T0297197

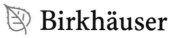

Bernd Schröder

Ordered Sets

An Introduction with Connections from Combinatorics to Topology

Second Edition

 Birkhäuser

Bernd Schröder
Department of Mathematics
University of Southern Mississippi
Hattiesburg, MS, USA

ISBN 978-3-319-29786-6   ISBN 978-3-319-29788-0
DOI 10.1007/978-3-319-29788-0

Library of Congress Control Number: 2016935408

Mathematics Subject Classification (2010): 06-01, 06-02, 05-01

© Springer International Publishing 2003, 2016
This work is subject to copyright. All rights are reserved by the Publisher, whether the whole or part of the material is concerned, specifically the rights of translation, reprinting, reuse of illustrations, recitation, broadcasting, reproduction on microfilms or in any other physical way, and transmission or information storage and retrieval, electronic adaptation, computer software, or by similar or dissimilar methodology now known or hereafter developed.
The use of general descriptive names, registered names, trademarks, service marks, etc. in this publication does not imply, even in the absence of a specific statement, that such names are exempt from the relevant protective laws and regulations and therefore free for general use.
The publisher, the authors and the editors are safe to assume that the advice and information in this book are believed to be true and accurate at the date of publication. Neither the publisher nor the authors or the editors give a warranty, express or implied, with respect to the material contained herein or for any errors or omissions that may have been made.

Printed on acid-free paper

This book is published under the trade name Birkhäuser.
The registered company is Springer International Publishing AG, CH

Preface to the Second Edition

Writing the second edition of this text was a wonderful opportunity for a more mature presentation than I was able to give 13 years ago. The intent still is to give a self-contained introduction to the theory of ordered sets and to its connections to other areas. I tried to shine a light on as many branches of the theory of ordered sets as possible, with the main obstacle being my limited cranial capacity. Indeed, although in this text I can present the most I have ever known about ordered sets, writing the text has also shown me how much I don't know. However, that should not be a problem for anyone, as that which we don't know will always exceed that which we do. Writing the second edition also was quite humbling in another way. Thorough re-reading revealed more typos than were reported on the posted errata, and some of them were rather embarrassing.

The presentation remains modular. Specifically, we have the following:

- Chapters 1, 2, 3, and Sections 4.1 and 4.2 form the core of the text. If you are new to ordered sets, you should read this part in the order in which it is presented here. (Skip Section 3.5 if you are not focusing on analysis.)
- The remaining chapters can be read in just about any order.

There have been some content rearrangements and additions. The automorphism problem (see Open Question 2.14), of which nothing appears to be known beyond the references given here, has been moved into Chapter 2 to feature it more prominently. Similarly, the chapter on algorithms has been "promoted" to Chapter 5 and expanded to focus even more strongly on constraint satisfaction problems. I believe that more results similar to the "$\Leftarrow$" direction of Theorem 5.57 can be proved for the fixed point property and for other constraint satisfaction problems. Chapter 6 is new and serves to separate the fixed clique property from the more fundamental fixed simplex property and to give an idea about graph homomorphisms and their connections to and differences from order-preserving maps. Finally, the future importance of discrete Morse functions for the fixed point property for ordered sets is indicated in Appendix B. Overall, I have shifted the primary focus toward finite ordered sets, with results on infinite ordered sets moved to the back of each chapter whenever possible. The references [21, 23, 25, 31, 41, 42, 54, 56, 91, 98, 102, 112,

120, 246, 247, 251, 252, 254, 311] should provide ample opportunity for further study of various aspects of ordered sets.

Although the goal was and is to have a self-contained exposition, the first edition's appendix on ordered L^p-spaces has been turned into Section 3.5 and Exercises 2-50, 3-32, 3-33, 3-34, 4-37, 8-11, 8-12, 8-13, and 8-18. If you are familiar with L^p-spaces, the exercises will be natural; if you are not, the appendix probably would have felt quite uncomfortable anyway. So the simple advice here is that if L^p-spaces are not part of your repertoire, then these exercises should probably be skipped.

Other additions are technical in nature, but important nonetheless. The text is now available as an ebook with live links for the internal references. This can make reading easier at times, but the usefulness of a paper copy should not be underestimated. I have read and written thousands of pages on screen. However, when I really want to learn something well, I read a paper copy.

I hope you will enjoy reading this text, in any of its forms, as much as I enjoyed writing it. Let me conclude with a final request/recommendation for readers who use this text as part of an effort to improve proof-writing skills. The transition to doing proofs is hard. To me, it is the hardest challenge that I have ever successfully met. Looking up solutions does not facilitate this transition. So, as you work the exercises, do *not* use the library, Google, or other resources to find solutions. (I maintain that I learned how to do proofs partly because I was too lazy to look up solutions, preferring hours of thinking over a half hour in the library.) Use your brain, your whole brain and nothing but your brain, aided by paper and pencil.

Hattiesburg, MS, USA   Bernd Schröder
January 7, 2016

Preface to the First Edition

Order theory can be seen formally as a subject between lattice theory and graph theory. Indeed, one can say with good reason that lattices are special types of ordered sets, which are in turn special types of directed graphs. Yet this would be much too simplistic an approach. In each theory, the distinct strengths and weaknesses of the given structure can be explored. This leads to general as well as discipline-specific questions and results. Of the three research areas mentioned, order theory undoubtedly is the youngest. The first specialist journal *Order* was launched in 1984, and much of the research that guided my own development started in the 1970s.

When I started teaching myself order theory (via a detour through category theory), I was only dimly aware of lattices and graphs. (I was working on a PhD in harmonic analysis and probability theory at the time.) I was attracted to the structure, as it apparently fits the way I think. It was possible to learn the needed basics from research papers as well as surveys. From there it was immensely enjoyable to start working on unexplored problems. This is the beauty of a fresh field. Interesting results are almost asking to be discovered. I hope that the reader will find the same type of attraction to this area (and will ultimately make interesting contributions to the field).

Yet there also is a barrier to entering such a new field. In new fields, standard texts are not yet available. I felt it would be useful to have a text that would expose the reader to order theory as a discipline without too quickly focusing on one specific subarea. In this fashion a broader picture can be seen. This is my attempt at such a text. It contains all that I know about the theory of ordered sets. From here, articles on ordered sets as well as the standard references I had available starting out (which are primarily Rival's Banff conference volumes [246, 247, 252], but also Birkhoff's classic [21] on lattice theory, Fishburn's text [91] on interval orders, and later Davey and Priestley's text [56] on lattices and Trotter's monograph [311] on dimension theory) should be easily accessible. The idea was to describe what I consider the basics of ordered sets without the work becoming totally idiosyncratic. Some of the salient features of this text are bulleted below.

- **Theme-driven approach.** Most of the topics in this text are introduced by
  investigating how they relate to research problems. We will frequently revisit
  the open problems that are explained early in the text. Further problems are
  added as we progress. In this fashion, I believe, the reader will be able to form
  the necessary intuitions about a new structure more easily than if there were no
  common undercurrent to the presentation. I have deliberately tried to avoid the
  often typical beginning of a text in discrete mathematics. There is no deluge of
  pages upon pages of basic definitions. New notions are introduced when they first
  arise, and they are connected with known ideas as soon as possible.
- **Connections between topics.** Paraphrasing W. Edwards Deming, one can say
  that *If we do not understand our work as part of a process, we do not understand
  our work at all.* This statement applies to industry, education, and also research.
  Many powerful results have been proved by connecting several seemingly
  different branches of mathematics. Consequently, I tried to show some of the
  connections between the different areas of ordered sets, as well as connections
  of the theory of ordered sets to areas such as algebraic topology, analysis, and
  computer science, to name several of them. These types of cross-connections
  have always been fascinating to me. Consider, for example, the use of algebraic
  topology in Chapter 9. Its connections to ordered sets yield results of a caliber
  that appears impossible to achieve by staying within a single discipline.

  Along the same lines, it must immediately be said that I cannot claim to know
  all connections between order and other fields. Thus this text is by no means a
  complete guide to interdisciplinary work that involves order.
- **Breadth and Flexibility.** One of my professors once said that mathematics is a
  field that one can study for 80 years without repeating a topic, but unfortunately
  also without contributing anything new. What we don't know will always exceed
  what we know, and order is no exception. On the other hand, there is much to be
  said for a broad education. The more one knows, the more connections one can
  make and the more potential one has to make good contributions.

  To allow for the benefits of broad training without the (very real) drawback of
  being overloaded with information, note that the core of the text was purposely
  kept lean and consists of Chapters 1, 2, 3, and Sections 4.1 and 4.2. The remaining
  chapters can be read in almost any order, if the reader is willing to follow the
  occasional backward reference. This organization makes it easier to tackle the
  breadth of the text and also makes the text more flexible. Follow your strongest
  interests first, then obtain more information about other things. So, I hope the
  reader will come back to this text frequently to learn more and to use this
  knowledge as a springboard towards new work in ordered sets.

  If the reader decides to stray from the linear presentation of the text, some
  results that use examples from earlier sections may need to be skipped (or
  better, acquired by selective reading).[1] Still, the reader should be able to acquire

---

[1] The index, which I tried to make as redundant as possible, should help in filling holes, clarifying
unknown notation, etc.

sufficient understanding of the chapter by relying on work that connects to topics already read. The reader can be guided by his/her primary interests and will still be exposed to many of the cross-connections mentioned above. (This remark should not dissuade the more intrepid readers from going cover-to-cover.)

- **Depth.** The greatest depth one can achieve in any research topic is to understand the open research problems and to be aware of most or all of the results pertaining to their solutions. For at least three problems, the fixed point property, the order reconstruction problem, and the automorphism problem, the reader is exposed to essentially all that I know about these topics. I will never claim comprehensive knowledge of a topic. For these problems, however, I am virtually certain that I have gone as far as the theory does to date. More open problems are given in the text as well as in the sections at the end of each chapter. In this fashion, the text allows the reader to attempt research even after reading a few chapters.

- **Open Problems.** The idea to work with a set of open problems certainly is inspired by the early issues of the journal *Order*. Open problems show the frontiers of a field and are thus in some way the life of a branch of mathematics. Therefore, I tried to at least show the reader what were and are considered major open problems in ordered sets. The main open problems that are part of the text come from *Order's* problem list or are inspired by it. I can safely say that they are of interest to the community at large. Naturally, the text is biased towards the problems that I know more about. For other problems in the text, I have tried to include references to more literature.

  The open problems at the end of each chapter and open problem 12.24 reflect my own curiosity. These problems range from special cases of the major problems, to simpler proofs of known results, to some things that "I simply would like to know." The only way I have to gauge their difficulty is to say that I have not solved them (yet?). I suspect that some may turn out to be quite difficult (material for MS or PhD theses). I sincerely hope that none of them turn out to be trivial or that their solutions were overlooked by me. Yet this possibility can never be excluded, especially since, to provide a more complete picture of order, I sometimes stretched myself into areas that I am not as familiar with. Either way, I would be interested in solutions as they arise.

  More open problems about order can be found in the collections [246, 247, 252].

- **Standard topics.** There are certain standard topics that mathematicians unfamiliar with order may automatically identify with order theory. The most worked-on parts of the theory of ordered sets appear to be lattice theory and dimension theory. Interval orders also have received a good bit of attention due to their applicability in modeling schedules. There are textbooks available in each of these areas (see [21, 56, 91, 311]) and any exposure of these areas as *part* of a text must necessarily be incomplete.[2] Therefore, in these areas, only the basics

---

[2]Recently, the text [102] approached order theory through the lens of formal concept analysis, which is interesting, too.

and some relations to the main themes of the text are explored. A deeper study
of these areas has to be relegated to further specialization. The same can be said
for the treatment of constraint satisfaction problems in Chapter 5.

- **Ordering of topics and history.** Like many authors, I want the text to be
  readable, not a list of achievements in exact chronological order. Thus, whenever
  necessary, logic supersedes historical order. Historically later results are used
  freely to prove historically earlier results when this appeared appropriate. The
  overall presentation is intended to be linear.

  The notes at the end of each chapter (if they refer to historical developments)
  reflect a very limited historical view at best. I could do no better, as many of
  the topics this text touches upon (such as reconstruction in general, lattices,
  dimension theory, constraint satisfaction, to name several) are in fact the tips
  of rather large icebergs, which in many cases I cannot fully fathom myself.[3]

  The reader who is interested in the history of a subject is advised to look
  at surveys I have referenced or to run searches of the *Mathematical Reviews*
  database with the appropriate key word. Electronic search tools are perfect for
  such tasks. The reader should, however, be aware that terminology is changing
  over time.[4]

- **Reading with a pencil** is mandatory, more so than in a regular mathematics text.
  The diagram of an ordered set is a very strong visual tool. Many proofs that
  appear difficult at first become clear when drawing diagrams of the described
  situations. Indeed, much of the appeal of the theory of ordered sets derives from
  its strongly visual character.

Mathematics is learned by doing, by confronting a topic, and by acquiring
the tools to master it. It is time to do so. I hope reading this text will be an
enjoyable and intellectually enriching contribution to the reader's mathematical life.
Readers interested in code for Chapter 5 or who have solutions to open problems
or suggestions for exercises are encouraged to contact the author. I am planning to
post updates on my web site.

Ruston, LA, USA   Bernd Schröder
October 24, 2002

---

[3]Recently, I did give a fairly chronological account of the fixed point property in the survey [285]
though.

[4]In fact, I have spent some time reinventing certain results because I was unaware of the particular
vocabulary of an area. While this appears inefficient, the only alternative would be to always just
stick with what I know well. This I consider an unacceptable proposition.

Acknowledgments

This book is dedicated to my family. It has been written largely on their time. Without the support of my wife Claire, my children Samantha, Nicole, Haven, and Mlle, and of my parents-in-law Merle and Jean, the writing would not have been possible. Moreover, without the help of my parents, Gerda and Siegfried, and the sacrifices they made, I would have never reached the starting point in the first place.

In addition to the immeasurable contribution of my family, many individuals made contributions of large magnitude. I would like to thank all of them here, and I hope I am not forgetting anyone.

Joseph Kung was a kind and resourceful editor. Ann Kostant and Tom Grasso, my contacts at Birkhäuser, gave me good insights into the publishing process, moral support, and a much-needed extension of a deadline; the same goes for Benjamin Levitt and Chris Tominich for the second edition. Elizabeth Loew was a great resource on typographical matters. Laura Ogden, Anita Dotson, and their staff at Louisiana Tech's Interlibrary Loan patiently and efficiently filled a never-ending stream of orders. Jean-Xavier Rampon was a wonderful host for a research visit to Nantes in the summer of 2000; aside from our work, much progress was made on this text. Jonathan Farley started the whole project by making me aware of Birkhäuser's search for authors. Without that one forwarded e-mail, the opportunity would have passed by. Along these lines, I owe special thanks to my Academic Director, Dr. Gene Callens, for the (very reasonable) advice against writing a book before being tenured and also for respecting my decision to disregard his advice.

Many people have helped me with questions regarding parts of the book. M. Roddy let me use an improvement of the proof of Theorem 12.17 and inspired part of the first edition's cover. Khaled Al-Agha helped with some parts of Chapter 10, and a number of questions were answered by Tom Trotter. Ralph McKenzie helped with Exercise 12-30 and Remark 15 in Chapter 12. Moreover, his help with the proof of Hashimoto's Theorem (Lemma 12.34!) was invaluable. My graduate student Joshua Hughes read the early chapters and eliminated a good number of typographical errors. Michał Kukieła read the first edition very thoroughly, made great suggestions, and answered questions, including one answered in [176]. Two reviewers of the first draft of the second edition provided further good suggestions.

Finally, I would like to thank the readers and the technical staff who helped this book through the final stages.

Contents

Chapter 1
Basics

Few prerequisites are needed to read this text. You should be familiar with real numbers, functions, sets, and relations. Moreover, the elusive property known as "mathematical maturity" should have been developed to the point that you can read and understand proofs and produce simple proofs. Texts that develop these skills are, for example, [117, 283]. A background in graph theory helps, but is not necessary.

1.1 Definition and Examples

This section introduces some basic definitions and notations used throughout the text. I have tried to keep this section short. Terminology will be introduced "in context" whenever possible. In case you don't remember notation, terminology, or symbols later in the text, consult the index to locate the definition.

The central concept in this book is that of an ordered set. An ordered set is a set equipped with a special type of binary relation. Recall that, abstractly, a **binary relation** on a set P is just a subset $R \subseteq P \times P = \{(p, q) : p, q \in P\}$. The statement $(p, q) \in R$ simply means that "p is related to q under R." A binary relation R thus contains all the pairs of points that are related to each other under R. For any binary relation R in this text, whenever p is related to q via R, we will write pRq instead of $(p, q) \in R$. The relations that interest us most are the order relations.

Definition 1.1. *An **ordered set** (or **partially ordered set** or **poset**) is an ordered pair $(P, \leq)$ of a set P and a binary relation $\leq$ contained in $P \times P$, called the **order** (or the **partial order**) on P, such that*

1. *The relation $\leq$ is reflexive. That is, each element of the ordered set $(P, \leq)$ is related to itself: $\forall p \in P : p \leq p$.*
2. *The relation $\leq$ is antisymmetric. That is, if p is related to q and q is related to p, then p must equal q: $\forall p, q \in P : [(p \leq q) \wedge (q \leq p)] \Rightarrow (p = q)$.*

© Springer International Publishing 2016

B. Schröder, *Ordered Sets*, DOI 10.1007/978-3-319-29788-0_1

3. *The relation $\leq$ is transitive. That is, if p is related to q and q is related to r, then*
   p is related to r: $\forall p, q, r \in P : [(p \leq q) \wedge (q \leq r)] \Rightarrow (p \leq r)$.

Similar to topology ([325]) or analysis ([62, 282]), where a set that is equipped with a structure is usually referred to by the name of the set only, we will usually say "Let P be an ordered set" instead of "Let $(P, \leq)$ be an ordered set." The order that is implicitly assumed to exist on P will usually be denoted $\leq$. When multiple orders exist on the same set, we distinguish them with indices. When we talk about orders as relations, we will automatically assume that a property of an order is defined in the same way as a property of the ordered set and vice versa.

The elements of P are called the **points** of the ordered set. Order relations introduce a hierarchy on the underlying set: The statement "$p \leq q$" is read "p is less than or equal to q" or "q is greater than or equal to p." The antisymmetry of the order relation ensures that there are no two-way ties ($p \leq q$ and $q \leq p$ for distinct p and q) in the hierarchy. Transitivity, in conjunction with antisymmetry, ensures that no cyclic ties ($p_1 \leq p_2 \leq \cdots \leq p_n \leq p_1$ for distinct $p_1, p_2, \ldots, p_n$) exist. We will also use the notation $q \geq p$ to indicate that $p \leq q$. In keeping with the idea of a hierarchy, we will say that $p, q \in P$ are **comparable** and write $p \sim q$ iff $p \leq q$ or $q \leq p$. We will write $p < q$ for $p \leq q$ and $p \neq q$. In this case we will say p is **(strictly) less than** q. An ordered set will be called finite (infinite) iff the underlying set is finite (infinite).

Example 1.2. As we will see in the following sections, ordered structures arise throughout mathematics. For starters, consider some examples of ordered sets.

1. The natural numbers $\mathbb{N}$, the integers $\mathbb{Z}$, the rational numbers $\mathbb{Q}$, and the real numbers $\mathbb{R}$, with their usual orders, are ordered sets.
2. Any set of sets is ordered by set inclusion $\subseteq$. Similarly, geometric figures (circles in the plane, balls, or simplices in higher-dimensional spaces) are ordered by inclusion. Note that, if we consider each geometric object as a set of points, we are back to sets ordered by set containment.
3. The simplest example of a set system ordered by inclusion is the **power set** $\mathcal{P}(X)$ of a set X.
4. Every ordinal number in set theory is an ordered set.
5. The vector space $C([0, 1], \mathbb{R})$ of continuous functions from $[0, 1]$ to $\mathbb{R}$ can be ordered as follows: For $f, g \in C([0, 1], \mathbb{R})$, we say $f \leq g$ iff, for all $x \in [0, 1]$, we have $f(x) \leq g(x)$, where the latter $\leq$ is the order of the real numbers.[1]
6. The natural numbers can also be ordered as follows: For $p, q \in \mathbb{N}$ we say $p \sqsubseteq q$ iff p divides q.
7. If J is a set of intervals of the real line, we can order J by $[a, b] \leq_{\text{int}} [c, d]$ iff $b \leq c$ or $[a, b] = [c, d]$.
8. If $(P, \leq_P)$ is an ordered set, the **dual** P^d of P is the set P with the dual order $\leq_{pd} := \{(a, b) : b \leq_P a\}$.

---

[1] Here it should be clear from the context which order (for numbers or for functions) is meant.

9. The lexicographic order on $\mathbb{N}^n$ is defined by $(x_1, \ldots, x_n) \leq (y_1, \ldots, y_n)$ iff $(x_1, \ldots, x_n) = (y_1, \ldots, y_n)$ or there is a $k \in \{1, \ldots, n\}$ with $x_i = y_i$ for $i < k$ and $x_k < y_k$.

10. (This example requires some background in analysis, as provided, for example, in [47] or [282].) Let (Ω, Σ, μ) be a measure space with Ω being the underlying set, Σ a σ-algebra of subsets of Ω, and $\mu : \Sigma \to \mathbb{R}$ a measure. Integration will be the usual (Lebesgue-) integration on measure spaces. For $p \geq 1$, the spaces $L^p(\Omega, \Sigma, \mu)$ are the usual spaces of all (equivalence classes of) functions $f : \Omega \to \mathbb{R}$ so that $|f|^p$ is μ-integrable with the usual L^p-norm $\|f\|_p = \left(\int_\Omega |f|^p d\mu \right)^{1/p}$. For two Σ-measurable functions $f, g \in L^p(\Omega, \Sigma, \mu)$, we define the relation $f \leq g$ iff $f(x) \leq g(x)$ holds μ-almost everywhere (μ-a.e.). Then $\leq$ is an order relation on $L^p(\Omega, \Sigma, \mu)$. $\qquad \square$

Proofs that the structures in Example 1.2 truly are ordered sets are not too hard. Here, we will only prove that the lexicographic order for $\mathbb{N}^n$ is indeed an order.

Proof that Example 1.2, part 9 is an ordered set. We must prove that the relation $\leq$ is reflexive, antisymmetric, and transitive. Let $(x_1, \ldots, x_n)$, $(y_1, \ldots, y_n)$, and $(z_1, \ldots, z_n)$ be elements of $\mathbb{N}^n$.

Reflexivity. By definition, we have that $(x_1, \ldots, x_n) \leq (x_1, \ldots, x_n)$.

Antisymmetry. Let $(x_1, \ldots, x_n) \leq (y_1, \ldots, y_n)$ and $(x_1, \ldots, x_n) \geq (y_1, \ldots, y_n)$ and suppose for a contradiction that $(x_1, \ldots, x_n) \neq (y_1, \ldots, y_n)$. Then there are $k_1, k_2 \in \{1, \ldots, n\}$ such that $x_i = y_i$ for $i < k_1$ and $x_{k_1} < y_{k_1}$, and $y_j = x_j$ for $j < k_2$ and $y_{k_2} < x_{k_2}$. Without loss of generality, we can assume that $k_1 \leq k_2$. Then $x_{k_1} < y_{k_1}$ and $y_{k_1} \leq x_{k_1}$, which is a contradiction. Thus we must have $(x_1, \ldots, x_n) = (y_1, \ldots, y_n)$, which proves antisymmetry.

Transitivity. Let $(x_1, \ldots, x_n) \leq (y_1, \ldots, y_n)$ and $(y_1, \ldots, y_n) \leq (z_1, \ldots, z_n)$. If any two of the three tuples are equal, then there is nothing to prove. Hence we can assume that the three tuples are mutually distinct. In this case, there are indices $k_1, k_2 \in \{1, \ldots, n\}$ such that $x_i = y_i$ for $i < k_1$ and $x_{k_1} < y_{k_1}$, and $y_j = z_j$ for $j < k_2$ and $y_{k_2} < z_{k_2}$. Let $k := \min\{k_1, k_2\}$. Then, for $i < k$, we have $x_i = y_i = z_i$, and, for the index k we have $x_k \leq y_k \leq z_k$, with at least one of the inequalities being strict. Thus $x_k < z_k$ and $(x_1, \ldots, x_n) \leq (z_1, \ldots, z_n)$, which concludes our proof of transitivity. $\qquad \blacksquare$

It is a good exercise to prove that every example in 1.2 is an ordered set. Not every relation that looks like it induces a hierarchy is an order relation however.

Proposition 1.3 (See [318]). *For a set of subsets $\mathcal{J}$ of an ordered set P, we define the relation $A \sqsubseteq B$ iff the following hold.*

1. *For all $a \in A$, there is an element b in B such that $a \leq b$, and*
2. *For all $b \in B$, there is an element a in A such that $b \geq a$.*

The relation $\sqsubseteq$ is reflexive and transitive, but not necessarily antisymmetric.

Proof. Let P be an ordered set, $\mathcal{J}$ a set of subsets of P and let $A, B, C \in \mathcal{J}$ be arbitrary. Because, for all $a \in A$, we have $a \leq a$, we obtain $A \sqsubseteq A$ for all $A \in \mathcal{J}$. Thus $\sqsubseteq$ is reflexive.

If $A \sqsubseteq B \sqsubseteq C$, then, for every $a \in A$ there is a $b \in B$ with $a \leq b$. For b, there is a $c \in C$ such that $b \leq c$ and hence $a \leq c$. Similarly, for every $c \in C$, there is a $b \in B$ with $c \geq b$. For b, there is an $a \in A$ such that $b \geq a$ and hence $c \geq a$. Thus $A \sqsubseteq C$ and $\sqsubseteq$ is transitive.

However $\sqsubseteq$ is in general not antisymmetric. Consider the sets $\{1, 2, 4\}$ and $\{1, 3, 4\}$ as subsets of $\mathbb{N}$ with the natural order. Then $\{1, 2, 4\} \sqsubseteq \{1, 3, 4\}$ and $\{1, 2, 4\} \sqsupseteq \{1, 3, 4\}$, but clearly the two sets are not equal. ∎

Exercises

1-1. Prove that the following are ordered sets.

    a. The set $\{a, b, c, d, e, f\}$ with the relation $\leq := \{(a, a), (a, c), (a, d), (a, e), (a, f), (b, b), (b, c), (b, d), (b, e), (b, f), (c, c), (c, e), (c, f), (d, d), (d, e), (d, f), (e, e), (f, f)\}$.

    b. Let L and U be disjoint ordered sets. Construct $(P, \leq)$ as follows. $P := L \cup U$ and $a \leq b$ iff $a \in L$ and $b \in U$ or $a, b \in L$ and $a \leq_L b$ or $a, b \in U$ and $a \leq_U b$.

    c. Consider the power set $\mathcal{P}(\{1, \ldots, n\})$ with the following relation. The set A is said to be **dominated** by the set B iff there is a k such that $|\{1, \ldots, k\} \cap A| < |\{1, \ldots, k\} \cap B|$ and, for all $l < k$, we have $|\{1, \ldots, l\} \cap A| = |\{1, \ldots, l\} \cap B|$. Define $A \leq B$ iff $A = B$ or A is dominated by B.

    d. (Requires graph-theoretical terminology.) Any tree $T = (V, E)$, ordered as follows: $x \leq y$ iff $x = y$ or there is a path from x to y that does not go through the root and the distance from x to the root is less than the distance from y to the root.

1-2. Let P be a set and let $\{\leq_i\}_{i \in I}$ be a family of order relations on P. Prove that $\bigcap_{i \in I} \leq_i$ is an order relation.

1-3. (Alternative definition of order.) Prove the following.

    a. If $\leq$ is an order relation on P, then $<$ is an antireflexive (that is, for all $p \in P$, we have $p \not< p$) and transitive relation.

    b. If $<$ is antireflexive and transitive, then $\leq := (< \cup =)$ is an order relation.

1-4. Let P be an ordered set and let $a_1, \ldots, a_n \in P$. Prove that, if $a_1 \leq a_2 \leq \cdots \leq a_n \leq a_1$, then $a_1 = a_2 = \cdots = a_n$.

1.2   The Diagram

After defining ordered sets as sets equipped with a certain type of relation, we are ready to investigate them. However, a visual aid would be helpful, because a picture often says more than a thousand words. One such visual aid is inspired by graph theory, so let us briefly review what a graph is.

Definition 1.4. *A **graph** G is a pair (V, E) of a set V (of **vertices**) and a set of doubleton sets $E \subseteq \{\{a, b\} : a, b \in V\}$ (of **edges**). If $\{a, b\} \in E$, we write $a \sim b$.*

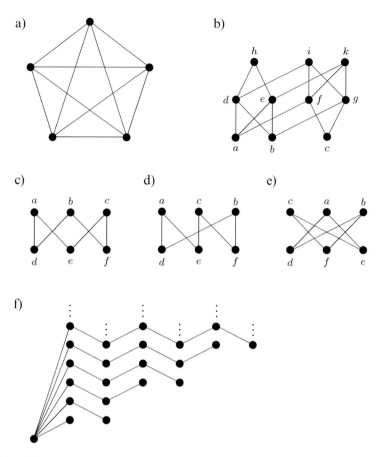

Fig. 1.1   a) The complete graph with five vertices K_5. **b)** The diagram of an ordered set (the set $P2$ in [264]). **c)–e)** Three diagrams for the same ordered set (points to be identified by an isomorphism have the same letters). **f)** The "diagram" of an infinite ordered set (this "spider," as Rutkowski named it, is taken from [318], Remark 5.1)

This is a perfectly fine definition. However, when working with graphs, most people do not think of the set theoretical entities of Definition 1.4. Instead, they visualize an entity such as shown in Figure 1.1 a). The connection is simple: For each vertex $v \in V$, we put a point in the plane (or in 3-space) to represent the vertex. For any two vertices $v, w \in V$, we join the corresponding points with a line (an edge) that does not touch any other points iff $\{v, w\} \in E$. These pictures provide a good visual tool for working with graphs as well as a way to translate real life problems into mathematics. Road networks, for example, can be modeled with graphs. We could do the same thing for orders: Put points in the plane or 3-space and join related points with edges. Arrows could indicate the way in which points are related. This idea would have two shortcomings. First, the hierarchical structure of the order may be hard to detect, and second, there will be many lines that can be considered superfluous because of transitivity. We tackle the second problem first.

Definition 1.5. *Let P be an ordered set. Then p ∈ P is called a **lower cover** of q ∈ P (and q is called an **upper cover** of p) iff p < q and, for all z ∈ P, we have that p ≤ z ≤ q implies z ∈ {p, q}. In this case, we write p ≺ q. Points p and q that satisfy p ≺ q or q ≺ p will also be called **adjacent**.*

Example 1.6. To become familiar with the covering relation, consider the following

1. In the power set $\mathcal{P}(\{1, \ldots, 6\})$, the set $\{1, 3\}$ is a lower cover of $\{1, 3, 5\}$, but it is not a lower cover of $\{1, 2, 3, 4\}$.
2. In $\mathbb{Z}$, each $k \in \mathbb{Z}$ has exactly one upper cover $(k+1)$ and one lower cover $(k-1)$.
3. Whether two elements are covers of each other depends on the surrounding universe. 2 is not an upper cover of 0 in $\mathbb{Z}$, but it is an upper cover of zero in the set of even numbers. Similarly $\{1, 2, 3, 4\}$ is an upper cover of $\{1, 3\}$ in the set of subsets of $\{1, \ldots, 6\}$ that have an even number of elements.
4. In infinite ordered sets, elements may or may not have covers. In $\mathbb{R}$ and $\mathbb{Q}$, no two elements are covers of each other.   □

The covering relation is the smallest relation that carries all the information for a given finite order. To visually incorporate the hierarchy, we must impose an up-down direction. The resulting tool is the Hasse diagram. Because of the difficulty indicated in Example 1.6, part 4, it is mainly used for finite ordered sets.

Definition 1.7. *The (**Hasse) diagram**[2] of a finite ordered set P is the ordered pair (P, ≺), where ≺ is the lower cover relation as defined in Definition 1.5.*

Again, we have a perfectly reasonable mathematical definition which gives rise to the following possibility for visualization (see Figure 1.1, b-e).

1. Draw the points of P as small disks in the plane (or as small balls in 3-space) such that any two disks (balls) are a positive distance apart from each other and so that, if $p \leq q$, then the center of the disk (ball) for q has a larger y-coordinate (z-coordinate) than the center of the disk (ball) for p.
2. If $p \prec q$, draw a line or a smooth curve (an edge) between the centers of the disks (balls) corresponding to p and q such that the following hold.

   - The slope of the edge (or $\frac{\partial}{\partial z}$ if we are in 3-space) does not change its sign.
   - The only disks (balls) touched by the edge correspond to p and q.

   Do not join any other pairs of disks (balls) by edges.

For an illustration, consider Figure 1.1, part b). We can easily read off the comparabilities in the diagram. For example, $a \leq d$, because there is an edge from a to d and a is lower than d. We also have $a \leq k$, because there is an edge going up from a to e and another going up from e to k. This means $a \prec e \prec k$ and, by transitivity, $a \leq k$. Transitivity allows us to avoid drawing some edges that

---

[2]Named after the German algebraic number theorist Helmut Hasse who used diagrams to picture the ordered sets of subfields or field extensions.

would only clutter the picture: Adding all the edges (a, h), (a, i), (a, k), (b, h), (b, i), (b, k), (c, i), and (c, k) would make the drawing confusing without adding useful information. Note that, from Figure 1.1 part b), we see that $c \not\leq h$: Indeed, even though there is an edge from c to f, an edge from f to a, an edge from a to d, and an edge from d to h, not all the edges are traversed in an upward direction in the trail just described. The edge from f to a is traversed downwards (from f to a), meaning that $c \prec f \succ a \prec d \prec h$. Transitivity cannot be applied to this sequence (or any other such sequence) and hence $c \not\leq h$.

Drawing diagrams is not canonical. An ordered set can be drawn in different ways according to the investigator's preferences or needs. Consider Figure 1.1, parts c)–e): Each picture depicts the same ordered set, yet they do look distinctly different.

Diagrams can also be drawn for infinite ordered sets, but must be explained. The infinite ordered set depicted in Figure 1.1 f) consists of one "zig-zag" with n elements for each $n \in \mathbb{N}$ such that all "zig-zags" have the same left endpoint. How to draw a diagram of a certain set is often a matter of taste. For some discussion on the subject, see [4]. For drawing diagrams on surfaces, see [83], and for a multitude of results regarding diagrams, see [248].

From a relation-theoretic point-of-view, the diagram is a certain subset of the order relation, formed according to the rule that only pairs (a, b) are selected for which there is no intermediate point i such that (a, i) and (i, b) are in the relation, too. To formalize how to recover the original relation from the diagram, we merely need to formulate the reading process indicated above as a mathematical construction.

Definition 1.8. *Let* $\sqsubset$ *be a binary relation on the finite set* P. *Then the* **transitive closure** $\sqsubset^t$ *of* $\sqsubset$ *is defined by* $a \sqsubset^t b$ *iff there is a sequence* $a = a_1, a_2, \ldots, a_n = b$ *such that* $a_1 \sqsubset a_2 \sqsubset \cdots \sqsubset a_n$.

The name is justified by the following result and by Exercise 1-7.

Proposition 1.9. *The transitive closure of a binary relation* $\sqsubset$ *on a finite set* P *is transitive.*

Proof. Let $a, b, c \in P$ with $a \sqsubset^t b \sqsubset^t c$. There are $a = a_1, a_2, \ldots, a_n = b$ with $a_1 \sqsubset a_2 \sqsubset \cdots \sqsubset a_n$ and $b = a_n, a_{n+1}, \ldots, a_{n+m} = c$ with $a_n \sqsubset a_{n+1} \sqsubset \cdots \sqsubset a_{n+m}$. Therefore $a = a_1 \sqsubset a_2 \sqsubset \cdots \sqsubset a_{n+m} = c$ and hence $a \sqsubset^t c$. ∎

For finite sets, the transitive closure allows us to translate back from the diagram to the order relation.

Proposition 1.10. *Let* P *be a finite ordered set with order* $\leq$ *and let* $\prec$ *be its lower cover relation. As usual,* $=$ *denotes the equality relation. Then* $\leq$ *is the transitive closure of the union of the relations* $\prec$ *and* $=$.

Proof. Let $\leq'$ denote the transitive closure of $\prec \cup =$. (Note that, despite the suggestive notation, at this stage, we do not even know if $\leq'$ is an order relation.) We will prove that $\leq = \leq'$. Because $\prec \cup =$ is contained in $\leq$ and because $\leq$ is transitive, we must have that $\leq' \subseteq \leq$ (see Exercise 1-7).

To prove the other inclusion, suppose for a contradiction that there are $a, b \in P$ with $a \leq b$ and $a \not\leq' b$. Because $a = b$ implies $a \leq' b$, we must have $a < b$.

Because P is finite, we can find points $c, d \in P$ with $c \leq d$ and $c \not\leq' d$ such that, for all $c < z < d$, we have $c \leq' z$ and $z \leq' d$. Because $c \prec d$ would mean $c \leq' d$, there must be a $z \in P$ with $c < z < d$. But then $c \leq' z \leq' d$ and, because the transitive closure of a relation is transitive, we infer $c \leq' d$, a contradiction. Thus a, b as described above cannot exist and we conclude that $\leq = \leq'$. ∎

We will discuss some algorithmic ramifications of computing transitive closures and the diagram in Sections 5.1 and 5.2.

Exercises

1-5. Draw the diagram of the ordered set.

    a. The ordered set in Exercise 1-1a.
    b. An ordered set as constructed in Exercise 1-1b with L and U isomorphic to ordered sets as in Exercise 1-1a.
    c. The set $\mathcal{P}(\{1, 2, 3, 4\})$ ordered as indicated in Exercise 1-1c.

1-6. Does the "spider" in Figure 1.1, part f) have a three-element subset C such that any two elements of C are comparable?

1-7. Let $\prec$ be a relation that is contained in the transitive relation $\leq$. Prove that the transitive closure $\prec^t$ of $\prec$ is contained in $\leq$. Conclude that the transitive closure of a relation $\prec$ is the intersection of all transitive relations that contain $\prec$.

1-8. Define the transitive closure of a relation on an arbitrary set. Then prove the analogue of Proposition 1.9.

1-9. Give an example of an infinite ordered set for which Proposition 1.10 fails. (Use the definition of the transitive closure from Exercise 1-8.)

1.3 Order-Preserving Mappings/Isomorphism

Figure 1.1 c), d), and e) shows three different pictures of the same ordered set. Had we assigned different labels to the points, we could have depicted three ordered sets that appear "different and yet the same." This phenomenon is similar to the topological result (which seems to have become folklore), that "a donut is homotopic/isomorphic to a teacup." Indeed, for Figure 1.1 c), d), and e), we can find a "continuous deformation" that has diagrams at each stage and, say, turns set c) into set d). The formal background lies in the investigation of structure-preserving maps (or "morphisms") which is strongly represented in algebra and topology. (Consider the strong role of structure homomorphisms in algebra and of continuous functions in topology.) Because the underlying structure we are interested in is the order, the following definition is only natural.

Definition 1.11. Let $(P, \leq_P)$ and $(Q, \leq_Q)$ be ordered sets and let $f : P \to Q$ be a map. Then f is called an **order-preserving function** iff, for all $p_1, p_2 \in P$, we have

$$p_1 \leq_P p_2 \implies f(p_1) \leq_Q f(p_2).$$

We will usually not index the orders in such a situation. Also, we will use the words "function" and "map" interchangeably. Finally, note that order-preserving maps are sometimes also called **isotone maps**.

Example 1.12. We continue with some examples of order-preserving and non-order-preserving maps. Note that the same map can be order-preserving or not, depending on the orders of domain and range.

1. The function $f : \mathbb{N} \to \mathbb{N}$, defined by $f(x) = 5x$, is an order-preserving map if both domain and range carry the natural order. It is also order-preserving if both domain and range carry the order $\sqsubseteq$ of part 6 in Example 1.2.
2. The function $f : \mathbb{N} \to \mathbb{N}$, defined by $f(x) = x + 1$, is order-preserving if both domain and range carry the natural order. However, it is not order-preserving if both domain and range carry the order $\sqsubseteq$ of part 6 in Example 1.2.
3. Let S be a set and let $\mathcal{P}(S)$ be its power set ordered by set inclusion. The map $\mathcal{I} : \mathcal{P}(S) \to \mathcal{P}(S)^d$ defined by $\mathcal{I}(X) := S \setminus X$ is an order-preserving map from $\mathcal{P}(S)$ to its dual.
4. Let B be the ordered set in part b) of Figure 1.1 and let C be the ordered set in part c) of Figure 1.1. Then the function $F : B \to C$ defined by $F(h) = a$, $F(a) = F(b) = F(d) = F(e) = d$, $F(f) = F(g) = F(i) = F(k) = b$, and $F(c) = f$ is order-preserving. We use the reflexivity of order relations here. Although reflexivity is essentially taken for granted and not noted on the diagram, it allows us to collapse related points into one.
5. Consider Figure 1.2, where maps are indicated by arrows. The sets where the arrows start are the domains of the maps. The arrows indicate where each individual point is mapped. All the maps thus given in Figure 1.2 are order-preserving. $\square$

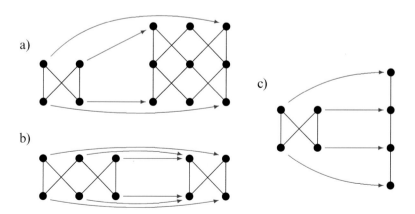

Fig. 1.2 Graphical representation of some order-preserving maps

If Definition 1.11 is a reasonable definition for structure-preserving maps on ordered sets, then the composition of two order-preserving maps should again be order-preserving.

Proposition 1.13. *Let P, Q, R be ordered sets and let $f : P \to Q$ and $g : Q \to R$ be order-preserving maps. Then the map $g \circ f : P \to R$ is order-preserving, too.*

Proof. Let $p_1, p_2 \in P$ with $p_1 \leq p_2$. Then, because f is order-preserving, we have $f(p_1) \leq f(p_2)$, and, because g is order-preserving, $g(f(p_1)) \leq g(f(p_2))$. Thus $g \circ f$ is order-preserving.   ∎

The above examples of order-preserving maps show that these maps preserve the order "one way." However, even the existence of a bijective order-preserving map between two sets, such as in part c) of Figure 1.2, is not a guarantee that both sets are "essentially the same." (This is similar to the situation in algebra or topology.) The problem in part c) of Figure 1.2 is that the inverse function is not order-preserving.

Suppose now that the two ordered sets P and Q have a bijection Φ between them such that Φ as well as Φ^{-1} preserve the order. When investigating an ordered structure, the underlying set normally only gives us the substance that we mold into structures. Which individual point is placed where in the structure is thus quite unimportant. Hence, for many purposes in order theory, ordered sets P and Q as indicated are indistinguishable: They have the same order-theoretical structure. This is the concept of (order-)isomorphism.

Definition 1.14. *Let P, Q be ordered sets and let $\Phi : P \to Q$. Then Φ is called an (**order-)isomorphism** iff the following hold.*

1. *Φ is order-preserving.*
2. *Φ has an inverse Φ^{-1}.*
3. *Φ^{-1} is order-preserving.*

*The ordered sets P and Q are called (**order-)isomorphic** iff there is an isomorphism $\Phi : P \to Q$.*

The following characterization of isomorphisms reinforces the notion that isomorphic ordered sets can be regarded as "the same."

Proposition 1.15. *Let P, Q be ordered sets. Then $f : P \to Q$ is an order-isomorphism iff the following hold.*

1. *f is bijective.*
2. *For all $p_1, p_2 \in P$, we have $p_1 \leq p_2 \Leftrightarrow f(p_1) \leq f(p_2)$.*

Proof. If $f : P \to Q$ is an order-isomorphism, then 1 and 2 are trivial. To prove that 1 and 2 imply that f is an isomorphism, we must prove that $f^{-1} : Q \to P$ is order-preserving. Let $q_1, q_2 \in Q$ be such that $q_1 \leq q_2$. Then there are $p_1, p_2 \in P$ such that $f(p_i) = q_i$ for $i = 1, 2$. Now $f(p_1) = q_1 \leq q_2 = f(p_2)$ implies $p_1 \leq p_2$, that is, $f^{-1}(q_1) \leq f^{-1}(q_2)$. Thus f^{-1} is order-preserving.   ∎

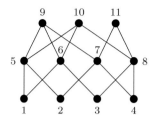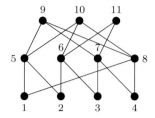

Fig. 1.3 Two isomorphic ordered sets

Proposition 1.16. *Let P, Q, R be ordered sets and let $\Phi : P \to Q$ and $\Psi : Q \to R$ be order-isomorphisms. Then $\Psi \circ \Phi$ is an order-isomorphism.*

Proof. Left as Exercise 1-13.   ∎

The strength of using isomorphisms is that structures that at first appear different can turn out to be equal for all intents and purposes. Consequently, even structures that appear different can have the same properties. For example, the ordered sets in Figure 1.3 appear quite different at first. However, they are isomorphic via the map (from the left set to the right set) $1 \mapsto 1$, $2 \mapsto 2$, $3 \mapsto 4$, $4 \mapsto 3$, $5 \mapsto 5$, $6 \mapsto 8$, $7 \mapsto 6$, $8 \mapsto 7$, $9 \mapsto 10$, $10 \mapsto 9$, $11 \mapsto 11$. Verification that the indicated map is an isomorphism is a good exercise. Although, with enough experience, we can (and should) classify this task as "tedious, but routine," it gives an indication how hard it is to find an isomorphism. How many calculations would have been necessary to find the indicated map, had it not been given above? Even worse, what if we start out with sets of which we do not know if they are isomorphic or not? How do we know how many and what types of checks to perform until we can be sure that no isomorphism exists? These questions show that it can be quite difficult to decide if two ordered sets are isomorphic or not. In fact it is still an open problem *how* difficult this decision is. We will discuss this issue in Remark 8 in Chapter 7.

Exercises

1-10. Prove that each of the following maps is order-preserving.

    a. With the natural numbers $\mathbb{N}$ ordered as described in Example 1.2, part 6, define
$$f : \mathbb{N} \to \mathbb{N} \text{ by } f(x) := \begin{cases} \frac{x}{2}; & \text{if } x \text{ is even,} \\ x; & \text{if } x \text{ is odd.} \end{cases}$$

    b. For the ordered set in Figure 1.1, part b) define $F : P2 \to P2$ by $F(a) = b$, $F(b) = a$, $F(c) = c$, $F(d) = e$, $F(e) = d$, $F(f) = g$, $F(g) = f$, $F(h) = h$, $F(i) = k$, $F(k) = i$.

    c. (Requires some notions from topology.) In the power set $\mathcal{P}(X)$ of any topological space (X, τ), the map $f : \mathcal{P}(X) \to \mathcal{P}(\mathcal{P}(X))$ that maps A to its closure.

1-11. Let P and Q be ordered sets and let $f : P \to Q$ be a function.

    a. Prove that, *if P and Q are finite*, then the following are equivalent.

        i. f is order-preserving.
        ii. For all $a, b \in P$, we have that $a < b$ implies $f(a) \leq f(b)$.
        iii. For all $a, b \in P$, we have that $a \prec b$ implies $f(a) \leq f(b)$.

    b. Prove that parts 1-11(a)i and 1-11(a)ii are equivalent for infinite ordered sets, too.
    c. Show that, in general, 1-11(a)iii is not equivalent to 1-11(a)i and 1-11(a)ii.

1-12. The relation between covers and isomorphisms.

    a. Let $f : P \to Q$ be an isomorphism. Prove that $x \prec_P y$ implies $f(x) \prec_Q f(y)$.
    b. Prove that, if P and Q are finite, then $f : P \to Q$ is an isomorphism iff f is bijective and $x \prec_P y$ is equivalent to $f(x) \prec_Q f(y)$.
    c. Give an example of a bijective function $f : P \to Q$ such that $x \prec_P y$ is equivalent to $f(x) \prec_Q f(y)$ and yet f is not an isomorphism.

1-13. Prove Proposition 1.16.
1-14. For two ordered sets $(P, \leq_P)$ and $(Q, \leq_Q)$, the ordered set $P \odot Q$ has the set $P \times Q$ as its points and $(p_1, q_1) \leq (p_2, q_2)$ iff $p_1 <_P p_2$ or $p_1 = p_2$ and $q_1 \leq_Q q_2$. Under what circumstances is $P \odot Q$ isomorphic to $Q \odot P$?
1-15. It is clear that, for every ordered set, the identity and the constant function are order-preserving self-maps. Call two order relations $\leq_1$ and $\leq_2$ on the same ground set P **perpendicular** iff the only order-preserving self-maps they have in common are the constant functions and the identity. Prove that, if two order relations $\leq_1$ and $\leq_2$ are perpendicular, then their intersection is $\leq_1 \cap \leq_2 = \{(x, x) : x \in P\}$.
1-16. (Also see [106].) Let $(P, \leq_1)$ and $(P, \leq_2)$ be ordered sets so that every function $f : P \to P$ that preserves $\leq_1$ also preserves $\leq_2$. Prove that one of the following holds: $\leq_2 = \leq_1$, or $\leq_2$ is the dual order $\leq_1^d$ of $\leq_1$, or no two points are $\leq_2$-comparable.

1.4 Fixed Points

A property that has attracted a good bit of attention in order theory is the fixed point property, defined as follows.

Definition 1.17. *We will call a function $f : D \to D$ whose domain and range are equal a **self-map** on D. A function $f : D \to D$ is an **order-preserving self-map** on D iff D carries an order $\leq$ and $a \leq b$ implies $f(a) \leq f(b)$.*

Definition 1.18. *Let P be an ordered set and let $f : P \to P$ be an order-preserving self-map. Then $p \in P$ is called a **fixed point** of f iff $f(p) = p$. If f has no fixed points, then f is called **fixed point free**. An ordered set P has the **fixed point property** iff each order-preserving self-map $f : P \to P$ has a fixed point. For any ordered set P and any order-preserving map $f : P \to P$, we define the set $\mathrm{Fix}(f) := \{p \in P : f(p) = p\}$.*

One original motivation for working with fixed points in ordered sets is a proof of the Bernstein–Cantor–Schröder Theorem, see Theorem 8.14. Subsequently, the property also became interesting in and of itself. For more background, consider Remark 2 in the "Remarks and Open Problems" section of this chapter.

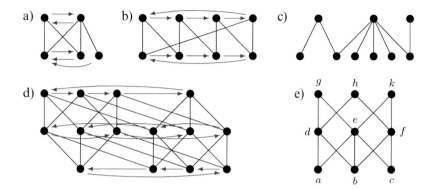

Fig. 1.4 Examples of ordered sets with and without the fixed point property. Fixed point free maps are indicated for sets without the fixed point property

We will use the fixed point property as a vehicle to introduce new order-theoretical notions. The fixed point property is well-suited for this task, because it combines properties of the set with properties of the maps on the set. For every new structure we introduce, the fixed point property can provide a familiar setting in which to investigate the structure. The underlying problem is the following.

> **Open Question 1.19.** *(See [250].) Characterize those (finite) ordered sets that have the fixed point property.*

We will frequently prove fixed point results to show how we can work with a certain class of ordered sets or a certain structure. Ultimately (see Theorem 7.32) we will prove that the NP version of this problem turns out to be co-NP-complete. At this stage, we only give a few examples of sets with and without the fixed point property (see Figure 1.4).

For the ordered sets in Figure 1.4 that do not have the fixed point property, a fixed point free order-preserving self-map is indicated. Yet how do we prove that such a map does *not* exist? It is instructive to try and do this for the remaining sets before reading on. We will soon have more sophisticated methods to prove that the ordered sets in Figure 1.4 c) and e) have the fixed point property. To become more acquainted with the basic notions of order theory, we give a quick proof that the ordered set in 1.4 e) has the fixed point property. The proof will be combinatorial, potentially with many cases to be treated separately. To reduce the number of cases to be treated, we can often use symmetry or, formally, the notion of an automorphism.

Definition 1.20. *Let P be an ordered set. The self-map $f : P \to P$ is called an* **(order-)automorphism** *iff f is an isomorphism.*

Example 1.21. For the ordered set in Figure 1.4 e), the map Φ that maps $a \mapsto c$, $b \mapsto b$, $c \mapsto a$, $d \mapsto f$, $e \mapsto e$, $f \mapsto d$, $g \mapsto k$, $h \mapsto h$, $k \mapsto g$ is an automorphism. We can see that this map "reflects the ordered set across the vertical axis through the middle" if the set is drawn as in Figure 1.4. This illustrates the common interpretation that automorphisms reveal the symmetries of a combinatorial structure.   □

Proposition 1.22. *The ordered set in Figure 1.4 e) has the fixed point property.*

Proof. Let's call the ordered set P and suppose for a contradiction that there is an order-preserving map $F : P \to P$ such that F has no fixed point. Then $F(b)$ cannot be related to b: Indeed, otherwise $b < F(b)$ and, because F has no fixed points, applying F twice to this inequality leads to $F(b) < F^2(b)$ and $F^2(b) < F^3(b)$. However, this is not possible, because there are no four distinct elements $w, x, y, z \in P$ such that $w < x < y < z$. Thus $F(b) \in \{a, c\}$. Because Φ in Example 1.21 is an automorphism, we can assume without loss of generality that $F(b) = a$. (Otherwise we would apply the whole following argument to $\Phi^{-1} \circ F \circ \Phi$, which is a fixed point free order-preserving self-map, too.)

Because $F(b) = a$, we have $F[\{d, e, g, h, k\}] \subseteq \{a, d, e, g, h, k\}$. Because $F(g)$ cannot be related to g (proved similar to $b \not\sim F(b)$), we must have $F(g) \in \{h, k\}$. If $F(g) = h$, then we must have that $a = F(b) \leq F(d) \leq F(g) = h$.

$F(d) = h$ would lead to $F(h) \geq F(d) = h$ and then $F(h) = h$, which is not possible. We exclude $F(d) = a$ in similar manner. This leaves $F(d) = d$, a contradiction.

Therefore we must have $F(g) = k$, which then leads to a contradiction in similar fashion. Thus P has no fixed point free order-preserving self-maps and hence P has the fixed point property.   ∎

With automorphisms being indicators of symmetry, ordered sets that have automorphisms without any fixed points should have a very high degree of symmetry, because they have at least one way to move every point and still retain the same order-theoretical structure. Asking about existence of fixed point free automorphisms is similar to asking about the existence of a fixed point free order-preserving self-map.

Definition 1.23. *Let P be an ordered set. P is called **automorphic** iff P has a fixed point free automorphism.*

> **Open Question 1.24.** *Characterize the (finite) automorphic ordered sets.*

Automorphic ordered sets play a role in the investigation of the fixed point property, as we will see in Theorem 4.8, part 2. The NP version of problem 1.24 turns out to be co-NP-complete as well (see [327]). We conclude this section by showing that, although the fixed point property implies that the ordered set is not automorphic, nonautomorphic ordered sets need not have the fixed point property.

Proposition 1.25. *The ordered set in Figure 1.4 a), call it Q, is not automorphic. However, it does not have the fixed point property.*

Proof. A fixed point free order-preserving map is indicated in the figure. To see that Q has no fixed point free automorphism, note that Q has exactly one point with exactly one upper cover. This point must be fixed by any automorphism.   ∎

Exercises

1-17. Let P be a finite ordered set and let $f : P \to P$ be order-preserving. Prove that, if there is a $p \in P$ with $p \leq f(p)$, then f has a fixed point. Then find an infinite ordered set in which this result fails.

1-18. Prove that the ordered set $(\emptyset, \emptyset)$ does not have the fixed point property.

1-19. Prove that the following ordered sets have the fixed point property.

    a. The ordered set in Figure 1.4 c).
    b. The ordered set in Figure 1.1 b).
    c. The ordered set in Figure 1.3.
    d. The ordered set in Figure 2.1 a).
    e. The ordered set in Figure 2.1 b_1/b_2).

1-20. Prove that the range of the map in Figure 1.2 a) does not have the fixed point property.

1-21. Let P be an ordered set and let $\Psi : P \to P$ be an automorphism.

    a. Prove that $f : P \to P$ has a fixed point iff $\Psi^{-1} \circ f \circ \Psi$ has a fixed point.
    b. What general hypotheses can be imposed on Ψ to assure that f has a fixed point iff $\Psi \circ f \circ \Psi$ has a fixed point?

1-22. Let P be an ordered set and let $f : P \to P$ be an injective order-preserving self-map.

    a. Prove that, if P is finite, then f is an automorphism.
    b. Show that, in general, f need not be an automorphism.

1.5   Ordered Subsets/The Reconstruction Problem

The last fundamental notion we need are the ordered subsets of an ordered set. They are defined analogous to subgroups in algebra or topological subspaces in topology.

Proposition 1.26. *Let $(P, \leq_P)$ be an ordered set and let $Q \subseteq P$. Then Q with the restriction $\leq_Q := \leq_P |_{Q \times Q}$ of the order on P to Q is an ordered set, too.*

Proof. The proofs of reflexivity, antisymmetry, and transitivity of $\leq_Q$ are trivial: Every property that holds for all elements of P will hold for all elements of Q. ∎

Knowing that the order properties are not destroyed when restricting ourselves to a subset, the following definition is sound.

Definition 1.27. *Let $(P, \leq_P)$ be an ordered set and let $Q \subseteq P$. If $Q \subseteq P$ and $\leq_Q = \leq_P |_{Q \times Q}$, then we call $(Q, \leq_Q)$ an **ordered subset** (or **subposet**) of P. Unless indicated otherwise, we will always assume that subsets of ordered sets carry the order induced by the surrounding ordered set.*[3]

Example 1.28. Every set of sets $\mathcal{S}$ that is ordered by inclusion is an ordered subset of $\mathcal{P}(\bigcup \mathcal{S})$. □

---

[3] So the notion of an ordered subset is similar to that of an *induced* subgraph in graph theory.

Order-theoretical properties may or may not carry over to ordered subsets. As we have not explored many properties yet, all we can do is record a negative example.

Example 1.29. Not every subset of an ordered set with the fixed point property has the fixed point property: Indeed, the subset $\{a, c, g, k\}$ of the ordered set in Figure 1.4 e) does not have the fixed point property. A fixed point free order-preserving map on $\{a, c, g, k\}$ would be $a \mapsto c, c \mapsto a, g \mapsto k, k \mapsto g$.   □

Connecting ordered subsets with order-preserving functions is simple: Naturally, for each $f : P \to Q$, the set $f[P]$ is an ordered subset of Q. Moreover, it is easy to see that, if P is finite and f is injective, then $f[P]$ has at least as many comparabilities as P. However, Figure 1.2 part c) shows that $f[P]$ need not be isomorphic to P, even if f is bijective. Order-preserving mappings for which P is isomorphic to $f[P]$ are called embeddings.

Definition 1.30. *Let P, Q be ordered sets. Then $f : P \to Q$ is called an (order) embedding iff the following hold.*

1. *f is injective.*
2. *For all $p_1, p_2 \in P$, we have $p_1 \leq p_2 \Leftrightarrow f(p_1) \leq f(p_2)$.*

Proposition 1.31. *Let P and Q be finite ordered sets and let $f : P \to Q$ be order-preserving. Then P is isomorphic to $f[P]$ iff f is an embedding. In this case, f is an isomorphism between P and $f[P]$.*

Proof. The direction "$\Leftarrow$" and the fact that f is the isomorphism follow immediately from Definition 1.30 and Proposition 1.15. For the direction "$\Rightarrow$," let P be isomorphic to $f[P]$ and let $g : P \to f[P]$ be an isomorphism. Because P is finite and $f[P]$ must have the same number of elements as P, f must be injective. Now $g^{-1}f : P \to P$ is an injective, and hence by finiteness of P bijective, order-preserving self-map. Because P is finite, this means that $g^{-1}f$ is an isomorphism. By Proposition 1.15 we have that $p_1 \leq p_2 \Leftrightarrow g^{-1}f(p_1) \leq g^{-1}f(p_2)$, which, after applying g, is seen to be equivalent to $f(p_1) \leq f(p_2)$. Hence f is an embedding. ∎

To further illustrate ordered subsets we introduce our second main research problem. (Characterizing the fixed point property was the first). Like the fixed point property, this problem will provide a familiar setting in which to investigate new structures that we introduce. Draw each ordered subset C of an ordered set P so that C has one point less than P on a card *without* labeling the points. Is it possible to take these pictures and reconstruct from them the original ordered set, up to isomorphism? Examples that this reconstruction does not always work are shown in Figure 1.5. However, these are the only examples known so far. The natural question that arises is if these are all such examples.

Definition 1.32. *For an ordered set P, we call the class of all ordered sets that are isomorphic to P the isomorphism class of P and we denote it $[P]$.*

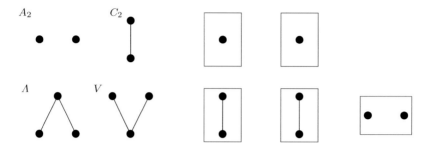

Fig. 1.5 Two pairs of nonisomorphic ordered sets, (A_2, C_2) and (Λ, V), with isomorphic decks, shown on the right. Are these the only examples of pairs of nonisomorphic ordered sets with isomorphic decks?

Definition 1.33. *Let P be a finite ordered set. For* $x \in P$*, the ordered subset* $P \setminus \{x\}$ *is called a* **card** *of P. Cards are considered to be unlabeled. That is, we have no way of determining which element of the card corresponds to which element of P.*

Let C be the set of all isomorphism classes of ordered sets with underlying set contained in $\mathbb{N}$*. The* **deck** *of P is the function* $\mathcal{D}_P : \mathcal{C} \to \mathbb{N}$ *such that, for each* $[C] \in \mathcal{C}$*, we have that* $\mathcal{D}_P([C])$ *is the number of cards of P that are isomorphic to the elements of* $[C]$*.*

> **Open Question 1.34.** *The Reconstruction Problem. Is every (finite) ordered set with at least four elements uniquely reconstructible from its deck? That is, is it true that, if* P, Q *are ordered sets with at least four elements such that* $\mathcal{D}_P = \mathcal{D}_Q$*, then P and Q must be isomorphic?*

For more background on the reconstruction problem, consider Remark 4 in the "Remarks and Open Problems" section of this chapter. We conclude this section with two simple-looking, but helpful, results on reconstruction. The first (Proposition 1.37) is a partial success, providing a positive answer to the problem in a restricted class of ordered sets. The second (Proposition 1.40) shows that a certain parameter can be reconstructed from the deck of any ordered set. These results provide some of the many indications that the answer might be positive in general. They will also be tools in later investigations. Both results are representative of possible approaches to this problem: We could try to prove reconstructibility for more and more special classes of ordered sets until every ordered set must belong to one class that has been proved to be reconstructible. Or, we could reconstruct more and more parameters of ordered sets until every ordered set is uniquely determined just by knowing a set of reconstructible parameters. So far, we are far from either of these goals.

Definition 1.35. *We will say an ordered set P is* **reconstructible** *from its deck* $\mathcal{D}_P$*, iff all ordered sets Q with* $\mathcal{D}_P = \mathcal{D}_Q$ *are isomorphic to P. A class* $\mathcal{K}$ *of ordered sets* is **reconstructible** iff each of its members $P \in \mathcal{K}$ is reconstructible. We will call a class $\mathcal{K}$ of ordered sets **recognizable** iff, for each ordered set $P \in \mathcal{K}$ and any ordered set Q, $\mathcal{D}_P = \mathcal{D}_Q$ implies that $Q \in \mathcal{K}$.

Definition 1.36. Let P be an ordered set. Then $s \in P$ is called the **smallest element** of P iff, for all $p \in P$, we have $s \leq p$.

Proposition 1.37. The class of finite ordered sets with at least 4 elements and a smallest element is reconstructible.

Proof. We will first prove that the class of ordered sets with at least 4 elements and a smallest element is recognizable. Indeed, if P has a smallest element s, then, for any $x \in P \setminus \{s\}$, we have that s is the smallest element of $P \setminus \{x\}$. Thus there is at most one card of P that does not have a smallest element. On the other hand, if Q does not have a smallest element, then there are at least two incomparable elements in Q that do not have any strict lower bounds. Therefore, ordered sets Q without a smallest element have at most two cards that have a smallest element and they have hence at least two cards that do not. (Cards are counted with multiplicity here.) Thus the ordered sets with a smallest element are recognizable as exactly those ordered sets whose deck has at most one card without a smallest element.

Having proved recognizability, we are ready to prove reconstructibility. Because we now know that ordered sets with a smallest element are recognizable from the deck, for the remainder of this proof, we can let P be an ordered set so that P has a smallest element s.

First consider the case that the deck of P has a card C that does not have a smallest element. Then, by the above, this card must be unique. Thus $C = P \setminus \{s\}$ and P is isomorphic to the ordered set obtained by attaching a smallest element to C. Formally, P is isomorphic to $C \cup \{s\}$, ordered by $\leq \cup \{(s, c) : c \in C \cup \{s\}\}$, where $\leq$ is the order on C.

If all cards of P have a smallest element, then removal of s must have introduced a new smallest element and we argue as follows. For every finite ordered set Q with a smallest element, there is a number $l_Q > 1$ such that the following hold.

1. For each $j \in \{1, \ldots, l_Q - 1\}$, there is exactly one element q_j of Q so that there are exactly j elements that are less than or equal to q.
2. Q contains zero or at least two elements that have l_Q elements that are less than or equal to each.

Essentially, $l_Q - 1$ is the number of times a smallest element can be removed before we arrive at an ordered set without a smallest element. For each card C of P that has a smallest element, find the number l_C. The cards C with $l_C \leq l_K$ for all cards K of P are all isomorphic. Moreover, $P \setminus \{s\}$ is isomorphic to one of them. Pick one and call it C_0. Then P is isomorphic to the ordered set obtained from C_0 by attaching a new smallest element below the smallest element of C_0. ∎

Definition 1.38. An **(order) invariant** $\alpha(\cdot)$ of ordered sets is a function from the class of all ordered sets to another set or class, such that, if P and Q are isomorphic, then $\alpha(P) = \alpha(Q)$. An order invariant will also be called a **parameter**.

Invariants are mostly numerical, but, for example, the deck is an invariant, too. The simplest invariant is probably the number of elements, closely followed by the number of comparabilities.

Definition 1.39. *An invariant α is called* **reconstructible** *iff, for all ordered sets P and Q, we have that $\mathcal{D}_P = \mathcal{D}_Q$ implies $\alpha(P) = \alpha(Q)$.*

The most easily reconstructed invariant is the number of elements. It is simply one more than the number of elements of any card. The number of comparabilities is to be reconstructed in Exercise 1-29a. If we could reconstruct a complete set of invariants, that is, a set of invariants, so that two ordered sets with the same invariants must be isomorphic, then the reconstruction problem would be solved. Unfortunately, no such complete set of invariants has materialized yet. A helpful invariant is the number of subsets of a certain type.

Proposition 1.40. *A* **Kelly Lemma** *for ordered sets, see [156], Lemma 1 and [174], Lemma 4.1. Let P be an ordered set with at least 4 elements and let Q be a finite ordered set with $|Q| < |P|$. Then the number $s(Q, P)$ of ordered subsets of P that are isomorphic to Q is reconstructible from the deck.*

Proof. Let $d_Q := \sum_C |\{S \subseteq C : S \text{ isomorphic to } Q\}|$, where the sum runs over all cards C of P, with multiplicity. Clearly d_Q can be computed from the deck. Let $P_Q \subseteq P$ be a fixed subset of P that is isomorphic to Q. Then P_Q is contained in exactly $|P| - |Q|$ cards of P, namely in exactly those cards obtained by removing an element outside P_Q. Because this is true for any subset of P that is isomorphic to Q, we have that $d_Q = s(Q, P)(|P| - |Q|)$ and we have reconstructed $s(Q, P)$ as $s(Q, P) = \frac{d_Q}{(|P| - |Q|)}$. ∎

Exercises

1-23. Let P, Q be ordered sets, let $f : P \to Q$ be order-preserving and injective and let P be finite. Prove that, if $f[P]$ contains as many comparabilities as P, then f is an embedding.
1-24. Prove that, in Definition 1.30, condition 2 implies condition 1.
1-25. Give an example that shows that Proposition 1.31 fails for infinite ordered sets.
1-26. Prove that each of the following ordered sets is reconstructible.

   a. The set $\{a, b, c, d\}$ with comparabilities $a < c > b < d$ and no further comparabilities.
      Note. This ordered set is also called "*N.*"
   b. The 5-element ordered set $\{a, b, c, d, x\}$ with comparabilities $a, b < x < c, d$ and the comparabilities dictated by transitivity.
   c. The ordered set in Figure 1.1 c).

1-27. Reconstruction of small ordered sets.

   a. Find all nonisomorphic ordered sets with up to five elements.
      Hint. The numbers of nonisomorphic sets are given in Remark 3 in Chapter 13.
   b. Verify that the reconstruction problem 1.34 is solvable for ordered sets with four and five elements.

    c. Attempt a positive solution of the reconstruction problem by hand or with a computer for
small sets with more than five elements.[4]

1-28. An **isolated point** is a point in an ordered set that is only comparable to itself. Prove that
ordered sets with an isolated point are reconstructible.

1-29. Some reconstructible parameters.

    a. Prove that the number of comparabilities in an ordered set is reconstructible,

    b. For $p \in P$, the **degree** $\deg(p)$ is the number of elements that are comparable to p. Prove
that, for every card $P \setminus \{x\}$ of P, the degree of the missing element x can be reconstructed.

Remarks and Open Problems

The "Remarks and Open Problems" sections provide more background on the
material that is discussed in the main body of the text. For this first section, it consists
solely of remarks. In later sections, we will list more and more open problems. To
keep things organized and easy to refer to, these sections are enumerated lists.

1. Anyone interested in the abstract underpinnings on objects and morphisms
   should look at category theory. A good introductory text is [3].
2. The fixed point property originated in topology using topological spaces and
   continuous functions. For a survey on the topological fixed point property,
   see [38]. It was apparently first studied for ordered sets by Tarski and Davis
   (see [58, 307]) and has since steadily gained in attention. The essentials of
   Tarski's result are already present in joint work with Knaster mentioned in
   [167]. A complete characterization of the ordered sets with the fixed point
   property might be beyond the realm of possibilities. The decisions whether an
   ordered set has a fixed point free order-preserving self-map or a fixed point
   free automorphism (also see problems 1.19 and 1.24) have been proved to be
   co-NP-complete, see [68, 327], as well as Theorem 7.32 here. Still, there are
   interesting connections to other fields (see [12, 219, 328]), explored here in
   Chapters 5 and 9 as well as the exercises on L^p-spaces. Moreover, it is possible to
   produce nice insights into the fixed point property for certain classes of sets (see
   [2, 58, 219, 244, 264, 270, 288, 307]), which can be used in applications (see
   [129, 130]) or to provide deeper structural insights into the theory of ordered
   sets (see [184, 187, 188] or Theorem 4.48 here). For a fairly chronological
   presentation, see [285].
3. The graph-theoretical analogue of an ordered subset is not the concept of a
   subgraph, but that of an *induced* subgraph. **Subgraphs** in graph theory are
   obtained by removing some vertices and possibly also some edges from the
   original graph. This works well, because, in graphs, there are no a priori

---

[4]In [144] it is shown via direct comparison of decks that ordered sets with up to 11 elements are
reconstructible. For reconstruction of small graphs, consider [202].

assumptions on the edges. In ordered sets, removal of comparabilities is not easy, because it might destroy transitivity. Thus generating a substructure by removing comparabilities is not a widely used notion in ordered sets.

4. The reconstruction problem has its origins in graph theory and was then also posed for ordered sets (see [145, 173, 174, 269]). According to "reliable sources" in [28] the problem for graphs is originally due to P.J. Kelly and S. Ulam who discovered it in 1942. The visualization via pictures on cards was suggested by Harary in 1964. For more background on the graph-theoretical reconstruction problem, check, for example, the survey article [28] by Bondy and Hemminger. For a thorough survey of order reconstruction, see [240].

5. Note that, in this section, we have frequently used a certain luxury theorists have. In Definitions 1.7 and 1.8, we defined the diagram of an ordered set and the transitive closure of a relation. We then showed how the diagram and the original order relation are linked via the taking of subsets and the formation of the transitive closure. Therefore, we can use these tools interchangeably. However, we did not provide step-by-step procedures to translate between diagrams and order relations. Similarly, whenever, in a proof, we say "one can find," we do not address the issue how to find the object we are looking for. These issues, together with the algorithmic ramifications of the fixed point property and order-isomorphism, are discussed in Chapter 5. Until then, we will freely use the mathematician's prerogative, to, if an object exists or a transformation can be made, assume that we can pick the object or make the transformation as necessary without worrying how long it might take for us to find or execute it.

Chapter 2
Chains, Antichains, and Fences

Chains and antichains are arguably the most common kinds of ordered sets in mathematics. The elementary number systems $\mathbb{N}, \mathbb{Z}, \mathbb{Q}$, and $\mathbb{R}$ (but not $\mathbb{C}$) are chains. Chains are also at the heart of set theory. The Axiom of Choice is equivalent to Zorn's Lemma, which we will adopt as an axiom, and the Well-Ordering Theorem. The latter two results are both about chains.

Antichains on the other hand are common when people are not talking about order, as an antichain is essentially an "unordered" set. Dilworth's Chain Decomposition Theorem (see Theorem 2.26) shows how these two concepts are linked together. Fences are not as widely known as either chains or antichains. Yet they do play a fundamental role, because they are the analogue of paths in graph theory.

2.1 Chains and the Rank of an Element

In a chain, any two elements are comparable. This means that the hierarchy is total. When people talk about ranking objects, they typically are talking about a chain, which may be why chains seem to be the most natural orders.

Definition 2.1. *An ordered set C is called a **chain** (or a **totally ordered set** or a **linearly ordered set**) iff, for all $p, q \in C$, we have $p \sim q$.*

Example 2.2. The following are all examples of chains.

1. The natural numbers, integers, rational and real numbers with their natural orders.
2. Every set of ordinal numbers.
3. The set $\mathbb{R} \times \mathbb{R}$ with the order $(a, b) \sqsubseteq (c, d)$ iff $a < c$ or ($a = c$ and $b \leq d$).
4. If $f : P \to P$ is order-preserving and $p \in P$ is such that $p \leq f(p)$, then $\{f^n(p) : n \in \mathbb{N}\}$ is a chain. $\qquad \square$

© Springer International Publishing 2016
B. Schröder, *Ordered Sets*, DOI 10.1007/978-3-319-29788-0_2

Proof. We will only prove 4. First note that, by induction on n, for all $n \in \mathbb{N}$, we have $p \leq f^n(p)$: The base step $n = 0$ is trivial, because $p = f^0(p)$. For the induction step $n \to (n+1)$, assume $p \leq f^n(p)$. Then $f(p) \leq f(f^n(p)) = f^{n+1}(p)$ and, by assumption, $p \leq f(p)$. Hence $p \leq f^{n+1}(p)$.

Now let $n, m \in \mathbb{N}$ with $m \leq n$. Then we have $p \leq f^{n-m}(p)$ and hence we conclude $f^m(p) \leq f^m(f^{n-m}(p)) = f^n(p)$. ∎

Remark 2.3. In relation to problems 1.19, 1.24, and 1.34 we can record:

1. Not every chain has the fixed point property. Simply consider $\mathbb{N}$ with the map $f(x) = x + 1$, which is order-preserving and fixed point free. It is possible to classify exactly those chains that have the fixed point property, and we will do so in Theorem 8.10.
2. Finite chains have exactly one automorphism, the identity. Infinite chains can have fixed point free automorphisms: Consider $\mathbb{Z}$ with the map $f(x) = x + 1$.
3. Finite chains with at least four points are reconstructible by Proposition 1.37.

Chains and their lengths are also closely intertwined with the rank, which indicates the "height" at which we can find an element in a finite ordered set.

Definition 2.4. *Let P be an ordered set. An element $m \in P$ is called **minimal** iff, for all $p \in P$ with $p \sim m$, we have $p \geq m$. We denote the set of minimal elements of P by* Min(P).

Note that we were already talking about minimal elements in the proof of Proposition 1.37.

Definition 2.5. *Let P be a finite ordered set. For $p \in P$, we define the **rank** rank(p) of p recursively as follows. If p is minimal, let* rank$(p) := 0$. *If the elements of rank $< n$ have been determined and p is minimal in $P \setminus \{q \in P : $ rank$(q) < n\}$, then we set* rank$(p) := n$.

The rank can be used to somewhat standardize the drawing of a finite ordered set. As was stated in the drawing procedure on page 6, $p < q$ forces that p is drawn with a smaller y-coordinate than q in the diagram. Using the rank function, a drawing of a diagram can be standardized by drawing all points of the same rank at the same height. In fact, this convention has been used in all drawings of diagrams of finite ordered sets so far. The ordered set in Figure 2.1 part a) shows once more that this convention can lead to good drawings of an ordered set. In part b) of Figure 2.1, we show that this convention should not be mechanically applied at all times. The down-up bias of ordering the points by rank can mask symmetries that are not compatible with the rank.

Example 2.6. It is possible for two points in an ordered set to be adjacent and yet have their ranks differ by more than 1. Just consider d and h in the ordered set in part b) of Figure 2.1. ☐

Returning to chains, we record the following.

Definition 2.7. *Let C be a finite chain. The **length of C** is $l(C) := |C| - 1$.*

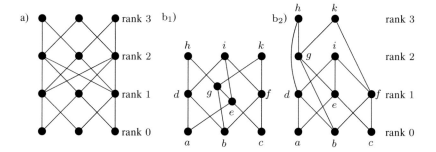

Fig. 2.1 Two ordered sets for which drawing points with the same rank at the same height has different effects. The set in a) has a very orderly drawing revealing several symmetries. The second set is drawn twice. Once (in b_1) to reveal its vertical symmetry (it is isomorphic to its own dual) and once (in b_2) by drawing points of the same rank at the same height

Proposition 2.8. *Let P be a finite ordered set and let $p \in P$. The rank of p is the length of the longest chain in P that has p as its largest element.*

Proof. The proof is an induction on the rank of p. For the base step, rank$(p) = 0$, there is nothing to prove.

For the induction step, let rank$(p) = k > 0$ and assume the statement is true for all elements of P of rank $< k$. Then p has a lower cover l of rank $k - 1$. By induction hypothesis, there is a chain C of length $k - 1$ that has l as its largest element. Thus $\{p\} \cup C$ is a chain of length k with p as its largest element.

Now suppose for a contradiction that there is a chain $K \subseteq P$ of length $> k$ such that p is the largest element of K. Let c be the unique lower cover of p in K. Then c is of rank $< k$ and yet $K \setminus \{p\}$ is a chain of length $\geq k$ with largest element c, a contradiction. Thus the largest possible length of a chain in P that has p as its largest element is k, and we are done. ∎

Proposition 2.9. *Let P, Q be finite ordered sets and let $\Phi : P \to Q$ be an isomorphism. Then, for all $p \in P$, we have* rank$_P(p) =$ rank$_Q(\Phi(p))$.

Proof. First note that the image of a chain under an order-preserving map is again a chain (details are left to Exercise 2-2). Thus, because Φ is injective, the image of any k-element chain under Φ is again a k-element chain. By Proposition 2.8, this means that rank$_P(p) \leq$ rank$_Q(\Phi(p))$.

Now Φ^{-1} is an isomorphism, too, so the result of the previous paragraph also applies to Φ^{-1}. This means that

$$\text{rank}_P(p) \leq \text{rank}_Q(\Phi(p)) \leq \text{rank}_P\left(\Phi^{-1}(\Phi(p))\right) = \text{rank}_p(p),$$

which implies rank$_P(p) =$ rank$_Q(\Phi(p))$. ∎

As we have seen, the elements of the largest rank in the ordered set determine how high the drawing of the order is. This motivates the following definition, which also makes sense for infinite ordered sets.

Definition 2.10. *Let P be an ordered set. The **height** of P is the length of the longest chain in P. If there are chains of arbitrary length in P, we will say that P is of **infinite height**.*

*The height of an ordered set is sometimes also called its **length**.*

Proposition 2.11. *The height of a finite ordered set with at least four elements is reconstructible from the deck of the ordered set.*

Proof. Let P be a finite ordered set with at least four elements. If P is a chain, then P is reconstructible by Proposition 1.37 and we are done. If not, the height of P is the maximum of the heights of the cards $P \setminus \{x\}$. ∎

We can now use the ideas of chains and the rank function to establish that every ordered set has an exponential number of order-preserving self-maps.

Theorem 2.12 (Compare with [75]). *Let P be a finite ordered set with $n > 1$ elements and height h. Then P has at least $2^{\frac{h}{h+1}n}$ order-preserving self-maps that are not automorphisms.*

Proof. The case that P is a chain is left to Exercise 2-4. (Also see Exercise 2-6 for a recursive formula for the number of order-preserving maps between chains.) If $h = 0$, there is nothing to prove. Thus we can assume in the following that P is not a chain, and $h > 0$. Let $C = \{c_0 < c_1 < \cdots < c_h\}$ be a chain with $h + 1$ elements.

Let $r_0, r_1, \ldots, r_h$ be the number of elements of rank h in P. Let $j \in \{0, \ldots, h\}$ be so that $r_j \leq r_i$ for all i. Let $R_0, \ldots, R_h$ be the sets of elements of rank h in P. Then, for any choices of numbers $u_i \in \{0, \ldots, r_i\}$ for $i < j$ and numbers $d_i \in \{0, \ldots, r_i\}$ for $i > j$, we have that the function that maps

- For each $i < j$, u_i elements of R_i to c_{i+1} and the remaining $r_i - u_i$ elements of R_i to c_i,
- For each $i > j$, d_i elements of R_i to c_{i-1} and the remaining $r_i - d_i$ elements of R_i to c_i,

is an order-preserving map that is not an automorphism. The number of such functions is $\prod_{i=0}^{j-1} 2^{r_i} \cdot \prod_{i=j+1}^{h} 2^{r_i} = 2^{n-r_j} \geq 2^{n-\frac{n}{h+1}} = 2^{\frac{h}{h+1}n}$. ∎

Because, for height 0, every non-bijective function is order-preserving, the above shows that every ordered set with more than one element has at least $2^{\frac{n}{2}}$ order-preserving maps that are not automorphisms. In [75] the better lower bound $2^{\frac{2}{3}n}$ is proved. You can do this later, in Exercise 4-14.

Because every automorphism is an order-preserving map, too, Theorem 2.12 shows that there will always be more order-preserving maps than automorphisms. The exponential lower bound on the difference seems substantial. Moreover, existence of a nontrivial automorphism, that is, symmetry, appears to be a rare property, a fact that we will substantiate later in Corollary 13.25. Hence, it is natural to ask if the ratio goes to zero.

Definition 2.13. *Let P be an ordered set. Then* Aut(P) *denotes the set of all automorphisms of P. Moreover,* End(P) *denotes the set of all order-preserving self-maps, or* **(order) endomorphisms** *of P.*

Open Question 2.14. *The automorphism problem. (See [253].) Is it true that*

$$\lim_{|P|\to\infty} \frac{|\mathrm{Aut}(P)|}{|\mathrm{End}(P)|} := \lim_{n\to\infty} \max_{|P|=n} \frac{|\mathrm{Aut}(P)|}{|\mathrm{End}(P)|} = 0 \, ?$$

The **automorphism conjecture** *states that the above limit is indeed zero.*

Natural stronger versions of the automorphism problem would include precise upper bounds on the quotient, even for restricted classes of ordered sets. The automorphism conjecture appears reasonable. Experiences with another problem, namely, how many maps of an ordered set are fixed point free, should caution us against jumping to conclusions, however.

Indeed, it seems that most endomorphisms of an ordered set should have a fixed point. It was even conjectured (very briefly) that the ratio of fixed point free endomorphisms to all endomorphisms should converge to zero. In [253], it is observed that, for an ordered set of height 0, that is, an ordered set in which no two elements are comparable, the limit of the quotient of the number of fixed point free order-preserving maps and the number of all order-preserving maps is $\frac{1}{e}$. It is not known if this is the largest possible quotient or not. Incidentally, the consideration of fixed point free endomorphisms versus all endomorphisms was the context in which the automorphism conjecture first arose. Please consider Open Problem 1 at the end of this chapter for the statement and references on this problem.

Exercises

2-1. Prove Proposition 2.11 using the Kelly Lemma.

2-2. Chains and order-preserving maps. Let P, Q be ordered sets, $f : P \to Q$ be order-preserving and let $C \subseteq P$ be a chain. Prove that $f[C]$ is a chain.

2-3. Prove that, for every finite ordered set P, there is an order-preserving map from P onto a $|P|$-element chain.

2-4. Let C be a chain of length ℓ. Prove that C has at least $2^{\ell} = 2^{\frac{\ell}{\ell+1}|C|}$ order-preserving self-maps that are not automorphisms.

2-5. Prove that, for any finite ordered set P and any two distinct elements $x, r \in P$, we have that $\mathrm{rank}_{P\setminus\{x\}}(r) \in \{\mathrm{rank}_P(r), \mathrm{rank}_P(r) - 1\}$. Give an example of an ordered set P and two points x and r such that the rank of r in $P \setminus \{x\}$ is less than the rank of r in P.

2-6. Let $MC_{n,m}$ be the number of order-preserving maps from an n-chain to an m-chain. Prove that $MC_{n,m} = MC_{n-1,m} + MC_{n,m-1}$.

2-7. Determine the largest class of ordered sets in which a sensible rank function can be defined.

2-8. Prove Proposition 1.10 for ordered sets that have no infinite chains.

2.2   A Remark on Duality

Definitions 2.4 and 2.5 clearly have a "down-up bias": We build the rank function by first defining rank zero, then rank one, etc. There is nothing wrong with this approach, as, for example, a house is built in the same "down-up" fashion. In order theory, however, going up and going down are closely related. Indeed, if we were to reverse all comparabilities, we would obtain equally sensible definitions. For example, if we reverse the comparabilities in Definition 2.4, we obtain the definition of a maximal element.

Definition 2.15.   *Let P be an ordered set. An element $m \in P$ is called **maximal** iff, for all $p \in P$ with $p \sim m$, we have $p \leq m$. We denote the set of maximal elements of P by* $\mathrm{Max}(P)$.

Note that the only difference between Definitions 2.4 and 2.15 is that the one inequality in the definition is reversed. This is why minimal and maximal elements are "at opposite ends" of an ordered set. For example, in the ordered set in Figure 2.1 part a), the elements of rank 0 are (naturally) the minimal elements, while the elements of rank 3 happen to be the maximal elements. Rank and maximality do not have a simple relationship, though. The maximal elements of the ordered set in Figure 2.1 part b) are h, i, k, which do not all have the same rank.

Taking an order-theoretical statement and reversing all inequalities as we did to obtain the definition of maximal elements from that of minimal elements is called **dualizing** the statement. Properties like minimality and maximality, that are obtained from each other by reversing all comparabilities are called **dual properties**. In part 8 of Example 1.2, we mentioned that, for each ordered set P, there is another ordered set P^d, called its dual ordered set, that is obtained by reversing all comparabilities. This means that the minimal elements of P are the maximal elements of P^d and vice versa.

Another example of dual properties are lower bounds and upper bounds.

Definition 2.16.   *Let P be an ordered set. If $A \subseteq P$, then $l \in P$ is called a **lower bound** of A iff, for all $a \in A$, we have $l \leq a$*

Definition 2.17.   *Let P be an ordered set. If $A \subseteq P$, then $u \in P$ is called an **upper bound** of A iff, for all $a \in A$, we have $u \geq a$.*

If l is a lower bound of A, we will also write $l \leq A$, and if L is a set of lower bounds of A we will write $L \leq A$. Similarly, or, better, dually, we can define $u \geq A$ and $U \geq A$.

Duality is a powerful tool when results are proved that are biased in one direction. Instead of re-stating and re-proving everything, one can simply invoke duality. For example, for a finite ordered set, we can also define a dual rank function that has a "top-down bias."

There are notions that are their own duals, such as the notion of being order-preserving. If $f : P \to Q$ is order-preserving, then so is $f : P^d \to Q^d$. Thus, for example, being an isomorphism is a self-dual notion and any result proved about the relation between isomorphisms and, say, upper bounds is also a result on isomorphisms and lower bounds. You are invited to state (and briefly prove) the duals of the results we have proved so far and in the future. In this fashion, when the need for dualization arises, it will not hold any surprises. We conclude this section with one simple example on the use of duality, complete with formal proof. Future uses of duality will not be elaborated as much.

Proposition 2.18. *Let P, Q be finite ordered sets and let $\Phi : P \to Q$ be an isomorphism. Then, for each $p \in P$, the image $\Phi(p)$ has as many upper bounds in Q as p has upper bounds in P. The same statement holds for lower bounds.*

Proof. Let $p \in P$ and let $x \in P$ be such that $x \geq p$. Then, because Φ is order-preserving, $\Phi(x) \geq \Phi(p)$. Moreover, because Φ is injective, no two upper bounds of p are mapped to the same point. Thus $\Phi(p)$ has at least as many upper bounds as p. Now suppose that $\Phi(p)$ has an upper bound q that is not the image of an upper bound of p. Then $\Phi^{-1}(q) \not\geq p$ even though $q \geq \Phi(p)$, a contradiction. This proves that the numbers of upper bounds of p and of $\Phi(p)$ are equal.

To prove the same statement for lower bounds, we could simply follow the above argument with reversed comparabilities. A formal proof using duality could work as follows. Note that Φ is an isomorphism between P^d and Q^d, too. Thus $\Phi(p)$ has as many Q^d-upper bounds in Q^d as p has P^d-upper bounds in P^d. For any ordered set R, an R^d-upper bound in R^d is an R-lower bound in R. Thus $\Phi(p)$ has as many lower bounds in Q as p has lower bounds in P. In the future, such arguments will be replaced with saying "By duality, the statement for the lower bounds holds." ∎

Exercises

2-9. Let P be a finite ordered set of height h. Let $C \subseteq P$ be a chain of length h. Prove each of the following.

  a. The largest element of C is maximal in P.
  b. The smallest element of C is minimal in P.
  c. If x is a lower cover of y in C, then x is a lower cover of y in P.

2-10. Find the maximal elements of each of the following ordered sets.

  a. The ordered set in Exercise 1-1a.
  b. The ordered sets in Figure 1.1.
  c. The ordered set in Figure 1.3.

2-11. Prove that, in a finite ordered set, every element is below at least one maximal element. Use this to prove quickly that, in a finite ordered set, every element is above at least one minimal element.

2-12. For the ordered set in Figure 2.1 part b), find

  a. The upper bounds of the set $\{e\}$,
  b. The upper bounds of the set $\{a, b, c\}$,
  c. The upper bounds of the set $\{f, g\}$,
  d. The upper bounds of the set $\{d, f, g\}$,
  e. The lower bounds of the set $\{g, h, k\}$,

2.3   Antichains and Dilworth's Chain Decomposition Theorem

Logically, antichains are the simplest possible ordered sets, because we impose no comparabilities at all on the points. If chains are totally ordered, one could say that antichains are totally unordered.

Definition 2.19. *An ordered set P is called an **antichain** iff, for all $p, q \in P$ with $p \neq q$, we have $p \not\sim q$.*

Proposition 2.20. *Let P be a finite ordered set and let $A \subseteq P$ be an antichain. Then there is an antichain $B \subseteq P$ that is maximal with respect to set inclusion and contains A. Antichains such as B are also called **maximal antichains**.*

Proof. Left to the reader as Exercise 2-13. Note that, for infinite ordered sets, this result is surprisingly challenging, requiring the use of Zorn's Lemma (see Exercise 2-39). ∎

Antichains can be used to pictorially characterize chains via a "forbidden subset" (also compare with Theorem 11.5).

Proposition 2.21. *An ordered set C is a chain iff C does not contain any two-element antichains.* ∎

Just as chains lead to a natural notion of height in ordered sets, antichains lead to a notion of width.

Definition 2.22. *Let P be an ordered set. We define the **width** $w(P)$ of P to be the size of the largest antichain in P, if such an antichain exists, and to be ∞ otherwise.*

Note that, although chains allow us to define a natural vertical ranking of ordered sets, there is no possibility to define a horizontal ranking using antichains. We can, however, reconstruct the width just as easily as the height.

Proposition 2.23. *The width of a finite ordered set with at least four elements is reconstructible from the deck.*

Proof. If P is not an antichain, then the width of P is the largest width of any card of P. If P is an antichain, then all its cards are antichains. Because only antichains can have a deck consisting of antichains, this would mean that P is reconstructible as the unique (up to isomorphism) antichain with $|P|$ elements. ∎

A nice connection between chains and antichains is provided by Dilworth's Chain Decomposition Theorem, which says that any ordered set of width k can be written as the union of k chains. This fact seems quite obvious. Call our ordered set P and consider the following.

Definition 2.24. *Let P be an ordered set. A chain C in P will be called a **maximal chain** iff, for all chains $K \subseteq P$ with $C \subseteq K$, we have $C = K$.*

Fig. 2.2 An ordered set for which there is a maximal chain (marked) whose removal does not decrease the width

Every chain in an ordered set is contained in a maximal chain (see Exercise 2-15 for the case of finite ordered sets and Proposition 2.55 for the surprisingly subtle infinite case). Let C be a maximal chain. Then $P \setminus C$ should have width $w(P) - 1$ and should thus be (if we argue inductively) the union of $w(P) - 1$ chains. Throw in C, and P is the union of $w(P)$ chains. Unfortunately what we just gave was a "poof," not a proof. Figure 2.2 shows an example of a maximal chain whose removal does not decrease the width. Of course, in this example, we simply did not remove the "right" maximal chain. A formalization of what the "right" maximal chain may be appears to be quite hard. For our proof (which is very similar to the proofs that can be found in [27], Chapter III.4, Theorem 11, [311], Theorem 3.3, or [314]; for another proof consider [220]), it will be useful to have the following notation at hand.

Definition 2.25. *Let P be an ordered set. For $p \in P$, we define the **filter** or **up-set** of p to be $\uparrow p := \{q \in P : q \geq p\}$ and the **ideal** or **down-set** of p to be $\downarrow p := \{q \in P : q \leq p\}$. Finally, we define $\updownarrow p :=\uparrow p \, \cup \downarrow p$ and call it the **neighborhood** of p.*

Theorem 2.26 (Dilworth's Chain Decomposition Theorem, see [63] or Exercise 2-43 for the proof for infinite ordered sets). *Let P be a finite ordered set of width k. Then P is the union of k chains. That is, there are chains $C_1, \ldots, C_k \subseteq P$ such that $P = \bigcup_{i=1}^{k} C_i$.*

Proof. The proof is an induction on $n := |P|$. For $n = 1$, the result is obvious.

For the induction step, assume Dilworth's Chain Decomposition Theorem holds for ordered sets with $\leq n$ elements and let P be an ordered set of width k with $(n+1)$ elements. Let us first assume there is an antichain $A = \{a_1, \ldots, a_k\}$ with k elements in P such that at least one a_j is not maximal and at least one a_j is not minimal. Then the sets $L := \bigcup_{i=1}^{k} \downarrow a_i$ and $U := \bigcup_{i=1}^{k} \uparrow a_i$ both have width k and $\leq n$ elements. Thus U is the union of k chains $U_1, \ldots, U_k$ and L is the union of k chains $L_1, \ldots, L_k$. Without loss of generality we can assume that a_i is the largest element of L_i and the smallest element of U_i. However then each $L_i \cup U_i$ is a chain and $P = \bigcup_{i=1}^{k} (L_i \cup U_i)$.

If the only antichain(s) with k elements in P are the antichain of maximal elements or the antichain of minimal elements, let $C \subseteq P$ be a maximal chain. Then C contains a maximal element and a minimal element. Thus we have $|P \setminus C| \leq n$ and $w(P \setminus C) = k - 1$. By the induction hypothesis, there are chains $C_1, \ldots, C_{k-1}$ such that $P \setminus C = \bigcup_{i=1}^{k-1} C_i$. Now we set $C_k := C$ and we are done. ∎

Dilworth's Chain Decomposition Theorem can be used to provide a surprisingly easy solution to the order-theoretical analogue of a hard graph-theoretical problem. The graph-theoretical analogue of a chain is a **complete graph**, that is, a graph in which any two vertices are connected by an edge. Complete graphs with s vertices are also denoted K_s. The natural analogue of antichains are graphs in which no two vertices are connected by an edge. These are called **discrete graphs**.

Definition 2.27 (For an introduction on Ramsey numbers see [27], Chapter VI; for an introduction to more sophisticated Ramsey Theory for ordered sets see [311], Chapter 10, Section 5). *Let $s, t \in \mathbb{N}$, $s, t \geq 2$. Then the **Ramsey number** $R(s, t)$ is the smallest natural number n such that any graph G with $\geq n$ vertices contains a complete subgraph K_s or a discrete subgraph with t elements.*

One of Ramsey's theorems (see [241]) states that the Ramsey numbers actually are finite. Very few Ramsey numbers are known. In fact, the precise value for $R(5, 5)$ is still unknown. Dilworth's Chain Decomposition Theorem allows us to easily solve the analogous problem for ordered sets.

Proposition 2.28. *Let $s, t \in \mathbb{N}$, $s, t \geq 2$ and let the **ordered set Ramsey number** $R_{\mathrm{ord}}(s, t)$ be the smallest natural number n such that every ordered set with $\geq n$ elements contains a chain with s elements or an antichain with t elements. Then $R_{\mathrm{ord}}(s, t) = (s - 1)(t - 1) + 1$.*

Proof. To see that $R_{\mathrm{ord}}(s, t) > (s - 1)(t - 1)$ consider the ordered set that consists of $t - 1$ pairwise disjoint chains with $s - 1$ elements each and no further comparabilities involved. It has $(s - 1)(t - 1)$ elements and no s-element chain and no t-element antichain.

Now suppose P has $(s - 1)(t - 1) + 1$ elements. We must show that P contains an s-element chain or a t-element antichain. If P contains an antichain with t elements, there is nothing to prove. If every antichain in P has at most $t - 1$ elements, then P has width at most $t - 1$. By Dilworth's Chain Decomposition Theorem, P is the union of at most $t - 1$ chains. However, then one of these chains must have more than $s - 1$ elements. ∎

We can immediately conclude that infinite size means we have an infinite chain or an infinite antichain.

Corollary 2.29. *Every infinite ordered set contains an infinite chain or an infinite antichain.*

Proof. Let P be an infinite ordered set. From Proposition 2.28, we conclude that P must contain chains of arbitrary length or antichains of arbitrary length. Indeed, if the longest chain in P was of length c and the longest antichain in P was of length a, then P would have at most ca elements.

If P contains an infinite chain, then there is nothing to prove. If P does not contain any infinite chains, we want to define a rank function. To make sure the rank function is well defined, we must first prove that there cannot be any elements that are top elements of chains of arbitrary length unless there are infinite antichains: Suppose there is an x so that, for every n, there is a chain X_n with at least n elements and top element x so that X_n is a maximal chain in $\downarrow x$. Then the chain $X_{\sum_{i=1}^{n}|X_n|+1}$ contains an element that is not in any of the chains $X_1, \ldots, X_n$. This fact can be used to inductively construct a nested sequence of antichains whose union is an infinite antichain. Hence, for every $p \in P$, the maximum length of a chain with top element p is finite. For each element $p \in P$, we define $\text{rank}_P(p)$ as the length of a longest chain that has p as its largest element. If, for any k, P has infinitely many elements of rank k, we are done.

Finally, in case P has no infinite chains and, for each k, only finitely many elements of rank k, there must be an element of rank k for each $k \in \mathbb{N}$. Then there must be an infinite sequence $k_1, k_2, \ldots$ of natural numbers, such that, for each i, there is a maximal element m_{k_i} of rank k_i: This sequence is constructed as follows. Start with a minimal element b_1. Because b_1 is not the bottom element of an infinite chain, there is a maximal element above b_1. The rank of this maximal element is k_1; we call the element m_{k_1}. Once k_i is found, let b_i be a non-maximal element of rank k_i. There is a maximal element above b_i. The rank of said maximal element is k_{i+1}; we call the element $m_{k_{i+1}}$.

Because no two maximal elements are comparable, the set $\{m_{k_i} : i \in \mathbb{N}\}$ is an infinite antichain. ∎

Exercises

2-13. Prove Proposition 2.20.

2-14. Prove Proposition 2.23 using the Kelly Lemma.

2-15. Let P be a finite ordered set and let $C \subseteq P$ be a chain. Prove that C is contained in a maximal chain.

2-16. Write the following ordered sets as a union of as few chains as possible.

    a. The ordered set in Figure 1.3,
    b. The ordered sets in Figure 1.4,
    c. The ordered sets in Figure 2.1.

2-17. A subset S of an ordered set P is called an **N** iff $S = \{a, b, c, d\}$ and $a \prec b \succ c \prec d$ with no further comparabilities. An ordered set that does not contain an N is called **N-free**.

    a. Prove that, if a finite ordered set is N-free, then the removal of any maximal chain decreases the width.
    b. Give an example that shows that the converse is not true.

2-18. In the following, we present an *incorrect proof of Dilworth's Chain Decomposition Theorem* (see Theorem 2.26). Find the mistake.

    Induction on $n := |P|$, $n = 1$ is obvious. For the induction step $n \to (n + 1)$, let us assume that Dilworth's Chain Decomposition Theorem holds for ordered sets with n elements and let P be an ordered set with $(n+1)$ elements. Let $m \in P$ be a maximal element. Then $|P \setminus \{m\}| = n$. By the induction hypothesis, there are chains $C_1, \ldots, C_{w(P \setminus \{m\})}$ such that $P \setminus \{m\} = \bigcup_{i=1}^{w(P \setminus \{m\})} C_i$. If $w(P \setminus \{m\}) = k - 1$, we set $C_k := \{m\}$ and we are done. Otherwise, m has strict lower bounds and thus, for some $i_0 \in \{1, \ldots, w(P \setminus \{m\})\}$, we have that m is an upper bound of C_{i_0}. Then $C_{i_0} \cup \{m\}$ is a chain and $C_1, \ldots, C_{i_0-1}, C_{i_0} \cup \{m\}$, $C_{i_0+1}, \ldots, C_{w(P \setminus \{m\})}$ are the desired chains. ∎??

Fig. 2.3 An ordered set (from [76]) without fibre of size $\leq \dfrac{|P|}{2}$

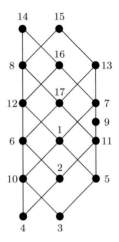

2-19. Consider the ordered set P_0 in Figure 2.3 (also see Remark 11 at the end of this chapter).

    a. Prove that $\{1, 9, 17\}$ is a maximal antichain in P_0.

    b. Find all two-element maximal antichains in P_0.

    c. Prove that every set in P_0 that intersects all maximal antichains and that contains an odd-numbered element must have at least 9 elements.

    d. Prove that every set in P_0 that intersects all maximal antichains must have at least nine elements.

2-20. Prove that, for every finite ordered set P of height h, there are antichains $A_0, \ldots, A_h \subseteq P$ such that $A_i \cap A_j = \emptyset$ for $i \neq j$ and such that $P = \bigcup_{i=0}^{h} A_i$.

2-21. Prove that, for an ordered set of width 3, we have that $|\mathrm{Aut}(P)| \leq 6^{\lceil \frac{n}{3} \rceil}$. Use the result for the number of endomorphisms for sets of height h in Theorem 2.12 to show that, for ordered sets of width 3, we have $\dfrac{|\mathrm{Aut}(P)|}{|\mathrm{End}(P)|} \to 0$ as $|P| \to \infty$.

2.4   Dedekind Numbers

Antichains also give rise to a simple-looking, but still baffling, counting problem.

> **Open Question 2.30 (See [232]).** *Dedekind's problem. What is the number of antichains in the power set $\mathcal{P}_n$ of an n-element set (ordered of course by inclusion)?*

    Although a formula for the number of antichains in $\mathcal{P}_n$ is given in [160], because the computational effort for this formula is so vast, the problem can still be considered unsolved. (To date, the numbers are only known up to $n = 8$, see [322] and Remark 4 at the end of this chapter.) In this section, we record two possible ways to start on this problem. One is to break the problem into smaller subproblems, the other is to translate the problem into another venue. These very common ideas can at least give us a flavor of the problem and of two common techniques. We start with a possible breakup of the problem into smaller problems.

Definition 2.31. *Let* $n \in \mathbb{N}$. *We define*

1. D_n *to be the number of antichains in* $\mathcal{P}(\{1,\ldots,n\})$; D_n *is also called the* n^{th} ***Dedekind number.***
2. T_n *to be the number of antichains in* $\mathcal{P}(\{1,\ldots,n\})$ *such that the union of these antichains is* $\{1,\ldots,n\}$.

Clearly $D_0 = 1$, $T_0 = 1$, because, vacuously, the empty set is an antichain, $D_1 = 2$ and $T_1 = 1$. The connection between the D_n and the T_n is the following.

Proposition 2.32. $D_n = \displaystyle\sum_{k=0}^{n} \binom{n}{k} T_k$.

Proof. Let $\mathcal{A}_k$ be the set of all antichains in $\mathcal{P}(\{1,\ldots,n\})$ whose union is a k-element set. Clearly $D_n = \sum_{k=0}^{n} |\mathcal{A}_k|$. For $B \subseteq \{1,\ldots,n\}$, let $\mathcal{A}_B$ be the set of all antichains in $\mathcal{P}(\{1,\ldots,n\})$ whose union is B. Then $|\mathcal{A}_k| = \sum_{|B|=k} |\mathcal{A}_B|$. Now if $B = \{b_1,\ldots,b_j\}$, then $\mathcal{P}(B)$ is order-isomorphic to $\mathcal{P}(\{1,\ldots,j\})$ via $\varphi : \mathcal{P}(\{1,\ldots,j\}) \to \mathcal{P}(B)$, $\{z_1,\ldots,z_i\} \mapsto \{b_{z_1},\ldots,b_{z_i}\}$. Hence $|\mathcal{A}_B| = T_{|B|}$. Because an n-element set has $\binom{n}{k}$ k-element subsets, we infer $|\mathcal{A}_k| = \binom{n}{k} T_k$, which directly implies the result. ∎

Thus the task of computing the D_n has been reduced to the task of computing the T_n, which is equally formidable. We can record the following.

Definition 2.33. *For* $n \geq 1$ *let* T_n^j *be the number of* j-*element antichains of* $\mathcal{P}(\{1,\ldots,n\})$ *whose union is* $\{1,\ldots,n\}$.

Proposition 2.34. *For* $n \geq 1$, *we have* $T_n = \displaystyle\sum_{j=1}^{n} T_n^j$. ∎

As is often done in a counting task, the above has reduced our task to a number of more specific counting tasks. Unfortunately, such reductions do not always lead to success, and Dedekind's problem is one example in which this is the case. The only further (very modest) advance I can report here is the following.

Proposition 2.35. *For* $n \geq 1$, *we have* $T_n^2 = \dfrac{1}{2}\left(3^n - 2^{n+1} + 1\right)$.

Proof. We will count the number of all pairs

$$(F, S) \in \mathcal{P}(\{1,\ldots,n\}) \times \mathcal{P}(\{1,\ldots,n\})$$

such that $\{F, S\}$ is an antichain and $F \cup S = \{1,\ldots,n\}$. Clearly this number is $2T_n^2$. We first count the ways in which F can be chosen and then we multiply, for each possible choice of F, with the number of ways S can be chosen.

If the first set F has k elements, then $k \neq 0$ and $k \neq n$. Hence we must sum from $k = 1$ to $n - 1$ and, for each k, there are $\binom{n}{k}$ possible choices for F. Once F

is chosen, the second set S must contain $\{1, \ldots, n\} \setminus F$. For $F \cap S$ there are $2^k - 1$ possibilities (only $F \cap S = F$ cannot happen). Hence

$$2T_n^2 = \sum_{k=1}^{n-1} \binom{n}{k} (2^k - 1) = \sum_{k=1}^{n-1} \binom{n}{k} 2^k 1^{n-k} - \sum_{k=1}^{n-1} \binom{n}{k} 1^k 1^{n-k}$$

$$= (3^n - 2^n - 1) - (2^n - 2) = 3^n - 2^{n+1} + 1.$$

∎

It is now easy to see that a similar process will give us the values for T_n^3, T_n^4, and so on. Unfortunately, the formulas become so unwieldy that they are not very useful any more. For example, I was unable to express the T_n^j recursively in terms of T_k^i with $k < n$ and $i < j$. For a computational approach to the problem and a list of the known Dedekind numbers so far (also given in Remark 4 at the end of this chapter), see [322].

Another time honored approach to difficult problems is to translate them into a different venue, in this case, order-preserving maps.

Proposition 2.36. *Let P be a finite ordered set. Then the number of antichains in P is equal to the number of order-preserving maps from P into the two-element chain.*

Proof. We will show that there is a bijective map B from the set $\mathcal{A}(P)$ of all antichains in P, to the set $\text{Hom}(P, \{0, 1\})$ of order-preserving maps from P into the two-element chain $\{0, 1\}$. Let $A = \{a_1, \ldots, a_m\} \in \mathcal{A}(P)$. Define $L(A) := \bigcup_{i=1}^{m} \downarrow a_i$ (note that, for $A = \emptyset$, we have that $L(A) = \emptyset$) and define

$$f_A(p) := \begin{cases} 1; & \text{if } p \notin L(A), \\ 0; & \text{if } p \in L(A). \end{cases}$$

Then $f_A : P \rightarrow \{0, 1\}$ is order-preserving. Indeed, if $p \leq q$ and $f_A(q) = 1$, there is nothing to prove, while if $p \leq q$ and $f_A(q) = 0$, then $q \in L(A)$, which implies $p \in L(A)$ and $f_A(p) = 0$.

Define $B(A) := f_A$. Then B is an injective function, because, for $A_1 \neq A_2$, we have that $(L(A_1) \setminus L(A_2)) \cup (L(A_2) \setminus L(A_1)) \neq \emptyset$, which means that there will be a point $p \in (L(A_1) \setminus L(A_2)) \cup (L(A_2) \setminus L(A_1))$ such that $f_{A_1}(p) \neq f_{A_2}(p)$. To prove that B is surjective, let $f : P \rightarrow \{0, 1\}$ be given. If f is identical to 1, then $f = B(\emptyset)$. Otherwise, let A be the set of maximal elements of $f^{-1}[0]$. Then $f^{-1}[0] = L(A)$ and $f = f_A = B(A)$. Thus B is the desired bijection. ∎

We will consider the problem of counting maps in more detail in Chapter 5. The translation in Proposition 2.36 reaches further than Dedekind's problem, because it applies to arbitrary finite ordered sets. Thus, a natural generalization of Dedekind's problem is to ask for the number of antichains in an arbitrary ordered set. Beyond what is presented in this text, I am unaware of any classes of ordered sets for which this question was asked or answered. However, it was proved in [230] that the computational enumeration must be considered "hard."

Exercises

2-22. Prove that the nth Dedekind number D_n is at least $2^{\binom{n}{\lfloor \frac{n}{2} \rfloor}}$.

2-23. a. Compute the number of all antichains in $\mathcal{P}_1, \mathcal{P}_2, \mathcal{P}_3$, and $\mathcal{P}_4$.
Hint. The numbers are given in Remark 4. (You need to justify why they are right.)
   b. Attempt to find the number of antichains in $\mathcal{P}_n$ for $n \geq 5$ via computer (see [322]).
[Publish the result if you get past 8.]

2.5   Fences and Crowns

The next-simplest ordered structures after chains and antichains are fences and crowns. Fences can be considered analogous to paths connecting distinct points in topology or graph theory. Crowns are very much analogous to closed, non-self-intersecting paths. We start with fences.

Definition 2.37. *Let P be an ordered set. An $(n + 1)$-**fence** (see Figure 2.4) is an ordered set $F = \{f_0, \ldots, f_n\}$ such that $f_0 > f_1, f_1 < f_2, f_2 > f_3, \ldots, f_{n-1} < f_n$ or $f_0 < f_1, f_1 > f_2, f_2 < f_3, \ldots, f_{n-1} > f_n$ if n is even, respectively, $f_0 < f_1, f_1 > f_2, f_2 < f_3, \ldots, f_{n-1} < f_n$ or $f_0 > f_1, f_1 < f_2, f_2 > f_3, \ldots, f_{n-1} > f_n$ if n is odd, and such that these are all comparabilities between the points. The **length** of the fence is n. The points f_0 and f_n are called the **endpoints** of the fence.*

Remark 2.38. Note that the "spider" in Figure 1.1 f) is made up of countably many fences with their left endpoints "glued together."

Proposition 2.39. *Every fence has the fixed point property.*

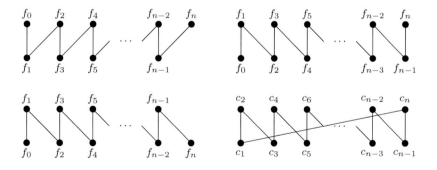

Fig. 2.4 Fences and crowns. On the left are the two nonisomorphic fences with an odd number of elements, top right is the unique (up to isomorphism) fence with an even number of elements. The bottom right shows the unique (up to isomorphism) crown with n elements (n even)

Proof. We will soon have more sophisticated tools to prove this result. However, the following direct proof nicely demonstrates the analogy between fences and intervals on the real line. It is also practice in working with order-preserving maps.

Let $P = \{p_0, p_1, p_2, \ldots, p_n\}$ be a fence and assume, without loss of generality, that $p_0 < p_1$. Suppose $f : P \to P$ was a fixed point free order-preserving map. For each $k \in \{0, \ldots, n\}$, let $g(k)$ be the number l such that $f(p_k) = p_l$. Then, for $|k_1 - k_2| \leq 1$, we have $|g(k_1) - g(k_2)| \leq 1$, because points in P are comparable iff their indices are adjacent. Moreover if $|k - g(k)| \leq 1$, then $p_k \sim f(p_k)$ and $f(p_k)$ is a fixed point of f. Thus $|k - g(k)| \geq 2$ for all k.

Let m be the smallest number such that $g(m) \leq m$. Then $g(m) \leq m - 2$ and $g(m - 1) \geq m + 1$, a contradiction to $|g(m) - g(m - 1)| \leq 1$. Thus P cannot have any fixed point free order-preserving self-maps f.   ∎

Just as we can turn a path from one point to another into a closed path by joining the endpoints, we can turn a fence with an even number of points into a crown by making the endpoints comparable in the appropriate manner.

Definition 2.40. *Let $n \in \mathbb{N}$ be even and ≥ 4. An n-**crown** (see Figure 2.4) is an ordered set C_n with point set $\{c_1, \ldots, c_n\}$ such that $c_1 < c_2, c_2 > c_3, c_3 < c_4, \ldots, c_{n-2} > c_{n-1}, c_{n-1} < c_n, c_n > c_1$ are the only strict comparabilities.*

Just as moving from injective paths with distinct endpoints to closed paths destroys the fixed point property in topology, moving from fences to crowns makes us lose the fixed point property, too.

Proposition 2.41. *Let $n \in \mathbb{N}$ be even and ≥ 4. Then C_n does not have the fixed point property.*

Proof. The map that maps $c_k \mapsto c_{k+2}$ for $k = 1, \ldots, n-2$, $c_{n-1} \mapsto c_1$, and $c_n \mapsto c_2$ is order-preserving and fixed point free.   ∎

Pictorially, it is easy to see that fences are parts of crowns and that crowns cannot be parts of fences. This observation can be formalized as follows.

Proposition 2.42. *Any fence of length k can be embedded into an l-crown with $l > k + 1$. However, no n-crown can be embedded into a fence of any length.*

Proof. Let $F = \{f_0, \ldots, f_k\}$ be a fence of length k and let $C = \{c_1, \ldots, c_l\}$ be an l-crown with $l > k + 1$. Assume without loss of generality that $f_0 < f_1$. Then the map $f_i \mapsto c_{i+1}$ is an embedding.

Now suppose for a contradiction that $C = \{c_1, \ldots, c_n\}$ is an n-crown and that the function $f : C \to F$ is an embedding. Let $f_k := f(c_1)$ and assume without loss of generality that $f(c_2) = f_{k+1}$. Then, for all $i \in \{1, \ldots, n\}$, we must have $f_{k+i-1} = f(c_i)$. In particular $f(c_n) = f_{k+n-1} \not\sim f_k = f(c_1)$ and f could not have been an embedding.   ∎

Fig. 2.5 A class of ordered sets that generalizes the ordered set in Figure 1.3

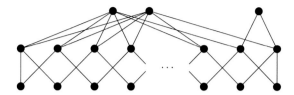

Exercises

2-24. Prove that fences are reconstructible.

2-25. The number of antichains in crowns and fences. For a finite ordered set P, we define $\#A(P)$ to be the number of antichains in P.

    a. (See [19].) Let P be a finite ordered set. Prove that $\#A(P) = \#A(P \setminus \{x\}) + \#A(P \setminus \updownarrow x)$.
    b. Prove that $\#A(F_n) = \#A(F_{n-1}) + \#A(F_{n-2})$.
    c. Prove that $\#A(F_n)$ is the $(n+2)^{\text{nd}}$ Fibonacci number.
    d. Prove that $\#A(C_n) = \#A(F_{n-1}) + \#A(F_{n-3})$.

2-26. Prove that the number of order-preserving self-maps of fences with n elements is bounded by $n3^{n-1}$.

2-27. Prove that crowns are reconstructible.

2-28. Prove that all fixed point free order-preserving self-maps of a crown are automorphisms.

2-29. Consider the ordered sets in Figure 2.5.

    a. Prove that each of these sets has the fixed point property.
    b. Find at least one more class of ordered sets that generalizes the ordered sets in Figure 2.5 and in which each ordered set has the fixed point property.

2-30. For $j \in \{1, \ldots, m\}$ let $C_j := \{c_1^j, \ldots, c_{2n}^j\}$ be crowns with $2n$ elements such that, for all $i \in \{1, \ldots, n\}$ and $j \in \{2, \ldots, m\}$, we have $c_{2i}^{j-1} = c_{2i-1}^j$. Let $\prec_j$ be the lower cover relation of C_j. Equip $CT_{2n}^m := \bigcup_{j=1}^m C_j$ with the order induced by the lower cover relation which is the union $\bigcup_{j=1}^m \prec_j$ of the individual lower cover relations. The ordered set CT_{2n}^m thus obtained is called a $2n$-**crown tower** with m crowns, or a $(2n\text{-})$**crown tower** when m or $2n$ are not needed. Prove that any $2n$-crown tower has a fixed point free automorphism.

2-31. Prove that a 6-crown tower with 3 crowns T_6^3 has more than two fixed point free order-preserving maps.

2-32. Prove that $\#A(CT_4^m) = 3m + 4$.

2.6 Connectivity

The notion of connectivity for ordered sets (as well as the analogous notion for graphs) is inspired by what is called pathwise connectivity in topology.

Definition 2.43. *Let P be an ordered set. P is called **connected** iff, for all $a, b \in P$, there is a fence $F \subseteq P$ with endpoints a and b. An ordered set that is not connected is called **disconnected**.*

    Similar to topology, the fixed point property implies connectivity.

Proposition 2.44. *Let P be an ordered set with the fixed point property. Then P is connected.*

Proof. Suppose for a contradiction that P is not connected. Let $a, b \in P$ be two points such that there is no fence with endpoints a and b. Define

$$f(p) := \begin{cases} a; & \text{if there is a fence with endpoints } b \text{ and } p, \\ b; & \text{otherwise.} \end{cases}$$

Then f is order-preserving and has no fixed points, a contradiction.   ∎

A related problem is the question what properties the set of fixed points of an order-preserving mapping has when it is nonempty. Although there are nice positive results (see, for example, Theorem 8.10 and Section 8.2.2), we present a negative result here.

Example 2.45. Even a finite ordered set P with the fixed point property can have a self-map $f : P \to P$ such that $Fix(f)$ is not connected. (See Exercise 2-36.)   □

Analogous to topology, we can define the connected components of an ordered set. Components and the metric notions defined thereafter will be useful for our proof that disconnected ordered sets are reconstructible.

Proposition 2.46. *Let P be an ordered set. If $S \subseteq P$ is a connected subset of P, then there is a maximal (with respect to inclusion) connected subset C of P such that $S \subseteq C$.*

Proof. Let $\mathcal{C}$ be the set of all connected subsets $C_x \subseteq P$ that contain S. Then $C := \bigcup \mathcal{C}$ is a connected subset of P: Indeed, for $a, b \in C$, let s be an arbitrary point in S. Let $C_a \in \mathcal{C}$ be such that $a \in C_a$ and let $C_b \in \mathcal{C}$ be such that $b \in C_b$. Then there is a fence F_a in C_a from a to s and there is a fence F_b in C_b from s to b. Thus, after possibly removing some elements from $F_a \cup F_b$, there is a fence in C from a to b and C is connected.

The definition of C shows that C is the largest connected subset of P that contains S.   ∎

Definition 2.47. *The maximal (with respect to inclusion) connected subsets of an ordered set are called the **(connected) components** of the ordered set.*

Metric notions such as distance and diameter arise in ordered sets through the lengths of fences.

Definition 2.48. *The **distance** dist(a, b) between two points $a, b \in P$ is the length of the shortest fence from a to b. If a and b are in different components of P, we will say the distance is infinite. The **diameter** diam(P) of an ordered set P is the largest distance between any two points in P. If P contains points that are arbitrarily far apart or if P is disconnected, we say the diameter of P is infinite.*

Example 2.49. 1. The "one-way infinite fence" $F_\omega := \{f_0 < f_1 > f_2 < f_3 > \cdots\}$ is connected and has infinite diameter and infinite width.
2. The "spider" in Figure 1.1 is an ordered set with infinite diameter that contains no infinite fences.
3. Let A be an antichain. Obtain the set V by adding an element to A that is an upper cover of all elements of A. Then V has width $|A|$, but its diameter is 2.   □

Diameter and width appear to be related notions, as they both measure how far away an ordered set is from being a chain. However, both notions are distinct, and there are no globally tight inequalities that relate either to each other. Indeed, Example 2.49, part 3 shows that there is no possibility to bound the width of an ordered set by using the diameter. Proposition 2.50 gives an inequality in the opposite direction. Although the inequality cannot be improved in general, Example 2.49, part 3 shows that it need not be very good in special cases.

Proposition 2.50. *Let P be an ordered set of finite diameter. Then the diameter of P is bounded by twice the width of P minus 1 and this inequality cannot be improved.*

Proof. A fence of length n contains an antichain of length $\left\lceil \frac{n+1}{2} \right\rceil$. Therefore $\left\lceil \frac{\text{diam}(P)+1}{2} \right\rceil \leq w(P)$, which implies $\text{diam}(P) \leq 2w(P) - 1$.
The fences with $2k$ elements all have length $2k - 1$ and width k. Thus the above inequality cannot be improved.   ∎

All the above notions merge nicely in the proof that disconnected ordered sets are reconstructible.

Proposition 2.51. *Let P be a disconnected finite ordered set with $|P| \geq 4$. Then P is reconstructible from its deck.*

Proof. First, we prove that disconnected ordered sets are recognizable. If P is disconnected, then all cards of P are disconnected unless P has a component with one point. In this case, all but one card of P are disconnected.
On the other hand, if Q is a connected ordered set, we can show that Q has at least two connected cards: Let $a, b \in P$ be such that $\text{dist}(a, b) = \text{diam}(P)$. Then $P \setminus \{a\}$ must be connected. Indeed, otherwise there is a $c \in P$ such that every fence from c to b goes through a, which means $\text{dist}(c, b) > \text{dist}(a, b)$, contradicting the choice of a and b. Similarly we prove that $P \setminus \{b\}$ is connected.
Thus disconnected ordered sets are recognizable: They are the ordered sets whose decks contain at most one connected card. For reconstruction, note that, if the deck of a disconnected ordered set D contains a connected card, then the components of D are the connected card and a singleton. This leaves the case in which all cards of D are disconnected. Find a card C of D such that the sum of the squares of the component sizes is maximal. This card must have been obtained by removing an element from a minimum-sized component. Thus C has a unique smallest component, which is the unique component of C that is not a component of D. Let s be the size of the unique smallest component of C. Then the last component of D is the unique connected ordered set S with $s + 1 < |D|$ elements such that D

contains more isomorphic copies of S than C. Because the number of isomorphic copies of S in D can be determined via the Kelly Lemma (see Proposition 1.40), we have reconstructed D.   ∎

Exercises

2-33. Prove that a one-way infinite fence does not have the fixed point property.

2-34. Prove that, if P is connected and $f : P \to Q$ is order-preserving, then $f[P]$ is connected.

2-35. Show that there are disconnected ordered sets P such that every automorphism of P has a fixed point. (Also see Exercise 12-25.)

2-36. For each of the following ordered sets, find an order-preserving self-map with a disconnected set of fixed points.

   a. The ordered set in Figure 1.1 b).
   b. The ordered set in Figure 1.3.
   c. The ordered sets in Figure 2.5.

2-37. The reconstruction problem has a negative answer for infinite ordered sets: Let P be a connected ordered set of height 1 such that P has no crowns, every maximal element has countably many lower covers and every minimal element has countably many upper covers. Let Q be an ordered set with two connected components that are isomorphic to P. Show that P and Q have the same (infinite) deck. (Of course the deck would have to be a function that assigns every class $[C]$ of ordered sets the cardinality of $\{p \in P : P \setminus \{p\} \in [C]\}$.)

2-38. Some notes on reconstruction of the diameter.

   a. Let P be an ordered set and let $A \subseteq P$ be a subset of P. Prove that $\mathrm{diam}(A) \geq \mathrm{diam}(P)$.
   b. Give an example of an ordered set P and a point $x \in P$ so that $\mathrm{diam}(P \setminus \{x\}) > \mathrm{diam}(P)$.
   c. Give an example of an ordered set P so that, for all $x \in P$, we have the strict inequality $\mathrm{diam}(P \setminus \{x\}) > \mathrm{diam}(P)$.

   Note. I do not know of any proof that the diameter is reconstructible. Exercise 2-38a gives an obvious starting point. Exercise 2-38c shows that a proof would not be totally trivial.

2.7   Maximal Elements in Infinite Sets: Zorn's Lemma

In a finite ordered set, every element is below at least one maximal element (see Exercise 2-11). For infinite ordered sets, as long as we don't have chains without upper bounds (like in the natural numbers $\mathbb{N}$), the same thing *should* hold. We can prevent the existence of chains without upper bounds by demanding that our set is inductively ordered.

Definition 2.52. *Let P be an ordered set. Then P is called **inductively ordered** iff every nonempty chain $C \subseteq P$ has an upper bound.*

Surprisingly, even when we exclude chains without upper bounds, existence of maximal elements cannot be derived from simpler axioms of set theory. Hence, the existence of maximal elements in inductively ordered sets actually has the status of an axiom.

Axiom 2.53. *Zorn's Lemma.* *Let P be an inductively ordered set. Then P contains a maximal element M.*

Zorn's Lemma is a standard tool for mathematicians who freely use the Axiom of Choice (to which Zorn's Lemma is equivalent, see Exercise 2-44). It is used, for example, in algebra to establish the existence of maximal ideals in rings with unity, or in functional analysis in the proof of the Hahn–Banach Theorem (see Exercise 2-41). On the other hand, the Axiom of Choice can be used to establish the existence of non-measurable sets and such counterintuitive things as the Banach–Tarski Paradox (see [317]). This is why some mathematicians decide to avoid its use. The philosophical issues that arise from using the infinitary generalization of a statement that is obvious in finite structures (see Exercise 2-11) are beyond the scope of this text. We will freely use Zorn's Lemma and anything equivalent here.

The structure of a proof involving Zorn's Lemma is fairly standard: The key is to design an appropriate nonempty set of objects that is inductively ordered and such that the desired element (if it exists) is maximal in it. The appropriate order often is set inclusion. We give examples in the following.

Our first example is the Axiom of Choice. The proof will feature all the characteristics of a proof using Zorn's Lemma without any additional order theory involved. Subsequent examples will (naturally) be examples that apply to order theory. These examples will often work with several orders at once and we will need to distinguish these orders carefully. Thus the Axiom of Choice is a good "warm-up" to the slightly more complex arguments that follow. Note that we do not give an *absolute* proof of the Axiom of Choice. All that is shown in the following is that the Axiom of Choice is true *if Zorn's Lemma is true.*

Theorem 2.54 (Axiom of Choice.). *If $\{P_\alpha\}_{\alpha \in I}$ is an indexed family of nonempty sets, then there is a function $f : I \to \bigcup_{\alpha \in I} P_\alpha$, also called a **choice function**, such that, for all $\alpha \in I$, we have $f(\alpha) \in P_\alpha$.*

Proof (using Zorn's Lemma). Let $\mathcal{P}$ be the set of all functions $f : D \to \bigcup_{\alpha \in I} P_\alpha$ such that $D \subseteq I$ and, for all $\alpha \in D$, we have $f(\alpha) \in P_\alpha$. Clearly this set is not empty, because, for any finite set $D \subseteq I$, such functions exist.

Because every function f is just the set $\{(\alpha, f(\alpha)) : \alpha \in \text{domain}(f)\}$, we have that $\mathcal{P}$ is ordered by set inclusion. If you prefer to think about restrictions to subsets we can say that, if $f_i : D_i \to \bigcup_{\alpha \in I} P_\alpha$, $i = 1, 2$ are functions in $\mathcal{P}$, then $f_1 \subseteq f_2$ iff $D_1 \subseteq D_2$ and $f_2|_{D_1} = f_1$. All following arguments that use unions can be re-written in language similar to the previous sentence. We choose not to do so, as we will see that the formation of unions is a much more compact way of arguing.

To show that $\mathcal{P}$ is inductively ordered, let $\mathcal{C} \subseteq \mathcal{P}$ be a nonempty chain. Define $g := \bigcup \mathcal{C}$. Then g is well-defined: Indeed, let $\alpha \in \text{domain}(g)$. Then, for all functions $f \in \mathcal{C}$ for which $f(\alpha)$ is defined, we have $(\alpha, f(\alpha)) \in g$, and these are all the pairs in g with first component α. If $f_1, f_2 \in \mathcal{C}$ satisfy $\alpha \in \text{domain}(f_1) \cap \text{domain}(f_2)$, then we can assume without loss of generality that $f_1 \subseteq f_2$. But that means $f_1(\alpha) = f_2(\alpha)$ and so g is well-defined.

Moreover, g is a (partial) choice function. For each $\alpha \in \text{domain}(g)$, there is an $f \in C$ with $\alpha \in \text{domain}(f)$. This means that $g(\alpha) = f(\alpha) \in P_\alpha$.

By Zorn's Lemma, we can now conclude that $\mathcal{P}$ has a maximal element F. To show that F is a choice function on I and not just on some subset $D \subset I$, suppose for a contradiction that $\text{domain}(F) \neq I$. Then there is an $\alpha \in I \setminus \text{domain}(F)$. Let $p_\alpha \in P_\alpha$ and define $G := F \cup \{(\alpha, p_\alpha)\}$. Then $G \in \mathcal{P}$ and $G \supsetneq F$ is a strict upper bound of F, contradicting the maximality of F. Thus $\text{domain}(F) = I$ and F is the desired choice function. ∎

Having warmed up to Zorn's Lemma we now focus on maximal chains once more.

Proposition 2.55. *Let $(P, \leq)$ be a nonempty ordered set and let $C_0 \subseteq P$ be a chain. Then there is a chain $M \subseteq P$ which is maximal with respect to set inclusion and such that $C_0 \subseteq M$. That is, there is a maximal chain that contains C_0.*

Proof. Let $\mathcal{C}$ be the (nonempty) set of all chains $C \subseteq P$ with $C_0 \subseteq C$. This set is ordered by set inclusion $\subseteq$.

To show that $\mathcal{C}$ is inductively ordered, let $\mathcal{K} \subseteq \mathcal{C}$ be a chain with respect to inclusion. Consider the set $K := \bigcup \mathcal{K}$ (with the induced order from P). To see that $K \in \mathcal{C}$, we must show that K is a chain. Let $x, y \in K$. Then there are $C_x, C_y \in \mathcal{K}$ such that $x \in C_x$ and $y \in C_y$. Because $\mathcal{K}$ is a chain, we can assume without loss of generality that $C_x \subseteq C_y$. Thus $x, y \in C_y$ and, because C_y is a chain, we have $x \sim_P y$. Thus K is a chain and hence K is in $\mathcal{C}$. Because, for all $C \in \mathcal{K}$, we trivially have $K \supseteq C$, we conclude that K is an upper bound of $\mathcal{K}$ in $\mathcal{C}$. Thus $\mathcal{C}$ is inductively ordered.

Therefore, by Zorn's Lemma, $\mathcal{C}$ has a maximal element M. By choice of $\mathcal{C}$, M is a chain that contains C_0 and is maximal with respect to inclusion. ∎

In the next section, we will see another application of Zorn's Lemma when we prove that every set can be well-ordered.

Exercises

2-39. Let P be an infinite ordered set and let $A \subseteq P$ be an antichain. Prove that there is an antichain $B \subseteq P$ that is maximal with respect to set inclusion and contains A.

2-40. Let V be a vector space. A **basis** of V is a subset $B \subseteq V$ such that any finite subset of B is linearly independent and such that any $v \in V$ has a (necessarily unique) representation as a finite linear combination of elements of B. Prove that every vector space has a basis.

2-41. Hahn–Banach Theorem. Let V be a normed vector space over $F = \mathbb{R}$ or $F = \mathbb{C}$. A **(continuous) linear functional** is a linear function $\Phi : V \to F$ such that there is a $c > 0$ such that, for all $v \in V$, we have $|\Phi(v)| \leq c\|v\|$.

Let $W \subset V$ be a linear subspace of V and let $v \in V \setminus W$. Prove that there is a continuous linear functional $\Phi : V \to F$ such that $\Phi|_W = 0$ and $\Phi(v) = 1$.

2-42. State the definition of a dual inductive order and state and prove the dual of Zorn's Lemma.

2-43. Let us now finally prove the general version of Dilworth's Chain Decomposition Theorem. It states that *any* ordered set of width k can be written as the union of k chains. We follow Dilworth's original idea (see [63], p. 163).

The idea is of course an induction on k with $k = 1$ still being trivial, so assume that the result holds for sets of width $k - 1$. Define a chain $C \subseteq P$ to be *strongly dependent* iff, for every finite subset $S \subseteq P$, there is a representation of $S = K_1 \cup \cdots \cup K_l$ as a union of chains K_i such that $S \cap C \subseteq K_i$ for some $i \in \{1, \ldots, l\}$.

a. Prove that there is a maximal strongly dependent chain C_1 in P.
b. Prove that $P \setminus C_1$ has width $k - 1$. To do so, assume that $\{a_1, \ldots, a_k\}$ is an antichain in $P \setminus C_1$.

- For each i find a finite set $S_i \subseteq P$ such that no chain decomposition of S_i contains $S_i \cap (C_1 \cup \{a_i\})$ in one chain.
- Apply the property of strong dependence of C_1 to $S = \bigcup_{i=1}^{k} S_i$.
- Use the insight gained to find a j such that S_j has a chain decomposition such that $S_j \cap (C_1 \cup \{a_j\})$ is contained in exactly one chain.

2-44. Prove that the Axiom of Choice is equivalent to Zorn's Lemma.
Note. This is quite challenging, see, for example, [283].

2.8 Well-Ordered Sets

Well-ordered sets are a particularly nice type of chain. Ordinal numbers in set theory are examples of well-ordered sets. In fact, they are, up to isomorphism, *all* well-ordered sets.

Definition 2.56. *Let P be an ordered set and let $S \subseteq P$. Then $s \in S$ is called the* **smallest element** *of S iff $s \leq S$.*

Definition 2.57. *An ordered set W is called* **well-ordered** *iff each nonempty subset $A \subseteq W$ has a smallest element.*

It is easy to see that well-ordered sets are chains and that not all chains are well-ordered. The next proposition shows that well-ordered sets are very closely related to the notion of counting. For every non-maximal element there will be a "next" element.

Proposition 2.58. *Let W be a well-ordered set. Then every non-maximal element $w \in W$ has an* **immediate successor**. *That is, for every non-maximal $w \in W$, there is an element w^+ such that $w < w^+$ and, for all $p > w$, we have $p \geq w^+$.*

Proof. Let $w \in W$ be not maximal. Then $\{p \in W : p > w\}$ is not empty and it therefore has a smallest element. This smallest element is w^+. ∎

Example 2.59. Every finite chain is well-ordered. The natural numbers $\mathbb{N}$ are the smallest infinite well-ordered set. It would be tempting to try to prove that, just as in $\mathbb{N}$, every element of a well-ordered set has an immediate predecessor, too. An immediate predecessor of w would of course be an element w^- such that $w > w^-$ and, for all $p < w$, we have $p \leq w^-$. The well-ordered set $\mathbb{N} \oplus \{\infty\}$, consisting of $\mathbb{N}$

with a largest element ∞ attached, shows that not every element of a well-ordered set has an immediate predecessor: In this example, ∞ does not have an immediate predecessor.   □

Informally speaking, well-ordered sets can be built as we just described. Pick a well-ordered set and attach a new largest element to get a new well-ordered set. This generates chains of well-ordered sets, which then can be united to form even bigger well-ordered sets. How far we can push this process will depend on how strong a version of the Axiom of Choice we are willing to accept. The standard class of examples of well-ordered sets (and, see Exercise 2-49, the only class of examples) is the class of ordinal numbers.

Example 2.60 (See [117], p. 75.).   An **ordinal number** is a well-ordered set α such that, for each $\xi \in \alpha$, we have that $\xi^+ = \{\eta \in \alpha : \eta \leq \xi\}$. That is, the immediate successor of each ordinal number ξ is the set of all ordinal numbers up to and including ξ. Because we can form sets of existing objects, this is a well-defined operation that allows formation of ordinal numbers. The set-theoretical problems start once we encounter infinite ordinals.

The **first infinite ordinal number** is (isomorphic to) the set of natural numbers. In ordinal arithmetic, it is denoted by ω. By simply defining the successor via the above equation, we obtain ω^+, $(\omega^+)^+$, and so on. In ordinal arithmetic, the nth successor of an ordinal α is called $\alpha + n$. Note that ω does not have an immediate predecessor. Indeed, any ordinal number that is less than ω is finite and thus its successor is finite, too, and not equal to ω. Ordinal numbers that do not have an immediate predecessor are also called **limit ordinals**.

The indicated counting process continues through all $\omega + n$ and the next limit ordinal is 2ω. Continuing in the above fashion we reach the limit ordinals 3ω, 4ω, ..., until we reach ω^2. What follows are $\omega^2 + 1$, $\omega^2 + 2$, ..., $\omega^2 + \omega$, ..., $\omega^2 + 2\omega$, ..., ω^3, ..., ω^4, ..., ω^ω. As in [325], p. 10 we will denote the **first uncountable ordinal number** by ω_1.   □

Proposition 2.58 and Examples 2.59 and 2.60 show that well-ordered sets are a natural generalization of the natural numbers. In a well-ordered set, there is a natural notion of counting (the immediate successor of each element is the "next" element as we count) and we can count "past infinity" if the well-ordered set is big enough. On the other hand, well-ordered sets can arise anywhere, if we believe Zorn's Lemma: With Zorn's Lemma, we can show that any set can be well-ordered.

Definition 2.61. *Let W be a well-ordered set. A well-ordered subset $V \subseteq W$ is called an **initial segment** of W iff $V \subseteq W$ and, for all $w \in W \setminus V$, we have that w is an upper bound of V.*

Theorem 2.62.   Well-Ordering Theorem. *For every set S, there is an order relation $\leq \subseteq S \times S$ such that $(S, \leq)$ is well-ordered.*

Proof (using Zorn's Lemma). Let $\mathcal{W}$ be the set of all pairs $(\leq, M)$ such that $M \subseteq S$ and $\leq \subseteq M \times M$ is a well-ordering. Then $\mathcal{W}$ is not empty, because, for each $s \in S$, the pair $(\{(s, s)\}, \{s\})$ is in $\mathcal{W}$. Order $\mathcal{W}$ by $(\leq_1, M_1) \sqsubseteq (\leq_2, M_2)$ iff

1. $M_1 \subseteq M_2$,
2. $\leq_1 \subseteq \leq_2$ (as relations),
3. M_1 is an initial segment of M_2 ordered by $\leq_2$.

To prove that $\mathcal{W}$ is inductively ordered, let $\mathcal{K}$ be a $\sqsubseteq$-chain in $\mathcal{W}$. Let

$$M_t := \bigcup \{M : (\leq, M) \in \mathcal{K}\}$$

and equip it with the relation

$$\leq_t := \bigcup \{\leq : (\leq, M) \in \mathcal{K}\}.$$

We must show that M_t is well-ordered by $\leq_t$ and that, for all $(\leq, M) \in \mathcal{K}$, we have that M is an initial segment of M_t. We shall first show that M_t is totally ordered. Let $x, y, z \in M_t$. Then there are M_x, M_y, and M_z so that $x \in M_x$, $y \in M_y$, and $z \in M_z$. Without loss of generality, assume that $(\leq_x, M_x) \sqsubseteq (\leq_y, M_y) \sqsubseteq (\leq_z, M_z)$. Thus $x \leq_x x$, which implies $x \leq_t x$ for all $x \in M_t$. If $x \geq_t y$ and $x \leq_t y$, then $x \geq_y y$ and $x \leq_y y$ (there is a small argument here, see Exercise 2-47), which implies $x = y$. If $x \leq_t y$ and $y \leq_t z$, then $x \leq_z y$ and $y \leq_z z$, which implies $x \leq_z z$ and then $x \leq_t z$. Finally, for all $x, y \in M$, we have either $x \geq_y y$ or $x \leq_y y$ which means $x \geq_t y$ or $x \leq_t y$ and $\leq$ is a total order on M.

Now we will show that, for every $(\leq, M) \in \mathcal{K}$, and every $b \in M_t \setminus M$, we have that b is an $\leq_t$-upper bound of M. Let $(\leq, M) \in \mathcal{K}$ and let $b \in M_t \setminus M$. Then there is a $(\leq', M') \in \mathcal{K}$ so that $b \in M'$. Because $b \in M' \setminus M$, we must have $(\leq, M) \sqsubseteq (\leq', M')$. By condition 3, we infer that b is an $\leq'$-upper bound of M and hence it is an $\leq_t$-upper bound of M.

To show that $\leq_t$ is a well-ordering, let $A \subseteq M_t$ be nonempty. Then there is a $(\leq, M) \in \mathcal{K}$ such that $A \cap M \neq \emptyset$. Let $a \in A$ be the $\leq$-smallest element of A. Then, for all $b \in A \cap M$, we have $b \geq a$, hence $b \geq_t a$. For $b \in A \setminus M$, the preceding paragraph shows $b \geq_t a$. Thus a is the $\leq_t$-smallest element of A. Hence $\leq_t$ is a well-ordering and, via the preceding paragraph, $(\leq_t, M_t)$ is a $\sqsubseteq$-upper bound of $\mathcal{K}$. This means $(\mathcal{W}, \sqsubseteq)$ is inductively ordered.

Let $(\leq, M) \in \mathcal{W}$ be a $\sqsubseteq$-maximal element as guaranteed by Zorn's Lemma. If $M = S$, we are done. Assume that there is an $s \in S$ that is not in M. Define

$$\leq' := \leq \cup \{(x, s) : x \in M \text{ or } x = s\} \subseteq (M \cup \{s\}) \times (M \cup \{s\}).$$

Then $\leq'$ is easily verified to be a well-ordering. But then $(\leq', M \cup \{s\}) \in \mathcal{W}$ is a $\sqsubseteq$-upper bound of $(\leq, M)$ that is not equal to $(\leq, M)$, which is a contradiction. Thus $\leq$ must be a well-ordering of $S = M$.   ∎

Note that, in the proof of the Well-Ordering Theorem, it is not possible to avoid the complicated definition of $\sqsubseteq$. Indeed, if we just used containment of the orders as $\sqsubseteq$, then the candidates for upper bounds of chains as defined in the proof need not be well-ordered. That is, they might not be in $\mathcal{W}$. For illustration, consider the set of chains $\mathcal{B} := \{\{\pm\frac{1}{k} : k = 1, \ldots, n\} : n \in \mathbb{N}\}$, with each individual chain ordered with the order inherited from $\mathbb{Q}$. Set containment induces a total order on this set of chains and any two chains in this set are well-ordered (after all, all chains in $\mathcal{B}$ are finite). Yet the union of these chains is the set $\bigcup \mathcal{B} = \{\pm\frac{1}{n} : n \in \mathbb{N}\}$, which is not well-ordered. So we need condition 3 in the definition of $\sqsubseteq$ to prevent chains that we unify from "filling in holes" instead of "building upwards in a well-ordered fashion."

A final observation about well-ordered sets is that they have many of the properties that increasing sequences have. In the near future (see Theorem 4.17), we will encounter situations in which we are only interested in "where the tops of certain chains go." In such situations it will be helpful to replace the chains we have with chains that have the same "growth" or "convergence behavior" and are otherwise well-behaved. The result that allows us to do so is the fact that any chain has a cofinal (see Definition 2.63) well-ordered subchain (see Proposition 2.64).

Definition 2.63. *Let P be an ordered set and let $A \subseteq B \subseteq P$. Then A is called* **cofinal (coinitial)** *in B iff, for every $b \in B$, there is an $a \in A$ such that $a \geq b$ ($a \leq b$).*

Proposition 2.64. *Let P be an ordered set. Then, for every chain $C \subseteq P$, there is a well-ordered cofinal subchain $W \subseteq C$.*

Proof. Left as Exercise 2-45. Hint: Order the well-ordered subchains of C with an order like in the proof of Theorem 2.62. ∎

Exercises

2-45. Prove Proposition 2.64.
2-46. Prove that a well-ordered set has the fixed point property iff it has a largest element.
2-47. In the proof of the Well-Ordering Theorem, we claim that, if $x \in M_x$ (ordered by $\leq_x$) and $y \in M_y$ (ordered by $\leq_y$) with $(\leq_x, M_x) \sqsubseteq (\leq_y, M_y)$ and $x \leq_t y$, then $x \leq_y y$. Formally we only know that $x \leq y$ for some $(\leq, M)$ where $x, y \in M$. Prove that $x \leq_y y$.
2-48. Prove that the Well-Ordering Theorem implies Zorn's Lemma.
     Hint. Use the result from Exercise 2-44.
2-49. Prove that every well-ordered set is isomorphic to one of the ordinal numbers in Example 2.60.
2-50. Let (Ω, Σ, μ) be a measure space and let $p \in [1, \infty)$. We only consider real valued functions. Prove that if $W \subseteq L^p(\Omega, \Sigma, \mu)$ is well-ordered without a largest element, then W is countable and there is a cofinal subchain $N \subseteq W$ that is isomorphic to $\mathbb{N}$.

Remarks and Open Problems

This is the first time in this text that the remarks section also includes open problems. The open problems presented in this section are special cases of the main open questions or modifications of them.

1. The original questions that motivated the automorphism problem: Let Fpf(P) be the set of fixed point free maps of the ordered set P.

   a. Find an overall bound for $\dfrac{|\text{Fpf}(P)|}{|\text{End}(P)|}$.

   b. Find $\limsup\limits_{|P|\to\infty} \dfrac{|\text{Fpf}(P)|}{|\text{End}(P)|}$.

   c. Find the above quantities when P is restricted to a special class of ordered sets.

   I am not aware of any progress in this direction beyond [253]. Natural candidates for first partial results might be ordered sets of small width, as the analogous automorphism problem also seems to be solvable there (see Exercises 2-21 and 9-7).

2. The ordered set in Figure 2.2 also appears in [64], where R. P. Dilworth recounts the history of Theorem 2.26. For more on the work of R. P. Dilworth, consider [25]. For connections between graph theory and Dilworth's Theorem, consider Section 2.1 in [319].

3. Dilworth's Theorem cannot be extended to ordered sets of infinite width. In [220], it is proved that, for every infinite cardinal c, there is an ordered set of cardinality c without infinite antichains, which cannot be decomposed into less than c chains. This construction is presented in Exercise 12-10.

4. The first eight Dedekind numbers have been found via computation. The list goes as follows (see [322]) 2; 3; 6; 20; 168; 7,581; 7,828,354; 2,414,682,040,998, and 56,130,437,228,687,557,907,788.

   Proofs of asymptotic formulas for Dedekind numbers can be found in [164, 171].

5. A conjecture on reconstruction of infinite ordered sets states that the example in Exercise 2-37 is characteristic for infinite ordered sets. The conjecture says that, if P, Q are infinite ordered sets that are not isomorphic and have the same decks, then there is a $p \in P$ such that $P \setminus \{p\}$ contains an isomorphic copy of Q or there is a $q \in Q$ such that $Q \setminus \{q\}$ contains an isomorphic copy of P.

6. Find formulas for T_n^j for $j \geq 3$ that allow (much) faster computation than [160].

7. Find easy-to-compute formulas for the number of antichains in classes of ordered sets other than fences or crowns. This might give ideas for the solution of Dedekind's problem. See Theorem 11.12 for a formula for interval ordered sets.

8. Characterize the ordered sets of height 2 or those of width 3 that have the fixed point property. We will consider width 2 in Theorem 4.34 and height 1 in Theorem 4.37. I conjecture that there is a polynomial algorithm to determine the fixed point property for ordered sets of width 3. In the light of the proof of Theorem 7.32, which says that it is NP-complete to decide if an ordered set of height 5 has a fixed point free order-preserving self-map, and in light of Exercise 7-27, I have no intuition what might happen for height 2.

9. Prove that ordered sets of width 3 or of height 1 are reconstructible. We will consider width 2 in Exercise 3-8. I believe that a proof of reconstructibility of ordered sets of width 3 should be possible with the reconstruction tools available today (also see [278] for a start). Reconstruction of ordered sets of height 1 on the other hand appears almost as hard as the reconstruction problem in general.

10. For Ramsey-type order-theoretical results beyond Proposition 2.28, see [97, 213, 214, 225]. One of Ramsey's theorems (the one which guarantees the existence of finite Ramsey numbers, see [241]) essentially says that certain structures (complete graphs and discrete graphs) are so plentiful, that a representative of a certain size can be found in any graph. Embed the graph $G = (V, E)$ into a complete graph with $|V|$ vertices and color the edges of G red and the remaining edges blue. Then Ramsey's theorem says that any such coloring will always allow for a monochromatic complete subgraph of a certain size.

   The mentioned papers investigate and prove the following. Fix $r, s \in \mathbb{N}$. For every ordered set P it is possible to find an ordered set P' such that for any r-coloring χ of the s-chains of P', there is an embedding e of P into P' such that the s-chains of $e[P]$ are monochromatic under χ. So there is an ordered set that contains so many copies of P that even a partition through coloring will still allow us to find a copy in one of the elements of the partition.

11. Define a **fibre** of an ordered set to be a subset $F \subseteq P$ such that F intersects every maximal antichain with at least two elements.[1] Pictorially, the maximal antichains of an ordered set represent "horizontal separators" in the order. That is, every element of the ordered set will be either above or below some element of the maximal antichain. The notion of a fiber is then a subset that "stabs through all the separators."

   It was conjectured in [193] that every finite ordered set has a fiber whose complement also is a fiber. Consequently, every ordered set would have a fiber of size at most $\frac{|P|}{2}$. This is not true, because, by Exercise 2-19 the smallest size for a fiber of the ordered set P_0 in Figure 2.3 is $\frac{9}{17}|P_0|$.

   The natural question that now arises is the following. What is the smallest λ such that any ordered set P is guaranteed to have a fiber of size at most $\lambda|P|$?

   In [199] an iterative construction using the set in Figure 2.3 is used to show that $\lambda \geq \frac{8}{15}$. In [69], Theorem 1, it is shown that the elements of an ordered set can be 3-colored so that all nontrivial maximal antichains receive at least two colors. This means $\lambda \leq \frac{2}{3}$. The exact value of λ remains unknown.

12. The automorphism conjecture can be settled trivially for classes of ordered sets for which $|\mathrm{End}(P)|$ grows faster than $n!$. Unfortunately the hope that this is true in general is false. Fences and crowns have $< O(3^n) \ll n!$ (for large n) endomorphisms.

---

[1] Sometimes, sets that intersect *every* maximal antichain are called fibers. The overall upper bound on the size of such a fiber is $|P|$, as can be seen considering chains. Upper and lower bounds on the fiber size for individual ordered sets in given classes of ordered sets can be interesting.

13. What general results are there to compute the number of homomorphisms from one ordered set to another? I am not aware of any results beyond [86, 280] and Exercise 2-6. A good list of earlier references is in [86]. Related to this, for each n, what is the ordered set P of size n with the fewest endomorphisms?

14. The **endomorphism spectrum** (see [111]) of a finite ordered set P is the set $S := \{s \in \{1, \ldots, |P|\} : (\exists f \in \mathrm{End}(P))|f[P]| = s\}$. That is, it is the set of possible range sizes for order-preserving self-maps. Aside from trivial choices such as $1, 2, |P|$, what numbers are guaranteed to always be in the spectrum? What are examples of ordered sets with "small" spectra? In [70], Theorem 1.1, it is shown that every ordered set has an endomorphism with $|P|^{\frac{1}{7}}$ elements in its image. It is also shown in [70], Theorem 1.2, that there is a $c > 0$ such that, for each n, there is an ordered set of size n such that every endomorphism that is not the identity has at most $c(n \log(n))^{\frac{1}{3}}$ elements in its image.

15. For every k in the spectrum of P, let e_k^P be the number of endomorphisms with image of size k. The spectrum analyzes the zeroes of this sequence. What more can be said about the properties of the sequence e_k^P?

Chapter 3
Upper and Lower Bounds

Upper and lower bounds have already been defined in Definitions 2.16 and 2.17. From their use in the proof of Dilworth's Chain Decomposition Theorem 2.26 and in the proof of Proposition 2.36, where, in each case, sets were defined in terms of their upper bounds, as well as from their role in Zorn's Lemma, we can infer that bounds of sets play an important role in ordered sets. In this chapter, we consider various types of bounds and relate them to open problems and to each other.

3.1 Extremal Elements

Minimal elements and their duals, maximal elements, were defined in Definitions 2.4 and 2.15. Zorn's Lemma shows how useful maximal elements can be in a variety of situations. Maximal elements and minimal elements are also called **extremal elements**. Extremal elements, if they exist, are absolute bounds of an ordered set: There are no points strictly above a maximal element or strictly below a minimal element. On the other hand, in infinite ordered sets, extremal elements need not exist. For example, the integers $\mathbb{Z}$ with their natural order have neither maximal nor minimal elements. In this section, we study the relation between extremal elements and the fixed point property, isomorphism, and the reconstruction problem, respectively. We start with some simple results on the fixed point property and on isomorphism. Then we will embark on the proof of a strong and useful result about the reconstruction of maximal cards, that is, cards obtained through the removal of a maximal element.

Proposition 3.1. *Let P be a finite ordered set. Then, for every fixed point free order-preserving function $f : P \to P$, there is a fixed point free order-preserving function $g : P \to P$ such that $g \leq f$ and g maps minimal elements to minimal elements.*

Proof. Let $f : P \to P$ be a fixed point free order-preserving self-map. For every minimal element $m \in P$, choose a minimal element $x_m \leq f(m)$ in P. Define

© Springer International Publishing 2016

B. Schröder, *Ordered Sets*, DOI 10.1007/978-3-319-29788-0_3

Fig. 3.1 An ordered set
without the fixed point
property for which every
rank-preserving map has a
fixed point

$$g(p) := \begin{cases} f(p); & \text{if } p \text{ is not minimal,} \\ x_p; & \text{if } p \text{ is minimal.} \end{cases}$$

If $x < y$, then $g(x) \le f(x) \le f(y) = g(y)$, so g is order-preserving. Clearly $g \le f$. Finally, g has no fixed point: Indeed, if g had a fixed point, there would be an $m \in P$ such that $m = g(m) \le f(m)$. By Exercise 1-17, this implies that f has a fixed point, a contradiction. ∎

Applying Proposition 3.1 and its dual, we see that, if an ordered set has a fixed point free order-preserving self-map, then it must also have a fixed point free order-preserving self-map that maps minimal elements to minimal elements and that maps maximal elements to maximal elements. Especially for ordered sets of small height, this insight can help reduce the number of candidates for fixed point free order-preserving self-maps in a direct proof of the fixed point property, such as in Proposition 1.22, or, in a brute-force computer search. Unfortunately, we cannot extend Proposition 3.1 to saying that there must even be a fixed point free order-preserving self-map that preserves the rank of each element: Figure 3.1 shows an ordered set P such that every order-preserving self-map that preserves the rank of all elements has a fixed point (Rutkowski calls this the **weak fixed point property**) and yet P does not have the fixed point property. Indeed, P has only one element of rank 2 and thus every map that preserves the rank must fix that element. A fixed point free order-preserving self-map is indicated in Figure 3.1.

Proposition 3.2. *Let P, Q be ordered sets and let $\Phi : P \to Q$ be an isomorphism. Then, for each maximal element $m \in P$, the image $\Phi(m)$ is maximal in Q.*

Proof. Let $m \in P$ be maximal and suppose for a contradiction that $\Phi(m)$ is not maximal. Then there is a $q \in Q$ such that $q > \Phi(m)$. However then we have $\Phi^{-1}(q) > \Phi^{-1}(\Phi(m)) = m$, which contradicts the maximality of m. ∎

Example 3.3. The **largest (or greatest) element** ℓ of an ordered set P is defined dually to the smallest element, that is, $\ell \ge P$. An infinite ordered set can have a unique maximal element and yet not have a largest element: Consider the ordered set that consists of the natural numbers $\mathbb{N}$ united with a singleton set $\{p\}$ and no further comparabilities added. Then the element p is the unique maximal element of the ordered set, yet it is not the greatest element. □

We will now prove some reconstruction results related to maximal elements. We start by reconstructing the decks of ideals and maximal ideals defined as follows.

Definition 3.4. *Let P be a finite ordered set. A **maximal ideal** is the ideal $\downarrow m$ of a maximal element $m \in P$.*

Definition 3.5. *Let P be an ordered set. For $[C] \in \mathcal{C}$ let $\mathcal{I}_P([C])$ denote the number of ideals $\downarrow_P p$ in P that are isomorphic to C and let $\mathcal{I}_P^M([C])$ denote the number of maximal ideals $\downarrow_P m$ in P that are isomorphic to C. $\mathcal{I}_P$ is called the **ideal deck** of P and $\mathcal{I}_P^M$ is called the **maximal ideal deck** of P.*

Theorem 3.6. *Let P be a finite connected ordered set with at least four elements. Then the ideal deck $\mathcal{I}_P$ of P is reconstructible from the deck.*

Proof. (The proof in its present form is influenced by suggestions from J.-X. Rampon, M. Ellingham, and an anonymous referee.) If P has a largest element, then, by the dual of Proposition 1.37, there is nothing to prove. Otherwise we argue as follows.

First note that, for given P, all functions considered in this proof will have only finitely many arguments for which they are not zero. Hence all sums and sets involved are finite. For any two finite ordered sets I, J that both have a largest element, let $s'(I, J)$ be the number of subsets of J that

- Contain the largest element of J and
- Are isomorphic to I.

Let I be a finite ordered subset of P with a largest element t. Then I is contained in the set $\downarrow t$ and I contains t. The number of isomorphic copies of I in P is the sum (over all ideals in P) of the number of copies of I that are contained in the ideal in such a way that the top point of the ideal is in the copy of I. Hence the number $s(I, P)$ (see Proposition 1.40) can be computed from the numbers $\mathcal{I}_P([J])$ as follows (the sum runs over all nonisomorphic ideals J in P with more elements than I):

$$s(I, P) = \sum_{|J| \geq |I|} \mathcal{I}_P([J])s'(I, J) = \mathcal{I}_P([I]) + \sum_{|J| > |I|} \mathcal{I}_P([J])s'(I, J),$$

because, if $|I| = |J|$, then $s'(I, J) = 1$ iff I is isomorphic to J and $s'(I, J) = 0$ otherwise. Solving for $\mathcal{I}_P([I])$, we obtain

$$\mathcal{I}_P([I]) = s(I, P) - \sum_{|J| > |I|} \mathcal{I}_P([J])s'(I, J).$$

This means that we can reconstruct $\mathcal{I}_P$ recursively from $s(\cdot, \cdot)$. To see this, first note that, essentially, we are solving a (large) system of linear equations. In the following we prove that the system really is uniquely solvable. Let

$$k_0 := \max\{|I| : s(I, P) > 0 \text{ and } I \text{ has a largest element}\}.$$

Then, by definition of k_0, there are no ideals of size $> k_0$. Moreover, for $|I| = k_0$, we have $s(I, P) = \mathcal{I}_P([I])$ by the above equation.

Having found $\mathcal{I}_P([I])$ for $|I| \geq k_i$, let

$$k_{i+1} := \max \left\{ |I| < k_i : s(I,P) - \sum_{|J|>|I|} \mathcal{I}_P([J])s'(I,J) > 0 \right.$$

$$\left. \text{and } I \text{ has a largest element} \right\},$$

where the sum runs over all nonisomorphic ideals J in P with $|J| > |I|$. If the set on the right is empty, stop, $\mathcal{I}_P$ has been reconstructed. Otherwise, there are no ideals with size in $\{k_{i+1} + 1, \ldots, k_i - 1\}$ and, for all I with $|I| = k_{i+1}$, we have

$$\mathcal{I}_P([I]) = s(I,P) - \sum_{|J|>|I|} \mathcal{I}_P([J])s'(I,J),$$

where the sum runs over all nonisomorphic ideals J in P with more elements than I. Because the right side has already been reconstructed, we have reconstructed the left side. Continue to k_{i+2}. This process eventually terminates with the set used for computing k_{i+1} being empty. We have thus reconstructed $\mathcal{I}_P$. ∎

Note that there are at least two ways to reconstruct a parameter of an ordered set. One is to prove equations that involve quantities available through the deck and obtain the parameter through these equations. This is what we have done above and what we will do mostly. Another way would be to consider two ordered sets with equal decks and show that the parameter must have the same value for either set.

In the proof of the following theorem, we will need to identify a card with certain properties. Similar to the above, there are at least two ways to look at such a proof. One way is to consider an ordered set P and its deck and show that the card can be identified from the deck without using any knowledge about P. This is the way we choose. Another way is to start with a deck $\mathcal{D}$ of an ordered set and let P be *any* ordered set with $\mathcal{D}_P = \mathcal{D}$. Then we would need to show that a card with the desired properties can be found in $\mathcal{D}$. This approach looks formally a little cleaner, but the proofs do not change except for more cumbersome language. (For example, instead of saying "$C = P \setminus \{x\}$ is a card such that x satisfies $\cdots$," we would need to say "C is a card such that for any P with $\mathcal{D}_P = \mathcal{D}$ there is an $x \in P$ such that $\cdots$.") This is why we choose the first approach throughout.

Theorem 3.7. *Let P be a finite connected ordered set with at least four elements. Then the maximal deck $\mathcal{I}_P^M$ of P is reconstructible from the deck.*

Proof. To reconstruct $\mathcal{I}_P^M$, let $C = P \setminus \{x\}$ be a card such that there is a set I with a largest element such that the following hold.

- $\mathcal{I}_C([J]) = \mathcal{I}_P([J])$ for all $[J] \neq [I]$,
- $\mathcal{I}_C([I]) = \mathcal{I}_P([I]) - 1$, and
- $|I|$ is as small as possible.

Because the ideal deck is reconstructible, such a card can be identified in the deck. First suppose for a contradiction that x had been below at least two maximal elements. If one of these maximal elements had more lower bounds than the others, then removal of a maximal element above x with the fewest lower bounds would have produced a card with the same properties as demanded above, but with a smaller I. Thus all maximal elements above x must have the same number of lower bounds. However, then, for all maximal elements m above x, the number $\mathcal{I}_C([\downarrow m])$ equals $\mathcal{I}_P([\downarrow m])$ minus the number of maximal elements above x whose ideal is isomorphic to $\downarrow m$. This contradicts the choice of C, which says that only one value of $\mathcal{I}_C$ is different from $\mathcal{I}_P$ and that difference is 1.

Thus x was below exactly one maximal element. By definition of C, x must have been below a maximal element m with as few lower bounds as possible and such that $[\downarrow_P m] = [I]$. This means that C contains all maximal ideals of P except for one copy of I. Moreover the maximal ideals of C with less than $|I|$ elements are exactly those maximal ideals of C that are not maximal ideals of P. We conclude that

$$\mathcal{I}_P^M([J]) = \begin{cases} \mathcal{I}_C^M([J]); & \text{for } |J| \geq |I| \text{ and } [J] \neq [I], \\ \mathcal{I}_C^M([I]) + 1; & \text{for } [J] = [I], \\ 0; & \text{for } |J| < |I|. \end{cases}$$

∎

Definition 3.8. *Let P be an ordered set. We denote the number of maximal elements of P by m_P.*

Corollary 3.9. *The number m_P of maximal elements of P is reconstructible.* ∎

Theorem 3.7 says we can reconstruct what is below single maximal elements. It seems natural to investigate what the set without the maximal element looks like. Maximal cards (see Definition 3.10) are essentially cards that were obtained by removing a maximal element. It is tempting to hope that a card is maximal iff it has fewer maximal elements than P. However, this is not the case. The number of maximal elements does not behave in a nice predictable way when an element is removed from the ordered set: It can increase, decrease, or stay the same. Consider, for example, the ordered set in Figure 3.2. The elements p_i are maximal and removal

Fig. 3.2 Illustration of how the removal of an element can increase, decrease, or not change the number of maximal elements (see this section) or the number of covers (see end of Section 3.2)

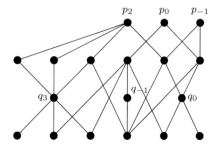

of p_i changes the number of maximal elements by i. Fortunately, the situation is not completely hopeless: The most that the number of maximal elements can decrease is 1. Still, recognition of maximal cards, or even better, the reconstruction of the maximal deck (see Definition 3.12), remains an open problem. We will reconstruct one maximal card here. The original definition of a maximal card is given below. Note that an ordered set can have more maximal cards than it has maximal elements (see Exercise 3-5). This is because of possible isomorphisms between $P \setminus \{m\}$ and $P \setminus \{x\}$ for m maximal and x not maximal. We will address this problem in our Definition 3.12 of the maximal deck.

Definition 3.10. *Let P be an ordered set and let C be an ordered set such that $\mathcal{D}_P([C]) > 0$. Then C is called a **maximal card** of P iff, for any Q with $\mathcal{D}_Q = \mathcal{D}_P$, there is a maximal element m of Q such that $[C] = [Q \setminus \{m\}]$. (This definition is equivalent to Kratsch and Rampon's definition in [174].)*

Note that Definition 3.10 specifically avoids calling a card maximal if it was obtained by removing a maximal element. In reconstruction, we start out with the deck and no knowledge of the set. So it is conceivable (if the reconstruction conjecture is false) to have two sets with equal decks for which a certain card is obtained by removing a maximal element in one set, whereas it can only be obtained by removing a non-maximal element in the other. Joshua Hughes observed that the 3-element nonreconstructible sets in Figure 1.5 are an example for this situation.

In the following result, the elements of the card that were not maximal in the ordered set turn out to be identifiable. This is because the removed element must have been maximal in *any* Q with $\mathcal{D}_Q = \mathcal{D}_P$.

Scholium 3.11 (See Theorem 6.3 in [174].). *Let P be an ordered set with at least four elements. Then either P is reconstructible or, from the deck, we can identify one maximal card C that was obtained by removal of a maximal element with as few lower bounds as possible. Moreover, the elements of C that are not maximal in P can be identified.*

Proof. Let $C = P \setminus \{x\}$ be a card as in the proof of Theorem 3.7. Then x was below exactly one maximal element m and said maximal element has as few lower bounds as possible. Thus we are done if we identify a maximal card among these cards. We will use notation from Theorem 3.7. If, for the numbers of maximal elements, we have $m_C \neq m_P$, then x must have been a maximal element. Also, in case $m_C = m_P$, if C has a maximal ideal of size $\leq |I| - 2$, then x must have been maximal. Thus, in either case, C is a maximal card. Moreover, because x was maximal with as few lower bounds as possible in P, the maximal elements of C that are not maximal in P can be identified as the maximal elements of C with fewer than $|I|$ lower bounds.

This leaves the case in which, for all the cards $C = P \setminus \{x\}$ as in the proof of Theorem 3.7, we have $m_C = m_P$ and C has a (necessarily unique) maximal ideal of size $|I| - 1$. In this case we will show that P is reconstructible.

Let m be the unique maximal element of P above x. Then m has a unique lower cover ℓ. (Otherwise there would be a card C as in the proof of Theorem 3.7 with a maximal ideal of size $< |I| - 1$.) Moreover, ℓ has only m as its unique upper cover.

(Otherwise there would be a card C as in the proof of Theorem 3.7 with $m_P - 1$ maximal elements.) This means that there are strict lower bounds y of m for which $\uparrow_P y$ is a chain and such that no strict upper bound of y has more than one lower cover.

Let the set J be obtained from I by removing the top element of I. We claim that, if y is such that the card $K = P \setminus \{y\}$ as in the proof of Theorem 3.7 has a maximal ideal that is isomorphic to the set J, then $(\uparrow_P y) \setminus \{y\}$ is a chain such that no element has more than one lower cover in P. To see this claim, suppose, for a contradiction, that $(\uparrow_P y) \setminus \{y\}$ was not a chain or that a strict upper bound of y has more than one lower cover. Let c be the largest point above y that has more than one lower cover. (Because $\uparrow_P y$ has a largest point, m, there is such a point $c > y$ in either case; moreover, by the above, $c \leq \ell < m$.) Then J would contain an ideal isomorphic to $\downarrow_P c$, while $\downarrow_{P\setminus\{y\}} m$ does not. Because $\downarrow_{P\setminus\{y\}} m$ would have to be isomorphic to J on any such card $P \setminus \{y\}$, this is a contradiction. To finish the proof, we use this fact to select one specific card now.

Let $K = P \setminus \{y\}$ be any card as in the proof of Theorem 3.7 so that the unique $(|I| - 1)$-element maximal ideal J' of K is isomorphic to J. Then, as shown above, $(\uparrow_P y) \setminus \{y\}$ must have been a chain such that no element has more than one lower cover. Moreover $y \leq m$ means that $\uparrow_P y$ is part of the chain at the top of J'. This however implies that P is isomorphic to K with a new top element attached to J' and P has been reconstructed.   ■

Although Scholium 3.11 is an extremely useful tool for order reconstruction, we have to (unfortunately) conclude this section on a lower note. Even if we could reconstruct the maximal deck (which would be quite an advance in reconstruction), we would still not be able to reconstruct all ordered sets as Example 3.13 shows. Note that in the definition of the maximal deck below, the deck contains only as many cards as there are maximal elements. Thus, even if we can find all maximal cards as defined in Definition 3.10, we might still not have the maximal deck, since several cards can be isomorphic and yet only some need be in the maximal deck.

Definition 3.12. *Let P be an ordered set. The **maximal deck** of P is the function $\mathcal{M}_P : \mathcal{C} \to \mathbb{N}$ such that $\mathcal{M}_P([C])$ is the number of cards $P \setminus \{x\}$ of P that are isomorphic to C and that were obtained by removal of a maximal element.*

Example 3.13 (See [173].).   Ordered sets are not reconstructible from their maximal decks. An example of two nonisomorphic ordered sets with the same maximal decks is given in Figure 3.3. Indeed, it is easy to see that the maximal decks of these two ordered sets are the same: For $i = 1, 2$ the sets $P \setminus \{m_P^i\}$ and $Q \setminus \{m_Q^i\}$ are isomorphic. On the other hand, the sets themselves are not isomorphic, because any isomorphism would have to map a_P to a_Q and hence b_P to b_Q and $\uparrow_P b_P$ to $\uparrow_Q b_Q$. This, however, is impossible, because $\uparrow_P b_P$ is not isomorphic to $\uparrow_Q b_Q$.   □

An example of a pair of ordered sets with equal maximal and minimal decks is given in Exercise 3-12.

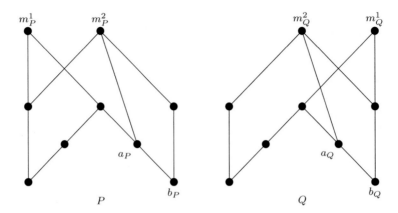

Fig. 3.3 Two nonisomorphic ordered sets with the same maximal decks (due to Kratsch and Rampon, see [173])

Exercises

3-1. Prove that, in a finite ordered set, a unique maximal element must be the greatest element of the set.

3-2. An ordered set satisfies the **descending chain condition** iff it contains no copies of the dual of $\mathbb{N}$. Prove that, in an ordered set with the descending chain condition, every element is above a minimal element.

3-3. Investigating the number of maximal elements cards can have.

    a. Construct an ordered set for which all cards have the same number of maximal elements.

    b. Prove that an ordered set cannot have one card with one maximal element, one card with three maximal elements, and one card with five maximal elements.

    c. Prove that, for every $n \in \mathbb{N}$ and every $(n + 1)$-element subset H of $\mathbb{N} \setminus \{1, \ldots, n - 1\}$, there is an ordered set P_H such that, for each $h \in H$, P_H has a card with h maximal elements.

3-4. Prove that, for every $k \in \mathbb{N}$ and every finite ordered set with at least four elements, the number of elements of rank k is reconstructible.

3-5. Prove that every card of a chain is a maximal card according to Definition 3.10.

3-6. (Extension of Scholium 3.11.) For an ordered set P, let s_P be the smallest size of a maximal ideal of P. Prove that, for every finite ordered set P with at least four elements, *all* cards obtained by removing a maximal element with at most s_P lower bounds are reconstructible.

3-7. Let P be a finite ordered set with at least four elements and let $C = P \setminus \{a\}$ be a maximal card that can be identified from the deck. A **down-set** is a set $H \subseteq P$ such that, for all $h \in H$, we have $\downarrow h \subseteq H$. Prove each of the following claims.

    a. It is possible to find a set A that is isomorphic to $\downarrow_P a$.

    b. Let $H_1, \ldots, H_k \subseteq P \setminus \{a\}$ be the distinct down-sets in $P \setminus \{a\}$ that are isomorphic to $A \setminus \{\bigvee A\}$. Let P_j be the ordered set obtained by attaching a to C as an upper bound of H_j with no further new comparabilities. Then $[P] \in \{[P_1], \ldots, [P_k]\}$. In particular, if $k = 1$, then P is reconstructible.

    c. Suppose the elements $l_1, \ldots, l_m$ in $P \setminus \{a\}$ have been identified as lower covers of a. Let $H_1, \ldots, H_k \subseteq P \setminus \{a\}$ be the distinct down-sets in $P \setminus \{a\}$ that are isomorphic to $A \setminus \{\bigvee A\}$ and contain $(\downarrow_C l_1) \cup \cdots \cup (\downarrow_C l_m)$. Let P_j be the ordered set obtained by attaching a to C as an upper bound of H_j with no further new comparabilities. Then $[P] \in \{[P_1], \ldots, [P_k]\}$. In particular, if $k = 1$, then P is reconstructible.

3-8. Prove that ordered sets of width 2 are reconstructible.

> *Hint.* Use Exercise 3-7.
>
> *Note.* The original solution is in [175].

3-9. Call a maximal element m of an ordered set P **dominating** iff m is above all non-maximal elements of P. For the following, we assume that the sets in question have at least four elements.

> a. Prove that ordered sets with a dominating maximal element are recognizable.
> b. Prove that cards obtained by removal of a dominating maximal element can be identified from the deck.
> c. Prove that ordered sets in which every element has at least two upper covers and which have a dominating maximal element are reconstructible.
> d. Prove that ordered sets of width 3 in which each element of rank 1 has at least two lower covers are reconstructible.

3-10. For an ordered set P, let $N_P(r, d, f, i, u, l)$ be the number of elements of rank r, of dual rank d (defined like the rank, only "dually"), with f upper bounds, ideal size i, u upper covers, and l lower covers. Give an example that shows that $N_P = N_Q$ does not imply that P is isomorphic to Q.

> *Hint.* Figure 3.3.

3-11. Prove that the ordered sets obtained from the sets in Figure 3.3 by deleting the element between a_P and m_P^1 in P and by deleting the element between a_Q and m_Q^1 in Q are also nonisomorphic with equal maximal decks.

3-12. Consider the ordered sets P and Q in Figure 3.4.

> a. Prove that P and Q are not isomorphic.
> b. Prove that P and Q have equal maximal and minimal decks.
>
> *Note.* For more examples, consider [277]. There are even nonisomorphic ordered sets for which the maximal deck, the minimal deck, and the deck obtained by removing all elements of a certain rank (that does not hold extremal elements) are equal, see [279].

3-13. Let P be an ordered set with a minimal element l and a maximal element h so that there is an isomorphism $\Phi : P \setminus \{h\} \to P \setminus \{l\}$. Let $K_l \subseteq P$ be the component of $P \setminus \{h\}$ that contains l and let $K_h \subseteq P$ be the component of $P \setminus \{l\}$ that contains h.

> a. Prove that K_l is isomorphic to K_h.
> b. Prove that the other components of $P \setminus \{h\}$ are isomorphic to components of $K_l \setminus \{l\}$.
> c. Prove that $K_l \setminus \{l\}$ is isomorphic to $K_l \setminus \{\Phi^{-1}(h)\}$.
>
> *Note.* For more on ordered sets with a minimal card that is isomorphic to a maximal card, consider [284].

3.2   Covers

Covers are a very local notion that was introduced in Definition 1.5 and was investigated thereafter. Proposition 1.10 shows that, if we know all the covers in a finite ordered set, then we also know all comparabilities. The relation between covers and isomorphisms was explored in Exercise 1-12. Also, in Section 3.1, we have seen the role of covers in the reconstruction of maximal cards. In this section, we will show that local knowledge about covers can have global consequences, such as the existence of a smallest element (see Theorem 3.16) and we will show that the number of covering relations is reconstructible (see Theorem 3.17).

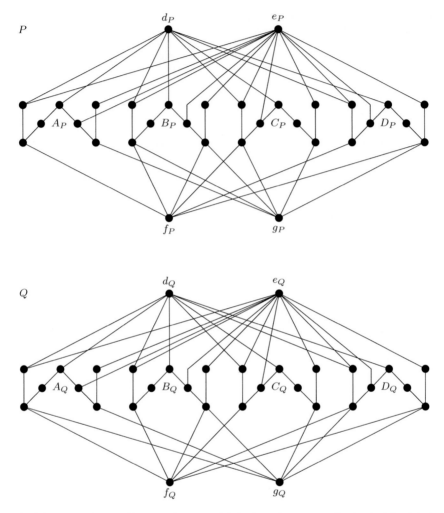

Fig. 3.4 Two nonisomorphic ordered sets P and Q that have the same maximal and minimal decks

Theorem 3.16 is inspired by the argument that is used in [85] to prove uniqueness of cores. (You can produce this argument in Exercise 4-30.)

Lemma 3.14. *Let P be a finite ordered set such that any two distinct elements that have a common upper cover have a common lower cover. Then any two elements of P that have a common upper bound have a common lower bound.*

Proof. Without loss of generality, we can assume that P is connected. (Otherwise we could work with the individual components.) For $x \leq y$, we define the cover-distance $c(x, y)$ from x to y to be the smallest number $n \in \mathbb{N}$ such that there is a chain $x = c_0 \prec c_1 \prec \cdots \prec c_n = y$ with each c_i being an upper cover of c_{i-1}. We claim that the following "parallelogram law" holds: If u is a common upper bound of $a, b \in P$, then there is a common lower bound l of a and b such that

$c(l, b) \leq c(a, u)$ and $c(l, a) \leq c(b, u)$. (Clearly this implies the result.) The proof of the parallelogram law is by induction on $n = c(a, u) + c(b, u)$.

Base step, $n = 0, 1, 2$. For $n = 0$ and $n = 1$, there is nothing to prove. In case $n = 2$, there is nothing to prove if $c(a, u) = 0$ or $c(b, u) = 0$. In the remaining case, $c(a, u) = 1$ and $c(b, u) = 1$, the claim follows directly from the hypothesis.

For the induction step $\{0, \ldots, n - 1\} \rightarrow n$ with $n > 2$, let $c(a, u) + c(b, u) = n$. We can assume without loss of generality that $c(b, u) > 1$. Because $c(b, u) > 1$, there is an upper cover b' of b such that $c(b', u) = c(b, u) - 1 > 0$. Thus there is a lower bound l' of a and b' such that we have the inequalities $c(l', a) \leq c(b', u)$ and $c(l', b') \leq c(a, u)$. Now $c(l', b') \leq c(a, u)$ and $c(b, b') = 1 < c(b, u)$. Thus, by induction hypothesis, there is a common lower bound l of l' and b such that we have the inequalities $c(l, l') \leq c(b, b')$ and $c(l, b) \leq c(l', b') \leq c(a, u)$. But then $c(l, a) \leq c(l, l') + c(l', a) \leq c(b, b') + c(b', u) = c(b, u)$ and we are done. ∎

Lemma 3.15. *Let P be a finite ordered set such that any two elements that have a common upper bound have a common lower bound. Then every component of P has a smallest element.*

Proof. Without loss of generality, we can assume that P is connected. Let $m \in P$ be minimal and suppose, for a contradiction, that m is not the smallest element of P. Then there is an $x \in P$ such that $m \nleq x$. Because P is connected, there is a $y \in P$ such that $m \nleq y$ and such that m and y have a common upper bound. However, then m and y have a common lower bound b. Because $m \nleq y$, we have $b < m$, a contradiction to the minimality of m. ∎

Theorem 3.16. *Let P be a finite ordered set such that any two elements that have a common upper cover have a common lower cover. Then every component of P has a smallest element.*

Proof. Simple combination of Lemmas 3.14 and 3.15. ∎

The above and previous results on covers show (unsurprisingly) that covers and bounds are tightly related to each other. In Exercise 1-29a, it was straightforward to reconstruct the number of comparability relations of an ordered set from the deck. The short proof of the reconstruction of the number of covering relations below should not lead us to believe that the reconstruction of the number of adjacencies is equally immediate. We are quoting a powerful lemma.

Theorem 3.17 (See [174], Theorem 7.1.). *The number of covering relations (or adjacencies) in a finite ordered set is reconstructible from its deck.*

Proof. Let a_P be the number of adjacencies of P. For any ordered set I with a largest element, let s_I be the number of lower covers of the largest element. Then we have that

$$a_P = \sum_{[I] \in \mathcal{C}} \mathcal{I}_P([I]) s_I,$$

which is reconstructible by Theorem 3.6. ∎

The number of covers is another tempting way to try to reconstruct the maximal deck. Removal of a maximal element decreases the number of covering relations. The temptation is to think that removal of a non-maximal element does not decrease the number of covering relations. To show this is not true, consider Figure 3.2 once more. Removal of any of the points q_k in Figure 3.2 changes the number of covering relations by k. Moreover, there is no bound on by how much the number of comparabilities can change by removal of one non-extremal point (see Exercise 3-14).

Exercises

3-14. The relation between removal of a point and the number of covering relations.

    a. For each $n \in \mathbb{N}$, construct an ordered set such that removal of a certain point of the ordered set decreases the number of covering relations by exactly n.

    b. For each $n \in \mathbb{N}$, construct an ordered set such that removal of a certain non-extremal point of the ordered set decreases the number of covering relations by exactly n.

    c. For each $n \in \mathbb{N}$, construct an ordered set such that removal of a certain point of the ordered set increases the number of covering relations by exactly n.

3-15. Find an ordered set that does not have the fixed point property and that is such that every cover-preserving map (that is, every map such that $x \prec y$ implies $f(x) \prec f(y)$) has a fixed point.

3-16. Recall from Exercise 2-17 that an ordered set P is called **N-free** iff P does not contain a subset $\{a, b, c, d\}$ such that $a \prec b \succ c \prec d$ and there are no further comparabilities between these elements. Let P be an N-free ordered set. Prove that

    a. If $a, b \in P$ have a common upper cover, then a and b have the same upper covers.

    b. If $a \in P$ has an upper cover u such that $u > b$, then $(\uparrow a) \setminus \{a\} \subseteq \uparrow b$.

3.3   Lowest Upper and Greatest Lower Bounds

The existence of lowest upper bounds and greatest lower bounds in an ordered set is a strong tool. For example, it is paramount to the development of analysis: The completeness of the real numbers is equivalent to the existence of a lowest upper (greatest lower) bound for every subset that has an upper (a lower) bound. In fact, our examples will show that, in infinite ordered sets, lowest upper bounds and greatest lower bounds can have a similar feel as limits of monotone sequences. This train of thought is followed in Section 3.4, where we investigate chain-completeness, and in Proposition 8.8, in which chain-completeness is related to lattices.

Definition 3.18. *Let P be an ordered set and let $A \subseteq P$. Then*

*1. The point u is called the **lowest upper bound** or **supremum** or **join** of A iff $u \geq A$ and, for all $p \in P$ with $p \geq A$, we have $p \geq u$.*

2. *The point l is called the **greatest lower bound** or **infimum** or **meet** of A iff $l \leq A$ and, for all $p \in P$ with $p \leq A$, we have $p \leq l$.*

We will denote the supremum of a set A (if it exists) by $\bigvee A$ and the infimum (if it exists) by $\bigwedge A$. For finite sets $A = \{a_1, \ldots, a_n\}$, we will also use the notation

$$a_1 \vee a_2 \vee \cdots \vee a_n := \bigvee \{a_1, a_2, \ldots, a_n\},$$

$$a_1 \wedge a_2 \wedge \cdots \wedge a_n := \bigwedge \{a_1, a_2, \ldots, a_n\}.$$

We have already made the acquaintance of lowest upper bounds when we were working with Zorn's Lemma. Although the hypothesis of Zorn's Lemma only requires the existence of some upper bound for every nonempty chain, in all our proofs, the bound that was constructed invariably was the lowest upper bound: You can check over the proofs of Theorems 2.54 and 2.62 and Propositions 2.55 and 2.64 to verify this claim. To justify our talking about *the* lowest upper bound, we note the following.

Proposition 3.19. *Let P be an ordered set and let $A \subseteq P$ be a subset that has a lowest upper bound. Then the lowest upper bound of A is unique.*

Proof. Let $A \subseteq P$ and let u and u' be upper bounds of A so that any upper bound of A must be above u and above u'. Then, in particular, because u is an upper bound of A, we have $u \geq u'$, and, by symmetry, we have $u' \geq u$, too. This implies $u = u'$ and we are done. ∎

There also is the possibility to iterate the formation of lowest upper bounds.

Proposition 3.20. *Let P be an ordered set and let A and B be subsets that have suprema. If $\bigvee A \vee \bigvee B$ exists, then it is the supremum of $A \cup B$.*

Proof. To abbreviate notation, let $s := \bigvee A \vee \bigvee B$. Then $s \geq A$ and $s \geq B$. Moreover, if $p \geq A \cup B$, then $p \geq \bigvee A$ and $p \geq \bigvee B$, hence $p \geq s$. ∎

Example 3.21. 1. In Figure 1.1 b), any two elements of rank 1 have a supremum or an infimum, but some sets of two elements of rank 1 do not have both. In Figure 1.1 c), any two minimal elements have a supremum and any two maximal elements have an infimum.
2. The ordered subset $\{x \in \mathbb{Q} : x \geq 0, x^2 \leq 2\}$ of $\mathbb{Q}$ has upper bounds but no lowest upper bound in $\mathbb{Q}$.
3. Every nonempty subset of $\mathbb{R}$ that has an upper bound has a lowest upper bound.
4. Let P be an arbitrary ordered set. The empty set $\emptyset$, as a subset of P, has a supremum in P iff P has a smallest element. The empty set $\emptyset$ has an infimum iff P has a largest element.
5. If X is a set and $\mathcal{P}(X)$ is its power set ordered by inclusion, then, for each set of subsets $A \subseteq \mathcal{P}(X)$, we have $\bigvee A = \bigcup A$ and $\bigwedge A = \bigcap A$.
6. Let $A \subseteq C([0, 2], \mathbb{R})$ be finite. Then $f(x) := \bigvee \{g(x) : g \in A\}$ is the supremum of A.

7. Let $p \geq 1$ and let $A \subseteq L^p(\Omega, \mathbb{R})$ be finite. Then $f(x) := \bigvee\{g(x) : g \in A\}$ is the supremum of A, where the supremum is taken pointwise almost everywhere.

8. For $n \in \mathbb{N}$, define $f_n \in C([0, 2], \mathbb{R})$ by

$$f_n(x) := \begin{cases} 1; & \text{for } x \in \left[0, 1 - \frac{1}{n}\right), \\ -nx + n; & \text{for } x \in \left[1 - \frac{1}{n}, 1\right], \\ 0; & \text{for } x \in (1, 2]. \end{cases}$$

Then $\{f_n : n \in \mathbb{N}\}$ has no supremum in $C([0, 2], \mathbb{R})$.

9. Monotone Convergence Theorem. In the space $L^1(\Omega, \Sigma, \mu)$, let $\{f_n\}_{n\in\mathbb{N}}$ be a sequence of functions such that $f_n \leq f_{n+1}$ a.e. Then $\{f_n : n \in \mathbb{N}\}$ has a lowest upper bound in $L^1(\Omega, \Sigma, \mu)$ iff it has an upper bound.   □

Proof[1] *1.* Part 1 allows a visualization of suprema and infima. Listing all sets that have suprema or infima and proving the claim is routine, if a little tedious. To see that all possibilities can occur, note that, in the set in Figure 1.1 b), the subset $\{c, d\}$ has a supremum and no infimum, and the subset $\{c, f\}$ has a supremum and an infimum. Moreover, the subset $\{i, k\}$ has neither a supremum nor an infimum.

For part 2, assume that u is the infimum of $\{x \in \mathbb{Q} : x \geq 0, x^2 \leq 2\}$. Because there is no rational number q with $q^2 = 2$, and because there are rational numbers whose squares are arbitrarily close to 2 but less than 2, we must have $u^2 > 2$. Let $n \in \mathbb{N}$ be such that $2\frac{u}{n} < u^2 - 2$. Then $\left(u - \frac{1}{n}\right) < u$ and

$$\left(u - \frac{1}{n}\right)^2 = u^2 - 2\frac{u}{n} + \frac{1}{n^2} > 2,$$

contradicting the choice of u.

Note that, in elementary set theory or analysis, part 2 normally is the motivation to go from the rational number system to the real number system. Analysis is concerned more with limits than with suprema and infima. The proof above suggests how to recursively construct a monotone Cauchy sequence that does not have a limit in $\mathbb{Q}$.

Part 3 is either part of the axiomatic definition of the real numbers or (if the reals were constructed from smaller number systems) must be proved in the construction of $\mathbb{R}$. Which it is depends on the author of the set theory chapter or text you are reading. Because we have not taken time to formally define the real numbers, we will not present a proof here. (If you are interested in the construction, consider, for example, [283].) For the development of order theory, this "hole" is not a problem. For examples that involve $\mathbb{R}$, use any definition of $\mathbb{R}$ you have seen in classes or other texts.

---

[1] This is a proof of the claims in the example. Such proofs will be given in this text for examples in which the claims are complex enough to require a proof.

The proof of part 4 goes back to the logical structure of the definition of suprema and upper bounds. The point $u \in P$ is an upper bound of $\emptyset$ iff u is above every element of $\emptyset$. Because $\emptyset$ has no elements, every $p \in P$ vacuously has this property. Therefore every element of P is an upper bound of $\emptyset$. This means that, if $\emptyset$ has a lowest upper bound, it would have to be below every element of P. Conversely, the smallest element of P, if P has one, would be the lowest upper bound of $\emptyset$. The infimum is treated dually.

To prove part 5, let $A \subseteq \mathcal{P}(X)$ be nonempty. Then $\bigcup A$ contains all sets in A and every set that contains all sets of A must contain $\bigcup A$. Because the union of an empty set of sets is the empty set, we have proved that unions are suprema. Infima are treated dually.

To see part 6, first note that, in any space of continuous functions, comparability is normally defined pointwise. Therefore, the supremum of a set of continuous functions must be at least above the pointwise supremum. It is a standard result of analysis that the pointwise supremum of finitely many continuous functions is again continuous (see Theorem 7.22 in [137], for example; they prove it for lower semicontinuous functions). Thus we are done.

The proof of part 7 is similar: The pointwise almost everywhere supremum of finitely many p-integrable functions is p-integrable, too; it is a pointwise a.e. upper bound for the set of functions; and any other such bound must be a.e. above it.

For part 8, first note that the pointwise supremum of the given set of functions is

$$f_{\text{ptws}}(x) = \begin{cases} 1; & \text{for } x \in [0, 1), \\ 0; & \text{for } x \in [1, 2]. \end{cases}$$

Now, for every continuous function $g \geq f_{\text{ptws}}$, we have $g(y) \geq 1$ for $y \in [0, 1)$, and hence $g(1) \geq 1$. Therefore there is an $x_\varepsilon \in (1, 2]$ such that $g(x_\varepsilon) = \varepsilon > 0$. Let

$$h(x) := \begin{cases} 1; & \text{for } x \in [0, 1], \\ 2 - x; & \text{for } x \in (1, 2]. \end{cases}$$

Then gh is an upper bound of all the f_n, $gh \leq g$ and $gh(x_\varepsilon) < \varepsilon = g(x_\varepsilon)$. Thus no upper bound of the given set of functions can be its lowest upper bound. This means that $\{f_n : n \in \mathbb{N}\}$ does not have a supremum in $C([0, 2], \mathbb{R})$.

The Monotone Convergence Theorem (part 9) is an important theorem in real analysis. In a sense it is what sets the spaces of Lebesgue integrable functions apart from spaces of continuous or Riemann integrable functions, where there is no analogue of the Monotone Convergence Theorem. For continuous functions, we have seen this in part 8. For Riemann integrable functions, note that the function on $[0, 1]$ that is 1 for rational numbers and zero for irrational numbers (Dirichlet's function) is not Riemann integrable. Yet it can be represented as the supremum of functions that are 1 in only finitely many places. A proof of the Monotone Convergence Theorem can be found, for example, in [47], Theorem 2.4.1, or in [282], Theorem 14.41. ∎

Suprema and infima are linked together, because the supremum of a set is the infimum of its set of upper bounds, as we are going to prove now.

Definition 3.22. *Let P be an ordered set and let $A \subseteq P$. Then we set*

$$\uparrow A := \{p \in P : p \geq A\},$$
$$\downarrow A := \{p \in P : p \leq A\}.$$

Lemma 3.23. *Let P be an ordered set and let $A \subseteq P$. Then*

1. *The equality $\bigvee A = \bigwedge \uparrow A$ holds in the sense that, if either of the involved two quantities exists, then both exist and the equality holds.*
2. *If $\bigwedge A$ exists in P, then $\downarrow A = \downarrow \bigwedge A$.*
3. *If $\bigvee A$ exists in P, then $\downarrow \uparrow A = \downarrow (\bigvee A)$.*

Proof. For part 1, first suppose that $\bigvee A$ exists. Then $\bigvee A$ is a lower bound of $\uparrow A$. Because $\bigvee A \in \uparrow A$, any lower bound of $\uparrow A$ is below $\bigvee A$, so $\bigvee A$ is the infimum of $\uparrow A$.

Now suppose that $\bigwedge \uparrow A$ exists. Because all elements of A are lower bounds of $\uparrow A$, $\bigwedge \uparrow A$ is an upper bound of A. By definition, it is below any upper bound of A, so it is the supremum of A.

In part 2, simply note that $x \leq A$ iff $x \leq \bigwedge A$.

To prove part 3, note that, by part 1, the upper bounds of A have an infimum. Thus $\downarrow \uparrow A = \downarrow (\bigwedge \uparrow A) = \downarrow (\bigvee A)$. ∎

As we have seen, existence of suprema and infima is by no means guaranteed in arbitrary ordered sets. Yet power sets (see part 5 of Example 3.21) show that, in the perhaps most natural example of an ordered set that is neither a chain nor an antichain, suprema and infima abound. In such a situation, mathematicians often strive to at least identify the general structures as substructures of nicer structures. Examples of this procedure include the completion of metric spaces (if one cares about convergence of Cauchy sequences) or the various ways to compactify topological spaces (if one wants small coverings using open sets). Thus it is only natural to wish to embed an ordered set into some power set (or a subset of a power set). The following two results show how this can be done. We will later consider the Dedekind–MacNeille completion, which is another, nicer, way to embed an ordered set into an ordered set with suprema and infima.

Proposition 3.24. *Let P be an ordered set and let $\mathcal{P}(P)$ be the power set of P ordered by set inclusion. Then the map $\Phi : P \to \mathcal{P}(P)$, $p \mapsto \downarrow p$ is an embedding of P into $\mathcal{P}(P)$.*

Proof. This is a very simple consequence of the definition of an embedding and the definition of Φ. ∎

The embedding Φ is also nice in the sense that existing infima are mapped to intersections, so at least the lower bound operations in the two sets correspond to each other.

Proposition 3.25 (See [56], Lemma 2.32.). *Let P be an ordered set and let $A \subseteq P$. If $\bigwedge A$ exists in P, then $\bigcap \{\downarrow a : a \in A\} = \downarrow (\bigwedge A)$.*

Proof. Note that

$$\bigcap \{\downarrow a : a \in A\} = \{p \in P : p \leq A\} = \left\{ p \in P : p \leq \bigwedge A \right\} = \downarrow \left(\bigwedge A \right).$$

∎

Exercises

3-17. Find a set S of rational numbers s with $s^2 > 2$, such that any rational number q with $q^2 > 2$ is above some $s \in S$.

3-18. Let P be an ordered set and let $A \subseteq B \subseteq P$. Prove that $\uparrow A \supseteq \uparrow B$ and $\downarrow A \supseteq \downarrow B$

3-19. Let P be an ordered set. Prove that $\uparrow \downarrow \uparrow A = \uparrow A$ for all $A \subseteq P$.

3-20. Let P, Q be ordered sets and let $f : P \to Q$ be order-preserving. Prove that, for any set A such that A and $f[A]$ have a supremum, we have the inequality $f[\bigvee A] \geq \bigvee f[A]$. Then give an example that shows that strict inequality can occur.

3-21. Prove that, in the space $C([-1, 1], \mathbb{R})$ of continuous functions on $[-1, 1]$, the set of monomials $\{x^n : n \in \mathbb{N}\}$ has a greatest lower bound. Subsequently prove that it does not have a greatest lower bound in the space $C([-2, 2], \mathbb{R})$ of continuous functions on $[-2, 2]$.

3-22. Prove that, if $c = a \vee b$ in P, then $c = a \wedge b$ in P^d, the dual of P.

3-23. On distributivity. The supremum $\vee$ and the infimum $\wedge$ of two elements can also be interpreted as algebraic operations. For algebraic operations, we can consider a variety of properties, including distributivity. In general, distributivity does not hold: Find an ordered set P and three points $a, b, c \in P$ such that $a \vee (b \wedge c) \neq (a \vee b) \wedge (a \vee c)$. That is, all involved quantities in the inequality exist and the inequality holds.

   This means that, in general, suprema do not distribute over infima. Use the above to conclude that, in general, infima do not distribute over suprema either.

   Note. This abstract distributivity is motivated by the fact that distributivity holds for unions and intersection. We shall investigate distributivity in more detail in Section 8.5.

3-24. (Jerzy Wojdylo.) Proposition 3.24 shows that, for every ordered set P, there is a set S with $|S| \leq |P|$ such that P can be embedded into the power set $\mathcal{P}(S)$ ordered by inclusion.

   a. Show that, if P can be embedded into $\mathcal{P}(S)$, then $|S| \geq \lceil \log_2 |P| \rceil$.
   b. Show that, if S is such that an n-element chain can be embedded into $\mathcal{P}(S)$, then $|S| \geq n - 1$ and that there is such an S with $|S| = n - 1$.
   c. Let n be an integer. Show that, for all integers k with $\lceil \log_2 n \rceil \leq k \leq n - 1$, there is an ordered set P with $|P| = n$ such that, for all sets S such that P can be embedded into $\mathcal{P}(S)$, we have that $|S| \geq k$ and there is such a set S of size k.
   d. Let n be an integer. Show that there is an ordered set P with $|P| = n$ such that, for all sets S such that P can be embedded into $\mathcal{P}(S)$, we have that $|S| \geq n$.

3.4   Chain-Completeness and the Abian–Brown Theorem

Clearly, existence of suprema and infima is a strong tool. It may not occur as
frequently as we want, but it is a fairly frequent occurrence in mathematics. Thus
a lot of work has been devoted to the study of ordered sets in which any finite
nonempty subset, or any subset in general, has a supremum and an infimum. Such
sets are called lattices and complete lattices, respectively. They are investigated in
lattice theory, which is a good-sized branch of discrete mathematics. We will devote
Chapter 8 to the study of lattices and Chapter 9 to truncated lattices. In this section,
we investigate a variation on the existence of suprema and infima that restricts its
attention to nonempty chains. This condition can be used to prove an important fixed
point result.

Definition 3.26. *Let P be an ordered set. Then P is called **chain-complete** iff each
nonempty subchain $C \subseteq P$ has a supremum and an infimum.*

Although the above definition of chain-completeness appears to be the most
common use of the term, some authors also demand that the empty chain has a
supremum and an infimum. (So, when in doubt, double check the definition used
by the author of what you're reading.) In this text, we will not assume that a chain-
complete ordered set has a largest or a smallest element unless we explicitly demand
the existence of these elements.

Example 3.27. 1. The power set of any set is chain-complete with respect to
   inclusion.
2. Every finite ordered set is chain-complete.
3. $C([0, 2], \mathbb{R})$ is not chain-complete.   □

Proof. By part 5 of Example 3.21, every subset of a power set has a supremum and
an infimum. Therefore, clearly, power sets must be chain-complete in particular.
This proves part 1. For part 2, note that finite chains have in fact a largest and a
smallest element. Finally, $C([0, 2], \mathbb{R})$ is not chain-complete, because, in part 8 of
Example 3.21, we have a nonempty chain of continuous functions on $[0, 2]$ that has
no supremum.   ∎

Proposition 3.28. *Let P be a chain-complete ordered set. For every $p \in P$, there is
a maximal element $M \in P$ and a minimal element $m \in P$ such that $m \leq p \leq M$.*

Proof. This is a consequence of Zorn's Lemma and its dual.   ∎

Because of the Abian–Brown Theorem, which we will prove below, chain-
completeness is a strong tool when working with the fixed point property. The idea
is very simple, and it was already used in the proof of Propositions 1.22 and 2.39: If
$p \leq f(p)$, then all the $f^n(p)$ form a chain. In a finite ordered set, this chain eventually
must stop, and the place where it stops must be a fixed point (see Exercise 1-17).
For infinite ordered sets, a similar argument will work. We just need some way
to push past limit ordinal numbers. Chain-completeness provides a vehicle to do
this: Whenever a chain goes up infinitely often, we can take the supremum and

continue. This is reflected in part 2 of the definition of f-chains below. Our proof
follows along the lines of the proof in [1]. The methods developed here will be used,
for example, in Exercises 12-19, 12-20, and 12-21 to give an alternative proof for
Roddy's product theorem (see Theorem 12.17) and in Section 3.5. (It should not
be surprising that a result this close to the Monotone Convergence Theorem has
applications in analysis. In fact, the motivation for the related work in [219] came
from the desire to solve certain types of integral equations.)

To allow for more versatility later on, we will not demand that our underlying set
be chain-complete until Theorem 3.32.

Definition 3.29 (See [1].).
*Let P be an ordered set, let $f : P \to P$ be an order-preserving function and let
$p \in P$ with $f(p) \geq p$. A well-ordered subset C of P is called an f-**chain starting at** p
if and only if the following hold.*

1. *The point p is the smallest element of C.*
2. *Using ordinal number terminology for elements of C, we have that, for every
$c \in C \setminus \{p\}$,*

$$c = \begin{cases} f(c^-); & \text{if } c \text{ has an immediate predecessor } c^-, \\ \bigvee_P \left[((\downarrow c) \cap C) \setminus \{c\} \right]; & \text{if } c \text{ is a limit ordinal and the supremum exists.} \end{cases}$$

Our first step is to find an f-chain that is maximal with respect to inclusion.
This is reminiscent of Zorn's Lemma. Indeed, a simple Zorn's Lemma argument
would prove the existence of maximal f-chains. However, with just a little more
effort, we can show the same fact without using Zorn's Lemma. This approach has
a theoretical and a practical aspect. From the theoretical angle, not needing Zorn's
Lemma might let us use a somewhat smaller system of set-theoretical axioms. From
a somewhat practical angle, showing that the f-chains are unique and only need to be
unified to obtain the maximal f-chain opens the door for an (admittedly transfinite)
algorithm to find fixed points in infinite sets. (For more on this idea, see Section 3.5.)

Lemma 3.30 (See [1], Lemma 1.). *Let P be an ordered set and let $f : P \to P$ be
order-preserving. If $p \in P$ is such that $p \leq f(p)$ and if C, K are two f-chains starting
at p, then either C is an initial segment of K or K is an initial segment of C. Hence
there is a unique maximal f-chain starting at p.*

Proof. Let C, K be f-chains starting at p and let

$$\mathcal{J} := \{ I \subseteq C \cap K : I \text{ is an initial segment of } C \text{ and of } K \}.$$

Then $\mathcal{J} \neq \emptyset$, because $\{p\}$ is in $\mathcal{J}$. Consider $H := \bigcup \mathcal{J}$. Clearly $H \subseteq C \cap K$. Let
$c \in C \setminus H = \bigcap_{I \in \mathcal{J}} C \setminus I$. Then c is an upper bound of all $I \in \mathcal{J}$ and thus it is an
upper bound of H. Hence H is an initial segment of C and similarly it is an initial
segment of K. Now assume that $K \setminus H \neq \emptyset$ and $C \setminus H \neq \emptyset$. If H has a largest element

h, then $f(h)$ is the immediate successor of h in C and in K. Hence $H \cup \{f(h)\}$ is an initial segment of C and of K, which is a contradiction. If H does not have a largest element, then $\bigvee H$ exists in P and it is the supremum of H in C and in K. Hence, in this case, $H \cup \{\bigvee H\}$ is an initial segment of C and of K, which is a contradiction. Thus $H = C$ or $H = K$ and we are done. The unique maximal f-chain starting at p is thus the union of all f-chains starting at p.   ∎

Lemma 3.31. *Let P be an ordered set and let $f : P \to P$ be an order-preserving self-map. If $p \in P$ is such that $f(p) \geq p$, then one of the following two holds.*

1. *The maximal f-chain starting at p has no supremum, or*
2. *There is a fixed point $q \in P$ of f such that $q \geq p$ and, for all fixed points $x \geq p$ of f, we have $x \geq q$.*

Proof. If the maximal f-chain C starting at p has no supremum, we are done. So let us assume that $q := \bigvee_P C$ does exist. By the definition of f-chains, $q \in C$. Moreover if $f(q) > q$, then $C \cup \{f(q)\}$ would be an f-chain that contains C, so $f(q) \not> q$.

On the other hand, we have $f(q) \geq q$, as the following shows. If q has an immediate predecessor q^-, then $f(q) = f(f(q^-)) \geq f(q^-) = q$. On the other hand, if q has no immediate predecessor, then $q = \bigvee_P \left[((\downarrow q) \cap C) \setminus \{q\} \right]$. Because the image of every element of C is greater than or equal to the element and again in C, we have

$$f(q) = f\left(\bigvee_P \left[((\downarrow q) \cap C) \setminus \{q\} \right] \right) \geq \bigvee_P \left[((\downarrow q) \cap C) \setminus \{q\} \right] = q.$$

Therefore $f(q) = q$.

Now let $x \in\uparrow_P p$ be a fixed point of f and suppose, for a contradiction, that $x \not\geq q$. Let $c \in C$ be the smallest element of C such that $x \not\geq c$. Then c cannot have an immediate predecessor c^-, because then we would have $x = f(x) \geq f(c^-) = c$. Thus $c = \bigvee_P \downarrow_C c$. Via $x \geq (\downarrow_C c) \setminus \{c\}$ we again infer $x \geq c$, a contradiction. Therefore we must have that $x \geq q$.   ∎

Note that the two possibilities in Lemma 3.31 are not mutually exclusive (see Exercise 3-30).

Theorem 3.32. *The **Abian–Brown Theorem** (see [2]). Let P be a chain-complete ordered set and let $f : P \to P$ be order-preserving. If there is a $p \in P$ with $p \leq f(p)$, then f has a smallest fixed point above p.*

Proof. Easy consequence of Lemma 3.31: Part 1 cannot occur because P is chain-complete.   ∎

Corollary 3.33. *Every chain-complete ordered set with a smallest element has the fixed point property.*   ∎

Exercises

3-25. On $\mathcal{P}(\{1, \ldots, n\})$, define the mapping $f(A) := A \cup \{1, \ldots, \min(|A| + 1, n)\}$.

   a. Show that $f : \mathcal{P}(\{1, \ldots, n\}) \to \mathcal{P}(\{1, \ldots, n\})$ is order-preserving,
   b. Find the fixed points of f,
   c. Find the largest number k such that there is an $A \in \mathcal{P}(\{1, \ldots, n\})$ such that $f^k(A)$ is not a fixed point of f.

3-26. The mapping $f : \mathbb{R} \to \mathbb{R}, x \mapsto x + 1$ is order-preserving and has no fixed points. Is this a contradiction to Theorem 3.32?

3-27. Let $f : P \to P$ be an order-preserving map. Show that, if $p \leq q$, $F(p)$ is the smallest fixed point of f above p and $F(q)$ is the smallest fixed point of f above q, then $F(p) \leq F(q)$.

3-28. Use Zorn's Lemma to prove that every f-chain is contained in a maximal f-chain.

3-29. Prove that a chain has the fixed point property iff it is chain-complete.

   Note. We will ultimately have the stronger Theorem 8.10.

3-30. Give an example of an ordered set P, an order-preserving map $f : P \to P$ and a $p \in P$ such that $p \leq f(p)$, the maximal f-chain starting at p has no supremum and f has a smallest fixed point $q \geq p$.

3-31. Let P and Q be ordered sets. Then the set of order-preserving maps $f : P \to Q$ can be ordered with the pointwise order. That is, $f \leq g$ iff $f(p) \leq g(p)$ for all $p \in P$. Show that, if Q is chain-complete, then so is the set of all order-preserving maps from P to Q. Does the reverse implication hold, too?

3-32. Let (Ω, Σ, μ) be a measure space and let $p \in [1, \infty)$. We only consider real valued functions. Prove that the unit ball $\{f \in L^p(\Omega, \Sigma, \mu) : \|f\|_p \leq 1\}$ is chain-complete.

   Hint. Exercise 2-50 and the Monotone Convergence Theorem.

3-33. **An Abian–Brown Theorem for L^p-spaces.** Let (Ω, Σ, μ) be a measure space, let $p \in [1, \infty)$, consider only real valued functions and let $F : L^p(\Omega, \Sigma, \mu) \to L^p(\Omega, \Sigma, \mu)$ be a mapping that is order-preserving on $\uparrow f \subseteq L^p(\Omega, \Sigma, \mu)$. If $F(f) \geq f$, then F has a smallest fixed point above f or F has no fixed points at all in $\uparrow f$.

   Hint. Exercise 2-50, Monotone Convergence Theorem, and the proof of the Abian–Brown Theorem.

3-34. Let (Ω, Σ, μ) be a measure space, let $p \in [1, \infty)$, consider only real valued functions and let $F : L^p(\Omega, \Sigma, \mu) \to L^p(\Omega, \Sigma, \mu)$ be a mapping that is order-preserving on $\uparrow f \subseteq L^p(\Omega, \Sigma, \mu)$. If $F(f) \geq f$, then the smallest fixed point of F above f can be found via a transfinite iteration scheme. Moreover, if F is continuous, then the fixed point can be found with an iteration that stops (at the latest) at the first infinite ordinal number.

   Hint. Extend the proof from Exercise 3-33.

3.5   The Abian–Brown Theorem in Analysis

Because L^p-spaces are ordered sets (see part 10 of Example 1.2) that almost satisfy the Abian–Brown Theorem (see Exercise 3-33), it is natural to use this fact to prove the existence or non-existence of fixed points for order-preserving operators T. The typical approach is to establish the existence of a function f so that $f \leq Tf$ and to then show that the Abian–Brown iteration does indeed converge. We exhibit this approach with the example below. What is presented here was inspired by Heikkilä's work (see, e.g., [40, 129, 130, 132, 133]). For a general overview of results like this, also consider [147].

Proposition 3.34 (For a more general context, see [129], Section 3.). *Consider the Hammerstein integral equation*

$$u(t) = v(t) + r \int_{\Omega} k(t,s) f(s, u(s)) \, ds, \qquad t \in \Omega,$$

where the set Ω is a closed and bounded subset of $\mathbb{R}^m$ and all functions assume values in $\mathbb{R}$. Assume that the following hold.

1. *$k : \Omega \times \Omega \to \mathbb{R}_+$ only assumes nonnegative values and is continuous.*
2. *$f : \Omega \times \mathbb{R}_+ \to \mathbb{R}_+$ only assumes nonnegative values and is such that*

   a. *$f(\cdot, u(\cdot))$ is measurable for each $u \in C(\Omega, \mathbb{R}_+)$,*
   b. *$f(t, \cdot)$ is increasing for almost every $t \in \Omega$,*
   c. *There are $h, h_0 \in L^1(\Omega, \mathbb{R}_+)$ such that, for all $x \in \mathbb{R}_+$ and almost every $t \in \Omega$, we have*

$$f(t, x) \le h_0(t) + h(t)x,$$

3. *$r > 0$ is such that $r \int_{\Omega} k(t,s) h(s) \, ds =: b < 1$ for each $t \in \Omega$.*

Then, for each $v \in C(\Omega, \mathbb{R}_+)$, the above integral equation has a solution.

Proof. We shall work with the operator

$$Gu(t) := v(t) + r \int_{\Omega} k(t,s) f(s, u(s)) \, ds, \qquad t \in \Omega$$

on $C(\Omega, \mathbb{R}_+)$ equipped with the L^1-norm. The fact that k and f map into $\mathbb{R}_+$ shows that $Gv \ge v$ and condition 2b assures that G is order-preserving. Now we need to prove that the G-chain starting at v has an upper bound. To do this, we consider the operator H defined as follows.

$$Hz(t) = v(t) + r \int_{\Omega} k(t,s) h_0(s) \, ds + r \int_{\Omega} k(t,s) h(s) z(s) \, ds.$$

By condition 3, we have that $\|Hz - Hz'\|_\infty \le b\|z - z'\|_\infty$, where the norm is the uniform norm $\|f\|_\infty = \sup\{|f(t)| : t \in \Omega\}$. Moreover H maps $\uparrow v$ to itself. By Banach's fixed point theorem (see [136], 111.11 "Banachscher Fixpunktsatz," [282], Theorem 17.64, or [62], Theorem 10.1.2; the result in Dieudonné would need to be adjusted slightly to fit our purposes), this means that there is a unique continuous function $w \in\uparrow v$ such that $Hw = w$. By condition 2c, we have $G \le H$, which means $Gw \le w$.

Thus G maps $[v, w] \subseteq C(\Omega, \mathbb{R}_+)$ to itself. Unfortunately this interval is not chain-complete in the L^1-norm, so we are not trivially done.

Let C be the maximal G-chain in $C(\Omega, \mathbb{R}_+)$ that starts at v. Let u_* be the pointwise a.e. supremum of C, which exists in $L^1(\Omega, \mathbb{R}_+)$. We shall now show that u_* is continuous. Because $C(\Omega, \mathbb{R}_+) \subseteq L^1(\Omega, \mathbb{R}_+)$, by Exercise 2-50, there is an increasing sequence $\{u_n\}_{n\in\mathbb{N}}$ that is cofinal in C. Then $\{G(u_n)\}_{n\in\mathbb{N}}$ is cofinal in C, too. Let $c_n := \int_\Omega f(s, u_n(s)) \, ds$ and let $c := \bigvee_{n\in\mathbb{N}} c_n$. Moreover, define $k_0 := \max\{rk(t, s) : t, s \in \Omega\}$. Then, for all $m > n$, we have

$$0 \le G(u_m) - G(u_n) \le k_0(c_m - c_n) \le k_0(c - c_n).$$

This means that $G(u_n)$ is an increasing uniform Cauchy sequence, which converges uniformly to its pointwise supremum u_*, which means that u_* is continuous.

Because u_* is the supremum of the maximal G-chain C, we must have that $G(u_*) = u_*$. Therefore u_* is a solution of the equation as desired. ∎

Note that conditions 2c and 3 are used exclusively to assure that the G-chain starting at v has an upper bound. Without them, the conclusion of Proposition 3.34 could be re-phrased as "the integral equation has a solution iff the G-chain starting at v has a supremum." This modification widens the scope of the result, while the verification if a solution exists becomes dependent on the outcome of a potentially transfinite (though countable) iteration scheme.

Remarks and Open Problems

1. Theorem 3.16 is useful when showing that a removal procedure such as dismantling (see Section 4.3) or a consistency enforcing algorithm for constraint networks (see Exercise 5-28) yields the same end result independent of the order in which the removal steps are being taken.
2. Are ordered sets reconstructible from their "ranked decks?" That is, does a deck such that, for each card, the rank of the removed element is known uniquely determine the ordered set?

   This problem is an extension/modification of trying to reconstruct an ordered set from the maximal deck or from the maximal deck and the minimal deck, both of which are not possible. More generally, we could ask "What additional information besides the deck is needed to reconstruct ordered sets?" Essentially, this amounts to a more focused quest for a complete set of invariants. My own motivation for these questions lies in the reconstruction of some further invariants such as the rank k neighborhood decks (that is, the decks that show all neighborhoods of points of rank k) and some maximal cards, see [275].

   In graph theory, similar questions have been asked for directed graphs. The reconstruction problem for directed graphs is completely analogous to that for ordered sets and graphs. However, there is an important family of directed graphs, namely, the family of tournaments, for which the reconstruction problem has a negative answer, see [298]. Moreover, in [299, 300], other examples of nonreconstructible digraphs have been constructed. This in turn has led to some fledgling efforts to reconstruct graphs from decks that contain the cards plus additional information, see, for example, [239].

3. Aside from reconstruction from the deck, which is also often called deck reconstruction, it is possible to consider reconstruction from the *set* of unlabeled, one-point deleted subsets. In graph theory, this idea is referred to as set reconstruction. The formal question for ordered sets is the following. Define the set of cards of an ordered set P as the function $\mathcal{S}_P : \mathcal{C} \to \{0, 1\}$ such that, for each $[C] \in \mathcal{C}$, we have that $\mathcal{S}_P([C]) = 1$ iff P has a card that is isomorphic to the elements of $[C]$. The **set reconstruction problem** asks the following. Is every (finite) ordered set with at least four elements uniquely reconstructible from its set of cards? That is, is it true that, if P, Q are ordered sets with at least four elements such that $\mathcal{S}_P = \mathcal{S}_Q$, then P and Q must be isomorphic? The set reconstruction problem for ordered sets is entirely unexplored.

   Obviously, the set reconstruction problem is harder than the (deck) reconstruction problem, because we have less information to work with. Similarly obviously, the problems are the same for any ordered sets for which $\mathcal{D}_P = \mathcal{S}_P$, that is, for any ordered sets so that no two distinct cards are isomorphic to each other.

   Exercise 3-13 exhibits an interesting class of ordered sets that have two distinct isomorphic cards. These sets are slightly more than mere examples. They play a key role in Theorem 2.4 in [284] and, together with ordered sets in which two distinct points have the same strict upper bounds and the same strict lower bounds, and the sets described in Theorem 2.4 and Proposition 2.5 in [284], they constitute *all* ordered sets that have two distinct isomorphic cards. If all these sets are set reconstructible, then set reconstruction and deck reconstruction of ordered sets are equivalent.

4. Which of the many finitary results available for the fixed point property can be translated to infinite ordered sets and then brought to bear on, for example, analysis (see Section 3.5)?

5. For $n \in \mathbb{N}$, consider all chains (with respect to inclusion) of order relations $\leq$ on $[n] := \{1, \dots, n\}$. Every such chain is contained in a chain of length $\frac{n}{2}(n-1)$. The greatest number of ordered sets with the fixed point property in such a chain is $\frac{n}{2}(n-1) - (n-1)$. (Create an ordered set with a point comparable to all others as fast as possible.) What is the smallest number of ordered sets with the fixed point property in such a chain? What is the largest number of adjacencies $\leq_1 \prec \leq_2$ in such a chain so that $([n], \leq_1)$ does not have the fixed point property and $([n], \leq_2)$ does?

6. For more on the connection between order and analysis, see [41, 131, 191].

7. For an overview of results that connect order-theoretic methods to Banach's fixed point theorem, see [147]. Note that these connections seem to move in the direction of using a graph as the underlying structure, rather than an ordered set.

Chapter 4
Retractions

Retractions are an important tool to explore the structure of ordered sets. For example, for the fixed point property, certain retractions can reduce the problem to a problem on smaller, easier to handle structures. Viewed "in the opposite direction," retractions can also be seen as a tool to build larger examples of ordered sets with certain properties. Similar statements hold for lexicographic sum decompositions, see Chapter 7, and for products and sets P^Q, see Chapter 12. Sometimes investigations via retractions can have surprising consequences, as can be seen in Theorem 4.48. For an excellent survey on retractions, see [245]. For a recent survey on the use of retractions in fixed point theory, see [289].

4.1 Definition and Examples

Definition 4.1. *Let P be an ordered set. Then an order-preserving map* $r : P \to P$ *is called a **retraction** iff* $r^2 = r$, *that is, iff r is **idempotent**. We will say that* $R \subseteq P$ *is a **retract** of P iff there is a retraction* $r : P \to P$ *with* $r[P] = R$.

Another way to describe retractions, which is less algebraic than idempotency, is to say that, after the first application of r, every point stays stationary. Although the two conditions are easily seen to be equivalent, the latter provides a more pictorial way of thinking about retractions.

Proposition 4.2. *Let P be an ordered set and let* $f : P \to P$ *be an order-preserving map. Then f is a retraction iff* $f|_{f[P]} = id_{f[P]}$.

Proof. Let $f^2 = f$ and let $q \in f[P]$. Then there is a $p \in P$ so that $q = f(p)$. Therefore, we have $f(q) = f(f(p)) = f(p) = q$. Conversely, let $f|_{f[P]} = id_{f[P]}$ and let $p \in P$. Then $f^2(p) = f|_{f[P]}(f(p)) = f(p)$. ∎

© Springer International Publishing 2016
B. Schröder, *Ordered Sets*, DOI 10.1007/978-3-319-29788-0_4

Example 4.3. Many of the examples that follow are indeed short theorems. Parts 3, 4, and 5 will be important tools when retractions are used to prove other results. Part 8 is the starting point for proving the Li–Milner Structure Theorem.

1. For every ordered set P, the identity id_P is a (trivial) retraction on P.
2. In the ordered set e) in Figure 1.4, the set $\{a, c, e, g, k\}$ is a retract.
3. (See [244, 270].) Let P be an ordered set and let $a, b \in P$. If $(\uparrow a) \setminus \{a\} \subseteq (\uparrow b)$ and $(\downarrow a) \setminus \{a\} \subseteq (\downarrow b)$, then

$$r(x) := \begin{cases} x; & \text{if } x \neq a, \\ b; & \text{if } x = a \end{cases}$$

   is a retraction. In this situation, a is called **retractable** to b and r is a **retraction that removes a retractable point**.
4. (See [244].) Let P be an ordered set and let $x \in P$ be such that $(\uparrow x) \setminus \{x\}$ has a smallest element u. Then x is retractable to u. In this situation (or its dual), we will say that x is **irreducible**[1] in P. The retraction r is referred to as a **retraction that removes an irreducible point**.
5. Let P be an ordered set and let $W \subseteq P$ be a well-ordered subchain such that there is no $p \in P \setminus W$ with $p > W$. Then W is a retract of P.
6. (Duffus, Rival, and Simonovits, see [74].) Let P be an ordered set and let $C \subseteq P$ be a maximal chain in P. Then C is a retract of P.
7. Let P be an ordered set and let $p \in P$. Then $\downarrow p$ is a retract of P.
8. For an ordered set P, the point $p \in P$ is said to be **funneled through** $q \in P$ iff all maximal chains through p also go through q. The subset Q of P is called a **good subset** iff no two elements of Q are funneled through each other (in P) and, for all $p \in P$, there is exactly one $q \in Q$ such that p is funneled through q. Every good subset of P is a retract of P.

Proof. The identity map trivially is idempotent, which establishes part 1.

For part 2, note that the map that maps $h \mapsto k$; $d, f \mapsto e$; $b \mapsto a$ and leaves all other points fixed is a retraction. Also note that this is not the only retraction onto the set $\{a, c, e, g, k\}$.

In part 3, it is clear that the proposed map is idempotent. To show that r is order-preserving, let $x < y$. If $a \notin \{x, y\}$, there is nothing to prove, so let us assume that $x = a$. (The case $y = a$ is handled dually.) Then $y \in (\uparrow a) \setminus \{a\} \subseteq (\uparrow b)$ and hence $r(a) = b < y = r(y)$. Thus r is order-preserving.

Part 4 follows directly from the definition of retractable points.

To prove 5, let $r(x) := \bigvee_W \{w \in W : w \leq x\}$. The map r is well-defined, because every initial segment of W that is not equal to W has a supremum and because, if W has an upper bound, it must be contained in W. (Also recall that the supremum of the empty set is the smallest element of W, so suprema of empty sets

---

[1]This choice of language comes from lattice theory and will be motivated in Proposition 8.29.

are accounted for, too.) Idempotency is again trivial. Finally, if $x \leq y$, then the
containment $\{w \in W : w \leq x\} \subseteq \{w \in W : w \leq y\}$ holds. Therefore we have
$r(x) = \bigvee_W \{w \in W : w \leq x\} \leq \bigvee_W \{w \in W : w \leq y\} = r(y)$.

The first step towards proving part 6 is to put another order on C. Let $\sqsubseteq$ be any
well-ordering of C. We define (see [245], p. 104)

$$r(x) := \min_{\sqsubseteq} \{c \in C : c = x \text{ or } c \nleq_P x\}$$

Note that the minimum in the definition is the minimum with respect to the well-
ordering, while the incomparability is with respect to the order C inherits from P.
Maximality of C guarantees that r is well-defined. It is clear that r is idempotent. To
see that r is order-preserving, let $x < y$. We must prove that $r(x) \leq r(y)$. If $x \in C$,
this is trivial, as $r(y)$ is an element of the set $\{c \in C : c \nleq y\} \cup \{y\}$, which consists
entirely of upper bounds of x. The case $y \in C$ is treated similarly. This leaves the
case in which $x, y \notin C$. If $r(x)$ is a $\leq$-lower bound of $\{c \in C : c \nleq y\}$ or if $r(y)$ is a
$\leq$-upper bound of $\{c \in C : c \nleq x\}$, then we are done. Now, $r(x)$ cannot be a strict
$\leq$-upper bound of $\{c \in C : c \nleq y\}$, because this would imply that $x < y \leq r(x)$,
which is not possible for $x \notin C$. Similarly $r(y)$ is not a strict $\leq$-lower bound of
$\{c \in C : c \nleq x\}$. This leaves the case in which $r(x), r(y) \in \{c \in C : c \nleq x, y\}$. In
this case, $r(x)$ and $r(y)$ are the $\sqsubseteq$-minimal element of $\{c \in C : c \nleq x, y\}$ and hence
they are equal. Thus r is order-preserving.

The retraction needed in part 7 is

$$r(x) := \begin{cases} x; & \text{if } x \leq p, \\ p; & \text{otherwise.} \end{cases}$$

The retraction for part 8 follows easily from the definition of the good subset Q.
(The following is adapted from [187], Lemma 2.5.) For each $p \in P$, we let $r(p)$
be the unique element of Q through which p is funneled. Clearly r is idempotent,
because, trivially, for each $q \in Q$, the point q itself is the unique element of Q
through which q is funneled.

To see that r is order-preserving, first note that, if we have $x < y \leq r(x)$, then
$r(x) = r(y)$: Indeed, let C be a maximal chain that contains y. Then the chain
$C' := [(\uparrow y) \cap C] \cup \{x\}$ is contained in a maximal chain K, which must contain $r(x)$.
Thus $[(\uparrow y) \cap C] \cup \{r(x)\}$ is a chain and, because $r(x) \geq y$, we have that $r(x) \cup C$ is a
chain. By maximality of C, we infer that $r(x) \in C$. Thus, because C was an arbitrary
maximal chain containing y, y is funneled through $r(x) \in Q$. By definition of good
subsets, this means $r(x) = r(y)$.

Now let $a \leq b$ in P. If $r(a) = r(b)$, then there is nothing to prove, so we can
assume that $r(a) \neq r(b)$. Because the set $\{a, b\}$ is a chain that is contained in a
maximal chain, the set $\{a, b, r(a), r(b)\}$ must be a chain. By the above, we cannot
have $a < b \leq r(a)$. Thus $r(a) < b$. Because $r(a) \neq r(b)$, by the dual of the previous
paragraph applied to $\{r(a), b\}$, we cannot have $r(b) \leq r(a) = r(a)) < b$. Thus the
only remaining possibilities are $r(a) < r(b) \leq b$ or $r(a) < b < r(b)$. In either case,
we have proved that r is order-preserving. ∎

Most of our examples above focused on the image set of the retraction. Although retracts of ordered sets are interesting subsets, we should not forget that retractions are functions. On a finite ordered set, any order-preserving map can be turned into a retraction via the following proposition.

Proposition 4.4. *Let P be a finite ordered set and let $f : P \to P$ be an order-preserving map. Then $f^{|P|!}$ is a retraction on P.*

Proof. To abbreviate notation, let $n := |P|$. Let $p \in P$. Because P has n elements, for the set $\{p, f(p), \ldots, f^n(p)\}$ there must be $0 \le k_p < l_p \le n$ such that $f^{k_p}(p) = f^{l_p}(p)$. In particular, this means that every point $f^j(p)$ with $j \ge k_p$ is $(l_p - k_p)$-periodic, that is, $f^{l_p - k_p}(f^j(p)) = f^j(p)$. Then, for all $j \ge l_p$, we have

$$f^j(p) = f^{k_p + [(j - l_p) \bmod (l_p - k_p)]}(p) \in \left\{ f^{k_p}(p), \ldots, f^{l_p - 1}(p) \right\}.$$

Now let $q \in f^{n!}[P]$. We must show that $f^{n!}(q) = q$. Let $p \in P$ be such that $f^{n!}(p) = q$. Then, by the above, there is an $i \le n$ such that $f^i(p) = q$ and such that $q = f^i(p)$ is periodic with a period $t = l_p - k_p \le n$. Thus $f^{n!}(q) = (f^t)^{\frac{n!}{t}}(q) = q$ and we are done.   ∎

With some examples at hand, let us now investigate some properties of retractions. Retractions behave "friendly" with respect to compositions and with respect to suprema and infima as evidenced by the next two propositions.

Proposition 4.5. *Let P be an ordered set and let $r : P \to P$ and $s : r[P] \to r[P]$ be retractions. Then $s \circ r : P \to P$ is a retraction onto $s \circ r[P]$.*

Proof. Clearly the composition $s \circ r$ is order-preserving. Now let $x \in s \circ r[P]$. Then $x \in s[r[P]] \subseteq r[P]$ and thus $s(r(x)) = s(x) = x$.   ∎

Proposition 4.6. *Let P be an ordered set, let $r : P \to P$ be a retraction, and let $A \subseteq r[P]$. If A has a supremum in P, then A has a supremum in $r[P]$ and $r\left[\bigvee_P A\right] = \bigvee_{r[P]} A$.*

Proof. Let the set $A \subseteq r[P]$ have a supremum a in P. Then $r(a)$ is an upper bound of $A = r[A]$. Moreover, if $x \in r[P]$ is an upper bound of A, then $x \ge a$ and hence $x = r(x) \ge r(a)$.   ∎

Note however that retractions do not preserve suprema of arbitrary subsets of P (see Exercise 4-12).

Exercises

4-1. Another definition for retractions. The ordered set P is called a **retract** of the ordered set Q iff there are order-preserving maps $r : Q \to P$ and $c : P \to Q$ such that $r \circ c = \mathrm{id}_P$. In this case, r is called the **retraction** and c is called the **coretraction**. Prove that $c[P]$ is a retract of Q in the sense of Definition 4.1 and that $c[P]$ is isomorphic to P.

This means that, although the above definition is more general than Definition 4.1, after the proper exchanges as necessary, we can use either description of retractions. We will commonly use Definition 4.1 throughout.

4-2. Find the isomorphism types of all retracts of fences and crowns.

4-3. Prove that every ordered set of height 1 with at least one comparability $x < y$ has at least $2^{\frac{n}{2}-1}$ retractions.

4-4. Prove that, if r is a retraction that removes a retractable point a and $r(x) \sim x$ for all $x \in P$, then a is in fact an irreducible point.

4-5. Let P be a finite ordered set such that there is a non-maximal $p \in P$ that has only one maximal upper bound. Prove that P has an irreducible point whose only upper cover is a maximal element.

4-6. Prove that, if r is a retraction of P onto $P \setminus \{a\}$, then a is retractable to $r(a)$.

4-7. Let P be a finite ordered set. Prove that a four-crown-tower that contains a maximal chain is a retract of P.

4-8. Give an example of an ordered set with more than one good subset.

4-9. Let P be an ordered set and let $f : P \to P$ be an order-preserving map such that there is a $k \in \mathbb{N}$ such that $|\{p \in P : f(p) \neq p\}| \leq k$. Prove that the function $f^{k!}$ is a retraction and that $|\{p \in P : f^{k!}(p) \neq p\}| \leq k$.

4-10. (Exploring Exercise 4-9.)

    a. Prove that, if P is an ordered set and $f : P \to P$ is an order-preserving mapping with $|\{p \in P : f(p) \neq p\}| = 1$, then f is a retraction.

    b. Find an ordered set P and an order-preserving map $f : P \to P$ such that f is not a retraction and $|P \setminus f[P]| = 1$.

    c. Find an ordered set P and an order-preserving map $f : P \to P$ such that f is not a retraction and $|\{p \in P : f(p) \neq p\}| = 2$.

4-11. Let P be an ordered set of finite width w.

    a. Prove that, for each $p \in P$, we have $f^{2w!}(p) \sim f^{w!}(p)$.

    b. Conclude that, if P is chain-complete and of finite width w, then, for every order-preserving self-map $f : P \to P$, there is a retraction r_f such that, for any two order-preserving self-maps f and g on P, if $f(p) \leq g(p)$ for all $p \in P$, then $r_f(p) \leq r_g(p)$ for all $p \in P$.

4-12. Find an example of an ordered set P, a retraction $r : P \to P$, and a subset $B \subseteq P$ such that B has a supremum in P, but $r[B]$ does not have a supremum in $r[P]$.

4-13. Prove that, for every ordered set of height r, there are at least $2^{\frac{r}{r+1}(n-r-1)}$ retractions onto an $(r + 1)$-chain.

4-14. Prove that every n-element ordered set of height 1 has at least $3^{\frac{n}{2}}$ order-preserving self-maps. Conclude that every n-element ordered set has at least $2^{\frac{2}{3}n}$ order-preserving self-maps.
    Hint. Retract onto suitable 3-fences.

4-15. Call a fence in an ordered set **spanning** iff it contains only extremal elements. Call a fence $F \subseteq P$ **isometric** iff, for any two points $a, b \in F$, we have $\mathrm{dist}_P(a, b) = \mathrm{dist}_F(a, b)$. Prove that an isometric spanning (finite or infinite) fence in an ordered set is a retract.
    Note. Part (2) of the main theorem in [74] gives a more general result.

4-16. Prove that any ordered set that contains no crowns, infinite chains, or infinite fences must have an irreducible point.

4.2 Fixed Point Theorems

This section presents two ways in which retractions are valuable tools for the fixed point property. Section 4.2.1 considers the removal of points, so that the task of determining whether an ordered set has the fixed point property is reduced to the, hopefully easier, task of answering the same question for a smaller set (or sets). The iteration of this process will be discussed in Section 4.3. Section 4.2.2 presents an extension of the Abian–Brown Theorem, see Theorem 3.32, which shows that finding a fixed point is closely related to finding a point comparable to its image in *any* ordered set, not just in the chain-complete ones.

We start with an elementary result that shows the close relation between the Open Questions 1.19 and 1.24.

Definition 4.7. *An ordered set P is called* **minimal automorphic** *iff P is automorphic and all proper retracts $R \neq P$ of P have the fixed point property.*

Theorem 4.8 shows that retractions preserve the fixed point property and that minimal automorphic sets are omnipresent in the class of finite fixed point free ordered sets.

Theorem 4.8. *Let P be an ordered set.*

1. *If P has the fixed point property, then every retract of P has the fixed point property.*
2. *If P is finite, then P does not have the fixed point property iff P has a minimal automorphic retract.*

Proof. To prove part 1, let P have the fixed point property, let $r : P \to P$ be a retraction, and let $f : r[P] \to r[P]$ be an order-preserving map. Then $f \circ r : P \to P$ is an order-preserving map and thus it has a fixed point p. Because p must be in $r[P]$, we have $r(p) = p$ and hence $f(p) = f(r(p)) = p$.

For part 2, first note that "$\Leftarrow$" is clear. To prove "$\Rightarrow$," let $n := |P|$. Then, by Proposition 4.4, for all order-preserving $f : P \to P$, we have that $f^{n!}$ is a retraction.

Now note that, for all order-preserving $f : P \to P$, the map $f|_{f^{n!}[P]}$ is an automorphism. Indeed, if $p \in f^{n!}[P]$, then

$$f(p) = f(f^{n!}(p)) = f^{n!}(f(p)) \in f^{n!}[P],$$

so $f|_{f^{n!}[P]}$ is an order-preserving self-map of $f^{n!}[P]$. The function $(f|_{f^{n!}[P]})^{n!-1}$ is the order-preserving inverse of $f|_{f^{n!}[P]}$.

The proof of part 2 is now an induction on n. For $n = 1$, there is nothing to prove. For the induction step $\{1, \ldots, n-1\} \to n$, assume that the result holds for all ordered sets with up to $n - 1$ elements. First note that, if $|P| = n$ and all fixed point free maps of P are automorphisms, then there is nothing to prove (use $Q = P$). Otherwise, let $f : P \to P$ be an order-preserving map that is not an automorphism and apply the induction hypothesis to $f^{n!}[P]$, which has fewer elements than P.

Let Q be the minimal automorphic retract obtained via the induction hypothesis and let $r : f^{n^1}[P] \to Q$ be a retraction. Then Q is a retract of P via $r \circ f^{n^1}$ and we are done. ∎

We will further investigate minimal automorphic sets in Section 4.4.3.

4.2.1 Removing Points

The initial work on removing a point is in the paper [244] by Rival, and it has since been extended in various directions. The idea is that, where fixed point theory is concerned, any map that maps points to their upper or lower bounds should not affect the fixed point property, because we have the Abian–Brown Theorem. Hence the strong interest in comparative retractions, defined below. We will present the best folklore version of Rival's original result in Theorem 4.11 and a modification in Theorem 4.12. Notice that chain-completeness is not needed in Theorem 4.12 and in Scholium 4.13.

Definition 4.9. *Let P be an ordered set. A retraction $r : P \to P$ is called*

1. *A **comparative retraction** iff, for all $p \in P$, we have $r(p) \sim p$.*
2. *An **up-retraction** or **closure operator** iff, for all $p \in P$, we have $r(p) \geq p$.*
3. *A **down-retraction** or **interior operator** iff, for all $p \in P$, we have $r(p) \leq p$.*

Example 4.10. Retractions that remove one irreducible point are always up- or down-retractions. Retractions onto good subsets are always comparative. Retractions that remove a retractable point are comparative iff the retractable point is in fact irreducible (see Exercise 4-31). □

Theorem 4.11 (See [266], Theorem 2.).
Let P be a chain-complete ordered set and let $r : P \to P$ be a comparative retraction. Then P has the fixed point property iff $r[P]$ has the fixed point property.

Proof. The direction "⇒" follows from part 1 of Theorem 4.8. To prove "⇐," let $r : P \to P$ be a comparative retraction and let $f : P \to P$ be order-preserving. Then $r \circ f|_{r[P]} : r[P] \to r[P]$ has a fixed point p and, because r is a comparative retraction, we have that $p = r \circ f|_{r[P]}(p) = r \circ f(p) \sim f(p)$. Now the Abian–Brown Theorem implies that f has a fixed point. ∎

When we restrict ourselves to only moving one point in the retraction, we obtain a similar result that holds for all ordered sets.

Theorem 4.12 (See [270], Theorem 3.3.).
Let P be an ordered set and let $a \in P$ be retractable to $b \in P$. Then P has the fixed point property iff

1. *$P \setminus \{a\}$ has the fixed point property, **and***
2. *One of $(\uparrow a) \setminus \{a\}$ and $(\downarrow a) \setminus \{a\}$ has the fixed point property.*

Proof. For "⇒," let P have the fixed point property. Condition 1 is a consequence of Example 4.3, part 3 and Theorem 4.8, part 1. To see 2, suppose for a contradiction that $(\uparrow a) \setminus \{a\}$ and $(\downarrow a) \setminus \{a\}$ do not have the fixed point property. Let the functions $g : (\uparrow a) \setminus \{a\} \to (\uparrow a) \setminus \{a\}$ and $h : (\downarrow a) \setminus \{a\} \to (\downarrow a) \setminus \{a\}$ be order-preserving maps without fixed points. Define

$$f(q) := \begin{cases} g(q); & \text{if } q \in (\uparrow a) \setminus \{a\}, \\ h(q); & \text{if } q \in (\downarrow a) \setminus \{a\}, \\ a; & \text{if } q \text{ is not comparable to } a, \\ b; & \text{if } q = a. \end{cases}$$

Clearly $f : P \to P$ has no fixed point. It is left to Exercise 4-17 to verify that f is order-preserving. We have arrived at a contradiction.

For "⇐," let $f : P \to P$ be order-preserving and let $r : P \to P \setminus \{a\}$ be as in Example 4.3, part 3. Then $r \circ (f|_{P \setminus \{a\}}) : P \setminus \{a\} \to P \setminus \{a\}$ is order-preserving and thus it has a fixed point $q \in P \setminus \{a\}$. In case $q \neq b$, we have that

$$q = r \circ (f|_{P \setminus \{a\}})(q) = f|_{P \setminus \{a\}}(q) = f(q)$$

and we are done. In case $q = b$, we either have $f(b) = b$ or $f(b) = a$. In the first case, we are done. In the second case, assume without loss of generality that $(\uparrow a) \setminus \{a\}$ has the fixed point property. Then $(\uparrow a) \setminus \{a\} \neq \emptyset$. Because $f(b) = a$, we have that $f[(\uparrow a) \setminus \{a\}] \subseteq f[\uparrow b] \subseteq \uparrow a$. In case $f[(\uparrow a) \setminus \{a\}] \subseteq (\uparrow a) \setminus \{a\}$ we have, by assumption, that f has a fixed point in $(\uparrow a) \setminus \{a\}$ and we are done. Otherwise, there is a $q \in (\uparrow a) \setminus \{a\}$ such that $f(q) = a < q$. Now, if $f(a) = a$ we are done. This leaves the case $d := f(a) < a$. In this case f maps $\downarrow d$ to itself. Yet $\downarrow d$ is a retract of $P \setminus \{a\}$ by part 7 of Example 4.3. Thus $\downarrow d$ has the fixed point property, which implies that f has a fixed point in $\downarrow d$. ∎

The difference between Theorem 4.12 and Scholium 4.13 is small but important. When our retractable point is in fact irreducible, then we do not need condition 2, which affects the complexity of iterating this process. (See open problem 9 at the end of Chapter 5.)

Scholium 4.13 (See [244], Proposition 1.).

Let P be an ordered set and let $a \in P$ be irreducible. Then P has the fixed point property iff $P \setminus \{a\}$ has the fixed point property.

Proof. As the direction "⇒" is trivial, we only need to consider "⇐." Suppose without loss of generality that b is the unique lower cover of a and let $f : P \to P$ be order-preserving. As seen in the proof of Theorem 4.12, the only interesting case is the case $f(b) = a > b$. If $f(a) = a$, we are done. Otherwise $f(a) > f(b) = a$ and f maps $\uparrow f(a)$ to itself. However $\uparrow f(a)$ is a retract of $P \setminus \{a\}$ and thus it has the fixed point property. Hence f has a fixed point in $\uparrow f(a)$. ∎

4.2.2 The Comparable Point Property

The Abian–Brown Theorem shows that, in chain-complete ordered sets, it is enough to find a point that is comparable to its image to establish existence of a fixed point. On the other hand, the proof of Theorem 4.12 shows that some fixed point arguments can avoid chain-completeness. This line of thought can be extended to show that understanding the fixed point property is essentially equivalent to understanding the comparable point property, see Theorem 4.17 below.

Until Section 4.6, this section will be the last section in this chapter that has a strong focus on infinite ordered sets.

Definition 4.14. *An ordered set P is said to have the **comparable point property** iff, for each order-preserving map $f : P \to P$, there is a $p \in P$ with $f(p) \sim p$.*

Definition 4.15. *Let P be an ordered set and let $C_l, C_u \subseteq P$ be chains in P such that $C_l \leq C_u$. Then the set*

$$K := \{p \in P : C_l \leq p \leq C_u\}$$

*will be called the (C_l, C_u)-**core**. (Note that C_l, C_u, or both could be empty).*

A word of caution is in order. The word "core" is also more commonly used for certain entities defined via retractions (see Definition 4.22) and in graph theory (see Definition 6.25). We will use (C_l, C_u)-cores only in this section, in the proof of Theorem 8.10, which is one of the seminal results in the fixed point theory for ordered sets, and in the proof of Theorem 8.15. The other notions of cores are not used in these contexts.

Lemma 4.16. *Let P be an ordered set and let $C_l \leq C_u$ be chains in P. If C_l is well-ordered and C_u is dually well-ordered, then*

$$C_l \cup \{p \in P : C_l \leq p \leq C_u\} \cup C_u$$

is a retract of P.

Proof. We define

$$r(x) := \begin{cases} x; & \text{if } C_l \leq x \leq C_u, \\ \sup_{C_l}\{c \in C_l : c \leq x\}; & \text{if } x \not\geq C_l, \\ \inf_{C_u}\{c \in C_u : c \geq x\}; & \text{if } x \geq C_l \text{ and } x \not\leq C_u. \end{cases}$$

The proof that r is a retraction is left as Exercise 4-19. ∎

Theorem 4.17 (See [263], Theorem 1.). *The following are equivalent for an ordered set P.*

1. *P has the fixed point property.*
2. *Every (C, K)-core of P has the fixed point property.*
3. *Every (C, K)-core of P has the comparable point property.*

Proof. To prove that 1 implies 2, let $H \subseteq P$ be a (C, K)-core of P. By Proposition 2.64 and its dual, we can assume without loss of generality that C is well-ordered and that K is dually well-ordered. By definition of (C, K)-cores, we know that the supremum of C, if it exists, is in H and the same goes for the infimum of K. Let $f : H \to H$ be order-preserving. For $c \in C \setminus H$ let c^+ denote the immediate successor of c in the well-ordering of C and for $k \in K \setminus H$, let k^- denote the immediate predecessor of k in the dual well-ordering of K. With r denoting the retraction from Lemma 4.16, we define

$$F(x) := \begin{cases} f(x); & \text{if } x \in H, \\ r(x)^+; & \text{if } x \notin H \text{ and } r(x) \in C, \\ r(x)^-; & \text{if } x \notin H \text{ and } r(x) \in K. \end{cases}$$

Then F is order-preserving (see Exercise 4-20a). Thus F has a fixed point, which, by definition of F, must be in H and must hence be a fixed point of f.

The implication "2⇒3" is trivial.

To prove "3⇒1," let every (C, K)-core of P have the comparable point property and let $f : P \to P$ be order-preserving. The set $\mathcal{H}$ of all (C, K)-cores H such that f maps H to itself is nonempty, because P is the $(\emptyset, \emptyset)$-core. This set is also inductively ordered by reverse inclusion $\supseteq$: The supremum of a chain of (C, K)-cores is its intersection, see Exercise 4-20b. Thus, by Zorn's Lemma, there is a maximal (C, K)-core H that is mapped to itself by f. Note that, because (C, K)-cores have the comparable point property, H is not empty.

If H is a singleton, we are done. So assume that H is not a singleton. Because H has the comparable point property, there is a $p \in H$ so that, without loss of generality, $f(p) \geq p$. If $f(p) = p$, we are done. If $f(p) > p$, then f must map $(\uparrow f(p)) \cap H$, which is a (C, K)-core, to itself and $(\uparrow f(p)) \cap H$ is a proper subset of H. This is a contradiction to the maximality of H. Thus, if H is a not singleton, then f must have a fixed point, too. ∎

Exercises

4-17. Prove that the function f in the "⇒" part of the proof of Theorem 4.12 is order-preserving.
4-18. Let P and Q be ordered sets with the fixed point property, let $m_P \in P$ and $m_Q \in Q$ be minimal elements. Define the ordered set R as the union of $P \setminus \{m_P\}$ and $Q \setminus \{m_Q\}$ with a new element m so that m is a minimal element of R that is less than all elements that are above m_P or m_Q. Prove that R has the fixed point property.
4-19. Finish the proof of Lemma 4.16 by proving that r is a retraction.
4-20. Finish the proof of Theorem 4.17 by proving the following.

a. The function F in the part "1⇒2" is order-preserving.
b. The set $\mathcal{H}$ in the part "3⇒1" is inductively ordered by $\supseteq$.

4-21. Call a set $A \subseteq P$ **retractable** to $b \in P \setminus A$ iff, for every $p \in P \setminus A$ with an $a \in A$ such that $p > a$ ($p < a$, respectively), we have $p \geq b$ ($p \leq b$, respectively).

    a. Prove that $P \setminus A$ is a retract of P.

    b. Now suppose that P is chain-complete and $P \setminus A$ has the fixed point property. Prove that, if, for every $a \in A$, every map from $(\uparrow a) \setminus A$ to $\uparrow a$ has a point that is comparable to its image, then P has the fixed point property.

4-22. Let P be an ordered set and let $a \in P$.

    a. Prove that, if P is chain-complete, then $(\updownarrow a) \setminus \{a\}$ has the fixed point property iff one of $(\uparrow a) \setminus \{a\}$ and $(\downarrow a) \setminus \{a\}$ has the fixed point property.

    b. Give an example that shows that the equivalence in part 4-22a does not hold for ordered sets that are not chain-complete.

4.3   Dismantlability

Theorem 4.11 and Scholium 4.13 suggest a reduction procedure to determine whether an ordered set has the fixed point property: Find a suitable retraction on the set and then determine if the retract has the fixed point property instead. Iterate this procedure until no suitable retraction can be found. In finite sets, this procedure is well understood (see e.g., [12, 71, 85, 244]). In infinite sets, we can sometimes iterate the above idea infinitely often, thus being faced with the problem how to get past the limit ordinal. This problem has been addressed successfully by Li and Milner (see [184–190]) for a large class of ordered sets.

The language surrounding dismantlability can be a little confusing, as essentially the same idea (which is invariably called "dismantlability") can be applied

- To finite and to infinite ordered sets.
- Using different types of retractions.
- With finitely as well as with infinitely many iterations.

In this text, we will apply dismantlability to finite as well as to infinite ordered sets, but we will always assume that the number of iterations is finite. For ways to work with infinitely many iterations, see [184], or, for a limited simpler approach, consider Exercise 4-36. The class of retractions used in the dismantling procedure will be indicated in the name.

Example 4.18. Some useful classes of retractions are the following.

1. $\mathcal{U}$: the class of all up-retractions.
2. $\mathcal{D}$: the class of all down-retractions.
3. $\mathcal{C}$: the class of all comparative retractions.[2]
4. $\mathcal{G}$: the class of all retractions onto good subsets (see [187, 188]).

---

[2]Although the letter $\mathcal{C}$ is used in Definition 1.33, there will be no confusion, as comparative retractions will not be used in the context of reconstruction.

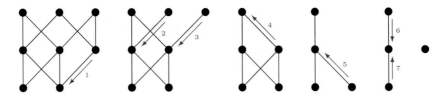

Fig. 4.1 An 8-point ordered set that is $\mathcal{I}$-dismantlable. Removals of irreducible points are indicated by arrows and the order of removal is numbered

5. $\mathcal{I}$: the class of retractions that remove an irreducible point.
6. $\mathcal{R}_k$: the class of all retractions $r : P \to P$ such that $|P \setminus r[P]| \leq k$.
7. $\mathcal{F} := \bigcup_{k \in \mathbb{N}} \mathcal{R}_k$.   □

Definition 4.19. *Let P be an ordered set, let $\mathcal{R}$ be a class of retractions, and let $Q \subseteq P$. Then P is called $\mathcal{R}$-**dismantlable to** Q iff there is a sequence*

$$P = P_0 \supseteq P_1 \supseteq \cdots \supseteq P_n = Q$$

of subsets of P and (with $i = 1, \ldots, n$) a sequence $r_i : P_{i-1} \to P_{i-1}$ of retractions in $\mathcal{R}$ such that $r_i[P_{i-1}] = P_i$. If Q is a singleton, we also say that P is $\mathcal{R}$-dismantlable.

For an example of an $\mathcal{I}$-dismantlable ordered set, see Figure 4.1.

Proposition 4.20. *Let P be an ordered set that is $\mathcal{R}$-dismantlable to Q. Then Q is a retract of P.*

Proof. The composition $r := r_n \circ r_{n-1} \circ \cdots \circ r_1$ is a retraction of P onto Q.   ∎

Some notions of dismantlability are actually equivalent for finite ordered sets. This is one of the reasons why the language is somewhat mixed in the literature. In Exercise 4-32, we will explore the differences between these notions when considering infinite ordered sets.

Proposition 4.21. *Let P be a finite ordered set. Then the following are equivalent.*

1. *P is $\mathcal{C}$-dismantlable to Q.*
2. *P is $(\mathcal{U} \cup \mathcal{D})$-dismantlable to Q.*
3. *P is $\mathcal{I}$-dismantlable to Q.*

Proof. Because $\mathcal{C} \supseteq (\mathcal{U} \cup \mathcal{D}) \supseteq \mathcal{I}$, the implications "3⇒2⇒1" are trivial.
To finish the proof, we must prove that part 1 implies part 3 for finite ordered sets. Let P be $\mathcal{C}$-dismantlable to Q. We proceed by induction on $|P|$. The base step $|P| = 1$ is of course trivial. For the induction step, assume that the result holds for all ordered sets S with $|S| < m$, let $|P| = m$ and let $c_i : P_{i-1} \to P_{i-1}$ be comparative retractions with $c_i[P_{i-1}] = P_i$, $P_0 = P$, and $P_n = Q$.
Let $U \subseteq P$ be the set of all points $p \in P$ such that $c_1(p) > p$ and assume without loss of generality that $U \neq \emptyset$. Because U is finite, we can let u be a maximal

element of U. Let $b > u$ be a strict upper bound of u. Then $b \geq c_1(b) \geq c_1(u) > u$, so $c_1(u)$ is the unique upper cover of u and hence u is irreducible. The map $c_1|_{P \setminus \{u\}}$ is a comparative retraction on $P \setminus \{u\}$ and $c_1|_{P \setminus \{u\}}[P \setminus \{u\}] = P_1$. Thus by induction hypothesis $P \setminus \{u\}$ is $\mathcal{I}$-dismantlable to Q. Because $P \setminus \{u\}$ is obtained from P by removal of an irreducible point, this finishes our proof. ∎

Of course the dismantling procedures would be quite uninteresting if all we could prove about them were their mutual equivalence (or lack thereof). The idea is to use dismantlings to shrink the sets we are considering. This reduces the difficulty of problems such as deciding if an ordered set has the fixed point property. Because we are working in finite sets, it is clear that the dismantling procedure must eventually stop. The sets at which the dismantlings must stop are called cores.

Definition 4.22. *Let P be an ordered set and let $\mathcal{R}$ be a class of retractions. Then P is called an $\mathcal{R}$-core iff the only retraction $r : P \to P$ that is in $\mathcal{R}$ is id_P.*

An ordered set can have many cores. For example, every singleton subset of a finite fence is a $\mathcal{C}$-core of that fence. However, we can show that the $\mathcal{C}$-core of a finite set is unique *up to isomorphism*. For fixed point investigations, this is exactly the result we need.

Definition 4.23. *Let P be a finite ordered set and let $\mathcal{R}$ be a class of retractions. We say that the $\mathcal{R}$-core of P is unique iff any two $\mathcal{R}$-cores C_1, C_2 to which P is $\mathcal{R}$-dismantlable are isomorphic.*

Lemma 4.24. *Let P be a chain-complete ordered set and let the order-preserving map $f \neq id_P$ be such that $f(p) \sim p$ for all $p \in P$. Then there is a retraction $r \neq id_P$ such that $r(p) \sim p$ for all $p \in P$.*

Proof. First note that $f \vee id_P$ and $f \wedge id_P$, with suprema taken pointwise, are both well-defined and order-preserving and that one of them, say, $f \vee id_P$, is not equal to id_P. For all $p \in P$, we have $f(p) \vee p \geq p$. Hence, by the Abian–Brown Theorem, $f \vee id_P$ has a smallest fixed point above p. Let $r(p)$ be the smallest fixed point of $f \vee id_P$ above p. Then $r : P \to P$ clearly is idempotent, not equal to id_P and $r(p) \geq p$ for all $p \in P$. What remains to be proved is that r is order-preserving. Let $p < q$ be given. Then $r(p)$ is the smallest fixed point of $f \vee id_P$ above p and $r(q)$ is a fixed point of $f \vee id_P$ above $q > p$. Thus $r(p) \leq r(q)$ and we are done. ∎

We are now ready to prove uniqueness of some types of cores in finite sets. The main idea for the proof is that, for some classes $\mathcal{R}$ of retractions, $\mathcal{R}$-cores have a certain resilience against $\mathcal{R}$-retractions. For these classes, $\mathcal{R}$-retractions will map any $\mathcal{R}$-core to which P is dismantlable to an isomorphic image of itself.

Theorem 4.25 (See [71, 85].). *Let P be a finite ordered set. Then the $\mathcal{C}$-, $\mathcal{U}$-, $\mathcal{D}$-, and $\mathcal{R}_k$-cores of P are unique.*

Proof. To prove uniqueness of $\mathcal{C}$-cores, let $K \subseteq P$ be a $\mathcal{C}$-core such that P is $\mathcal{C}$-dismantlable to K. Let $S : P \to K$ be a composition of comparative retractions as in Definition 4.19 and in the proof of Proposition 4.20. Now consider a nested sequence $P = P_0 \supset P_1 \supset \cdots \supset P_n$ so that, for $i = 1, \ldots, n$, there is a comparative retraction $c_i : P_{i-1} \rightarrow P_{i-1}$ with $c_i[P_{i-1}] = P_i$. For $j = 1, \ldots, n$, we define $C_j := c_j \circ \cdots \circ c_1$ and we let $C_0 = \mathrm{id}_P$.

We will prove inductively that, for all j, we have $SC_j|_K = \mathrm{id}_K$. For $j = 0$, this is trivial, so let $j > 0$. First note that $SC_{j-1}|_K = \mathrm{id}_K$. For all $p \in K$, we have that $c_j(C_{j-1}(p)) \sim C_{j-1}(p)$ and thus $S(c_j(C_{j-1}(p))) \sim S(C_{j-1}(p)) = p$. Thus, because K is a C-core, by Lemma 4.24, we infer that $\mathrm{id}_K = Sc_jC_{j-1}|_K = SC_j|_K$.

Thus we have that $SC_n|_K = \mathrm{id}_K$. Now let the sets K_1, K_2 be C-cores of P with $S_i : P \rightarrow K_i$ being compositions of the retractions in the definition of dismantlability. Then, by the above, $S_1|_{K_2}S_2|_{K_1} = \mathrm{id}_{K_1}$ and $S_2|_{K_1}S_1|_{K_2} = \mathrm{id}_{K_2}$, and hence K_1 and K_2 are isomorphic.

To prove uniqueness of $\mathcal{U}$-cores, denote by F_P the set of all $p \in P$ that are funneled through a $u_p > p$. Note that a finite ordered set is a $\mathcal{U}$-core iff it does not have any points p that are funneled through a $u_p > p$. Moreover if $r : P \rightarrow P$ is a $\mathcal{U}$-retraction and p is funneled through u_p, then $r(p)$ is funneled through $r(u_p)$ in $r[P]$ (proved similar to the proof of Example 4.3, part 8).

Now let P be a finite ordered set. The above implies that no $\mathcal{U}$-core can contain any points $p \in F_P$. Thus all $\mathcal{U}$-cores are contained in $P \setminus F_P$. However $P \setminus F_P$ is itself a $\mathcal{U}$-core, see Exercise 4-25. Thus, for a finite ordered set, the $\mathcal{U}$-core itself is unique and not just up to isomorphism.

Proving uniqueness of $\mathcal{D}$-cores is the dual of the above.

Proving uniqueness of $\mathcal{R}_k$-cores is left as Exercise 4-26. ∎

We can thus talk about *the* $\mathcal{R}$-core of an ordered set.

Definition 4.26. *Whenever the $\mathcal{R}$-core of an ordered set P is unique, we denote the set of all isomorphic $\mathcal{R}$-cores of P (or sometimes, by abuse of notation, one of its representatives) as $\mathcal{R}$-core(P).*

Theorem 4.27. *A finite ordered set P has the fixed point property iff C-core(P) has the fixed point property.* ∎

4.3.1 (Connectedly) Collapsible Ordered Sets

When using retractable points and Theorem 4.12 to prove fixed point results, we not only have to contend with the set $P \setminus \{a\}$, but also with the sets $(\uparrow a) \setminus \{a\}$ and $(\downarrow a) \setminus \{a\}$. Thus, although $\mathcal{R}_1$-dismantlability is an interesting structural property, it is a priori not sufficient to determine whether an ordered set has the fixed point property: We also must know if the sets $(\updownarrow a) \setminus \{a\}$ that we encounter along the way have the fixed point property (also see Exercise 4-22a). It is thus natural to consider a recursively defined class of ordered sets for which any such decision can be made through successive $\mathcal{R}_1$-dismantlings.

Definition 4.28. *A finite ordered set P is called **collapsible** iff $|P| \in \{0, 1\}$ or there is a point $x \in P$ such that*

1. $P \setminus \{x\}$ is a retract of P,
2. $P \setminus \{x\}$ is collapsible,
3. $\updownarrow x \setminus \{x\}$ is collapsible.

Definition 4.29. *A finite ordered set P is called **connectedly collapsible** iff $|P| = 1$ or there is a point $x \in P$ such that*

1. $P \setminus \{x\}$ is a retract of P,
2. $P \setminus \{x\}$ is connectedly collapsible,
3. $\updownarrow x \setminus \{x\}$ is connectedly collapsible.

The definitions of collapsibility and connected collapsibility are very similar and, clearly, connected collapsibility implies collapsibility. The following result shows the distinct difference between collapsibility and connected collapsibility.

Theorem 4.30. *A collapsible ordered set has the fixed point property iff it is connectedly collapsible.*

Proof. The proof is an induction on $n = |P|$, with the base steps 0 and 1 being trivial. For the induction step, let $n > 1$ and assume that the result holds for all ordered sets of size less than n. Now let P be a collapsible ordered set of size $|P| = n$. Because P is collapsible, P has a retractable point a as in Definition 4.28. By Theorem 4.12 and Exercise 4-22a, P has the fixed point property iff $P \setminus \{a\}$ and $(\updownarrow a) \setminus \{a\}$ have the fixed point property. Because P is collapsible, $P \setminus \{a\}$ and $(\updownarrow a) \setminus \{a\}$ are collapsible. By induction hypothesis this means that both sets have the fixed point property iff they are connectedly collapsible. Thus P has the fixed point property iff P is connectedly collapsible. ∎

Another nice property of the classes of $\mathcal{I}$-dismantlable ordered sets, collapsible ordered sets, and connectedly collapsible ordered sets is that each class is closed under retractions.

Theorem 4.31 (See Lemma 5 in [71] and Theorem 2 in [272].). *Let P be an ordered set and let $r : P \to P$ be a retraction. If P is $\mathcal{I}$-dismantlable, then so is $r[P]$. If P is collapsible, then so is $r[P]$. If P is connectedly collapsible, then so is $r[P]$.*

Proof. We will only prove the result for connectedly collapsible ordered sets. The proof that $r[P]$ is connectedly collapsible is an induction on the size $n := |r[P]|$. The base step $n = 1$ is trivial. For the induction step $\{1, \ldots, n-1\} \to n$, assume that the result holds for all connectedly collapsible ordered sets Q and all retractions $r' : Q \to Q$ with $|r'[Q]| < n$, let P be a connectedly collapsible ordered set and let $r : P \to P$ be a retraction with $|r[P]| = n$. Let $P = \{a_1, \ldots, a_{|P|}\}$ be a sequence such that, for $k \in \{1, \ldots, |P| - 1\}$, the point a_k is retractable in the set $P_k := P \setminus \{a_1, \ldots, a_{k-1}\}$ and the subsets P_k and $\updownarrow_{P_k} a_k \setminus \{a_k\}$ are connectedly collapsible.

Let m be the smallest number such that $a_m \in r[P]$. Let $s : P_m \to P_m$ be a retraction that maps a_m to $b_m := s(a_m)$ and that fixes all other points of P_m. If $r(b_m) = a_m$, then, for all $x \in r[P] \setminus \{a_m\}$, we have $x > a_m$ iff $x > b_m$ and $x < a_m$ iff $x < b_m$. Thus $r[P] \cup \{b_m\} \setminus \{a_m\}$ is (via sr) a retract of P that is isomorphic to $r[P]$ and which is contained in $\{a_{m+1}, \ldots, a_{|P|}\}$. We can apply this argument to the new retract until we eventually have a retract with $r(b_m) \neq a_m$. Thus we can assume that $r(b_m) \neq a_m$ in the following.

Now $r \circ s|_{r[P]} : r[P] \to r[P]$ is a retraction on $r[P]$ that maps a_m to $r(b_m)$ and leaves all other points fixed. Because $|rsr[P]| < n$ and rsr is a retraction, by induction hypothesis, $rsr[P]$ is connectedly collapsible. There are now two cases to distinguish: First, if $r(b_m) \not\sim a_m$, then, for all $x \in \updownarrow_{P_m} a_m \setminus \{a_m\}$, we have that $r(x) \sim a_m, r(b_m)$ and thus $r(x) \in \updownarrow_{r[P]} a_m \setminus \{a_m\}$. Thus, the center-deleted $r[P]$-neighborhood of a_m is equal to $\updownarrow_{r[P]} a_m \setminus \{a_m\} = r[\updownarrow_{P_m} a_m \setminus \{a_m\}]$, which is, by induction hypothesis, connectedly collapsible. Second, if $r(b_m) \sim a_m$, then all points in the center-deleted neighborhood $\updownarrow_{r[P]} a_m \setminus \{a_m\}$ are comparable to $r(b_m)$ and hence the center-deleted neighborhood $\updownarrow_{r[P]} a_m \setminus \{a_m\}$ is dismantlable and thus connectedly collapsible. Thus $r[P]$ is connectedly collapsible in either case.

The proof for collapsible ordered sets is the same and the proof for $\mathcal{I}$-dismantlable ordered sets is left for Exercise 4-23.   ∎

Exercises

4-23. Prove directly that every retract of a finite $\mathcal{I}$-dismantlable set is again $\mathcal{I}$-dismantlable.

4-24. Prove that an ordered set P is $\mathcal{C}$-dismantlable to Q iff P is $(\mathcal{U} \cup \mathcal{D})$-dismantlable to Q.

4-25. Finish the proof of the uniqueness of $\mathcal{U}$-cores (see Theorem 4.25) by showing that $P \setminus F_P$ is a $\mathcal{U}$-core.

4-26. Prove that $\mathcal{R}_k$-cores are unique.

4-27. Let P be a finite ordered set and let $a \in P$ be retractable. Prove that P is connectedly collapsible iff $P \setminus \{a\}$ and $\updownarrow a \setminus \{a\}$ are connectedly collapsible.

4-28. Let P be a connectedly collapsible ordered set with more than 1 point. Prove that P contains a non-maximal chain that is contained in exactly one maximal chain. Then give an example that this result does not hold for collapsible ordered sets.

4-29. a. Let P be a finite ordered set and let the point $a \in P$ be retractable. Prove the inequality $|\mathrm{End}(P \setminus \{a\})| < |\mathrm{End}(P)|$.

   b. Give an example of a finite ordered set P and an element $p \in P$ such that the inequalities $|\mathrm{End}(P \setminus \{a\})| > |\mathrm{End}(P)|$ and $|\mathrm{Aut}(P \setminus \{a\})| < |\mathrm{Aut}(P)|$ hold.

   c. Show that, for any $n \in \mathbb{N}$, there is a P as in 4-29b such that $|P| > n$.

4-30. (An alternative proof for Theorem 4.25, see [85].) Let P be a finite ordered set. Let $\mathcal{Q}$ be the set of isomorphism classes of retracts of P that were obtained by repeated removal of irreducible points.

   a. Prove that $\leq$ defined as follows is an order for $\mathcal{Q}$. $[A] \leq [B]$ iff some representative of $[A]$ is obtained from some representative of $[B]$ by repeated removal of irreducible points.

   b. Use Theorem 3.16 to prove that $\mathcal{Q}$ has a smallest element.

   c. Conclude that $\mathcal{C}$-cores are unique for finite ordered sets.

   d. Use the same technique to prove that $\mathcal{R}_1$-cores are unique for finite ordered sets.

   e. Try to use the same technique to prove that $\mathcal{R}_k$-cores are unique for finite ordered sets.

4-31. Prove that, for finite ordered sets, $\mathcal{R}_1 \cap \mathcal{C} = \mathcal{I}$.

4-32. This exercise investigates how different notions of dismantlability are related in infinite ordered sets.

    a. Prove that P is $\mathcal{C}$-dismantlable to Q iff P is $(\mathcal{U} \cup \mathcal{D})$-dismantlable to Q. (Note that P or Q or both could be infinite.)

    b. Construct an $\mathcal{I}$-core that has a nontrivial good subset, that is, a nontrivial $\mathcal{G}$-retraction.

    c. Construct a $\mathcal{G}$-core that has a nontrivial $\mathcal{C}$-retraction.

4-33. Prove that a connectedly collapsible ordered set of height 1 is $\mathcal{C}$-dismantlable.

4-34. (See [264].) Prove that all ordered sets with the fixed point property and at most nine elements are connectedly collapsible.

    Note. Also consider Remark 3 at the end of this chapter.

4-35. See [20] and [308]. Let P_1 and P_2 be disjoint ordered sets and let $a \in P_1$. Define the **up-1-sum** of P_1 and P_2 with respect to a as the ordered set $P_1]_a P_2$ whose ground set is $P_1 \cup P_2$ and for which the order is the union of the orders on P_1 and P_2 with all elements of P_2 being above $\downarrow a$. The **down-1-sum** of P_1 and P_2 with respect to a is the ordered set $P_1]^a P_2$ whose ground set is $P_1 \cup P_2$ and for which the order is the union of the orders on P_1 and P_2 with all elements of P_2 being below $\uparrow a$. If $a < b$ and b is not an upper cover of a, we define the **2-sum** of P_1 and P_2 with respect to a and b as the ordered set $P_1]_a^b P_2$ whose ground set is $P_1 \cup P_2$ and for which the order is the union of the orders on P_1 and P_2 with all elements of P_2 being above $\downarrow a$ and below $\uparrow b$.

    a. Let P_1 and P_2 be ordered sets and let $a, b \in P_1$ be so that $a < b$ and b is not an upper cover of a. Prove that the order on $P_1]_a^b P_2$ is the union of the order on $P_1]_a P_2$ with the order on $P_1]^b P_2$.

    b. Let P_1 and P_2 be finite and let $a \in P_1$. Prove that $P_1]_a P_2$ has the fixed point property iff P_1 does.

    c. Let P_1 and P_2 be finite and let $a \in P_1$. Prove that $P_1]^a P_2$ has the fixed point property iff P_1 does.

    d. Let P_1 and P_2 be finite and let $a, b \in P_1$ be so that $a < b$ and b is not an upper cover of a. Prove that $P_1]_a^b P_2$ has the fixed point property iff P_1 does.

4-36. (Folklore.) Let λ be an ordinal and let P be a set. Let $\{f_\alpha : P_\alpha \to P_{\alpha+1}\}_{\alpha < \lambda}$ be a family of surjective functions such that $P_{\alpha+1} \subseteq P_\alpha \subseteq P$, $P_0 = P$ and such that, for each limit ordinal, $\gamma \leq \lambda$ we have $P_\gamma = \bigcap_{\alpha < \gamma} P_\alpha$. Inductively define $F_\alpha : P \to P_\alpha$ by

$$F_0 := id_P,$$

$$F_{\alpha+1} := f_\alpha \circ F_\alpha,$$

$$F_\gamma(p) := F_\beta(p), \text{ if } \gamma \text{ is a limit ordinal and } \beta \text{ is such that}$$

$$F_\alpha(p) = F_\beta(p) \text{ for all } \beta \leq \alpha < \gamma.$$

Stop when F_γ is not totally defined for some limit ordinal γ or after having defined F_λ for all $p \in P$. If F_λ can be totally defined, we will call $\{f_\alpha\}_{\alpha < \lambda}$ **infinitely composable** and F_λ is the **infinite composition** of $\{f_\alpha\}_{\alpha < \lambda}$.

    Let $\mathcal{R}$ be a class of retractions and let P be a chain-complete ordered set. Then P is called $\mathcal{R}$-**infinite-dismantlable to** $Q \subseteq P$ iff there is an infinitely composable family $\{r_\alpha\}_{\alpha < \lambda}$ of retractions in $\mathcal{R}$ such that $R_\lambda[P] = Q$.

    a. Prove that, if P is $\mathcal{C}$-infinite-dismantlable to a finite ordered set with the fixed point property, then P has the fixed point property.

    b. Find an ordered set for which nontrivial $\mathcal{C}$-retractions can be iterated indefinitely, but which is not $\mathcal{C}$-infinite-dismantlable to any $\mathcal{C}$-core.

c. Prove that, for any ordered set P, any two C-cores C_1, C_2 to which P is C-infinite-dismantlable must be isomorphic.

   Hint. Push the proof in Theorem 4.25 past the limit ordinal.

4-37. (Also see [41], Theorem 2.44.) Let (Ω, Σ, μ) be a measure space and let $p \in [1, \infty)$. We only consider real valued functions. Prove that the unit ball $\{f \in L^p(\Omega, \Sigma, \mu) : \|f\|_p \le 1\}$ is C-dismantlable and hence has the order-theoretical fixed point property.

   Hint. Exercise 3-32 and the fact that every function has a supremum with the constant function $f(x) = 0$.

   Note. It is worth emphasizing that we are talking about the *order-theoretical* fixed point property. The L^p-unit ball does <u>not</u> have the *topological* fixed point property, as Exercise 4-38 shows. For more, consider Remark 19 at the end of this chapter.

4-38. Consider the space ℓ^1 of sequences of real numbers with the norm $\left\|\{x_n\}_{n=1}^\infty\right\|_1 := \sum_{n=1}^\infty |x_n|$. On the unit ball $B := \{x \in \ell^1 : \|x\|_1 \le 1\}$ of ℓ^1, define the operator $T : B \to B$ as follows.

   - For each $x = \{x_n\}_{n=1}^\infty$ in $\ell^1 \setminus \{0\}$, let n_x be the smallest $n \in \mathbb{N}$ so that $x_n \ne 0$.

   - Define $S : B \to B$ by $S(0) := 0$ and, for $x \ne 0$, by $(Sx)_n := \begin{cases} 0; & \text{for } n \le n_x, \\ \frac{1}{2}x_{n_x}; & \text{for } n = n_x + 1, \\ x_{n-1}; & \text{for } n > n_x + 1. \end{cases}$

   - Define $T : B \to B$ by $(Tx)_n := \begin{cases} 1 - \|Sx\|_1; & \text{for } n = 1, \\ (Sx)_n; & \text{for } n > 1. \end{cases}$

   a. Prove that T maps B to B.
   b. Prove that T is continuous.
   c. Prove that T has no fixed points.
      Hint. Prove that, if $x = \{x_n\}_{n=1}^\infty$ is a fixed point, then, because $\{x_n\}_{n=1}^\infty$ must converge to zero, all x_n are equal to zero.
   d. Translate this example to an example on the unit ball in ℓ^p, the space of sequences of real numbers with the norm $\left\|\{x_n\}_{n=1}^\infty\right\|_p := \sqrt[p]{\sum_{n=1}^\infty |x_n|^p}$.

4.4   The Fixed Point Property for Ordered Sets of Width 2 or Height 1

By Scholium 4.13, the presence or absence of irreducible points does not affect the fixed point property. Via $\mathcal{I}$-dismantlability, we can now characterize the fixed point property for ordered sets of width 2 or height 1.

4.4.1   Width 2

The characterization of the fixed point property for ordered sets of width 2 is quickly proven, because ordered sets of width 2 have only two possible kinds of $\mathcal{I}$-cores.

Definition 4.32. *A* ***four-crown-tower*** *is an ordered set* $\{a_0, b_0, a_1, b_1, \ldots, a_t, b_t\}$ *such that, for all* $k \in \{0, \ldots, t\}$ *we have that* a_k *is not comparable to* b_k, *and, for all* $k \in 1, \ldots, t$, *the points* a_{k-1} *and* b_{k-1} *are the only lower covers of* a_k *and* b_k.

Proposition 4.33. *Let P be a finite ordered set of width* 2 *with no irreducible points. Then P is either a singleton, an antichain, or a four-crown-tower.*

Proof. If P is a singleton, there is nothing to prove, so assume that P is not a singleton. Let $t \in \mathbb{N}$ be the height of P. If $t = 0$, then P is the 2-antichain and we are done. Hence we can assume that $t > 0$.

We will first prove, by induction on k, that, for all $k \in \{0, \ldots, t\}$, all points of rank k are upper bounds of all points of rank $< k$. For the base step $k = 0$, the claim is vacuously true. For the induction step, assume that the statement is proved for all natural numbers less than $k \in \{0, \ldots, t\}$. Let p be an element of rank k. Then p has at least one lower cover of rank $k - 1$. By induction hypothesis, this means that p is not adjacent to any elements of rank less than $k - 1$. Therefore, because p cannot have a unique lower cover, p must have at least two elements of rank $k - 1$ as lower covers. Because P can have at most two elements of rank $k - 1$, p is an upper bound of all elements of rank $k - 1$.

Now first assume that $k < t$. By the above, every element of rank $k + 1$ is above all elements of rank k and not adjacent to any elements of rank less than k. Because each element of rank $k + 1$ has at least two lower covers, this means that there must be two elements of rank k. For $k = t$, note that neither element of rank $t - 1$ has a unique upper cover. Thus P must have two elements of rank t.

Therefore P must be a four-crown-tower.   ∎

Theorem 4.34 (See Theorem 1 in [94] for the version for chain-complete ordered sets.).

Let P be a finite ordered set of width 2. *Then P has the fixed point property iff P is* $\mathcal{I}$*-dismantlable to a singleton.*

Proof. Every finite ordered set of width 2 is $\mathcal{I}$-dismantlable to an $\mathcal{I}$-core. By Proposition 4.33, this $\mathcal{I}$-core is either a singleton, an antichain or a four-crown-tower. By Scholium 4.13, P has the fixed point property iff its $\mathcal{I}$-core does. Because four-crown-towers and antichains do not have the fixed point property (for a fixed point free map, simply switch the a_i and b_i at every level), this proves the result.   ∎

4.4.2   Height 1

For ordered sets of height 1, the situation is not quite as easy as for width 2. Intuition dictates that crowns should be part of the picture. Knowing this, we need the right types of crowns and the right idea what to do with them.

Lemma 4.35. *Let P be an ordered set of height 1 and let $C = \{c_0, \ldots, c_{n-1}\}$ be a crown of minimal size in P. Then C is a retract of P.*

Proof. (Compare with [245], p. 118, 119.) Assume without loss of generality that c_0 is minimal in P. Let P' be the ordered set obtained from P by erasing the cover (c_0, c_{n-1}). Define

$$r(x) := \begin{cases} c_{n-1}; & \text{if } \operatorname{dist}_{P'}(x, c_0) \geq n, \\ c_i; & \text{if } \operatorname{dist}_{P'}(x, c_0) = i < n. \end{cases}$$

Because C is a crown of smallest possible size in P, we have $\operatorname{dist}_{P'}(c_i, c_0) = i$ for all i and the map r is idempotent. Now let $x < y$ (in P). Because r clearly preserves the comparability $c_0 < c_{n-1}$, we can assume that $x \neq c_0$ or $y \neq c_{n-1}$. Then $\operatorname{dist}_{P'}(x, c_0)$ is even and $\operatorname{dist}_{P'}(y, c_0) \in \{\operatorname{dist}_{P'}(x, c_0) - 1, \operatorname{dist}_{P'}(x, c_0) + 1\}$. Thus, if $\operatorname{dist}_{P'}(x, c_0) \geq n$, then $r(x) = r(y) = c_{n-1}$, while, if $\operatorname{dist}_{P'}(x, c_0) < n$, then $r(x)$ is a minimal element of C and $r(y)$ is one of its maximal upper covers in C. Thus r is order-preserving and we are done. ∎

So, in ordered sets of height 1, the presence of a crown destroys the fixed point property. Next, we consider what happens in the absence of crowns.

Lemma 4.36. *Let P be a finite connected ordered set of height 1 that does not contain any crowns. Then P must contain an irreducible point.*

Proof. We prove the contrapositive. Suppose P has no irreducible points and let $p_0 \in P$ be minimal. Let p_1 be an upper cover of p_0. Then p_1 is maximal and it has at least two lower covers. Thus p_1 has a lower cover $p_2 \neq p_0$. Suppose mutually distinct points $p_0, \ldots, p_n$ such that $\{p_0, \ldots, p_n\}$ is a fence have been constructed already and, without loss of generality, assume that p_n is minimal. Then p_n has at least two upper covers, so p_n has a maximal upper cover $p_{n+1} \neq p_{n-1}$. If $\{p_0, \ldots, p_n, p_{n+1}\}$ is a fence, continue this process.

Because P is finite, this construction must ultimately produce a p_{n+1} that is comparable to an earlier p_j with $j < n$. Let j_m be the largest $j < n$ such that p_{n+1} is comparable to p_j. Then $\{p_{j_m}, \ldots, p_{n+1}\}$ is a crown. ∎

Theorem 4.37 (See [218, 244].).

Let P be a connected finite ordered set of height 1. Then the following are equivalent.

1. *P has the fixed point property.*
2. *P contains no crowns.*
3. *P is $\mathcal{I}$-dismantlable to a singleton.*

Proof. "1⇒2" follows directly from Lemma 4.35, "2⇒3" follows from Lemma 4.36, and "3⇒1" follows from Theorem 4.27. ∎

4.4.3 Minimal Automorphic Ordered Sets

For finite ordered sets of width 2 as well as for ordered sets of height 1, we have
seen that $\mathcal{I}$-dismantlability to a singleton is equivalent to the fixed point property.
Proposition 1.22 shows that there are ordered sets with the fixed point property
that are not $\mathcal{I}$-dismantlable. Further simple characterizations of the fixed point
property, say, for width 3, are not available. This section will show why this is
the case. Proposition 4.38 below still gives a reasonably simple characterization
of automorphic ordered sets of width 3, but Proposition 4.39 shows that there does
not seem to be a simple characterization of minimal automorphic ordered sets of
width 3.

Proposition 4.38. *(See [216], parts (1) and (3) of Theorem 10.) Let P be a finite
ordered set of width $\leq$ 3. Then P has a fixed point free automorphism iff, for all
$k < \ell$, we have that the set $\{x \in P : \text{rank}(x) \in \{k, \ell\}\}$ has a fixed point free
automorphism.*

Proof. The direction "$\Rightarrow$" is trivial. We prove the direction "$\Leftarrow$" by induction on
the height h of the ordered set P. The base step for $h \leq 1$ is trivial.

   For the induction step, let P be an ordered set of width $\leq$ 3 and height $h > 1$ and
assume that the result holds for all ordered sets of width $\leq$ 3 and height $\leq h - 1$.
Because, for all $k < \ell$, the set $\{x \in P : \text{rank}(x) \in \{k, \ell\}\}$ has a fixed point free
automorphism, for all $0 \leq k \leq h$, the set $\{x \in P : \text{rank}(x) = k\}$ has at least 2
elements. If, for some $0 \leq k \leq h - 1$, the set $\{x \in P : \text{rank}(x) \leq k\}$ lies entirely
below the set $\{x \in P : \text{rank}(x) \geq k + 1\}$, then, by induction hypothesis, both these
sets have a fixed point free automorphism and hence so does P. Therefore, for the
following, we can assume that no two consecutive ranks of P form a subset as in
Figures 4.2b, 4.2c, 4.2d, or 4.2f.

   First, consider the case that, for all $0 \leq k \leq h$, the set $\{x \in P : \text{rank}(x) = k\}$
has exactly 2 elements. By assumption, for all $0 \leq k < \ell \leq h$, we have that the
set $\{x \in P : \text{rank}(x) \in \{k, \ell\}\}$ is a 4-crown or a disjoint union of two 2-chains (see
Figures 4.2a and 4.2b). Clearly, the function that maps each element of rank k to

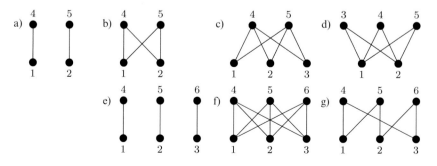

Fig. 4.2 All ordered sets of width $\leq$ 3 and height 1 that have a fixed point free automorphism

the other element of rank k is the only fixed point free automorphism for this set. Therefore, in the following, we can assume that, for some ℓ with $0 \le \ell \le h$, the set $\{x \in P : \text{rank}(x) = \ell\}$ has 3 elements.

In case there is a $0 \le k \le h$ so that $\{x \in P : \text{rank}(x) = k\}$ has 2 elements, assume without loss of generality that $k < h$ and that $\{x \in P : \text{rank}(x) = k + 1\}$ has 3 elements. Then, because $\{x \in P : \text{rank}(x) \in \{k, k+1\}\}$ has a fixed point free automorphism, the set $\{x \in P : \text{rank}(x) \in \{k, k+1\}\}$ is so that every element of $\{x \in P : \text{rank}(x) = k\}$ is below every element of $\{x \in P : \text{rank}(x) = k + 1\}$ (see Figures 4.2c and 4.2d), which we have already excluded.

Thus, we are left to consider the case that, for all $0 \le k \le h$, we have that the set $\{x \in P : \text{rank}(x) = k\} =: \{a_{k,0}, a_{k,1}, a_{k,2}\}$ has exactly 3 elements. In this case, the set $\{x \in P : \text{rank}(x) \ge 1\}$ has a fixed point free automorphism Φ. Because every rank has exactly 3 elements, we can assume that P is the union of three chains $C_j := \{a_{0,j}, a_{1,j}, \ldots, a_{h,j}\}$ and that $\Phi[C_j \setminus \{a_{0,j}\}] = C_{j+1} \setminus \{a_{0,j+1}\}]$, with arithmetic for the index j running modulo 3.

Extend Φ to become a bijective function $\Psi : P \to P$ by setting

$$\Psi(a_{k,j}) := \begin{cases} \Phi(a_{k,j}); & \text{for } k \ge 1, \\ a_{0,j+1}; & \text{for } k = 0, \text{ arithmetic for } j \text{ modulo } 3. \end{cases}$$

To prove that the function Ψ is order-preserving, which implies that Ψ is an automorphism and hence a fixed point free automorphism, let $k \ge 1$ and consider the set $\{x \in P : \text{rank}(x) \in \{0, k\}\}$. By assumption, this set has a fixed point free automorphism, which means that $\{x \in P : \text{rank}(x) \in \{0, k\}\}$ is either the disjoint union of the three 2-chains $\{a_{0,0}, a_{k,0}\}$, $\{a_{0,1}, a_{k,1}\}$, and $\{a_{0,2}, a_{k,2}\}$, or every element of rank 0 is below every element of rank k or $\{x \in P : \text{rank}(x) \in \{0, k\}\}$ is a 6-crown (see Figures 4.2e, 4.2f, and 4.2g). In the first two cases, Ψ trivially is order-preserving on $\{x \in P : \text{rank}(x) \in \{0, k\}\}$. In case $\{x \in P : \text{rank}(x) \in \{0, k\}\}$ is a 6-crown, the comparabilities are $a_{0,0} < a_{k,0} > a_{0,1} < a_{k,1} > a_{0,2} < a_{k,2} > a_{0,0}$ or $a_{0,0} < a_{k,0} > a_{0,2} < a_{k,2} > a_{0,1} < a_{k,1} > a_{0,0}$. In either case, it is easy to see that Ψ is order-preserving on $\{x \in P : \text{rank}(x) \in \{0, k\}\}$. Hence Ψ is order-preserving.

We conclude that, in all cases, P has a fixed point free automorphism. ∎

Proposition 4.38 is an efficient characterization of automorphic ordered sets of width 3, because the number of automorphic ordered sets of height 1 and width ≤ 3 is small, see Figure 4.2: Simply check each set $\{x \in P : \text{rank}(x) \in \{k, \ell\}\}$ whether it is or isn't one of the sets in Figure 4.2 to determine if P is automorphic. Figure 4.3 shows that there is no simple analogue of Proposition 4.38 for ordered sets of width 4.

Regarding *minimal* automorphic sets, we can note that graded minimal automorphic sets of width ≤ 4 have been characterized in [88] and we will see a large number of examples of minimal automorphic sets in Section 9.5. (We will consider graded ordered sets in Section 13.1.) The example that follows is one of the examples mentioned in [88]. It is not graded and seems to indicate that, even for width 3, a characterization of all minimal automorphic sets would be quite difficult.

Fig. 4.3 An ordered set of width 4 that does not have a fixed point free automorphism, but so that, for all $k < \ell$, the set $\{x \in P : \operatorname{rank}(x) \in \{k, \ell\}\}$ has a fixed point free automorphism

Fig. 4.4 A minimal automorphic set of width 3

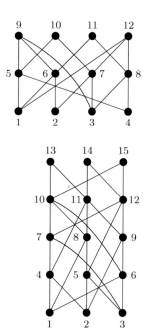

Proposition 4.39. *The ordered set M^{33333} in Figure 4.4 is minimal automorphic.*

Proof. Clearly the map Φ with cycles $(1, 2, 3)$, $(4, 5, 6)$, $(7, 8, 9)$, $(10, 11, 12)$, and $(13, 14, 15)$ is an automorphism of M^{33333}.

To show that M^{33333} is minimal automorphic, let $f : M^{33333} \to M^{33333}$ be a fixed point free order-preserving map.

First note that $M^{33333} \setminus \{1\}$ is connectedly collapsible and hence it has the fixed point property: Indeed, $4, 6$, and 8 have unique lower covers in $M^{33333} \setminus \{1\}$. After removing them, 7 and 9 are retractable to 5 in $M^{33333} \setminus \{1, 4, 6, 8\}$ and their sets of strict upper bounds are 5-fences. The resulting set is such that 5 is comparable to all other elements. Thus, $M^{33333} \setminus \{1\}$ is connectedly collapsible and has the fixed point property. This means that we must have $1 \in f[M^{33333}]$. Repeated application of Φ shows that we must also have $2 \in f[M^{33333}]$ and $3 \in f[M^{33333}]$.

Now note that $(\uparrow 1) \cap (\uparrow 2) \cap (\uparrow 3) = \{7, \ldots, 15\}$. By the above

$$(\uparrow f(1)) \cap (\uparrow f(2)) \cap (\uparrow f(3)) = (\uparrow 1) \cap (\uparrow 2) \cap (\uparrow 3) = \{7, \ldots, 15\},$$

that is, $f[\{7, \ldots, 15\}] \subseteq \{7, \ldots, 15\}$. But $\{7, \ldots, 15\}$ is a stack of two 6-crowns, which is minimal automorphic (see Exercise 4-40). Thus $\{7, \ldots, 15\} \subseteq f[M^{33333}]$.

Finally, none of $f(4), f(5)$, and $f(6)$ is minimal or in $\{7, \ldots, 15\}$, because, otherwise, f would not be surjective on the minimal elements or on $\{7, \ldots, 15\}$. Thus $f(4)$ will be the unique lower cover of $f(7)$ that has rank 1, $f(5)$ will be the unique lower cover of $f(8)$ that has rank 1, and $f(6)$ will be the unique lower cover of $f(9)$ that has rank 1. This means that f must be surjective, hence injective and hence an automorphism.   ∎

Exercises

4-39. Prove that an ordered set of height 1 has the fixed point property iff it is C-infinite-dismantlable to a singleton (see Exercise 4-36), which is the case iff it is connected and does not contain any crowns or infinite fences.

4-40. Prove that a stack of two six crowns (see Exercise 2-30) is minimal automorphic.

4-41. Prove that a collapsible ordered set is minimal automorphic iff it is an antichain or a four-crown-tower.

4-42. Prove that the ordered sets in Figure 4.2 are indeed all ordered sets of width ≤ 3 and height 1 that have a fixed point free automorphism.

4-43. (Also consider Exercise 2-29.) Let P be a finite ordered set without retractable points so that the following hold. The subset $P \setminus \max(P)$ is minimal automorphic and no automorphism of $P \setminus \max(P)$ can be extended to an order-preserving map on P. For each maximal element m of P, the ordered set $\downarrow m \setminus \{m\}$ has the fixed point property or is minimal automorphic. For each maximal element x of $P \setminus \max(P)$, we have that $\downarrow\uparrow x \supseteq P \setminus \max(P)$. Prove that P has the fixed point property.

4.5 Isotone Relations

Isotone relations are a variation on the order-preserving mapping theme. In the finite case, there is a complete solution to the corresponding version of the fixed point problem (see Theorem 4.47). Most results in this section are due to or inspired by Walker (see [318]). The idea behind isotone relations (which originally occurred in [294]) is to consider multifunctions, that is, functions for which the images are sets instead of single points, and to define a suitable notion of preservation of order. The ordering for the images that turns out to be the right one is the relation $\sqsubseteq$ of Proposition 1.3. Although this relation does not define an order relation for sets, it does induce a hierarchy, which is all that is needed in the following.

Definition 4.40. *Let P be an ordered set and let $\mathcal{P}(P)$ be its power set. A function $\phi : P \to \mathcal{P}(P) \setminus \{\emptyset\}$ is called an **isotone relation** iff $p \leq q$ implies $\phi(p) \sqsubseteq \phi(q)$ in the sense of Proposition 1.3. That is, if $p \leq q$, then*

1. *For all $a \in \phi(p)$, there is a $b \in \phi(q)$ with $a \leq b$ and*
2. *For all $d \in \phi(q)$, there is a $c \in \phi(p)$ with $d \geq c$.*

Definition 4.41. *An ordered set P is said to have the **relational fixed point property** iff, for every isotone relation $\phi : P \to \mathcal{P}(P) \setminus \{\emptyset\}$, there is a fixed point $p \in P$, that is, a $p \in P$ such that $p \in \phi(p)$.*

We start our exploration of this property with a negative result in Example 4.42. This example, together with Example 4.3, part 6 and Proposition 4.43, part 2, shows that a meaningful analysis of the relational fixed point property can only be carried out in ordered sets that do not contain a bi-infinite chain. Hence our later restriction to finite ordered sets is not as severe as it may first look.

Example 4.42. The set $\mathbb{Z} \cup \{\pm\infty\}$ with its natural order does not have the relational fixed point property.

Proof. Let $E \subseteq \mathbb{Z}$ be the set of even integers and let $O \subseteq \mathbb{Z}$ be the set of odd integers. The mapping $f : \mathbb{Z} \cup \{\pm\infty\} \to \mathcal{P}(\mathbb{Z} \cup \{\pm\infty\}) \setminus \{\emptyset\}$ defined by

$$f(x) := \begin{cases} E; & \text{if } x \in O, \\ O; & \text{if } x \in E, \\ \mathbb{Z}; & \text{if } x \in \{\pm\infty\} \end{cases}$$

clearly is an isotone relation with no fixed point. ∎

Proposition 4.43. *Let P be an ordered set with the relational fixed point property. Then the following hold.*

1. *P has the fixed point property.*
2. *Every retract of P has the relational fixed point property.*

Proof. For part 1, note that, if $f : P \to P$ is an order-preserving map, then $\phi_f : P \to \mathcal{P}(P) \setminus \{\emptyset\}$ defined by $\phi_f(p) := \{f(p)\}$ is an isotone relation.

Part 2 is analogous to part 1 of Theorem 4.8, see Exercise 4-44. ∎

As for order-preserving maps, there is a version of the Abian–Brown Theorem for isotone relations.

Lemma 4.44 (see [318], Proposition 5.2; compare with Theorem 3.32). *An Abian–Brown Theorem for isotone relations. Let P be an ordered set with no infinite chains and let $\phi : P \to \mathcal{P}(P) \setminus \{\emptyset\}$ be an isotone relation. If there are $p, q \in P$ such that $q \in \phi(p)$ and $p \leq q$, then ϕ has a fixed point.*

Proof. Exercise 4-45. ∎

Because the Abian–Brown Theorem was the key ingredient in the proof of Theorem 4.11, it should now be easy to prove the following.

Theorem 4.45. *Every finite ordered set P that is $\mathcal{I}$-dismantlable to a singleton has the relational fixed point property.*

Proof. Let P be a finite ordered set and let $p \in P$ be irreducible. We will show that, if $P \setminus \{p\}$ has the relational fixed point property, then P has the relational fixed point property, which clearly implies the result.

To do so, let $\phi : P \to \mathcal{P}(P) \setminus \{\emptyset\}$ be an isotone relation. Assume without loss of generality that $p \in P$ has a unique upper cover u and let $r : P \to P \setminus \{p\}$ be the retraction that maps p to u. Then $x \mapsto r[\phi(x)]$ is an isotone relation on $P \setminus \{p\}$, which must have a fixed point y. If $y \in \phi(y)$, then we are done. Otherwise $y = u$ and $u \notin \phi(u)$. However, because $u \in r[\phi(u)]$, we must have that $p \in \phi(u)$. By Lemma 4.44, ϕ has a fixed point. ∎

The most natural candidate for an isotone relation without a fixed point is obtained by mapping every point to the set of all points that are not comparable to it. If this candidate turns out to not be an isotone relation, the underlying set must already have a very special structure.

Lemma 4.46 (Compare with [318], Theorem 5.6.). *Let P be a finite ordered set with more than one element. If $\Phi(p) := \{x \in P : x \not\sim p\}$ does not define an isotone relation, then P has an irreducible point.*

Proof. If Φ is not an isotone relation, then either there is a $p \in P$ such that $\Phi(p) = \emptyset$ or there are $a, b \in P$ such that $a \leq b$ and $\Phi(a) \not\subseteq \Phi(b)$.

In the first case, p is comparable to all elements of P. Without loss of generality assume that p has upper covers. Then, for any upper cover u of p, p is the unique lower cover of u.

In the second case, we can assume without loss of generality that there is an $x \in \Phi(a)$ such that there is no $y \in \Phi(b)$ that is above x. Then $x \sim b$ and, because $x \not\sim a$, we have $x < b$. Let z be an upper bound of x such that $z \prec b$. Then no upper bound of z is in $\Phi(b)$, so all upper bounds of z are comparable to b. Because $z \prec b$, this means that all strict upper bounds of z are upper bounds of b and b is the unique upper cover of z.   ∎

Lemma 4.46 already was the last ingredient needed to characterize the finite ordered sets that have the relational fixed point property.

Theorem 4.47 (Compare with [318], Theorem 5.7.). *A finite ordered set P has the relational fixed point property iff P is $\mathcal{I}$-dismantlable to a singleton.*

Proof. By Theorem 4.45, the direction "⇐" holds. To prove "⇒," we proceed by induction on $n = |P|$. The base case $n = 1$ is trivial. Now assume that $|P| = n$, that the result holds for ordered sets with fewer than n elements, and that P has the relational fixed point property. Then Φ from Lemma 4.46 is not an isotone relation on P. Therefore P has an irreducible point p. By part 2 of Proposition 4.43, this means that $P \setminus \{p\}$ has the relational fixed point property. By induction hypothesis, $P \setminus \{p\}$ is $\mathcal{I}$-dismantlable to a singleton, and therefore so is P.   ∎

Exercises

4-44. Prove part 2 of Proposition 4.43.
4-45. Prove Lemma 4.44.
4-46. Prove that every retract of a finite $\mathcal{I}$-dismantlable set is again $\mathcal{I}$-dismantlable by combining two results in this section.
4-47. $\mathcal{C}$-infinite-dismantlability (see Exercise 4-36) and the relational fixed point property.

   a. Prove that, if P has no infinite chains and is $\mathcal{C}$-infinite-dismantlable to a singleton, then P has the relational fixed point property.
   b. Conclude that the "spider" in Figure 1.1 part f) has the relational fixed point property.

4-48. Prove directly that, if a finite ordered set of width 2 has a fixed point free isotone relation, then it also has a fixed point free order-preserving map. Use this fact and Theorem 4.47 to give an alternative proof of Theorem 4.34.

4-49. (A partial converse to Lemma 4.46.) Let P be a finite ordered set such that the set-valued function $\Phi(p) := \{x \in P : p \not> x\}$ is an isotone relation.

    a. Prove that P contains no points a, b, c such that b is the unique upper cover of c, $a < b$, and $c \not> a$.

    b. Give an example that shows that P could contain points c and b, such that b is the unique upper cover of c and c is the unique lower cover of b.

4-50. Let P be an ordered set with more than one element. Prove that, if $\Phi(p) := \{x \in P : x \not> p\}$ does not define an isotone relation, then P has a comparative retraction.

4-51. Let P be a finite ordered set. Then an isotone relation $\phi : P \to \mathcal{P}(P) \setminus \{\emptyset\}$ is called a **connected isotone relation** iff, for all $p \in P$, the set $\phi(p)$ is connected. P is said to have the **connected relational fixed point property** iff, for every connected isotone relation $\phi : P \to \mathcal{P}(P) \setminus \{\emptyset\}$, there is a fixed point $p \in P$, that is, a $p \in P$ such that $p \in \phi(p)$.

    a. Prove that, if $a \in P$ is retractable to $b \in P$, then P has the connected relational fixed point property iff

        i. $P \setminus \{a\}$ has the connected relational fixed point property, and

        ii. One of $(\uparrow a) \setminus \{a\}$ or $(\downarrow a) \setminus \{a\}$ has the connected relational fixed point property.

    b. Conclude that connectedly collapsible ordered sets have the connected relational fixed point property.

    c. Take two finite ordered sets P and Q without any irreducible points. Choose a minimal element $m_P \in P$ and a minimal element $m_Q \in Q$ and form the set R by identifying m_P and m_Q. Show that $\Phi(p) := \{x \in R : x \not> p\}$ is an isotone relation on R, but not a connected isotone relation.

        Note. This shows that Φ cannot be used as the canonical candidate for a fixed point free connected isotone relation.

4-52. Define a "chain isotone relation" to be an isotone relation ϕ so that all $\phi(x)$ are nonempty chains. Prove that, for a finite ordered set P, every chain isotone relation has a fixed point iff P has the fixed point property.

4.6 Li and Milner's Structure Theorem

Li and Milner's theorem (Theorem 4.48) is a surprising result. It says that the class of chain-complete ordered sets without infinite antichains is in fact very close to the class of finite ordered sets: Each of its members can be $\mathcal{G}$-dismantled to a finite ordered set in finitely many steps. We thus have a family of infinite sets for which the fixed point property can be reduced to arguments on finite sets. The following proof of the Li–Milner Theorem on chain-complete ordered sets with no infinite antichains is due to J. D. Farley. Many of the concepts we have introduced so far are exhibited nicely in this proof. We start by stating the theorem, which we will finish proving on p. 107.

Theorem 4.48 (Li–Milner Structure Theorem, see [188].). *Any chain-complete ordered set with no infinite antichain is $\mathcal{G}$-dismantlable to a finite $\mathcal{C}$-core, which is unique up to isomorphism.*

The proof of Theorem 4.48 has several stages, which we will formulate as lemmas and propositions, before finally putting them together to finish the proof. We must prove that we can reach finite sets in finitely many steps. To do this, we embark on an argument by contradiction that assumes that we cannot. Both published proofs of Theorem 4.48 (Li and Milner's original argument in [188] as well as Farley's proof in [87]) go this route. We start with some notation.

Definition 4.49. *A **generalized perfect sequence** of the ordered set P is a sequence $\{P_\alpha\}_{\alpha<\lambda}$ indexed by the ordinal numbers before λ such that*

1. *$P_0 = P$,*
2. *If $\alpha < \lambda$, then $P_{\alpha+1}$ is a good subset of P_α,*
3. *For all limit ordinals $\mu < \lambda$, we have $P_\mu = \bigcap_{\alpha<\mu} P_\alpha$.*

*A **perfect sequence** is a strictly decreasing generalized perfect sequence such that the last set in the sequence has no nontrivial good subset.*

Although perfect sequences are defined for arbitrary ordered sets, in general, they can be quite trivial and need not end with a finite set. For example, a four-crown-tower of infinite height has no nontrivial good subsets, so its perfect sequence has only one element. Perfect sequences can also have infinite length, as can be seen by considering the perfect sequence of the "spider" in Figure 1.1, part f).

To show that no perfect sequence of a chain-complete ordered set without infinite antichains is infinite and that the last element of the perfect sequence is finite, we will *fix a chain-complete ordered set P without infinite antichains and a generalized perfect sequence $\{P_\alpha\}_{\alpha\le\omega}$ of P.* (Recall that ω is the first infinite ordinal number.) Because no P_α has infinite antichains, by Corollary 2.29, we would be done if we could show that some P_α with $\alpha < \omega$ had no infinite chains. We start with some general lemmas on chains.

Lemma 4.50 (See [87], Lemma 6.8.). *Let $\{p_n\}_{n<\omega}$ be a sequence in an ordered set without infinite antichains. Then there is an infinite sequence of natural numbers $n_0 < n_1 < n_2 < \cdots < \omega$ such that we have $p_{n_0} \le p_{n_1} \le p_{n_2} \le \cdots$ or $p_{n_0} \ge p_{n_1} \ge p_{n_2} \ge \cdots$.*

Proof. First note that, by Corollary 2.29, we can assume without loss of generality that $\{p_n : n < \omega\}$ is a chain. Suppose for a contradiction that $\{p_n\}_{n<\omega}$ has no infinite increasing or decreasing subsequences.

Find a maximal increasing subsequence of $\{p_n : n < \omega\}$ that starts at p_1 and call its top element p_{k_1}. Then no element with index greater than k_1 is $\ge p_{k_1}$. Now suppose we have already found $k_1 < k_2 < \cdots < k_m$ such that

1. If i is odd, no element with index greater than k_i is $\ge p_{k_i}$,
2. If i is even, no element with index greater than k_i is $\le p_{k_i}$.

Without loss of generality assume that m is odd. Find a maximal decreasing subsequence of $\{p_n : k_m \leq n < \omega\}$ that starts at p_{k_m} and call its smallest element $p_{k_{m+1}}$. Then no element with index greater than k_{m+1} is $\leq p_{k_{m+1}}$.

The inductive procedure above produces an infinite sequence of indices k_i satisfying parts 1 and 2 above. However, this means that $\{p_{k_{2i}} : i \in \mathbb{N}\}$ is an infinite increasing sequence and $\{p_{k_{2i+1}} : i \in \mathbb{N}\}$ is an infinite decreasing sequence, which is a contradiction. ∎

Definition 4.51 (See [187], p. 328.). *Let P be an ordered set and let $X, Y \subseteq P$. Then $y \in Y$ is called the **supremum of X in Y**, denoted $\sup_Y(X)$ iff y is an upper bound of X and, for all $z \in Y$ with $z \geq X$, we have $z \geq y$.*

Lemma 4.52 (Special case of [187], Lemma 4.1 (1).). *Let $\{P_n\}_{n<\omega}$ be a decreasing sequence of chain-complete subsets of P and let $P_\omega := \bigcap_{n<\omega} P_n$. If $X \subseteq P$ is such that $\sup_{P_n} X$ exists for all $n < \omega$, then $\sup_{P_\omega} X$ exists, too. In particular, P_ω is chain-complete.*

Proof. Let $s_n := \sup_{P_n} X$. Because $P_n \supseteq P_{n+1}$, we have that $s_{n+1} \geq s_n$. If $s_m = s_n$ for some n and all $m > n$, then $s_n = \sup_{P_\omega} X$ and we are done. Otherwise, every upper bound of X in P_ω is also an upper bound of the s_n. Thus, in P_ω, the chain of elements s_n has the same upper bounds as X. Therefore we can assume that X is a well-ordered chain C without a largest element.

Now, for every well-ordered chain K with $C \subseteq K \subseteq P$ without a largest element, either $\sup_{P_\omega} K$ exists, or, as above, we can construct a countable chain C' such that $K \cup C'$ is well-ordered, has no largest element and all upper bounds of K in P_ω are also upper bounds of $K \cup C'$ (and hence they have the same upper bounds in P_ω). Consider the set of all well-ordered chains K with $C \subseteq K \subseteq P$ such that K and C have the same upper bounds in P_ω. Order this set by $K \sqsubseteq K'$ iff K is an initial segment of K'. Then this set is inductively ordered and must thus have a maximal element K_M. By maximality and the above, K_M must have a largest element, which is $\sup_{P_\omega}(C)$. ∎

Lemma 4.53. *If $x < p$ is funneled through p and $x < y < p$, then y is funneled through p.*

Proof. Exercise 4-53. ∎

For the rest of this proof, we focus on a special kind of chain.

Definition 4.54. *A nonempty chain $C \subseteq P$ is called an **approaching chain** iff*

1. *C has no greatest element,*
2. *For all $n < \omega$ the set $C \cap P_n$ is cofinal in C.*

Now note that, by definition of approaching chains and by Lemma 4.52, every approaching chain has a P_ω-supremum. We will also say that an approaching chain C **approaches** its P_ω-supremum $\sup_{P_\omega} C$.

meaning all elements of $C \cap (\uparrow c) \cap P_n$ would be funneled
through p and thus would not be in P_{n+1}, a contradiction.

If the sequence $\{p_n^c\}_{n\in\mathbb{N}}$ has an increasing subsequence $\{p_{n_k}^c\}_{k\in\mathbb{N}}$, then the desired
element is $p^c := \sup_{P_\omega}\{p_{n_k}^c : k \in \mathbb{N}\}$. Because p^c can serve as the p^d for
all $d \in C \cap (\downarrow c)$, we are done unless there is a $b \in C$ such that for all
$c \geq b$ in C the sequence $\{p_n^c\}_{n\in\mathbb{N}}$ has no increasing subsequence. For the remainder
of this proof, we shall assume that this is the case.

In this case, for all $c \in C \cap (\uparrow b)$, the sequence $\{p_n^c\}_{n\in\mathbb{N}}$ has a decreasing
subsequence $\{p_{n_k}^c\}_{k\in\mathbb{N}}$. Now, for some k_c, we must have that $p_{n_k}^c \not\geq C$ for $k \geq k_c$,
because otherwise $\inf_{P_\omega}\{p_{n_k}^c : k \in \mathbb{N}\} \geq p$, which would imply $p_{n_k}^c \geq p$, a
contradiction.

Therefore, for any $c \in C \cap (\uparrow b)$, we can construct strictly increasing sequences
$c = c_1 < c_2 < \cdots$ in $C \cap (\uparrow b)$ and $k_1 < k_2 < \cdots$ in $\mathbb{N}$ such that $p_{k_n}^{c_n} \not\geq c_{n+1}$. Now, if
$m < n$, then $p_{k_m}^{c_m} \not\geq p_{k_n}^{c_n}$, because otherwise $c_{m+1} \leq c_n \leq p_{k_n}^{c_n} \leq p_{k_m}^{c_m}$, a contradiction.
Thus $\{p_{k_n}^{c_n}\}_{n\in\mathbb{N}}$ has an increasing subsequence whose P_ω-supremum p^c is not less
than or equal to p. Because this construction can be started at any $c \in C \cap (\uparrow b)$, we
are done. ∎

Before we continue, we need to insert a proof of the Dushnik–Miller–Erdös
Theorem for graphs, which will allow us to finish the proof of the Li–Milner
Theorem.

Definition 4.57. *For every set S, we define the **cardinality** of S to be the smallest
ordinal number α such that there is a bijection between S and α. We define $|S| = \alpha$.
Ordinal numbers α such that there is no bijection between α and any ordinal number
strictly less than α are called (**regular**) **cardinal numbers**.*

Theorem 4.58 (Dushnik–Miller–Erdös, see [78], Theorem 5.23.). *Let $G = (V, E)$ be a graph of infinite cardinality $|V| = \alpha$ such that every infinite subset
contains two adjacent elements. Then G contains a complete subgraph of cardinality
α.*

Proof. Suppose, for a contradiction, that G has no complete subgraph of cardinality
α and let $M \subseteq V$ be a maximal complete subgraph. Then $|M| < \alpha$ and hence
$|V \setminus M| = \alpha$. Now, for each $v \in V \setminus M$, there is an $f(v) \in M$ such that v is

not adjacent to $f(v)$. Thus $V \setminus M = \bigcup_{m \in M} f^{-1}(m)$, which means that, for some $m \in M$, we have $|f^{-1}(m)| = \alpha$. This means that m is not adjacent to any element of $f^{-1}(m)$ and the induced subgraph $(f^{-1}(m), \{e \in E : e \subseteq f^{-1}(m)\})$ of G satisfies the assumption above. Iteration of this step produces a countable set of vertices in G such that no two of them are adjacent. This contradicts the hypothesis that any infinite set of vertices contains two adjacent vertices. ∎

Corollary 4.59. *Let P be an ordered set without any infinite antichains, let α be a cardinal, and let $X := \{x_\beta : \beta < \alpha\}$ be a family of elements of P. Then there exists a chain $C \subseteq X$ such that $\{\beta < \alpha : x_\beta \in C\}$ is cofinal in α.*

Proof. If α is finite, there is nothing to prove. Define the graph $G = (V, E)$ by

$$V = \{(x_\beta, \beta) : \beta < \alpha\},$$

$$E = \{\{(x_\beta, \beta), (x_\gamma, \gamma)\} : \beta, \gamma < \alpha \text{ and } x_\beta \sim x_\gamma\}.$$

By the Dushnik–Miller–Erdös Theorem, the graph G contains a complete subgraph $H = (W, F)$ of cardinality α. Define $I := \{\beta < \alpha : (x_\beta, \beta) \in W\}$. Then I has cardinality α. Because no ordinal number before α has cardinality α, I must be cofinal in α. Thus $C = \{x_\beta : (x_\beta, \beta) \in W\}$ is a chain as desired. ∎

Lemma 4.60 (See [87], Lemma 6.10 or [188], Lemma 3.2.). *For all $p \in \mathcal{A}$ there is a $q \in \mathcal{A}$, such that $p < q$.*

Proof. Let C be an approaching chain that approaches p. Without loss of generality, assume that C is isomorphic to a cardinal. (This can be achieved by going to an appropriate subchain as necessary, see Exercise 4-55.) By Lemma 4.56, for each $c \in C$, there is a $p^c \in \mathcal{A}$ such that $c < p^c$ and $p^c \not\leq p$. By Corollary 4.59, there is a nonempty chain $D \subseteq \{p^c : c \in C\}$ such that $\{c \in C : p^c \in D\}$ is cofinal in C. Because $D \subseteq \mathcal{A}$, D is an approaching chain that approaches, say, $q \in P_\omega$. Because $p^c \not\leq p$ for all $c \in C$ we must have $p < q$. ∎

Proof of Theorem 4.48. (See proof of [87], Theorem 6.11 or [188], Corollary 3.4.) Suppose for a contradiction that $\mathcal{A}$ is not empty. Because every chain in $\mathcal{A}$ is a chain in P_ω, every infinite chain $C \subseteq \mathcal{A}$ without a largest element is an approaching chain, whose P_ω-supremum is an upper bound of C in $\mathcal{A}$. Thus $\mathcal{A}$ is inductively ordered. (Note that the only reason we needed to work with uncountable approaching chains was to show that $\mathcal{A}$ is inductively ordered.) However, by Lemma 4.60, $\mathcal{A}$ has no maximal elements, in contradiction to Zorn's Lemma. Thus $\mathcal{A}$ is empty.

This implies that, for some finite $n \in \mathbb{N}$, the set P_n must be finite: Indeed, otherwise we could find a sequence of mutually distinct $p_n \in P_n$ that must, by Corollary 2.29, contain an infinite chain, which in turn would contain an approaching chain or a dual approaching chain, a contradiction.

Therefore chain-complete ordered sets that do not contain infinite antichains are $\mathcal{G}$-dismantlable to a finite $\mathcal{G}$-core. For finite ordered sets, the $\mathcal{G}$-core is the $\mathcal{C}$-core. By Theorem 4.25, the $\mathcal{C}$-core is unique. ∎

Fig. 4.5 The "Tower of Doom"

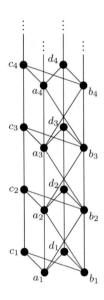

We conclude this section by showing that the chain-completeness assumption in the Li–Milner Theorem is vital. Indeed, the following example shows that even though every infinite ordered set without infinite antichains has a comparative retraction (see Exercise 4-57), the chain-completeness assumption cannot be relaxed.

Example 4.61. There is an ordered set of finite width that cannot be dismantled to a finite set in finitely many steps, see Figure 4.5. Let $A := \{a_n : n \in \mathbb{N}\}$, $B := \{b_n : n \in \mathbb{N}\}$, $C := \{c_n : n \in \mathbb{N}\}$, and $D := \{d_n : n \in \mathbb{N}\}$ be countable sets and let $T := A \cup B \cup C \cup D$ be ordered via $a_n \leq a_k, b_{k+1}, c_k, d_k$ for all $k \geq n$, $b_n \leq a_{k+1}, b_k, c_k, d_k$ for all $k \geq n$, $c_n \leq c_k$ for all $k \geq n$, and $d_n \leq d_k$ for all $k \geq n$. Let $R := A \cup B \cup \{c_{n_k} : k \in \mathbb{N}\} \cup \{d_{m_j} : j \in \mathbb{N}\}$ be a subset of T, where n_k and m_j are increasing sequences of natural numbers. We will show that, for every retraction $r : R \to R$ such that $r(p) \sim p$ for all $p \in R$, the image $R[T]$ must contain $A \cup B$ and infinite subsequences of C and of D.

Let $r : R \to R$ be such a retraction. Then there is no point $p \in R$ with $r(p) < p$: Indeed, otherwise, let p be a minimal point in the set of points q with $r(q) < q$. Then $r(p)$ must be the unique lower cover of p, and no point in R has a unique lower cover, a contradiction.

Moreover, $r[A \cup B] = A \cup B$: First notice that $r(a_n) = c_{n_k}$ for any $n \leq n_k$ would mean that, for all $m_j \geq n$, we have $r(d_{m_j}) \geq c_{n_k}$, that is, $r(d_{m_j}) \in C$ and hence $r(d_{m_j}) \not\sim d_{m_j}$, which is not possible. Similarly we show $r(a_n) \notin D$ and $r(b_n) \notin C \cup D$. Thus $r[A \cup B] \subseteq A \cup B$. Now if $r(a_n) > a_n$ for some $n \in \mathbb{N}$, then an easy inductive argument shows that, for each $l > n$, we would have $r(a_l) > a_l$ or $r(b_l) > b_l$. This implies that, for all $k, j \in \mathbb{N}$ so that $n_k > n$ and $m_j > n$, we would have $r(c_{n_k}) > c_{n_k}$ and $r(d_{m_j}) > d_{m_j}$. Because $r[\{c_{n_k} : k \in \mathbb{N}\}] \subseteq \{c_{n_k} : k \in \mathbb{N}\}$ and $r[\{d_{m_j} : j \in \mathbb{N}\}] \subseteq \{d_{m_j} : j \in \mathbb{N}\}$, r would not be idempotent. Hence $r[A \cup B] = A \cup B$ and r is the identity on $A \cup B$.

Finally, because $r(p) \geq p$ for all p and $r[\{c_{n_k} : k \in \mathbb{N}\}] \subseteq \{c_{n_k} : k \in \mathbb{N}\}$ and $r[\{d_{m_j} : j \in \mathbb{N}\}] \subseteq \{d_{m_j} : j \in \mathbb{N}\}$, $r[T]$ must contain infinite subsequences of C and of D.

Thus if $T_0 := T$ and $r_k : T_k \to T_{k+1}, k = 0, \ldots, n-1$ are retractions such that every point is comparable to its image under r_k, then T_n is of the same form as the set R above. Thus T cannot be dismantled to a finite set in finitely many steps. ∎

Exercises

4-53. Prove Lemma 4.53.

4-54. Let α be a cardinal and let β be an ordinal number with $\beta < \alpha$. Prove that $\alpha \setminus \beta$ is isomorphic to α.

4-55. Prove that every infinite chain has a cofinal well-ordered subchain that is isomorphic to a cardinal.

4-56. Give a direct proof that, if P is a $\mathcal{G}$-core that is chain-complete with no infinite antichains, then P must be finite.

4-57. Prove that every infinite ordered set of finite width has a comparative retraction.

Hints. First prove the following. Let P be an ordered set and let $C := \{c_\alpha : \alpha < \xi\}$ be an increasing well-ordered chain in P. Assume that there is a $y \in P$ such that there is an $\alpha_0 < \xi$ with $y < c_{\alpha_0}$, $y \notin C$ and such that for all $d \geq y$ we have $d \geq C$ or $\exists c \in C : c \geq d$. Then there is a retraction $r : P \to P$ such that $r \neq \mathrm{id}_P$ and for all $p \in P$ we have $r(p) \geq p$.

Then show that if the only comparative retraction is the identity, then the set must be finite.

Remarks and Open Problems

Most of the open problems in this section are motivated by my own strong interest in the fixed point property and retractions.

1. For more on retractions, see [245, 289].

2. Although irreducible points are presented here as special retractable points, they are actually the more widely investigated notion. We will see more on irreducible points when we discuss their role in lattices in Section 8.4 and, subsequently, in the correspondence between ordered sets and distributive lattices in Section 8.5.

3. Irreducible and retractable points suffice to characterize the fixed point property for ordered sets with up to 9 points (see [264]). After that, the proverbial "combinatorial explosion" sets in: Up to isomorphism and duality, there are 4 ordered sets with 10 points that have the fixed point property and no retractable point (this can also be verified by checking the sets in [264]). For 11 points, there are 8 such ordered sets, as is verified in [270]. For 12 points, there are 102; for 13 points, there are 964; and for 14 points, there are 24,546. (Verified with [286] and lists of nonisomorphic ordered sets up to size 14.)

4. The idea of minimal automorphic sets has been present since the characterization of the fixed point property for ordered sets of height 1. Yet it is not widely used. One interesting use related to the product problem is presented in [50]. (Also see Remark 13 in Chapter 12.)
5. Some generalizations of isotone relations and their connection to the product problem are presented in [258].
6. It is tempting to hope that ordered sets that are defined recursively via removals of single points, such as ordered sets that are dismantlable to a singleton and collapsible ordered sets are easily proved to be reconstructible. Unfortunately, this hope so far has not produced any fruit, leading to our first open problems.

    Are ordered sets that are $\mathcal{I}$-dismantlable to a singleton reconstructible? What about collapsible ordered sets?
7. Let P be chain-complete with no infinite antichains. What is the smallest number of retractions onto good subsets that will lead to a finite ordered set?

    In [87], Figure 5, Farley shows a chain-complete set with no infinite antichains and an infinite good subset. However the good subset of the good subset already is finite. Rutkowski told me ([265]) that perfect sequences in general can be arbitrarily long, but his examples contain infinite antichains. A related question would be how many $\mathcal{C}$-retractions it would take to reach a finite ordered set. (In Farley's example, this is possible in the first step.)
8. Is infinite dismantlability equivalent to the relational fixed point property for ordered sets with no infinite chains?
9. Is there a polynomial algorithm that decides if a finite ordered set, all of whose subsets are $\mathcal{R}_1$-dismantlable to a singleton, has the fixed point property? I conjecture that this is true.
10. For a given finite ordered set P, find the smallest number k such that, for all order-preserving maps $f : P \to P$, the map f^k is a retraction.
11. Let Q be a fixed, given ordered set. How hard is it to determine if Q is isomorphic to a retract of an arbitrary given ordered set? (The answer will depend on Q.)
12. Characterize those ordered sets for which, for every order-preserving self-map $f : P \to P$, the set Fix(f) is connected. Is this "connected fixed point property" preserved by products (see Section 12.2.1)?
13. For infinite ordered sets, is it true that the $\mathcal{R}_k$-core is unique or doesn't exist? Is it true that the $\mathcal{F}$-core is unique or doesn't exist?

    Uniqueness of cores is subtle in infinite ordered sets. In [274], it is shown that, in chain-complete ordered sets, the $\mathcal{C}$-core is unique if it exists, while the $\mathcal{I}$-core need not be unique.
14. Characterize the connected relational fixed point property.

    I do not know how hard this problem is. Exercise 4-51c shows that there is no analogue of Lemma 4.46. Indeed, the connected relational fixed point property is not equivalent to connected collapsibility as Exercise 7-5 shows.
15. Find a nontrivial characterization of all minimal automorphic sets of width 3.
16. Is there a way to prove the Li–Milner Theorem without using the Dushnik–Miller–Erdös Theorem?

17. Are cores in the sense of Li and Milner (see [184]) unique?

18. Define the **retract spectrum** analogous to the endomorphism spectrum, only for retractions. For the retract spectrum, what can we say about the questions analogous to those in Remark 14 in Chapter 2?

19. An example in which, at least to my knowledge, Exercise 4-37 or mimicking its proof is the only possible option. For applications of the theorem in Exercise 4-37 to differential equations, see [41].

    Let ℓ^1 be the space of sequences $\{a_n\}_{n=1}^\infty$ of real numbers so that we have $\sum_{n=1}^\infty |a_n| < \infty$, equipped with the usual ℓ^1-norm $\|\{a_n\}_{n=1}^\infty\|_1 := \sum_{n=1}^\infty |a_n|$. Let $\mathrm{Hom}\,(\mathbb{R}, (-1, 1))$ be the set of order-preserving functions from the real numbers $\mathbb{R}$ to the interval $(-1, 1)$ and let $\Phi : \ell^1 \to \mathrm{Hom}\,(\mathbb{R}, (-1, 1))$ be an order-preserving function. As is customary, let a $\widehat{}$ denote the absence of the element in the sequence. Define the function $g : \ell^1 \to \ell^1$ componentwise as follows.

    $$\pi_m \circ g(\{a_n\}_{n=1}^\infty) := \frac{1}{2^m}\Phi(a_1, \ldots, a_{m-1}, \widehat{a_m}, a_{m+1}, \ldots)[a_m]$$

    That is, the mth component of $g(\{a_n\}_{n=1}^\infty)$ is obtained by applying to a_m the function that Φ returns for the sequence obtained from $\{a_n\}_{n=1}^\infty$ by deleting the mth entry and then multiplying by $\frac{1}{2^m}$.

    Clearly, each function g as defined above is order-preserving. Because each codomain is contained in the unit ball, each function g maps the unit ball of ℓ^1 to itself. By Exercise 4-37, each function g as defined above has a fixed point.

20. In [290], the author has generalized Exercise 4-37 to show that, for $p > 1$, closed neighborhoods of line segments in L^p are $\mathcal{C}$-dismantlable. In L^1, they need not be $\mathcal{C}$-dismantlable. Moreover, in Banach lattices, any increasing operator from a closed r-neighborhood of a line segment to a closed $(r - \varepsilon)$-neighborhood of the line segment has a fixed point.

Chapter 5
Constraint Satisfaction Problems

When proving theorems, we have a luxury that is so fundamental, we often take it for granted: If we need to work with an object, we say "let x be ⟨the object in question⟩" and we move on with the proof. Especially in the finite setting it is obvious that, given enough patience, we should be able to find the object: Simply try out all possibilities and, if there is an object as desired, at least one of them will work. As long as we are not interested in the object itself, this approach is very efficient for developing a theory.

In practice, the situation is a bit different. When I schedule class assignments, I know that there is an assignment that will satisfy all faculty requests and budgetary constraints (or at least I hope so). However, that is not enough. By the deadline, I must have an actual schedule in hand. In this text, asking for the existence or non-existence of a fixed point free order-preserving self-map is the theoretical side of actually trying to find such a map, which can require great effort. Proposition 1.22 shows this effort for a comparatively small and well-behaved set. Similarly, although it is not impossible to show by hand that the ordered set in Figure 1.1 b) has the fixed point property (see Exercise 1-19b), it is quite tedious. Finally, "in principle," counting tasks for a fixed ordered set (for example, to check the quotient in the automorphism problem 2.14) only require enough patience. In practice, the effort may be insurmountable.

In a class of problems that are "too tedious to solve by hand," we have "truly insurmountable" problems, but we usually also have problems that can be solved with a fast enough computer and enough time. For example, with the program [286], it takes less than a second to show that the ordered set in Figure 1.1 b) has the fixed point property. It also takes less than a second to enumerate the 30,126 order-preserving self-maps that this set has.

In this chapter, we will investigate algorithms which, for a given problem, construct solutions or which show that there is no solution. The most basic question in this setting is that of algorithm correctness. This is a reliability question that must have an affirmative answer: If we want to be sure that an algorithm produces the desired results, we better prove it. Once correctness is assured, the main practical

© Springer International Publishing 2016
B. Schröder, *Ordered Sets*, DOI 10.1007/978-3-319-29788-0_5

question is how long an algorithm may run. The precise formulation of this question leads to the idea of algorithm complexity, which provides a *worst case* upper bound on run time or storage requirements.

The philosophical issues that arise from the use of computers in mathematics are complex and numerous. For example, is a proof by computer really a proof? How do we know a proof is correct if we did not personally check the millions of cases that were checked by a computer? Do we really know that the number mentioned above is 30,126? Is it enough to know that a program, which is assumed to do the right thing, was (hopefully) correctly executed on a computer?

Practically speaking, computers have become a part of daily life, and it only took the time since the first edition of this text for this change to occur. Your smartphone has more computing power than the supercomputers of my youth. Algorithms are used and their results are trusted or at least accepted in branches of research that were greatly enhanced by the use of computers or that could literally not exist without computers. So, in many ways, paraphrasing my former Academic Director, Gene Callens, the question is not *if* to use technology, but *how*.

5.1 Algorithms

The notion of an algorithm is central to the use of computers. We will use the definition given in a mainstream computer science text.

Definition 5.1 (See [5], p. 2). *An **algorithm** is a finite sequence of instructions, each of which has a clear meaning and can be performed with a finite amount of effort in a finite length of time.*

Note that the definition of an algorithm does not state what the algorithm should produce. All instructions must merely be intelligible and doable in finite time. It is left up to us to make sure the algorithm does what it is supposed to do.

Let us first consider a few examples. In Proposition 1.10, we have seen that finite ordered sets can be represented by their cover relation or by the order relation itself. In many situations, one representation is preferable over the other: To determine if a point is irreducible, the cover relation is a much better tool than the order relation. On the other hand, to determine if two points are comparable, the order relation is a much better tool than the cover relation. Consequently, it is often useful to have both representations available. Therefore, when we have one of them, we want to be able to produce the other. This process has likely become automatized in your mind by now. How would we "explain" it to a computer, though?

Example 5.2. Simple algorithm to compute the cover relation from the order relation.
   Given. An order relation $\leq$ on a finite set P.
   Task. Find the corresponding cover relation $\prec$.

The intuitive idea is that, to go from the order relation to the diagram, we erase superfluous connections. The algorithm below spells out exactly this idea as a sequence of executable instructions.

We will describe algorithms in English, thus circumventing a long (and tedious) process of defining a language in which algorithms will be formulated. The "**For** ··· **do**" structures we use are to be understood as follows. We assume that the set in question has been linearly ordered in some fashion. The operations between "For ··· do" and "end for ···" are executed exactly once for each element of the set and in the order that was imposed.

Start with $\prec^*$ being equal to $\leq$,
Erase all pairs (x, x) from $\prec^*$,
For every pair $(x, y) \in \leq$ do

    For every $z \in P \setminus \{x, y\}$ do

        If $x < z$ and $z < y$, then

                Erase (x, y) from $\prec^*$ and continue to the next pair in $\leq$,
    end for z,

end for (x, y),
The covering relation will be stored in $\prec^{*,\text{end}}$, which is the relation $\prec^*$ after the above loops terminate.

It is a small, but important, issue to verify that the above algorithm really produces the covering relation for $\leq$ in its output $\prec^{*,\text{end}}$. To see this, we must show $\prec = \prec^{*,\text{end}}$.

For $\prec \subseteq \prec^{*,\text{end}}$, note that, if $(x, y) \in \prec$, then (x, y) is in $\prec^*$ at the start. Because $x \neq y$ and there are no elements z of P strictly between x and y, (x, y) will never be erased.

For $\prec \supseteq \prec^{*,\text{end}}$, let $(x, y) \notin \prec$. Then either $x = y$ and (x, y) will be erased before the For-loop starts or there is a z with $x < z < y$ and (x, y) will be erased during the For-loop. Either way, $(x, y) \notin \prec^{*,\text{end}}$.

We have proved that $\prec = \prec^{*,\text{end}}$.   □

Conversely, Warshall's algorithm (see below) computes the transitive closure of any finite relation. It can be used to compute the order relation from the diagram.

Example 5.3 (Warshall's algorithm). for computing the transitive closure of a relation. In the following, we use notation that suggests order and cover relations. However, as noted, the algorithm in fact computes *any* transitive closure.

Given. A relation $\prec$ on a finite set P.

Task. Find the transitive closure of the relation.

Definition 1.8 of the transitive closure is not very efficient for designing an algorithm: It essentially says to try out all possible chains of adjacencies and to add the pairs of endpoints for each chain. Warshall's algorithm shortens this straightforward, but inefficient, approach. Note that, to compute the order relation that belongs to a given cover relation, we would precede Warshall's algorithm by adding all relations (x, x) to $\prec$.

> Set $\leq^*$ equal to $\prec$,
> For every $z \in P$ do
>> For every $x \in P$ do
>>> For every $y \in P$ do
>>>> If $x \leq^* z$ and $z \leq^* y$, add (x, y) to $\leq^*$,
>>> end for y,
>> end for x,
> end for z,

The transitive closure $\leq$ is stored in the relation $\leq^{*,\text{end}}$, which is the relation $\leq^*$ at the end of all the loops.

At every stage of the algorithm, the relation $\leq^*$ is a subset of the transitive closure $\leq$ of $\prec$. This is because all added relations arise by transitivity. Hence $\leq^{*,\text{end}} \subseteq \leq$. The proof of the inclusion $\leq \subseteq \leq^{*,\text{end}}$ is harder. Let

$$x = z_0 \prec z_1 \prec \cdots \prec z_n \prec z_{n+1} = y.$$

Now let σ be a permutation of $\{1, \ldots, n\}$ such that the order in which the z_k occur in the outer (z-) loop is $z_{\sigma(1)}, z_{\sigma(2)}, \ldots, z_{\sigma(n)}$. After the execution of the z-loop for $z = z_{\sigma(1)}$ we have

$$x = z_0 \leq^* \cdots \leq^* z_{\sigma(1)-1} \leq^* z_{\sigma(1)+1} \leq^* \cdots \leq^* z_{n+1} = y.$$

That is, $\leq^*$ has been extended in such a way that $z_{\sigma(1)}$ can be eliminated from the original progression of $\leq^*$-relations that go from x to y. Similarly, for every $k \in \{1, \ldots, n\}$, the execution of the z-loop for $z_{\sigma(k)}$ will extend $\leq^*$ in such a way that $z_{\sigma(k)}$ can be erased from the connection between x and y. Therefore, after the execution of the z-loop for $z_{\sigma(n)}$ we must have $x \leq^* y$ and hence $\leq \subseteq \leq^{*,\text{end}}$.   □

Exercises

5-1. Turn the description of a correctness proof in Example 5.3 into a solid proof by induction on n.

5-2. Prove that any algorithm that computes the transitive closure of a relation can be extended to compute the Hasse diagram of an order.

Hint. Erase all comparabilities between points whose ranks differ by k. Compute the transitive closure of this relation. All edges in this relation that are between points whose ranks differ by k are not in the Hasse diagram.

5-3. Consider the following task.
      Given. Two finite ordered sets P and Q and a function $f : P \to Q$.
      Task. Determine if f is order-preserving.

    a. Write an algorithm that solves the task.
    b. Prove that the algorithm solves the task correctly.

5.2 Polynomial Efficiency

From a theoretical and from a practical point-of-view, a correct algorithm is nice if it terminates quickly. However, actual run time will depend on the input. Hence, to estimate the run time of an algorithm, we typically consider upper bounds on the run time for input of a given size.

Definition 5.4. *An algorithm is said to be of **polynomial efficiency** iff there is a polynomial p such that, for any allowed input of size n, the algorithm terminates after at most $p(n)$ steps.*

For Definition 5.4 to be sensible, we must specify what is counted as a step in an algorithm. Ultimately we could define a "step" as one clock cycle on a computer. Although this definition of a "step" would give the most accurate run-time estimates, it would be rather unwieldy. Any analysis would depend on what language, what compiler, and what computer we use. For theoretical results of lasting value, such a dependency is unacceptable. Therefore, the steps we count are normally the most elementary operations that occur in an algorithm. These operations can usually be performed in a polynomial number of machine cycles. Moreover, to keep the theoretical picture uncluttered, we do not explicitly count the overhead the computer encounters for controlling a loop or for the branching associated with an if-then statement. With this underlying idea, once an algorithm is proved to be of polynomial efficiency, it is polynomially efficient on any machine and in any language. (We will not delve into more subtle details, such as Turing machines and quantum computers here.)

The algorithms in Section 5.1 are of polynomial efficiency.

Proposition 5.5. *Let P be an ordered set with n elements, let $\leq$ be its order relation, and let $\prec$ be its covering relation. Let a step be the erasure of a pair in a relation or the checking if two elements are related. Then the simple algorithm to compute $\prec$ from $\leq$ in Example 5.2 terminates in at most $|P| + 3|P|^3$ steps.*

Proof. Erasing all pairs (x, x) takes $|P|$ steps. There are at most $|P| \cdot |P|$ pairs in any relation on P and there are $|P| - 2$ elements in any set $P \setminus \{x, y\}$. Thus the instructions inside the nested loops are executed at most $|P|^2(|P| - 2) < |P|^3$ times. These instructions include two comparisons and possibly one erasure, for a total of up to three steps. Thus the total number of steps is at most $|P| + 3|P|^3$. ∎

The estimates in Proposition 5.5 are not terribly sophisticated. Crude estimates are adequate as long as we only want to establish that a problem can be solved in polynomial time. For development of faster polynomial algorithms, it is useful to know the smallest degree of a polynomial p such that a given task can be solved in $p(\langle\text{input size}\rangle)$ steps. These bounds, together with testing on cases that occur frequently, can then be used to decide if an algorithm should be replaced by another. The significance of such improvements is normally measured in economic terms. For example, the faster your VLSI layout algorithms are, the faster chips can be designed.

Also note that we chose $|P|$ as our underlying variable, even though, because we work with the order relation $\le$, we could argue that the input size is in fact $|P|^2$. As long as the relation between the sizes of the objects in question is polynomial, this is not a problem.

Keeping track of all the different terms can be tedious. When we are only interested in the degree of the bounding polynomial, it is helpful to use "**big-O notation**."

Definition 5.6. *Let $\{a_n\}_{n\in\mathbb{N}}$ and $\{b_n\}_{n\in\mathbb{N}}$ be sequences of real numbers. Then we say $a_n = O(b_n)$ (read "a_n is big-oh of b_n") iff $\lim\sup_{n\to\infty}\frac{a_n}{b_n} < \infty$.*

The above means that, if $a_n = O(b_n)$, then a_n grows at most at a rate comparable to b_n. Note that, for example, any sequence that is $O(n)$ is eventually much smaller than n^3. Hence, when faster growing terms are present, terms of size $O(n)$ can be ignored. In particular, by Proposition 5.5, the algorithm in Example 5.2 requires $O\left(|P|^3\right)$ steps.

Definition 5.7. *Let $k > 0$ and let $\mathcal{A}$ be an algorithm. We will say that $\mathcal{A}$ is $O(n^k)$ iff the maximum number of steps s_n that the algorithm requires for input of size n is $s_n = O\left(n^k\right)$. We do not specify what a step is, but it is assumed that a step can be executed in an amount of time independent of the input size.[1]*

In this language, Proposition 5.5 says that the algorithm in Example 5.2 is $O(|P|^3)$, which also happens to be the complexity of Warshall's algorithm.

Proposition 5.8. *Warshall's algorithm is $O\left(|P|^3\right)$, with a step being the check if two elements are related or the adding of an ordered pair to a relation.*

Proof. Because P has $|P|$ elements, the nested loops in Warshall's algorithm are such that the instructions inside are executed exactly $|P|^3$ times. Thus Warshall's algorithm is $O(|P|^3)$. ∎

Other parameters of an ordered set that we often take for granted are the height and the width. In Exercise 5-6, you can show that the height of an ordered set is computable in polynomial time. What about the width? We might be tempted to

---

[1]This is a possible trap when implementing algorithms or analyzing their complexity. Simple-looking steps can become quite complex and orders of magnitude of the actual complexity can be overlooked because they are hidden in sub-steps whose length does depend on the input size.

assume that the width of an ordered set is the size of the largest set of points of the same rank. Exercise 5-7 shows that this is not the case. For references to an algorithm that computes the width of an arbitrary ordered set, see Remark 2. In this text, we will consider interval orders as a special case in Proposition 11.25.

Exercises

5-4. Give a linear time algorithm to check if a function $f : P \to P$ has a fixed point.

5-5. Prove that your algorithm from Exercise 5-3 is of polynomial efficiency.

5-6. Prove that the algorithm to compute the rank of each element that suggests itself from the definition of the rank (see Definition 2.5) takes $O\left(|P|^2\right)$ steps. Then prove that the height of an ordered set can be determined in $O\left(|P|^2\right)$ steps.

5-7. Prove that the width of a finite ordered set P need *not* be given by the size of the largest set of points with the same rank max $\left\{|S| : (\forall s, t \in S)\ \mathrm{rank}_P(s) = \mathrm{rank}_P(t)\right\}$.

5-8. Prove that, in an ordered set of width w, the transitive closure of the Hasse diagram is computable in $O\left(w|P|^2\right)$ steps.

    Hint. Construct from the top down. Use that each element has at most w upper/lower covers.

5-9. Let P be a finite ordered set. Give an $O\left(|P|^2\right)$ algorithm that computes $|\uparrow x|$ and $|\downarrow x|$ for all $x \in P$.

5-10. Let P be a finite ordered set of which we have the Hasse diagram and the order relation.

    a. Prove that, for any $x \in P$, it can be checked in $O\left(|P|\right)$ time if x is irreducible.

    b. Prove that it can be verified in $O\left(|P|^3\right)$ time if P is $\mathcal{I}$-dismantlable.

    c. Conclude that, for finite ordered sets of width 2 or height 1, it can be determined in polynomial time if the ordered set has the fixed point property.

    d. Prove that, for fixed k, there is a polynomial algorithm that verifies if a given finite ordered set is $\mathcal{R}_k$-dismantlable.

5-11. State a polynomial algorithm that checks if an ordered set of width 3 has a fixed point free automorphism.

    Note. In [216], a linear time algorithm is given.

5.3 NP Problems and Constraint Satisfaction Problems

Although polynomial efficiency is nice, not every problem can be solved with a polynomially efficient algorithm. By Theorem 2.12, an ordered set with n elements has at least $2^{\frac{n}{2}}$ order-preserving self-maps. Therefore, if we want a list of all order-preserving self-maps of an ordered set with n elements, then any algorithm will take at least $2^{\frac{n}{2}}$ steps, one for each map. Interestingly enough, beyond examples similar to the above, it is usually hard to prove that a problem can*not* be solved in polynomial time, even if the answer will be either "yes" or "no."

Definition 5.9. *A decision problem is a problem for which the answer will be "yes" or "no."*

The problem "Does a given ordered set have the fixed point property?" definitely is a decision problem. Just because there are exponentially many order-preserving self-maps does not mean that establishing the fixed point property always takes exponential time.[2] For example, Exercise 5-10c provides a polynomial algorithm to determine if an ordered set of height 1 or width 2 has the fixed point property. A correct answer to the question "Does a given ordered set have the fixed point property?" is equally valuable to us as a correct answer to the question "Does a given ordered set *not* have the fixed point property?" That is, we could consider the following decision problem.

Given. A finite ordered set P.

Question. Is there a fixed point free order-preserving map $f : P \to P$?

Proving that a given ordered set does not have the fixed point property would be easy if some "oracle" were to provide us with the right conjecture for a fixed point free order-preserving self-map. Indeed, it takes $|P|$ steps to check if a given self-map of P has a fixed point (see Exercise 5-4) and another $O(|P|^2)$ steps to check if the map is order-preserving (see Exercise 5-5). Although it is not realistic to hope for such an oracle, the idea that the discovery of a suitable object will solve a problem quickly is at the heart of the idea of nondeterministically polynomial problems.

Definition 5.10. *A decision problem for input of size n is said to be **nondeterministically polynomial** or **NP** iff there is an algorithm $\mathcal{A}$ whose run time is polynomial in the size of its input and, for each input I, a certain structure C(I), called a **certificate**, of size polynomial in n such that running $\mathcal{A}$ on C(I) "proves" that the answer is "yes."*

Formally (see [103], p. 156), because NP problems focus on an answer of "yes," this means that the decision if an ordered set has the fixed point property is a **co-NP** problem. That is, it is the complement of an NP problem. Determining if an ordered set has the fixed point property would only be in NP if there was a way to design a "certificate" that proves in polynomial time that the set has the fixed point property. A decision problem that directly translates to an NP problem is the question "Are two given ordered sets isomorphic?" because we can formulate the problem as follows.

Given. Two finite ordered sets P and Q.

Question. Is there an order-isomorphism $\Phi : P \to Q$?

The problem above is in NP, because, for any given map, it can be verified in polynomially many steps if it is an order-isomorphism.

Unfortunately, the polynomial verifiability of the answer being "yes" only helps if we guess right. Moreover, there is no provision for verification of a negative answer. Thus it may not be a surprise that there are no known polynomial algorithms which solve the above problems in general. However, the framework of NP problems so far is the best framework in which to analyze the level of difficulty

---

[2]Generating every possible order-preserving self-map and checking it for a fixed point would be an exceedingly naive idea.

of many decision problems. The crucial advantage of working with NP problems is that there are results that allow us to formally distinguish those problems that are "genuinely hard." This idea of NP-completeness (see Definition 7.30) will be discussed in Section 7.6, while we now turn to the main topic of this chapter.

Constraint satisfaction problems, sometimes also called constraint networks, provide a framework for problems in which values are assigned to variables subject to certain constraints. This is a widely applicable idea, see [60, 177, 198, 212, 260, 313]. From our point-of-view, a fixed point free order-preserving self-map certainly assigns values (elements of the ordered set) to variables (which are the elements of the ordered set, too) subject to constraints (namely, the preservation of order and the avoidance of fixed points).

As is mentioned in the introduction to [313], the wide applicability of constraint satisfaction problems has led to many rediscoveries of the setup, rediscoveries of solution algorithms, and to a multitude of terminologies. The presentation here should give an overview of this area and how it interfaces with ordered sets and other branches of mathematics.

Definition 5.11. *A **binary constraint satisfaction problem (CSP)** (compare [313], Section 1.2; or **binary constraint network**, compare [60], p. 276) consists of the following.*

1. *A set of variables $x_1, \ldots, x_r$.*
2. *A set of domains $D_1, \ldots, D_r$, one for each variable. Because an instantiation of the variables (see Definition 5.12 below) is, mathematically speaking, a function from the domain set $\{x_1, \ldots, x_r\}$ to the codomain set $\bigcup_{i=1}^{r} D_i$, we will also refer to the D_i as **value domains**.*
3. *A set $\mathcal{C}$ of unary and binary constraints.[3]*

   - *Each **unary constraint** consists of a variable x_i and a set $C_i \subseteq D_i$.*
   - *Each **binary constraint** consists of a set of two variables $\{x_i, x_j\}$ and a binary relation $C_{ij} \subseteq D_i \times D_j$, where we assume that $i < j$.[4]*
   - *For each set of two variables, we have at most one constraint.*

It is possible to define higher order CSPs using k-ary constraints. For our purposes, binary CSPs will be sufficient. Moreover, there is a translation process that turns higher order constraint satisfaction problems into binary ones (see [61], p. 355).

Note that Definition 5.11 does not say anything about which assignments are allowed and which are not. Allowed assignments for some or all variables are defined as follows.

---

[3] Again, there will be no confusion with the use of the letter $\mathcal{C}$ here.

[4] The ambiguity resulting from specification of constraints for *sets* $\{x_i, x_j\}$ of variables (the order of the variables matters in the specification of C_{ij}) is removed by demanding $i < j$. The alternative, which makes things unnecessarily technical, would be to specify constraints for ordered pairs of variables and demand the appropriate symmetry for the constraints C_{ij} and C_{ji}.

Definition 5.12 (Compare [60], p. 276).   *For a given CSP, let $Y \subseteq \{1, \ldots, r\}$. Any set $\{(x_i, a_i) : a_i \in D_i, i \in Y\}$ is an **instantiation** of the variables $\{x_j : j \in Y\}$. An instantiation of the variables $\{x_j : j \in Y\}$ is called **consistent** iff for all $i, j \in Y$ we have that $a_i \in C_i$ and $(a_i, a_j) \in C_{ij}$ (assuming $i < j$).*[5]

A consistent instantiation for all variables $\{x_i : i \in \{1, \ldots, r\}\}$ is called a **solution**.

Hence the unary constraints encode the consistent instantiations for single variables and the binary constraints encode the consistent instantiations for sets of two variables. (These ideas are easily generalized to ternary and higher order constraints.) Because, for any consistent instantiation, we want that $a_i \in C_i$, we could simply start the problem with $D_i := C_i$ and **omit any further unary constraints**. This is what we will do in this text.

Of course we want to know if there is a solution for a given binary CSP. That is, the central problem in binary constraint satisfaction is the following.

Given. A binary CSP.

Question. Is there a solution for the given binary CSP?

Clearly the above is another NP problem: For a given instantiation of all variables, it can be verified in $O(n^2)$ time if it is a solution. In the appropriate context, the following questions/tasks can also be addressed.

- How many solutions are there?
- List all solutions.
- In case a weight function for the instantiations is given, find a solution with lowest possible weight.

The fixed point property was first cast into the framework of CSPs in [328], though apparently without any connection to the main body of the literature on constraint satisfaction. (The setting in [328] is formal concept analysis.)

Example 5.13. Let P be a finite ordered set. The constraint satisfaction problem $FPF(P)$ has $\{x_1, \ldots, x_r\} := P$ as its variable set. (Assume that, if $x_i < x_j$, then $i < j$.) For each x_i, the value domain D_i is $D_i := \{p \in P : p \not> x_i\}$. The constraints C_{ij} are

- If $x_i \not\sim x_j$, there is no constraint between x_i and x_j.
- If $x_i < x_j$, let $C_{ij} := \{(y_i, y_j) \in D_i \times D_j : y_i \leq y_j\}$.

The above gives at most one constraint for each set of two variables. Any consistent instantiation $\{(x_{i_1}, y_{i_1}), \ldots, (x_{i_k}, y_{i_k})\}$ of k variables $x_{i_1}, \ldots, x_{i_k}$ corresponds to an order-preserving map from $\{x_{i_1}, \ldots, x_{i_k}\}$ to P, namely, the map that maps each x_{i_j} to y_{i_j}.   □

---

[5]Careful with notation here. In some works, an instantiation is called consistent if $(a_i, a_j) \notin C_{ij}$, which is consistent with a constraint being something that forbids configurations.

Exercises 5-12 and 5-13 show other problems that can be translated into binary CSPs. Another way to translate the problem if a given ordered set has a fixed point free order-preserving self-map is outlined in Exercise 5-47. Aside from giving a first idea of the versatility of constraint satisfaction problems, translating a given problem into the framework of another problem is a standard technique in complexity theory. Once a problem is embedded into the framework of CSPs, we can apply the tools that exist for CSPs. These tools will be the subject of the following sections.

From an order-theoretical point-of-view, we may ask if there is a way to express CSPs as problems of certain ordered sets having the fixed point property. There is such a way. Essentially we could combine Cook's Theorem (see Theorem 7.29) and Theorem 7.32 to embed CSPs into fixed point problems. However, this approach is very cumbersome, and it is therefore not used.

Exercises

5-12. For each of the following decision problems, formulate a CSP for which the solutions are the maps in question.

a. Let P and Q be finite ordered sets. Is there an isomorphism from P to Q? We will call this constraint satisfaction problem $ISO(P, Q)$.

b. Let P be a finite ordered set. Does P have a fixed point free automorphism? We will call this constraint satisfaction problem $FPFAUT(P)$.

c. Let P be a finite ordered set and let $A \subseteq P$. Is A a retract of P? We will call this constraint satisfaction problem $RETR(P, A)$.

d. Let G be a graph and let v be a vertex of G. Does G have a Hamiltonian cycle that starts at v? We will call this constraint satisfaction problem $HC_v(G)$.

e. Let G be a graph and let v be a vertex of G. Does G have a Hamiltonian path that starts at v? We will call this constraint satisfaction problem $HP_v(G)$.

f. Given sets P and Q and first-order-logic propositions $\sigma(x, y)$ and $\tau(x, y, u, v)$, with two and four unquantified variables, respectively, is there a function $f : P \to Q$ such that $\forall x \in P : \sigma(x, f(x))$ and $\forall x, u \in P : \tau(x, f(x), u, f(u))$?

g. Given a finite ordered set P, find all antichains in P.
Hint. Use Proposition 2.36.

5-13. Some translations into a CSP are not immediate. Let P and R be ordered sets with $R \cap P = \emptyset$. Translate the question if P has a retract that is isomorphic to R into a binary CSP. We will call this constraint satisfaction problem $ISORETR(P, R)$, where the distinction to Exercise 5-12c is that $R \cap P = \emptyset$.
Hint. Set up a CSP whose variable set is $P \cup R$ and whose solution is a retraction-coretraction pair as in Exercise 4-1.

5-14. Consider the following two problems.

a. **Given.** Two finite graphs G and H.
Question. Is there an isomorphism between G and H?

b. **Given.** Two finite ordered sets P and Q.
Question. Is there an isomorphism between P and Q?

Prove that these two problems have the same complexity. That is, every algorithm that solves one can be turned into an algorithm that solves the other via a polynomial translation.

5.4   Search Algorithms

To solve a constraint satisfaction problem, we can always run a search algorithm. Unfortunately, any of the search algorithms discussed in this section, and any of their refinements that could be used instead (see, for example, [170]), requires, in the worst case, exponential time. In fact, unless every NP problem can be solved in polynomial time, running such a search algorithm is the best we can do for "hard" problems. We will discuss what "hard" means in Section 7.6. The question if, for every NP problem, there is a polynomially efficient algorithm that solves it is known as "is P=NP?" and it is one of the Clay Millennium Problems (see [46]).

The **search space** for solving a CSP in n variables is the set of all possible instantiations of the variables. Clearly, this space contains the solutions of the CSP, if there are any. The task of every search algorithm is to find these solutions in the search space, or to report that there are none. A very primitive (and inefficient) idea is to first linearly order the variables and their value domains[6], then recursively compute all possible instantiations for $\{x_1, \ldots, x_k\}$ from the set of all possible instantiations for $\{x_1, \ldots, x_{k-1}\}$ and, finally, when all possible instantiations for all variables are available, we could check each one if it is a solution or not.

You can immediately check that the set thus computed has $\prod_{i=1}^r |D_i|$ elements, which, assuming we have $|D_i| \geq 2$, are at least 2^r.[7] Moreover, indiscriminately generating instantiations would force us to check many instantiations that we could have ruled out with less effort. To wit, if $\{(x_1, y_1), (x_2, y_2)\}$ is inconsistent, then none of its extensions can be consistent and we need not check any of them. Therefore, the search space generally is a "background entity" that only arises in the theoretical analysis of search algorithms as in, say, [293] or [326]. We include its formal definition for completeness' sake.

Definition 5.14. *The **search tree** or **search space** of a CSP in r variables is the set*

$$\left\{ \{(x_1, y_1), \ldots, (x_k, y_k)\} : y_i \in D_i, 0 \leq k \leq r \right\}$$

of all instantiations of the first k variables, ordered by reverse inclusion.

The search tree as defined above is indeed a **tree**, that is, it is a connected ordered set T so that, for all $t \in T$, we have that $\uparrow t$ is a chain (see Exercise 5-15). Note that Definition 5.14 assumes a static ordering of the variables. Variable ordering heuristics, which can speed up algorithms, will be touched only briefly in this text.

There are many ideas for searching trees in a more efficient manner than the naive exhaustive search that was just mentioned (see [170]). In this text, we will focus on the two main paradigms for search algorithms, backtracking and forward checking.

---

[6]For our purposes we can always assume that the order of the variables $x_1, \ldots, x_n$ is the order of the indices. We will not explicitly specify the order of the value domains.

[7]I have seen a colleague, who was not a mathematician, use a similarly inefficient approach in his research. This is another instance that shows how mathematicians can help in collaborations.

Throughout this section, we will assume that the variables are ordered as indicated by their indices and that there is a fixed value order on each value domain.

Algorithm 5.15 (Backtracking). The backtracking algorithm maintains a consistent instantiation of the first k variables at all times. At the start, with $k = 0$, this is the empty set. Given a consistent instantiation CI of the first k variables, backtracking instantiates x_{k+1} to the first value $y_{k+1,1}$ of D_{k+1}. If $CI \cup \{(x_{k+1}, y_{k+1,1})\}$ is consistent, then CI is replaced with $CI \cup \{(x_{k+1}, y_{k+1,1})\}$ and backtracking tries to instantiate x_{k+2}. If not, the next value in D_{k+1} is tried. If backtracking does not find any instantiation of x_{k+1} that allows a consistent extension of the current instantiation, then x_k is uninstantiated. That is, CI is replaced with $CI \setminus \{(x_k, y_{k,current})\}$. The search then resumes as above by instantiating x_k to the next element of the kth value domain, $y_{k,current+1}$. When all instantiations for x_k have been explored, x_k is uninstantiated.

If backtracking encounters a solution, necessarily when $k + 1 = r$, the algorithm stops if only one solution is to be found. It continues with a backtrack if all solutions are to be found. If backtracking terminates without finding any solutions, then there is no solution for the CSP (see Exercise 5-17). □

Pseudocode for a recursive backtracking algorithm is given in Figure 5.1. Note that, although recursion is a very efficient way to talk about backtracking, a non-recursive implementation normally has shorter run times, because it requires less overhead in the control structures. Backtracking terminates when a solution is found

---

recursive algorithm `backtrack`(CI, depth)
If there is another instantiation y for x_{depth} that was not checked against CI,

　If $CI \cup \{ (x_{\text{depth}}, y) \}$ is consistent and depth $< r$,
　　Replace depth with depth+1, CI with $CI \cup \{ (x_{\text{depth}}, y) \}$.
　　Call `backtrack`(CI, depth).
　If $CI \cup \{ (x_{\text{depth}}, y) \}$ is consistent and depth $= r$,
　　Output $CI \cup \{ (x_{\text{depth}}, y) \}$ as a solution.
　　If the algorithm only needs to find the first solution, stop.
　　If the algorithm is supposed to find all solutions,
　　　Go back to the start of this routine.
　If $CI \cup \{ (x_{\text{depth}}, y) \}$ is not consistent,
　　Go back to the start of this routine.

If all instantiations for x_{depth} have already been tested,

　If depth$= 1$, stop. Otherwise,
　Remove the instantiation of $x_{\text{depth}-1}$ from CI.
　Replace depth with depth-1.
　Return to the previous level of execution.

Fig. 5.1 Pseudocode for a recursive backtracking algorithm. The algorithm is called with $CI = \emptyset$ and depth $= 1$. Instantiations for x_{depth} that have not been checked against CI are normally "detected" by checking D_{depth} in a fixed order. If the recently checked or removed instantiation is the last in this order, we backtrack

or after the whole tree of instantiations generated this way has been searched. In the latter case, either there is no solution or all solutions have been listed. The tree generated by backtracking is easily characterized.

Definition 5.16 (Compare [170], Figure 3). *For a CSP with variables $\{x_1, \ldots, x_r\}$ with value domains $D_1, \ldots, D_r$, we define the **backtracking tree** to be the set of all instantiations $\{(x_1, y_1), \ldots, (x_k, y_k)\}$ such that the parent instantiation $\{(x_1, y_1), \ldots, (x_{k-1}, y_{k-1})\}$ is consistent. The backtracking tree is assumed to be ordered by reverse inclusion.*

For an example of a backtracking tree, see Figure 5.2. We say that a search algorithm **visits** an instantiation I iff, during the execution, the algorithm checks if I can become the algorithm's current instantiation CI. You can prove in Exercise 5-17 that the instantiations of the backtracking tree are exactly the instantiations visited by backtracking.

The crucial step in backtracking considers the *past*, not the future: Backtracking only checks if the new instantiation for x_{k+1} is consistent with the already recorded instantiations for $x_1, \ldots, x_k$. The paradigm for algorithms that consider the *future* is forward checking.

Definition 5.17 (Compare [170], Definition 2). *For a CSP with r-element variable set $\{x_1, \ldots, x_r\}$ and value domains $D_1, \ldots, D_r$, let A and B be two consistent instantiations of disjoint sets of variables. Then A is called **consistent with** B iff $A \cup B$ is consistent.*

Definition 5.18. *A consistent instantiation $\{(x_1, y_1), \ldots, (x_k, y_k)\}$ of the first k variables in a CSP is called **forward consistent** iff, for every $i > k$, there is a $y_i \in D_i$ such that $\{(x_i, y_i)\}$ is consistent with $\{(x_1, y_1), \ldots, (x_k, y_k)\}$.*

Example 5.19. To describe the connection between consistent and forward consistent instantiations refer to the ordered set in Figure 1.4 e). For parts 1 through 3 below, we shall consider the task of enumerating all order-preserving maps of this set. We assume that the variables (points) are ordered alphabetically.

1. The instantiation $\{(a, a), (b, c), (c, b), (d, e)\}$ is consistent.
2. The instantiation $\{(a, d), (b, e), (c, f)\}$ is consistent, but not forward consistent. Indeed, there is no instantiation for the variable e that is consistent with $\{(a, d), (b, e), (c, f)\}$. There is one for d, though. This is the strength of forward consistency. Problems that are "(far) ahead" can be recognized "early."
3. Every forward consistent instantiation of $\{a, b, c\}$ actually is part of an order-preserving map, because we could map all elements to the value y so that $\{(e, y)\}$ is consistent with the instantiation. (This does not always happen.)
4. Forward consistent instantiations need not be part of solutions. For example, consider the problem of finding automorphisms for the ordered set in Figure 1.4 e), with the variables (points) ordered in reverse alphabetical order. The instantiation $\{(g, h), (h, k), (k, g)\}$ is forward consistent for this problem, but it is not part of a solution. (That is, it is not part of an automorphism.)

You can generate similar examples for the search for fixed point free order-preserving maps in Exercise 5-19. □

Algorithm 5.20 (Forward Checking.). The forward checking algorithm maintains, at all times, a forward consistent instantiation CI of the first k variables and a list of remaining possible extensions $(x_i, y_{i,j})$, $i > k$ that are consistent with CI. (We will assume that this list is updated as needed as we describe the algorithm.) Given a forward consistent instantiation CI of the first k variables, forward checking instantiates x_{k+1} to the first value $y_{k+1,1}$ of D_{k+1} that is consistent with CI. If $CI \cup \{(x_{k+1}, y_{k+1,1})\}$ is forward consistent, then CI is replaced with $CI \cup \{(x_{k+1}, y_{k+1,1})\}$ and forward checking instantiates x_{k+2}. If not, the next consistent value in D_{k+1} is tried. If forward checking does not find any instantiation of x_{k+1} that allows us to extend the current instantiation to a forward consistent instantiation of $x_1, \ldots, x_k, x_{k+1}$, then x_k is uninstantiated. That is, CI is replaced with $CI \setminus \{(x_k, y_{k,\text{current}})\}$. The search then resumes as above by instantiating x_k to the next consistent element of D_k, $y_{k,\text{current}+1}$. When all instantiations for x_k have been explored, x_k is uninstantiated.

If forward checking encounters a solution, the algorithm stops if only one solution was to be found. It continues with a backtrack if all solutions were to be found. If forward checking terminates without finding any solutions, then there is no solution for the CSP (see Exercise 5-20). □

Definition 5.21 (Compare [170], Figure 3). *For a CSP with variables $\{x_1, \ldots, x_r\}$ with value domains $D_1, \ldots, D_r$ we define the* **forward checking tree** *to be the set of all consistent instantiations $\{(x_1, y_1), \ldots, (x_k, y_k)\}$ such that the parent $\{(x_1, y_1), \ldots, (x_{k-1}, y_{k-1})\}$ is forward consistent, ordered by reverse inclusion.*

The forward checking tree is a subtree of the backtracking tree (see Exercise 5-21a). In Figure 5.2, the forward checking tree is the backtracking tree without the leaves. Indeed, you can prove in Exercise 5-21b that all nodes of the forward checking tree are interior nodes of the backtracking tree. The removal of the leaves of the backtracking tree is already a large reduction in the number of nodes visited. However, forward checking usually prunes the search tree even more than that. In Exercise 5-21c, you can order the variables for the 5-fence in such a way that the backtracking tree has an interior node that is not in the forward checking tree.

If the primary concern is the size of the tree searched, then forward checking is a better algorithm than backtracking. However, when visiting a node that is in both search trees, each algorithm requires a different amount of processing at this node. To check if an extension $CI \cup \{(x_{k+1}, y_{k+1})\}$ of a consistent instantiation CI is still consistent, backtracking has to check k pairs of instantiations $\{(x_i, y_i), (x_{k+1}, y_{k+1})\}$, $i \leq k$ for consistency. In contrast, to check if a consistent extension $CI \cup \{(x_{k+1}, y_{k+1})\}$ of a forward consistent instantiation CI is still forward consistent, forward checking, for each variable x_i with $i > k + 1$, has to check if $\{(x_{k+1}, y_{k+1})\}$ is consistent with at least one of the consistent extensions $CI \cup \{(x_i, y_{i,j})\}$ of CI. In fact, to update the list of remaining possibilities, forward checking has to check consistency between (x_{k+1}, y_{k+1}) and all consistent extensions $CI \cup \{(x_i, y_{i,j})\}$ of

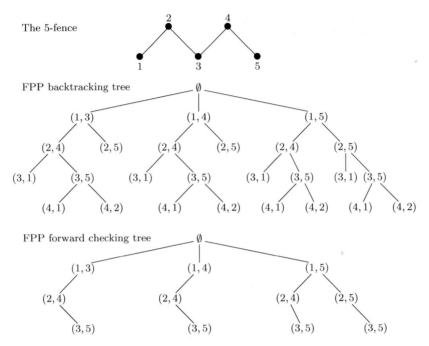

Fig. 5.2 The backtracking and forward checking trees for determining if the 5-fence has a fixed point free order-preserving self-map. The unary constraint is that no element is comparable to its image

CI, erasing those that are not consistent with (x_{k+1}, y_{k+1}). That is, the overhead that forward checking encounters *decreases* as we go deeper into the tree, while the overhead that backtracking encounters *increases*.

The large initial overhead appears to make forward checking slow. However, the reduction in the tree size generally makes up for the larger overhead encountered for instantiations of the first few variables (see [229]).[8] There is an implementation of forward checking that reduces the overhead at every extension $CI \cup \{(x_{k+1}, y_{k+1,c})\}$ to $n - k - 1$ bitwise ANDs. For details, see [201], Sec. 4 for the representation idea, and [119], p. 270–271 for the application to forward checking. We outline this idea in Exercise 5-23. This bit-parallel forward checking algorithm is very efficient.

An example of a forward checking tree is given in Figure 5.2. You should also check the proof of Proposition 1.22 and your solution to Exercise 1-19b to see how backtracking and forward checking ideas are present in these proofs. With the tool [286], you can apply backtracking and forward checking, in their regular and in their bit-parallel versions, as well as other search algorithms, to a wide variety of CSPs and compare their performance.

---

[8]Personally, I prefer forward checking over backtracking and I don't think I'm the only one.

Unfortunately even the most efficient implementations of either algorithm are still, in the worst case, exponential. Although this problem cannot be resolved by the preprocessing algorithms in the next section, the idea of reducing the CSP before starting a search can greatly reduce the search time.

Exercises

5-15. Prove that the search space (see Definition 5.14) is a tree.

5-16. Draw the backtracking and forward checking trees for the following problems.

    a. Decide if the six crown has a fixed point free order-preserving self-map.

    b. Find all fixed point free order-preserving self-maps of the six crown.

    c. Find all order-preserving self-maps of the four crown.

5-17. Nodes visited by backtracking.

    a. Prove that a backtracking algorithm that is set to find all solutions of a CSP visits exactly the instantiations that are vertices of the backtracking tree.

    b. Prove that backtracking set to find all solutions will visit all solutions of a CSP. (This means the algorithm can be called "correct.")

5-18. Provide a pseudocode description of forward checking that is similar to that for backtracking in Figure 5.1.

5-19. For the ordered set in Figure 1.4 e), consider the task of searching for fixed point free order-preserving self-maps of this set.

    a. Find a consistent instantiation of $\{a, b, c\}$.

    b. Find a consistent instantiation of $\{a, b, c, d\}$ that is not forward consistent.

    c. Is there a consistent instantiation of $\{a, b, c\}$ that is not forward consistent?

    d. Find a forward consistent instantiation of $\{a, b, c\}$ that is not part of a solution. (Why is this trivial?)

5-20. Nodes visited by forward checking.

    a. Prove that a forward checking algorithm that is set to find all solutions of a CSP visits exactly the instantiations that are vertices of the forward checking tree.

    b. Prove that forward checking set to find all solutions will visit all solutions of a CSP. (This means the algorithm can be called "correct.")

5-21. Comparing forward checking and backtracking.

    a. Prove that the forward checking tree for a CSP is always a subtree of the corresponding backtracking tree.

    b. Prove that the forward checking tree for a CSP is contained in the interior nodes of the corresponding backtracking tree.

    c. Find an ordering of the variables in the 5-fence P such that the $FPF(P)$ backtracking tree as in Figure 5.2 contains an interior node that is not in the $FPF(P)$ forward checking tree.

5-22. Give an example of a CSP for which the backtracking trees corresponding to two different orderings of the variables do not have the same size. Do the same for forward checking trees.

    (Variable pre-ordering is an important topic in CSPs, because a better variable order can sometimes drastically speed up computations, see [226].)

5-23. **Bit-parallel implementation of backtracking and forward checking**, see [119, 201].
Given a CSP, for each instantiation (x_i, y) and each variable $x_j \neq x_i$, let $b(x_i, y, x_j)$ be a
vector of $|D_j|$ zeroes and ones such that a 1 in the ℓth place indicates that $\{(x_i, y), (x_j, y_{j,\ell})\}$
is consistent, where $y_{j,\ell}$ is the ℓth element of D_j. (A zero indicates that $\{(x_i, y), (x_j, y_{j,\ell})\}$ is
inconsistent.)

  a. Prove that the componentwise AND of $b(x_i, y, x_j)$ and $b(x_k, z, x_j)$ encodes exactly those
     instantiations (x_j, v) that are consistent with (x_i, y) and with (x_k, z).
  b. Bit-parallel backtracking. Use part 5-23a to write a backtracking algorithm that, for
     each consistent instantiation CI, computes the instantiations for the next variable that
     are consistent with CI using at most $|CI|$ componentwise ANDs.
  c. Bit-parallel forward checking. Use part 5-23a to write a forward checking algorithm that,
     for each visited instantiation CI, computes the instantiations for all future variables that
     are consistent with CI using at most $n - |CI|$ componentwise ANDs.
       Hint. At each level, store the consistent instantiations for the future variables in a
     separate vector.
  d. Bit-parallel **Backjumping**. Modify the algorithm in part 5-23b as follows. Whenever
     the bitwise ANDs show that there is no consistent instantiation at the next level,
     backtrack not to the previous level, but to the first level k for which, with compAND
     denoting the componentwise AND operation, $b(x_1, y_1, x_{\text{current}})$ compAND $\cdots$ compAND
     $b(x_k, y_k, x_{\text{current}})$ is a vector of zeroes. Prove that this algorithm does not miss any
     solutions.
  e. A comparison between bit-parallel forward checking and regular backtracking. Prove
     that, for CSPs with r variables and value domain sizes at least r, bit-parallel forward
     checking performs fewer componentwise AND operations than regular backtracking
     visits nodes when searching for all solutions.

5.5   Expanded Constraint Networks and Local Consistency

The idea for backtracking is that, if an instantiation $\{(x_1, y_1), (x_2, y_2)\}$ is inconsistent, then it need not be further considered in the search for a solution. Forward
checking extends this idea by noting that, if an instantiation of the first k variables is
not forward consistent, it need not be further considered in the search for a solution.
Consequently, if we instantiate the variables x and u first, and $\{(x, y), (u, v)\}$ is not
forward consistent, then it need not be further considered. In fact, independent of
the ordering of the variables in a search algorithm, if, for $\{(x, y), (u, v)\}$, there is a
variable $z \notin \{x, u\}$ so that $\{(x, y), (u, v)\}$ is not consistent with any instantiation of z,
then no instantiation of variables that contains $\{(x, y), (u, v)\}$ needs to be considered,
because no solution can contain $\{(x, y), (u, v)\}$.

The idea for enforcing local consistency *before* running a search algorithm is to
modify the constraints by disallowing certain instantiations that cannot be contained
in a solution. In the language of Definition 5.11, to disallow an instantiation
$\{(x, y), (u, v)\}$, we simply modify the constraint C_{xu} by removing the pair (y, v).
If there was no constraint between the variables x and u, we would assume that
$C_{xu} = D_x \times D_u$. Although such operations modify the problem, we will make
sure that we only use operations that neither destroy solutions nor induce new ones.
When this is given, we might as well work with the new, more constrained, problem.

Definition 5.22. *Two CSPs are called **equivalent** iff they have the same solutions.*

The power of enforcing local consistency can be seen when we repeatedly check all consistent instantiations of two variables for forward consistency (in a variable order that puts these two variables first) and remove those that are not forward consistent: As long as some instantiations were removed on an earlier pass through all remaining consistent instantiations, further instantiations can possibly become inconsistent and can then be removed in the next pass. The expanded constraint network, defined below, is a very useful tool for encoding the original CSP and for the tracking of removals of consistent instantiations.

Definition 5.23. *The **(binary) expanded constraint network** $C^{\exp}$ for a binary constraint satisfaction problem C (see p. 195 of [212]) is a graph whose vertices are the set*

$$V = \{(x, y) : x \in X, y \in D_x\}$$

of consistent instantiations of single variables. For any $x \in X$ and $y, v \in D_x$, there is no edge between the vertices (x, y) and (x, v). For any distinct $x, u \in X$ and any $y \in D_x$, $v \in D_u$, there is an edge between (x, y) and (u, v) iff assigning y to x is consistent with assigning v to u, that is,

$$E = \{\{(x, y), (u, v)\} : x \neq u \text{ and } \{(x, y), (u, v)\} \text{ is consistent}\}.$$

*See Figure 5.3 for a visualization. We will also refer to a graph whose vertices are pairs (x, y) so that no two pairs with the same first component are adjacent as an **expanded constraint network with variable set** X, where $X := \{\pi_1(s) : s \in V\}$ and π_i is the projection onto the ith component. In an expanded constraint network H with variable set X, for each $x \in X$, we set $D_x := \{\pi_2(s) : s \in V, \pi_1(s) = x\}$.*

Clearly, the assignment $C \mapsto C^{\exp}$ is a bijective correspondence between binary constraint satisfaction problems C and expanded constraint networks $C^{\exp}$. Hence, the two encodings are completely equivalent. However, I am not aware of much work that explicitly refers to the expanded constraint network. Moreover, the only work that gives explicit theorems regarding the effect of enforcing local consistency seems to be in [65, 315, 328]. So, on one hand, we will work with terminology that may not quite be mainstream, but, on the other hand, I think the following sections will show the potential for nice results in specific contexts.

To start, we should note that solutions in an expanded constraint network with variable set X are in bijective correspondence with cliques of size $|X|$.

Definition 5.24. *A **clique** in a graph $G = (V, E)$ is a nonempty set of vertices K such that, for all distinct $x, y \in K$, we have $\{x, y\} \in E$.*

Proposition 5.25. *A clique of size $|X|$ in an expanded constraint network with variable set X is a complete assignment of values to all variables so that any two individual assignments are consistent with each other and vice versa. That is, a clique of size $|X|$ is a solution of the corresponding constraint satisfaction problem, and conversely.* ∎

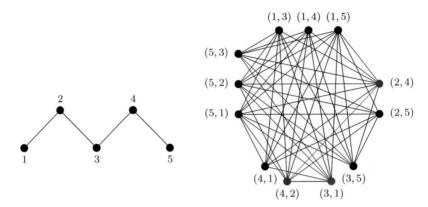

Fig. 5.3 The expanded constraint network for finding fixed point free order-preserving self-maps of a 5-fence. The vertex $(3,1)$ is not $(1,1)$-consistent, see Definition 5.27, because there is no edge of the form $\{(3,1),(2,y)\}$. The edge $\{(4,2),(2,4)\}$ is not $(2,1)$-consistent, see Definition 5.26, even though both vertices are $(1,1)$-consistent, because there is no instantiation of 3 that is consistent with $\{(4,2),(2,4)\}$

In the same fashion, consistent instantiations of variables for the constraint satisfaction problem $\mathcal{C}$ are cliques in the expanded constraint network $\mathcal{C}^{\exp}$. The advantage of using expanded constraint networks is that disallowing an instantiation $\{(x,y),(u,v)\}$ corresponds to removing an edge in a graph. We are left with a graph of the same type and there is no further bookkeeping regarding which of the many constraints were modified. Moreover, the absence of a constraint between two variables is a non-issue, because, if there is no constraint between variables x and u, then there is an edge between any two vertices (x,y) and (u,v).

Before we remove edges, we should think about which edges are sensible to remove. In keeping with earlier (and standard) terminology, we identify the entities that we cannot rule out.

Definition 5.26 (See [59, 260, 313]). *Let H be an expanded constraint network with variable set X and let $n \geq 1$. For $x, u \in X$ and $y \in D_x$, $v \in D_u$, the edge $\{(x,y),(u,v)\}$ is $(2,n)$-consistent iff, for any n variables $u_1, \ldots, u_n \in X \setminus \{x,u\}$, there are $v_i \in D_{u_i}$ so that $\{(x,y),(u,v),(u_1,v_1),(u_2,v_2),\ldots,(u_n,v_n)\}$ is an $(n+2)$-clique in H.*

When an edge is $(2,n)$-consistent, it may or may not be part of an $|X|$-clique. However, an edge that is *not* $(2,n)$-consistent (see Figure 5.3) is definitely not part of an $|X|$-clique. We can define $(1,n)$-consistency for vertices similarly.

Definition 5.27 (See [59, 260, 313]). *Let H be an expanded constraint network with variable set X and let $n \geq 1$. For $x \in X$ and $y \in D_x$, the vertex (x,y) is called $(1,n)$-consistent iff, for any n variables $u_1, \ldots, u_n \in X \setminus \{x\}$, there are values $v_i \in D_{u_i}$ so that $\{(x,y),(u_1,v_1),(u_2,v_2),\ldots,(u_n,v_n)\}$ is an $(n+1)$-clique in H.*

As for $(2, n)$-consistency for edges, when a vertex is $(1, n)$-consistent, it may or may not be part of an $|X|$-clique. On the other hand, a vertex that is *not* $(1, n)$-consistent (see Figure 5.3) is definitely not part of an $|X|$-clique.

Remark 5.28. The difference between $(1, n)$-consistency and $(2, n)$-consistency is that $(1, n)$-consistency focuses on vertices, whereas $(2, n)$-consistency focuses on edges. Hence, many arguments for $(2, n)$-consistency translate easily to corresponding results for $(1, n)$-consistency. $\qquad\qquad\qquad\qquad\qquad\qquad\qquad\qquad\quad\square$

It should not be surprising that $(1, n)$-consistency and $(2, n)$-consistency are special instances of a more general notion of (m, n)-consistency, which you can define in Exercise 5-25. However, to disallow an edge that is not $(2, n)$-consistent, we only need to remove the edge. To disallow an m-clique C that is not (m, n)-consistent, we would need to remove the clique C, but each edge could still be part of a solution that does not contain C. The only way to handle this situation is to work with a hypergraph, that is, a graph in which hyperedges can have more than 2 ends. For hypergraphs, the theoretical analysis, as well as the programming, becomes quite messy. Because, for our purposes, removal of edges will provide several good results, we will focus on $(1, n)$-consistency and $(2, n)$-consistency. To unify both ideas under the umbrella of removing edges, we define the following.

Definition 5.29. *Let H be an expanded constraint network, let e be an edge of H, and let $n > 0$. We say that e is $(1, n)$-**consistent** iff both its incident vertices are $(1, n)$-consistent.*

We will work with different levels of $(2, n)$- and $(2, m)$-consistency, so it is natural to establish a relation between these notions.

Proposition 5.30. *Let H be an expanded constraint network with variable set X and let $n \geq 1$ and $j > 0$ be so that $|X| \geq n + j + 2$. If the edge $\{(x, y), (u, v)\}$ is not $(2, n)$-consistent, then it is not $(2, n + j)$-consistent either. The same result holds for edges that are not $(1, n)$-consistent.*

Proof. Because the edge $\{(x, y), (u, v)\}$ is not $(2, n)$-consistent, there are mutually distinct variables $u_1, \ldots, u_n \in X \setminus \{x, u\}$ so that there is no $(n + 2)$-clique $\{(x, y), (u, v), (u_1, v_1), \ldots, (u_n, v_n)\}$ with $v_i \in D_{u_i}$. If we arbitrarily pick mutually distinct $u_{n+1}, \ldots, u_{n+j} \in X \setminus \{x, u, u_1, \ldots, u_n\}$, then there is no $(n + 2 + j)$-clique $\{(x, y), (u, v), (u_1, v_1), \ldots, (u_n, v_n), (u_{n+1}, v_{n+1}), \ldots, (u_{n+j}, v_{n+j})\}$ with $v_i \in D_{u_i}$. Hence $\{(x, y), (u, v)\}$ is not $(2, n + j)$-consistent either.

As indicated in Remark 5.28, the argument for edges that are not $(1, n)$-consistent is similar (see Exercise 5-26). $\qquad\qquad\qquad\qquad\qquad\qquad\qquad\qquad\blacksquare$

There is a simple relationship between $(1, n)$-consistency and $(2, n)$-consistency.

Proposition 5.31. *Let H be an expanded constraint network with variable set X and let $n \geq 1$. If e is an edge of H that is not $(1, n)$-consistent, then e is not $(2, n)$-consistent either.*

Proof. Exercise 5-27.   ∎

The idea for enforcing local consistency is now to remove edges that are not $(2, n)$-consistent for some n.

Definition 5.32. *Let H be an expanded constraint network, let K be a subgraph of H with the same vertex set as H, and let $n \geq 1$. Then we say that the network H can be **reduced, by enforcing** $(2, n)$-**consistency, to the network K** iff the edges that are in H, but not in K, can be enumerated as a sequence of edges $e_1, \ldots, e_m$ so that, if, for $i \in \{0, \ldots, m\}$, we define H_i as obtained from H by removing the edges $e_1, \ldots, e_i$, then, for $i \in \{1, \ldots, m\}$, each edge e_i is not $(2, n)$-consistent in H_{i-1}. Reduction by enforcing $(1, n)$-consistency is defined similarly.*

Algorithmically, the idea to remove edges until this is no longer possible is quite simple.

Algorithm 5.33 (Enforcing $(2, 1)$-consistency.).
   Given. An expanded constraint network with variable set X.
   Task. Compute an equivalent $(2, 1)$-consistent expanded constraint network.

The algorithm in Figure 5.4 simply removes inconsistent edges until no further inconsistent edges can be found. It runs in polynomial time. Indeed, all for-loops go over polynomially many objects and the repeat-loop is carried out at most as many times as there are edges in the expanded constraint network. Moreover, as you will show in Exercise 5-28, the reduced network the algorithm produces does not depend on any of the orderings in the for-loops. It is also easy to see that the removals do not affect any solutions that might be present. Indeed, if $\{(x, y), (u, v)\}$ is an edge in an r-clique, then for every $z \in X \setminus \{x, u\}$ there is a $w \in D_z$ so that (z, w) is in the r-clique. This means that $\{(x, y), (u, v), (z, w)\}$ is consistent and hence $\{(x, y), (u, v)\}$ is not removed.

For more efficient implementations of $(2, 1)$-consistency algorithms see [118, 201, 210] or Section 4.3 of [313].   □

```
Repeat

    changed := FALSE
    For every edge {(x, y), (u, v)}
        For every z ∈ X \ {x, u}
            Check if there is an instantiation {(z, w)} such that
            {(x, y), (u, v), (z, w)} is consistent. If there is no such
            instantiation, remove {(x, y), (u, v)} from the expanded
            constraint network and set changed := TRUE
        end for z
    end for {(x, y), (u, v)}

until changed = FALSE.
```

Fig. 5.4 A simple algorithm to enforce $(2, 1)$-consistency

Remark 5.34. Descriptions of $(1, n)$-consistency enforcing algorithms typically focus on the removal of vertices rather than edges. For our purposes, it is more uniform to turn the discussion of $(1, n)$-consistency enforcement into a discussion of edge removals, as we did via Definitions 5.29 and 5.32. Where a standard $(1, n)$-consistency enforcing algorithm would remove a vertex (x, y), in our discussions, the vertex (x, y) will remain as an isolated vertex and vice versa. Hence the descriptions are equivalent.   □

As we remove edges, we change the expanded constraint network and its corresponding CSP. These changes lead to equivalent CSPs (see Exercise 5-29). It is also true that "reduction until no further edges can be removed" always leads to the same CSP, but the proof is a little less immediate.

Lemma 5.35. *Let H be an expanded constraint network, let H' be a subgraph of H with the same vertex set as H, and let $n \geq 1$. If the edge e is not $(2, n)$-consistent in H and e is an edge in H', then e is not $(2, n)$-consistent in H' either. The same result holds for edges that are not $(1, n)$-consistent.*

Proof. The condition for $(2, n)$-consistency in H' is the same as the condition for $(2, n)$-consistency in H, except that H' has even fewer edges available to satisfy the condition. Hence, if e is not $(2, n)$-consistent in the larger graph H, then it cannot be $(2, n)$-consistent in H'.   ∎

Theorem 5.36. *Let H be an expanded constraint network, let H' and K be subgraphs of H with the same vertex set as H, and let $n \geq 1$. If H can be reduced, by enforcing $(2, n)$-consistency, to the network K, then H' can be reduced, by enforcing $(2, n)$-consistency, to a network K' that is a subgraph of K and that can be obtained from H' by removing the edges that are in H', but not in K. The same result holds for $(1, n)$-consistency.*

Proof. Let $e_1, \ldots, e_m$ be a sequence of edges in H, so that, if, for $i \in \{0, \ldots, m\}$, we define H_i as obtained from H by removing the edges $e_1, \ldots, e_i$, then, for $i \in \{1, \ldots, m\}$, each edge e_i is not $(2, n)$-consistent in H_{i-1}. Then, by Lemma 5.35, if, for $i \in \{0, \ldots, m\}$, we define H_i' as obtained from H' by removing the edges among $e_1, \ldots, e_i$ that are in H', then, for $i \in \{1, \ldots, m\}$, if the edge e_i is present in H_{i-1}', then it is not $(2, n)$-consistent in H_{i-1}'. Let $K' := H_m'$. By construction, H' can be reduced, by enforcing $(2, n)$-consistency, to K'. Moreover, K' is a subgraph of K. Because the above process does not remove any edges of H' that are in K, K' can be obtained from H' by removing the edges of H' that are not in K.

As indicated in Remark 5.28, the argument for $(1, n)$-consistency is similar (see Exercise 5-30).   ∎

In particular, Theorem 5.36 shows that, if an expanded constraint network H can be reduced, by enforcing $(2, n)$-consistency, to a network K, then the order in which edges are removed is immaterial: Indeed, if, in Theorem 5.36, we fix H and K, we see that, independent of which edges we remove from H to obtain, by enforcing $(2, n)$-consistency, a network H' that contains K, the network H' can still be reduced, by enforcing $(2, n)$-consistency, to K.

Every expanded constraint network H can, by enforcing $(2, n)$-consistency, be reduced to a network K so that all remaining edges in K (if there are any) are $(2, n)$-consistent. Theorem 5.36 says, in particular, that this remaining network K is unique.

Definition 5.37. *Let H be an expanded constraint network. The unique network K so that all edges in K are $(2, n)$-consistent to which H can be reduced by enforcing $(2, n)$-consistency is called the $(2, n)$-**core** of H. The $(1, n)$-**core** of H is defined similarly.*

Clearly, an expanded constraint network without edges contains no $|X|$-cliques. Consistent with notation in graph theory, we define empty networks.

Definition 5.38. *An **empty expanded constraint network** or **empty network** is an expanded constraint network without any edges.*

So, if the $(2, n)$-core of an expanded constraint network is empty, then the corresponding constraint satisfaction problem has no solution. Because the $(2, n)$-core can be computed in polynomial time (see Exercise 5-32), this means that certain constraint satisfaction problems without a solution can be identified in polynomial time.

Remark 5.39. We know that, if an ordered set has the fixed point property, then all its automorphisms have fixed points, too. Consider the expanded constraint networks for $FPF(P)$ (see Example 5.13) and for $FPFAUT(P)$ (see Exercise 5-12b). Then $FPFAUT(P)^{\exp}$ is a subgraph of $FPF(P)^{\exp}$ with the same vertex set. Hence, by Theorem 5.36, if the $(2, n)$-core of $FPF(P)^{\exp}$ is empty, then the $(2, n)$-core of $FPFAUT(P)^{\exp}$ is empty, too. The converse is false, of course. This is an algorithmic analogue of the note at the start of this paragraph.   □

Exercises

5-24. Define m-ary constraint satisfaction problems and m-ary expanded constraint networks and establish a bijective correspondence between them.

5-25. Let H be an expanded constraint network with variable set X. Define what it means for an m-clique to be (m, n)-consistent.

5-26. Prove the part of Proposition 5.30 that applies to edges that are not $(1, n)$-consistent.

5-27. Prove Proposition 5.31.

5-28. Prove that the simple $(2, 1)$-consistency algorithm in Example 5.33 always produces the same reduced CSP independent of the ordering of the edges and the variables in the loops.

5-29. Let H be an expanded constraint network with variable set X that can be reduced, by enforcing $(2, n)$-consistency, to the expanded constraint network K. Prove that H and K have the same $|X|$-cliques.

5-30. Prove the part of Theorem 5.36 that applies to $(1, n)$-consistency.

5-31. Let H be an expanded constraint network with variable set X that can be reduced, by enforcing $(1, n)$-consistency (on edges), to the expanded constraint network K. Prove that H can be reduced, by enforcing $(2, n + j)$-consistency, to the expanded constraint network K.

5-32. Prove that the natural algorithm to compute the $(2, n)$-core of an expanded constraint network takes at most $O\left(M^4 \cdot M^4 \cdot M^{2n}\right)$ steps, where $M = \max\left\{|D_x| : x \in X\right\} \cup \left\{|X|\right\}$.

5-33. **Bit-parallel $(2, 1)$-consistency.** Use Exercise 5-23a to write a $(2, 1)$-consistency enforcing algorithm that uses componentwise ANDs to check, for each edge $\{(x, y), (u, v)\}$, if there is an instantiation of $z \notin \{x, u\}$ that is consistent with $\{(x, y), (u, v)\}$.

5-34. Devise an algorithm that reduces a given expanded constraint network to an equivalent $(1, 1)$-consistent expanded constraint network.

5.6   Retractions and $FPF(P)^{exp}$

Let P be an ordered set and let $n \in \mathbb{N}$. If the $(2, n)$-core of $FPF(P)^{exp}$ is empty, then P has the fixed point property. Algorithms that enforce $(2, n)$-consistency will be good tools for investigating the fixed point property if there are large classes $\mathcal{C}$ of ordered sets P that have the fixed point property if *and only if* the $(2, n)$-core of $FPF(P)^{exp}$ is empty. Experiments with [286] suggest that this is the case.

Definition 5.40. *Let $\mathcal{C}$ be a class of ordered sets and let $n \in \mathbb{N}$. We say that* **enforcing $(2, n)$-consistency decides the fixed point property** *for $\mathcal{C}$ iff, for all $P \in \mathcal{C}$, we have that P has the fixed point property iff the $(2, n)$-core of $FPF(P)^{exp}$ is empty. The definition for $(1, n)$-consistency is similar.*

To identify classes for which enforcing $(2, n)$-consistency decides the fixed point property, it is natural to start by determining the $(2, n)$-core of $FPF(P)^{exp}$ for ordered sets P for which there is a theorem that guarantees the fixed point property. For classes of ordered sets that can be identified in polynomial time, such as $\mathcal{I}$-dismantlable ordered sets (see Exercise 5-10b), it should not be surprising if the $(2, n)$-core is empty. The class of connectedly collapsible ordered sets, see Definition 4.29, is not likely to be identifiable in polynomial time, even within the class of collapsible ordered sets, see Definition 4.28. (Also consider Remark 9 at the end of this chapter.) If enforcing $(2, n)$-consistency decides the fixed point property in the class of connectedly collapsible ordered sets, we could decide the fixed point property in polynomial time in a rather unwieldy class. The following results indicate that this may well be the case, but a complete proof has eluded me so far.

We start by establishing an analogue of part 1 of Theorem 4.8 in Theorem 5.43 below. This result will be useful for the equivalences in Theorems 5.55 and 5.60.

Definition 5.41. *A constraint satisfaction problem with variable set X is called* **endomorphic** *iff, for all $x \in X$, we have that $D_x \subseteq X$. An expanded constraint network H is called* **endomorphic** *iff the corresponding constraint satisfaction problem is endomorphic.*

Definition 5.42. *Let H be an endomorphic expanded constraint network with variable set X and let $R \subseteq X$. Then we denote by $H|_R$ the induced subgraph of H with vertex set $\left\{(z, w) : z \in R, w \in D_z \cap R\right\}$.*

Theorem 5.43 (Compare with part 1 of Theorem 4.8). *Let P be an ordered set and let $n \geq 1$ be so that the $(2,n)$-core of $FPF(P)^{\exp}$ is empty. Let $r : P \to P$ be a retraction. Then the $(2,n)$-core of $FPF(r[P])^{\exp}$ is empty, too. The same result holds for the $(1,n)$-core.*

Proof. Because the result is trivial for $|r[P]| < n + 2$, we can assume that $|r[P]| \geq n + 2$.

Let $x \in P$ and let $y \in r[P]$. Note that, if $x \leq y$, then $r(x) \leq r(y) = y$ and the dual of this implication holds, too. Hence, if $(r(x), y)$ is a vertex of $FPF(r[P])^{\exp}$, then $r(x)$ is not comparable to y, which means that x is not comparable to y, and therefore (x, y) is a vertex of $FPF(P)^{\exp}$, too.

Now let $x, u \in P$ and $y, v \in r[P]$ be so that $\{(r(x), y), (r(u), v)\}$ is an edge of $FPF(r[P])^{\exp}$. In particular, this means that $r(x) \neq r(u)$. Then, by the preceding paragraph, (x, y) and (u, v) are vertices of $FPF(P)^{\exp}$. Moreover, if $r(x) \leq r(u)$, then $y \leq v$ and, because $r(x) \neq r(u)$, $x \not\geq u$. Hence, in case $r(x) \leq r(u)$, we have that $\{(x, y), (u, v)\}$ is an edge of $FPF(P)^{\exp}$. The case $r(x) \geq r(u)$ is handled similarly. Finally, in case $r(x)$ is not comparable to $r(u)$, we have that x is not comparable to u and hence, in this case, too, $\{(x, y), (u, v)\}$ is an edge of $FPF(P)^{\exp}$. Overall, for $x, u \in P$ and $y, v \in r[P]$ so that $\{(r(x), y), (r(u), v)\}$ is an edge of $FPF(r[P])^{\exp}$, we have that $\{(x, y), (u, v)\}$ is an edge of $FPF(P)^{\exp}$.

Construct the subgraph H of $FPF(P)^{\exp}$ as follows. The vertices of H are the vertices of $FPF(P)^{\exp}$. For $x, u \in P$ and $y, v \in r[P]$, so that $r(x) \neq r(u)$, we let $\{(x, y), (u, v)\}$ be an edge of H iff $\{(r(x), y), (r(u), v)\}$ is an edge of $FPF(r[P])^{\exp}$. For $x, u \in P$ and $y, v \in r[P]$, so that $r(x) = r(u)$, we let $\{(x, y), (u, v)\}$ be an edge of H iff we have $y = v$ and y is not comparable to $r(x) = r(u)$. By the above and by definition of $FPF(P)^{\exp}$, H is a subgraph of $FPF(P)^{\exp}$ that contains $FPF(r[P])^{\exp}$ as the induced subgraph $H|_{r[P]}$. Finally, by hypothesis on $FPF(P)^{\exp}$ and by Theorem 5.36, the $(2,n)$-core of H and of all subgraphs of H with the same vertex set is empty.

Let K be a subgraph of H so that the following hold. For $x, u \in P$ and $y, v \in r[P]$ so that $r(x) \neq r(u)$, we have that $\{(x, y), (u, v)\}$ is an edge of K iff $\{(r(x), y), (r(u), v)\}$ is an edge of $K|_{r[P]}$. For $x, u \in P$ and $y, v \in r[P]$ so that $r(x) = r(u)$ we have that $\{(x, y), (u, v)\}$ is an edge of K iff $y = v$ and y is not comparable to $r(x) = r(u)$ and $(r(x), y)$ is not isolated in K. In particular, H itself satisfies these hypotheses: For edges with $r(x) = r(u)$, note that, for $y \not\geq r(x)$, the set $\{(r(x), y), (y, r(x))\}$ is an edge of $FPF(r[P])^{\exp}$, so that $(r(x), y)$ is not isolated in H.

Let $\{(x, y), (u, v)\}$ be an edge of K that is not $(2,n)$-consistent. Then there are mutually distinct $u_1, \ldots, u_n \in P \setminus \{x, u\}$ so that K does not contain a clique of the form $\{(x, y), (u, v), (u_1, v_1), \ldots, (u_n, v_n)\}$ with $v_j \in D_{u_j}$. Sort the points $u_1, \ldots, u_n$ so that the points $r(u_1), \ldots, r(u_k)$ are mutually distinct, so that none of $r(u_1), \ldots, r(u_k)$ is equal to $r(x)$ or $r(u)$, and so that we have $\{r(u_{k+1}), \ldots, r(u_n)\} \subseteq \{r(x), r(u), r(u_1), \ldots, r(u_k)\}$. Suppose, for a contradiction, that there is a clique $\{(r(x), y), (r(u), v), (r(u_1), v_1'), \ldots, (r(u_k), v_k')\}$ in $K|_{r[P]}$. For $j \in \{k + 1, \ldots, n\}$ so that $r(u_j) = x$, let $v_j' := y$. For $j \in \{k + 1, \ldots, n\}$ so that $r(u_j) = u$, let

$v'_j := v$. For the remaining $j \in \{k+1, \ldots, n\}$, there is a unique $i_j \in \{1, \ldots, k\}$ so that $r(u_j) = u_{i_j}$ and, for all these j, let $v'_j := v'_{i_j}$. By assumption on K, $\{(x,y),(u,v),(u_1,v'_1),\ldots,(u_k,v'_k),(u_{k+1},v'_{k+1}),\ldots,(u_n,v'_n)\}$ is a clique in K, a contradiction. Hence, in the case that $r(x) \neq r(u)$, the edge $\{(r(x),y),(r(u),v)\}$ is not $(2,k)$-consistent in $K|_{r[P]}$. Therefore, by Proposition 5.30, it is not $(2,n)$-consistent in $K|_{r[P]}$. In case $r(x) = r(u)$, the same argument shows that the vertex $(r(x),y) = (r(u),v)$ is not $(1,n)$-consistent in $K|_{r[P]}$.

Now let $\{(r(x),y),(r(u),v)\}$ be an edge of $K|_{r[P]}$ that is not $(2,n)$-consistent in $K|_{r[P]}$. Then there are mutually distinct points $u_j = r(u_j) \in r[P] \setminus \{r(x), r(u)\}$ ($j = 1, \ldots, n$) so that $K|_{r[P]}$ does not contain any $(n+2)$-cliques of the form $\{(r(x),y),(r(u),v),(r(u_1),v_1),\ldots,(r(u_n),v_n)\}$. Because all second components of vertices of K are in the retract $r[P]$, K does not contain any such cliques either. Therefore, by assumption on K, K does not contain any $(n+2)$-cliques of the form $\{(x,y),(u,v),(r(u_1),v_1),\ldots,(r(u_n),v_n)\}$. That is, the edge $\{(x,y),(u,v)\}$ is not $(2,n)$-consistent in K.

Therefore, an edge $\{(x,y),(u,v)\}$ of K with $r(x) \neq r(u)$ is not $(2,n)$-consistent in K iff $\{(r(x),y),(r(u),v)\}$ is an edge of $K|_{r[P]}$ that is not $(2,n)$-consistent in $K|_{r[P]}$.

With the above, we will show that a reduction of H, by enforcing $(2,n)$-consistency, to an empty graph induces a reduction of $FPF(r[P])^{\exp} = H|_{r[P]}$, by enforcing $(2,n)$-consistency, to an empty graph. So consider a sequence of edges $e_1, \ldots, e_m$ in H so that, if, for $i \in \{0, \ldots, m\}$, we define H_i as obtained from H by removing the edges $e_1, \ldots, e_i$, then, for $i \in \{1, \ldots, m\}$, each edge e_i is not $(2,n)$-consistent in H_{i-1}, and H_m is empty.

Define $K_0 := H$, let $i > 0$ and assume that K_{i-1} is a subgraph of H_{i-1} so that the following hold. For all $x, u \in P$ and $y, v \in r[P]$ with $r(x) \neq r(u)$, we have that $\{(x,y),(u,v)\}$ is an edge of K_{i-1} iff $\{(r(x),y),(r(u),v)\}$ is an edge of $K_{i-1}|_{r[P]}$. For all $x, u \in P$ and $y, v \in r[P]$ with $r(x) = r(u)$, we have that $\{(x,y),(u,v)\}$ is an edge of K_{i-1} iff $y = v$ and y is not comparable to $r(x) = r(u)$ and $(r(x),y)$ is not isolated in K_{i-1}.

In case e_i is not an edge of K_{i-1}, we let $K_i := K_{i-1}$ and we let $f_i := \emptyset$. Clearly, K_i satisfies the hypotheses originally stated for K_{i-1} and the process can continue.

In case $e_i = \{(x,y),(u,v)\}$ is an edge of K_{i-1}, first consider the case that $r(x) \neq r(u)$. We know that e_i is not $(2,n)$-consistent in K_{i-1}. Therefore the edge $e'_i := \{(r(x),y),(r(u),v)\}$ is not $(2,n)$-consistent in $K_{i-1}|_{r[P]}$ and we set $f_i := \{e'_i\}$. By the above, all edges $\{(x',y),(u',v)\}$ of K_{i-1} with $\{(r(x'),y),(r(u'),v)\} = \{(r(x),y),(r(u),v)\}$ are not $(2,n)$-consistent in K_{i-1}. Let K'_i be the network obtained from K_{i-1} by removing these edges. If the vertex $(r(x),y)$ is isolated in K'_i, remove from K'_i all edges $\{(x',y),(u',y)\}$ of K_{i-1} with $r(x') = r(u') = r(x)$ to obtain K''_i, otherwise let $K''_i := K'_i$. If the vertex $(r(u),v)$ is isolated in K'_i, remove from K''_i all edges $\{(x',v),(u',v)\}$ of K_{i-1} with $r(x') = r(u') = r(u)$ to obtain K_i, otherwise let $K_i := K''_i$. This newly obtained network K_i satisfies the hypotheses originally stated for K_{i-1} and the process can continue.

Now consider the case that $e_i = \{(x,y),(u,v)\}$ is an edge of K_{i-1} so that $r(x) = r(u)$. By the above, the vertex $(r(x),y)$ is not $(1,n)$-consistent in $K_{i-1}|_{r[P]}$. Hence none of the edges a in $K_{i-1}|_{r[P]}$ that are incident with $(r(x), y)$ are $(2, n)$-consistent in $K_{i-1}|_{r[P]}$. Therefore, none of these edges a are $(2, n)$-consistent in K_{i-1} either. We conclude that the procedure from the preceding paragraph can be applied to each of these edges. This produces a network K_i that satisfies the same hypotheses as K_{i-1} and that does not contain $\{(x, y), (u, v)\}$. It also produces a set f_i of edges a' of $K_{i-1}|_{r[P]}$ that are not $(2, n)$-consistent in $K_{i-1}|_{r[P]}$. (The set f_i happens to be the set of all edges in $K_{i-1}|_{r[P]}$ that are incident with $(r(x), y)$.)

This process continues until all e_i are processed and we have a sequence of sets f_i ($i = 1, \ldots, m$) of sets of edges that are not $(2, n)$-consistent in $K_{i-1}|_{r[P]}$ and K_m is empty.

From the sequence of sets $f_1, \ldots, f_m$ generate a sequence $d_1, \ldots, d_p$ of edges of $FPF(r[P])^{\exp} = H|_{r[P]}$ by deleting the empty sets among the f_i and by consecutively listing the edges in the nonempty sets f_i so that, for $i < j$, edges in f_i occur before edges in f_j. Then $d_1, \ldots, d_p$ is a sequence of edges in the network $FPF(r[P])^{\exp} = H|_{r[P]}$ so that, if, for $i \in \{0, \ldots, p\}$, we define L_i as obtained from $FPF(r[P])^{\exp} = H|_{r[P]}$ by removing the edges $d_1, \ldots, d_i$, then, for $i \in \{1, \ldots, p\}$, each edge d_i is not $(2, n)$-consistent in L_{i-1}, and L_p is empty. Hence, the $(2, n)$-core of $FPF(r[P])^{\exp}$ is empty.

As indicated in Remark 5.28, the argument for $(1, n)$-consistency is similar. The (substantial) details are left as Exercise 5-35. ∎

Aside from Theorem 5.43, we can present some results that show that enforcing $(2, n)$-consistency eliminates some edges that a human investigator reasonably would eliminate, too. These results make for good exercises in working with expanded constraint networks. Because we will not use them in the following, we will leave them as exercises. Proposition 5.44 below shows that, from a constraint propagation point-of-view it does not matter that, in the definition of $FPF(P)$, we immediately disregarded vertices (x, y) with $x \sim y$.

Proposition 5.44. *Let P be a finite ordered set. Consider the constraint satisfaction problem $FPF^*(P)$ that is defined as follows.*

1. *The set of variables is the set P.*
2. *For each $x \in P$, we set $D_x := \{y \in P : y \neq x\}$.*
3. *Assigning y to x is consistent with assigning v to u iff*

$$(x \leq u \Rightarrow y \leq v) \wedge (u \leq x \Rightarrow v \leq y)$$

is a true logical statement.

Denote the expanded constraint network for $FPF^*(P)$ by $FPF^*(P)^{\exp}$. Then, by enforcing $(1, 1)$-consistency, $FPF^*(P)^{\exp}$ can be reduced to the subgraph K that is obtained from $FPF^*(P)^{\exp}$ by removing all edges incident with a vertex (x, y) so that x is comparable to y.

Proof. See Exercise 5-36.   ∎

Similarly, edges that should not be present in $FPF(P)$ because of distance considerations are removed by enforcing $(2, 1)$-consistency.

Proposition 5.45. *Let P be a finite ordered set. By enforcing $(2, 1)$-consistency, the network $FPF(P)^{\exp}$ can be reduced to the subgraph K that is obtained from $FPF(P)^{\exp}$ by removing all edges $\{(x, y), (u, v)\}$ of $FPF(P)^{\exp}$ so that* $\mathrm{dist}(x, u) < \mathrm{dist}(y, v)$.

Proof. See Exercise 5-37.   ∎

There also is a constraint satisfaction version of Proposition 3.1.

Proposition 5.46 (Compare with Proposition 3.1). *Let P be a finite ordered set and let $n \geq 1$. Let K be the network obtained from $FPF(P)^{\exp}$ by removing all edges incident with vertices (x, y) so that x is minimal and y is not, or so that x is maximal and y is not. Then the $(2, n)$-core of $FPF(P)^{\exp}$ is empty iff the $(2, n)$-core of K is empty. The same result holds for $(1, n)$-cores.*

Proof. See Exercise 5-38.   ∎

Exercises

5-35. Prove Theorem 5.43 for $(1, n)$-cores.
5-36. Prove Proposition 5.44.
       Hint. Induction on the rank of x in vertices (x, y) with $x > y$ and duality.
5-37. Prove Proposition 5.45.
       Hint. Induction on the distance between x and u.
5-38. Prove Proposition 5.46.
       Hint. The direction "$\Rightarrow$" follows from Theorem 5.36. For the direction "$\Leftarrow$," eliminate the vertices (x, y) with x minimal and y not minimal first. Use that, if x is minimal, y is not, and $\{(x, y), (u, v)\}$ is an edge, then, for every minimal point $m \leq y$ we have that $\{(x, m), (u, v)\}$ is an edge. (This property must be preserved throughout the construction.)

5.7 Essentially Controlled Networks

Theorem 5.43 and its cousin, part 1 of Theorem 4.8, show that, if we have a large structure with good properties, then it is possible to carve out smaller structures with the same good properties. This should not be surprising: Large structures with good properties *should* induce smaller structures with good properties. The strength of the results in Section 4.3 stems from the fact that good properties of *smaller* sets are used to establish good properties for *larger* sets. We want to do the same thing for algorithms that enforce local consistency.

Definition 5.47. *In the context of endomorphic constraint satisfaction problems, a* **retraction** *is a function* $r : X \to X$ *that satisfies* $r^2 = r$. *The set* $r[X]$ *is called a* **retract**.

Note that, because we are working in the context of a general constraint satisfaction problem, idempotency is the only property we demand of r.

Suppose we have a retract $r[P]$ of an ordered set so that the $(2,1)$-core of $FPF(r[P])^{\exp}$ is empty. To establish that the $(2,1)$-core of $FPF(P)^{\exp}$ is empty, we need some connection between the edges of $FPF(r[P])^{\exp}$ and the edges of $FPF(P)^{\exp}$. The notion of essential control below provides such a connection in the context of general constraint satisfaction problems. The idea for essential control is implicit in [328].

Definition 5.48. *Let H be an endomorphic expanded constraint network with variable set X, let $R \subseteq X$, and let $r : X \to R$ be a retraction onto R. Then H is called* **essentially controlled (via r)** *by $H|_R$ iff, for all $x, u \in R$ and all $y \in D_x$, $v \in D_u$, we have that, if $\{(x, y), (u, v)\}$ is an edge of H, then $\{(x, r(y)), (u, r(v))\}$ is an edge of $H|_R$.*

It is tempting to hope that, for finding fixed point free order-preserving maps, any retraction will confer essential control of $FPF(r[P])^{\exp}$ over $FPF(P)^{\exp}$. The following shows why this is not the case.

Remark 5.49. Let P be an ordered set and let $r : P \to P$ be a retraction. In $FPF(P)^{\exp}$, for $x, u \in P$, $y \in D_x$, and $v \in D_u$, we have that $\{(x, y), (u, v)\}$ is an edge iff $(x \leq u \Rightarrow y \leq v) \wedge (u \leq x \Rightarrow v \leq y)$ is a true logical statement. Because r is order-preserving, we infer that, if $\{(x, y), (u, v)\}$ is an edge of $FPF(P)^{\exp}$, then $(x \leq u \Rightarrow r(y) \leq r(v)) \wedge (u \leq x \Rightarrow r(v) \leq r(y))$ is a true logical statement, too. However, $FPF(P)^{\exp}$ need not be, and *typically is not*, essentially controlled via r by $FPF(r[P])^{\exp}$. The reason is that, for an edge $\{(x, y), (u, v)\}$ of $FPF(P)^{\exp}$, neither $(x, r(y))$ nor $(u, r(v))$ needs to be a vertex of $FPF(r[P])^{\exp}$: The retraction r can, and typically does, for some points $z \in r[P]$ and $w \in P$ that are not comparable, map w to a point $r(w)$ that is comparable to z. In such a case, (z, w) is a vertex of $FPF(P)^{\exp}$, but $(z, r(w))$ is not a vertex of $FPF(r[P])^{\exp}$. $\qquad\qquad\square$

The following lemma shows how essential control enables us to translate removal of edges in $H|_R$ to removal of edges in H.

Lemma 5.50. *Let H be an endomorphic expanded constraint network with variable set X, let $R \subseteq X$, let $r : X \to R$ be a retraction so that H is essentially controlled via r by $H|_R$, let $n > 0$, and let $\{(x_1, y_1), (x_2, y_2)\}$ be an edge of $H|_R$ that is not $(2, n)$-consistent in $H|_R$. Then there is a set F of edges of H so that the following hold.*

1. *None of the edges in F is $(2, n)$-consistent in H.*
2. *The only edge in F that is also an edge of $H|_R$ is the edge $\{(x_1, y_1), (x_2, y_2)\}$.*
3. *Let K be the network obtained from H by removing the edges in F. Then K is essentially controlled via r by $K|_R$.*

Proof. Let F be the set of all edges $\{(x_1, y'), (x_2, v')\}$ of H so that the equation $\{(x_1, r(y')), (x_2, r(v'))\} = \{(x_1, y_1), (x_2, y_2)\}$ holds. In particular, we have that $\{(x_1, y_1), (x_2, y_2)\} \in F$.

For part 1, let $\{(x_1, y'), (x_2, v')\} \in F$. Because $\{(x_1, y_1), (x_2, y_2)\}$ is not $(2, n)$-consistent in $H|_R$, there are mutually distinct $u_1, \ldots, u_n \in R \setminus \{x_1, x_2\}$ so that there is no $(n+2)$-clique $\{(x_1, y_1), (x_2, y_2), (u_1, v_1), \ldots, (u_n, v_n)\}$ in the network $H|_R$, where each v_i is in $D_{u_i} \cap R$. Suppose, for a contradiction, that there is an $(n+2)$-clique in H that is of the form $\{(x_1, y'), (x_2, v'), (u_1, v_1'), \ldots, (u_n, v_n')\}$, with each v_i' being in D_{u_i}, but not necessarily in R. Then, because H is essentially controlled by $H|_R$, the set $\{(x_1, r(y')), (x_2, r(v')), (u_1, r(v_1')), \ldots, (u_n, r(v_n'))\}$ would be an $(n+2)$-clique in $H|_R$ with each $r(v_i')$ being in $D_{u_i} \cap R$, contradicting the fact that there is no $(n+2)$-clique in $H|_R$ that is of the form $\{(x_1, y_1), (x_2, y_2), (u_1, v_1), \ldots, (u_n, v_n)\}$ with each v_i in $D_{u_i} \cap R$. Hence every edge $\{(x_1, y'), (x_2, v')\} \in F$ is not $(2, n)$-consistent in H.

For part 2, let $\{(x_1, y'), (x_2, v')\} \in F$ be so that $\{(x_1, y'), (x_2, v')\}$ is an edge of $H|_R$. Because $\{(x_1, y'), (x_2, v')\}$ is an edge of $H|_R$, we have that $y' \in R$ and $v' \in R$. However, then $y' = r(y') = y_1$ and $v' = r(v') = y_2$, which means that $\{(x_1, y'), (x_2, v')\} = \{(x_1, y_1), (x_2, y_2)\}$.

For part 3, let $\{(x, y'), (u, v')\}$ be an edge of K. By definition of K, we have $\{(x, r(y')), (u, r(v'))\} \neq \{(x_1, y_1), (x_2, y_2)\}$. By part 2 and the hypothesis that H is essentially controlled by $H|_R$, we conclude that $\{(x, r(y')), (u, r(v'))\}$ is an edge in $K|_R$. Hence K is essentially controlled by $K|_R$. ∎

Lemma 5.51. *Let H be an endomorphic expanded constraint network with variable set X, let $R \subseteq X$, and let $r : X \to R$ be a retraction so that H is essentially controlled via r by $H|_R$. Let (x_1, y_1) be a vertex of $H|_R$ that is not $(1, n)$-consistent in $H|_R$. Then there is a set F of edges of H so that the following hold.*

1. *Each edge in F is incident with at least one vertex that is not $(1, n)$-consistent in H.*
2. *All edges of H that are incident with (x_1, y_1) are in F. Moreover, if an edge in F is also in $H|_R$, then it is incident with (x_1, y_1).*
3. *Let K be the network obtained from H by removing the edges in F. Then K is essentially controlled via r by $K|_R$.*

Proof. The proof is similar to that of Lemma 5.50 (recall Remark 5.28), see Exercise 5-39. ∎

Note that, both in Lemma 5.50 and in Lemma 5.51, the network H can be reduced, by enforcing $(2, n)$-consistency or $(1, n)$-consistency, respectively, to the network K. The strength of Theorem 5.52 below now lies in the fact that $H|_R$ has a smaller variable set than the original network H: If a large endomorphic expanded constraint network is essentially controlled by a well-behaved smaller network, then the large network is well-behaved, too. With smaller networks usually being easier to analyze, this makes for a promising tool.

Theorem 5.52. *Let H be an endomorphic expanded constraint network with variable set X, let $R \subseteq X$, and let $r : X \to R$ be a retraction so that H is essentially* *controlled via r by $H|_R$. Let $n \geq 1$ be so that $H|_R$ can be reduced, by enforcing $(2, n)$-consistency, to the network C. Then H can be reduced, by enforcing $(2, n)$-consistency, to a network K, so that $K|_R = C$ and K is essentially controlled via r by $K|_R = C$. In particular, if the $(2, n)$-core of $H|_R$ (or $H|_R$ itself) is empty, then the $(2, n)$-core of H is empty. The same result holds for $(1, n)$-consistency and $(1, n)$-cores.*

Proof. Consider a sequence of the edges $e_1, \ldots, e_m$ that are in $H|_R$ but not in C so that, if, for $i \in \{0, \ldots, m\}$, we define C_i as obtained from $H|_R$ by removing the edges $e_1, \ldots, e_i$, then, for $i \in \{1, \ldots, m\}$, each edge e_i is not $(2, n)$-consistent in C_{i-1}. Note that $C_m = C$. Define $H_0 := H$. Let $j \in \{1, \ldots, m\}$. Inductively, assume that, for $i \in \{0, \ldots, j - 1\}$, the network H_i is so that $H_i|_R = C_i$ and so that H_i is essentially controlled (via r) by $H_i|_R = C_i$. By Lemma 5.50, we can, by enforcing $(2, n)$-consistency, reduce H_{j-1} to a new network H_j so that $H_j|_R = C_j$ and so that H_j is essentially controlled (via r) by $H_j|_R = C_j$. Thus, by enforcing $(2, n)$-consistency, H can be reduced to the network $K := H_m$, which is essentially controlled (via r) by the network $K|_R = C$.

In case the $(2, n)$-core of $H|_R$ is empty, the above shows that the network H can be reduced, by enforcing $(2, n)$-consistency, to a network K that is essentially controlled (via r) by the empty network $K|_R$. Hence, for all $x \in R$ and $y \in D_x$, in K there is no edge incident to a vertex (x, y). Thus, K (and hence H) can be reduced, by enforcing $(2, n)$-consistency, to an empty network.

The proof for $(1, n)$-consistency and $(1, n)$-cores is left as Exercise 5-40.   ∎

As for dismantlability, composition of retractions passes the control to the smaller retract.

Corollary 5.53. *Let H be an endomorphic expanded constraint network with variable set X, let $R \subseteq X$, and let $r : X \to R$ be a retraction so that H is essentially controlled via r by $H|_R$. Let $S \subseteq R$, let $s : R \to S$ be a retraction onto S, and let $n \geq 1$ be so that $H|_R$ can be reduced, by enforcing $(2, n)$-consistency, to a network C that is essentially controlled via s by $C|_S$. Then H can be reduced, by enforcing $(2, n)$-consistency, to a network K that is essentially controlled via $s \circ r$ by $K|_S$. The same result holds for $(1, n)$-consistency.*

Proof. By Theorem 5.52, H can be reduced, by enforcing $(2, n)$-consistency, to a network K that is essentially controlled via r by $K|_R = C$.

Now, let $x, u \in S$, $y \in D_x$, and $v \in D_u$ and let $\{(x, y), (u, v)\}$ be an edge of K. Because K is essentially controlled via r by $K|_R = C$, the set $\{(x, r(y)), (u, r(v))\}$ is an edge of $K|_R = C$. Because C is essentially controlled via s by $C|_S$, the set $\{(x, s(r(y))), (u, s(r(v)))\}$ is an edge of $C|_S = (K|_R)|_S = K|_S$. Hence K is essentially controlled via $s \circ r$ by $K|_S$.   ∎

Exercises

5-39. Prove Lemma 5.51.
5-40. Prove Theorem 5.52 for $(1, n)$-consistency and $(1, n)$-cores.

5.8 Retractable Points and Constraint Propagation for FPF(P)

We will now apply the idea of essential control to expanded constraint networks $FPF(P)^{\exp}$. First, we focus on irreducible points, then on retractable points.

Proposition 5.54. *Let P be an ordered set and let $a \in P$ be irreducible. If a has a unique upper cover, let b be the unique upper cover of a. Otherwise, a has a unique lower cover and we let b be said lower cover. Let $r : P \to P \setminus \{a\}$ be the retraction that maps a to b. Then $H := FPF(P)^{\exp}$ can be reduced, by enforcing $(1, 1)$-consistency, to a network K that is essentially controlled via r by $K|_{P \setminus \{a\}}$, which is equal to $FPF(P \setminus \{a\})^{\exp}$.*

Proof. Without loss of generality, assume that $b = r(a)$ is the unique upper cover of a. Let $x \in P$ be so that $x < b$, but $x \not\leq a$. Suppose for a contradiction that there is a vertex (b, y) in $FPF(P)^{\exp}$ so that (x, a) is consistent with (b, y). Then $x < b$ implies $a \leq y$. However, the strict upper bounds of a are $\geq b$ and a itself satisfies $a < b$. That is, every upper bound of a is comparable to b. We conclude that y is comparable to b, which is not possible by definition of $FPF(P)^{\exp}$. Hence, all vertices (x, a) so that $x < b$, but $x \not\leq a$, are not $(1, 1)$-consistent and all edges incident with them can be removed by enforcing $(1, 1)$-consistency. Call the thus obtained network K. Note that $K|_{P \setminus \{a\}} = FPF(P)^{\exp}|_{P \setminus \{a\}} = FPF(P \setminus \{a\})^{\exp}$.

To prove that K is essentially controlled via r by $FPF(P \setminus \{a\})^{\exp}$, we must show that we do not encounter the problem noted in Remark 5.49. Let $x, u \in P \setminus \{a\}$ and let $y, v \in P$ be so that $\{(x, y), (u, v)\}$ is an edge in K. If $y = a$, then, by construction of K, x is not comparable to b and $(x, b) = (x, r(y))$ is a vertex of $FPF(P \setminus \{a\})^{\exp}$. Similarly, if $v = a$, then u is not comparable to b and $(u, b) = (u, r(v))$ is a vertex of $FPF(P \setminus \{a\})^{\exp}$. By definition of the expanded constraint network for the fixed point property, $x \leq u$ implies $y \leq v$, which, because r is a retraction, implies $r(y) \leq r(v)$, and similarly $x \geq u$ implies $r(y) \geq r(u)$. Hence $\{(x, r(y)), (u, r(v))\}$ is an edge of $FPF(P \setminus \{a\})^{\exp}$. Thus K is essentially controlled via r by $FPF(P \setminus \{a\})^{\exp}$. ∎

Theorem 5.55. *Let P be an ordered set and let $a \in P$ be irreducible. Then, for any $n \geq 1$, the $(2, n)$-core of $FPF(P)^{\exp}$ is empty iff the $(2, n)$-core of the network $FPF(P \setminus \{a\})^{\exp}$ is empty. The same result holds for $(1, n)$-cores.*

Proof. The direction "$\Rightarrow$" follows from Theorem 5.43.

For the direction "$\Leftarrow$," assume that $FPF(P \setminus \{a\})^{\exp}$ can be reduced, by enforcing $(2, n)$-consistency, to an empty graph. By Exercise 5-31 and Proposition 5.54, $FPF(P)^{\exp}$ can be reduced, by enforcing $(2, n)$-consistency, to a network K that is essentially controlled by $K|_{P \setminus \{a\}} = FPF(P \setminus \{a\})^{\exp}$. Therefore, by Theorem 5.52, K can be reduced, by enforcing $(2, n)$-consistency, to an empty graph, too. Hence, $FPF(P)^{\exp}$ can be reduced, by enforcing $(2, n)$-consistency, to an empty graph.

The proof for $(1, n)$-cores is left as Exercise 5-41.   ∎

So, similar to how the presence or absence of an irreducible point does not affect the fixed point property (see Theorem 4.11), the presence or absence of an irreducible point does not affect whether enforcing local consistency decides the fixed point property either. Moreover, the following shows that ordered sets with an empty $(1, 1)$-core can be completely characterized.

Lemma 5.56. *Let P be an ordered set. Then $FPF(P)^{\exp}$ has a vertex that is not $(1, 1)$-consistent iff P has an irreducible point.*

Proof. The direction "$\Leftarrow$" is proved in the proof of Proposition 5.54.

For "$\Rightarrow$," we prove the contrapositive. So let P be so that no point in P is irreducible. Let (x, y) be an arbitrary vertex of $FPF(P)^{\exp}$. To prove that (x, y) is $(1, 1)$-consistent, let $u \in P \setminus \{x\}$. If x and u are not comparable, then, for all points $v \in D_u$, the set $\{(x, y), (u, v)\}$ is an edge of $FPF(P)^{\exp}$. Now consider the case that $x < u$. If y is not comparable to u, then $\{(x, y), (u, y)\}$ is an edge of $FPF(P)^{\exp}$. If y is comparable to u, then $y < u$. Because P has no irreducible points, y must have an upper bound v that is not comparable to u. Now $\{(x, y), (u, v)\}$ is an edge of $FPF(P)^{\exp}$. This means that (x, y) is $(1, 1)$-consistent. Because (x, y) was arbitrary, we conclude that all vertices of $FPF(P)^{\exp}$ are $(1, 1)$-consistent.   ∎

Theorem 5.57 (See [328], proof of Theorem 5). *Let P be an ordered set. Then the $(1, 1)$-core of $FPF(P)^{\exp}$ is empty iff P is $\mathcal{I}$-dismantlable.*

Proof. We prove the direction "$\Leftarrow$" by induction on $|P|$. The cases $|P| = 1, 2$ are trivial, because, in these cases, $FPF(P)^{\exp}$ is empty. For the induction step, assume that the result holds for $\mathcal{I}$-dismantlable ordered sets with fewer than $|P|$ elements and let $a \in P$ be irreducible. Then, by Theorem 4.31, $P \setminus \{a\}$ is $\mathcal{I}$-dismantlable. By induction hypothesis, the $(1, 1)$-core of $FPF(P \setminus \{a\})^{\exp}$ is empty. Therefore, by Theorem 5.55, the $(1, 1)$-core of $FPF(P)^{\exp}$ is empty, too.

For "$\Rightarrow$," we also proceed by induction on $|P|$, with $|P| = 1, 2$ being trivial. For the induction step, suppose the result holds for ordered sets with fewer than $|P|$ elements and assume that the $(1, 1)$-core of $FPF(P)^{\exp}$ is empty. By Lemma 5.56, P has an irreducible point a. By Theorem 5.55, the $(1, 1)$-core of the network $FPF(P \setminus \{a\})^{\exp}$ is empty. By induction hypothesis, $P \setminus \{a\}$ is $\mathcal{I}$-dismantlable and hence P is $\mathcal{I}$-dismantlable.   ∎

Theorem 5.57 is the only result I am aware of that completely characterizes empty (m, n)-cores for any type of constraint satisfaction problem. The value of such characterizations is that they clearly identify where enforcing a certain type of local consistency will decide the problem: Enforcing $(1, 1)$-consistency will decide the fixed point property within any class in which the fixed point property is equivalent to $\mathcal{I}$-dismantlability. This includes the classes of ordered sets of width 2 (see Theorem 4.34) or of height 1 (see Theorem 4.37). Enforcing $(1, 1)$-consistency will not decide the fixed point property in any class that contains non-$\mathcal{I}$-dismantlable ordered sets with the fixed point property.

Open Question 5.58. *For which constraint satisfaction problems and what level of consistency are there satisfactory characterizations of empty $(2, n)$-cores similar to Theorem 5.57?*

Now we can turn to retractable points. Note that, although we will be able to establish Theorem 5.60 below, it is unknown if there is any level of consistency for which enforcing this level of consistency on $FPF(P)^{\exp}$ produces an empty network when P is connectedly collapsible.

Proposition 5.59. *Let P be an ordered set, let $a \in P$ be retractable to $b \in P$, and let $r : P \to P \setminus \{a\}$ be the retraction that maps a to b. Let the network H be obtained from $FPF(P)^{\exp}$ by, if the vertex (b, a) is present, removing all edges incident with the vertex (b, a). Then H can be reduced, by enforcing $(1, 1)$-consistency, to the network K that is obtained from $FPF(P)^{\exp}$ by removing all edges incident with vertices (x, a) so that x is comparable to b, but not comparable to a. Moreover, the network K is essentially controlled via r by $FPF(P \setminus \{a\})^{\exp} = K|_{P \setminus \{a\}}$.*

Proof. If a is irreducible in P, by Proposition 5.54, then $H = FPF(P)^{\exp}$ can be reduced, by enforcing $(1, 1)$-consistency, to a network K that is essentially controlled via r by $FPF(P \setminus \{a\})^{\exp} = K|_{P \setminus \{a\}}$.

If a is not irreducible, then a is retractable to b, a is not comparable to b, and the vertex (b, a) is an isolated vertex in H. (Note that, in this case, $H \neq FPF(P)^{\exp}$.) Let $x \in P \setminus \{b\}$ be comparable to b and not comparable to a. Suppose, for a contradiction, that there is an edge from the vertex (x, a) to a vertex (b, v). Then, because (b, a) is an isolated vertex, v is not equal to a. Hence v is comparable to a and not equal to a, which means v is comparable to b, a contradiction. Therefore, all vertices (x, a) so that x is comparable to b, but not comparable to a, are not $(1, 1)$-consistent. By definition, all edges incident with these vertices are not $(1, 1)$-consistent. Let K be the network obtained from H by removing these edges. Note that $K|_{P \setminus \{a\}} = FPF(P)^{\exp}|_{P \setminus \{a\}} = FPF(P \setminus \{a\})^{\exp}$.

To prove that K is essentially controlled via r by $FPF(P \setminus \{a\})^{\exp}$, we must show that we do not encounter the problem in Remark 5.49. Let $x, u \in P \setminus \{a\}$ and let $y, v \in P$ be so that $\{(x, y), (u, v)\}$ is an edge in K. If neither of y and v is equal to a, then $\{(x, r(y)), (u, r(v))\} = \{(x, y), (u, v)\}$ is an edge of the network $FPF(P \setminus \{a\})^{\exp}$. If $y = a$, then, by construction of K, x is not comparable to b and hence (x, b) is a vertex of $FPF(P \setminus \{a\})^{\exp}$. Similarly, if $v = a$, then u is not comparable to b and (u, b) is a vertex of $FPF(P \setminus \{a\})^{\exp}$. Hence, by definition of the expanded constraint network for the fixed point property, the edge $\{(x, r(y)), (u, r(v))\}$ is an edge of $FPF(P \setminus \{a\})^{\exp}$. Thus K is essentially controlled via r by $FPF(P \setminus \{a\})^{\exp}$. ∎

Proposition 5.59 shows that, if a is retractable to b in P, then, for analysis of $FPF(P)^{\exp}$, the vertex (b,a) is a natural target. Proposition 5.59 is also consistent with the fact that, if a is retractable to b and not irreducible, then there is either a fixed point free order-preserving self-map on the retract or there is such a map that switches a and b (see the proof of Theorem 4.12).

Theorem 5.60. *Let P be an ordered set, let $a \in P$ be retractable to $b \in P$, and let $n \geq 1$. Then the $(2,n)$-core of $FPF(P)^{\exp}$ is empty iff*

5-1. If the vertex (b,a) is present, then $FPF(P)^{\exp}$ can be reduced, by enforcing $(2,n)$-consistency, to a graph that has no edges incident with the vertex (b,a).
5-2. The $(2,n)$-core of $FPF(P \setminus \{a\})^{\exp}$ is empty.

The same result holds for $(1,n)$-consistency.

Proof. In case a is comparable to b, the result follows from Theorem 5.55. Hence, we can assume that a and b are not comparable.

For the "$\Rightarrow$" direction, note that part 5-1 holds trivially and that part 5-2 follows from Theorem 5.43.

For the "$\Leftarrow$" direction, we argue as follows. By part 5-1, $FPF(P)^{\exp}$ can be reduced, by enforcing $(2,n)$-consistency, to a graph H' which is contained in the graph H that is obtained from $FPF(P)^{\exp}$ by removing all edges incident with the vertex (b,a). By Proposition 5.59, H can be reduced, by enforcing $(1,1)$-consistency, and hence (via Exercise 5-31) by enforcing $(2,n)$-consistency, to the network K that is obtained from $FPF(P)^{\exp}$ by removing all edges incident with vertices (x,a) so that x is comparable to b, but not comparable to a. Moreover, the network K is essentially controlled via r by $FPF(P \setminus \{a\})^{\exp} = K|_{P \setminus \{a\}}$. By part 5-2, the $(2,n)$-core of $FPF(P \setminus \{a\})^{\exp}$ is empty. Now, by Theorem 5.52, the $(2,n)$-core of K is empty, and hence the $(2,n)$-core of H is empty. Therefore, by Theorem 5.36, the $(2,n)$-core of H' is empty and hence the $(2,n)$-core of $FPF(P)^{\exp}$ is empty.

The proof for $(1,n)$-consistency is left as Exercise 5-42. ∎

Although Theorem 5.60 is an "if and only if," it is unsatisfying, because there is no guarantee that the removal of edges incident with (b,a) can be achieved. This leaves us with some interesting open questions.

Open Question 5.61. *Let P be a connectedly collapsible ordered set. Are there $m,n \in \mathbb{N}$ so that the (m,n)-core of $FPF(P)^{\exp}$ empty? Is there an $n \in \mathbb{N}$ so that the $(2,n)$-core of $FPF(P)^{\exp}$ empty? Is the $(2,1)$-core of $FPF(P)^{\exp}$ empty?*

There is hope for a positive answer to the questions in 5.61. In [288], it is shown that it can be determined in polynomial time if an ordered set of interval dimension 2 (see Definition 11.18) has the fixed point property. Moreover, the result below shows that, for connectedly collapsible ordered sets of height at most 2, enforcing $(2,1)$-consistency will suffice.

Proposition 5.62. *Let P be a connectedly collapsible ordered set of height at most 2. Then the $(2,1)$-core of $FPF(P)^{\exp}$ is empty.*

Proof. If the height of P is 1, then P is $\mathcal{I}$-dismantlable and the result follows from Theorem 5.57.

If the height of P is 2, we proceed by induction on $|P|$. Let $a \in P$ be retractable to $b \in P \setminus \{a\}$. If a is irreducible, then, by Theorem 5.55, the $(2,1)$-core of $FPF(P)^{\exp}$ is empty iff the $(2,1)$-core of $FPF(P \setminus \{a\})^{\exp}$ is empty. If a is not irreducible, then, using Theorem 5.57, it can be shown (see Exercise 5-43) that enforcing $(2,1)$-consistency removes all edges incident with (b,a). In either case, the result is established by induction. ∎

Exercises

5-41. Prove Theorem 5.55 for $(1,n)$-cores.

5-42. Prove Theorem 5.60 for $(1,n)$-consistency.

5-43. Provide a detailed proof of Proposition 5.62.

5-44. $(2,1)$-consistency is also called **path consistency**. This exercise explains why. In a $(2,1)$-consistent CSP, let $\{(x,u),(y,v)\}$ be a consistent instantiation of two variables that share a constraint. Let $x = x_0, \ldots, x_k = y$ be a path from x to y in the constraint graph (see Exercise 5-45). Prove that there are instantiations $\{(x_i, y_i)\}$ (with $y_0 = u$ and $y_k = v$) such that, for all $i \in \{1, \ldots, n\}$, we have that $\{(x_{i-1}, y_{i-1}), (x_i, y_i)\}$ is consistent.

5-45. (See [99], p. 26.) A CSP is called k-**consistent** iff, for each consistent instantiation of $(k-1)$ variables and any choice of a kth variable, there is an instantiation of the kth variable such that the k instantiations together are consistent, too. A CSP is **strongly k-consistent** iff, for all $1 \le j \le k$, the CSP is j-consistent.

A nontrivial constraint between two variables is a constraint that declares at least one instantiation of the two variables as inconsistent. The **constraint graph** of a binary CSP with variables $x_1, \ldots, x_r$ has vertex set $V = \{x_1, \ldots, x_r\}$ and there is an edge between x_i and x_j iff there is a nontrivial constraint between these two variables, that is, $E = \{\{x_i, x_j\} : C_{ij} \text{ is not trivial }\}$.

The **width** of the constraint graph is the smallest integer k such that there is a linear ordering $x_{i_1}, \ldots, x_{i_r}$ of the vertices such that, for each d, the vertex x_{i_d} is adjacent to at most k vertices x_{i_j} with $j < d$.

Prove that, if, for a CSP, the level of strong k-consistency is larger than the width of the constraint graph, then there exists a backtrack-free search order for the CSP.

Note. The enforcing of consistency changes the constraints and hence, typically, the constraint graph and its width. Therefore, the hope that, if we start with a constraint graph of width k, then, after enforcing $(k+1)$-consistency, there is a backtrack-free search order, is a false one: This will only work if the width of the constraint graph did not increase during the processing, which typically does not happen.

5-46. For any finite ordered set P we define the ordered set P_{2lc} as follows.

a. For every $x \in P$, let c_x be the number of lower covers of x and let $l_1^x, \ldots, l_{c_x}^x$ be an enumeration of these lower covers.

b. To build the underlying set for P_{2lc}, for every $x \in P$ add points $d_1^x, \ldots, d_{c_x-2}^x$ to P.

c. To build the order relation for P_{2lc}, add the following comparabilities to the comparabilities in P:

    i. The unique upper cover of d_1^x is x.

    ii. For $i \in \{2, \ldots, c_x - 2\}$ the unique upper cover of d_i^x is d_{i-1}^x.

    iii. For $i \in \{1, \ldots, c_x - 3\}$, d_i^x has exactly two lower covers, namely, l_{i+1}^x and, as forced by part 5-46(c)ii, d_{i+1}^x.

    iv. $d_{c_x-2}^x$ has exactly two lower covers, namely, $l_{c_x-1}^x$ and $l_{c_x}^x$.

    v. Plus all the comparabilities forced by transitivity and reflexivity.

    Prove each of the following.

    a. P is an ordered subset of P_{2lc}.

    b. P has the fixed point property iff P_{2lc} has the fixed point property.

    c. Every element of P_{2lc} has at most two lower covers.

    d. $|P_{2lc}| \leq |P|^2$.

5-47. (Another CSP for FPP.) For determining if the ordered set P has a fixed point free order-preserving self-map, let the variable set be equal to P, that is, $\{x_1, \ldots, x_n\} := P$. For each x_i the value domain D_i is $D_i := \{p \in P : p \not\leq x_i \text{ and } p \not\geq x_i\}$.

    The constraints C_{ij} are

- If $x_i \not< x_j$ and $x_i \not> x_j$, there is no constraint between x_i and x_j.
- If $x_i \prec x_j$, let C_{ij} be $C_{ij} := \{(y_i, y_j) \in D_i \times D_j : y_i \leq y_j\}$.

    Prove each of the following.

    a. Any consistent instantiation of all the variables corresponds to a fixed point free order-preserving self-map of P.

    b. When applied to the set P_{2lc} of Exercise 5-46, the constraint graph (see Exercise 5-45) of the above CSP has at most width 2.

5-48. Explain why going from P to P_{2lc} as in Exercise 5-46, then enforcing 3-consistency for the corresponding CSP from Exercise 5-47, and then using backtracking do not lead to a polynomial algorithm to determine the fixed point property.

Remarks and Open Problems

Open questions abound in an area as vast as algorithms in general and constraint satisfaction in particular.

1. For a start on the body of work on transitive closures, consider [116].
2. There is an $O(n^{2.5})$ algorithm to compute the width of an ordered set. This is done via Dilworth's Theorem through a translation to a matching problem as in [319], p. 274 and then using an $O(n^{2.5})$ algorithm for matching, see [143]. (This is also mentioned in [30]. p. 259.) [319] provides many connections between graph-theoretical and order-theoretical parameters.
3. The current "standard reference" on CSPs appears to be [260].
4. For an excellent exposition on search algorithms for CSPs, consider [170].
5. Results and references in [9, 170, 229], as well as, in some measure, Exercise 5-23e, indicate why forward checking is generally (though not always) preferable to backtracking.
6. Another search paradigm has gained in importance. It is called Maintaining Arc Consistency, see [104, 201, 267]. In this algorithm, the consistent future assignments are generated and stored just as in forward checking. However, to achieve further pruning of the tree, a $(1, 1)$-consistency algorithm is run on the set of consistent future assignments every time we go deeper into the tree.

7. For some results on variable ordering heuristics, see [119, 226, 313].

8. Is there an algorithm that determines in polynomial time whether an ordered set of width 3 has the fixed point property? I conjecture that a combination of local consistency enforcement and Proposition 4.38 should point the way towards such an algorithm.

9. Is it possible to determine in polynomial time if a given finite ordered set is (connectedly) collapsible? I was only able to show this for fixed height (see [272]). Moreover, the natural way to check (connected) collapsibility can lead to unavoidable exponential effort (see [289]).

10. In [142] and in other papers in the same journal volume, the **phase transition** for constraint satisfaction problems is described. For the following, it is helpful to assume that there is a parameter that is linked to how constrained a problem is. Say, the "average instance" of the problem becomes more constrained as the parameter grows. (Such a parameter often exists.)

    In essence, for any type of constraint satisfaction problem, problems with "few" constraints or "many" constraints are easy to solve: In the former case, usually a solution is found almost immediately, in the latter case, the problem usually is so overconstrained, that the backtracking tree is of very small size. Hard problems can be found in abundance in the region in between, where the problem changes from highly underconstrained to highly overconstrained. This is the region of the phase transition. It often coincides with the region where the probability of being solvable changes from 1 to 0. (Plotting the probability of being solvable against the parameter shows a steep drop.) Problems in this region have few enough solutions that search algorithms are not likely to find a solution fast and their backtracking trees are large.

    What is the region of the phase transition for determining if a given ordered set has a fixed point free order-preserving self-map? What parameter would we use?

11. For any finite ordered set P, the following is true. Let G be a $(1, 1)$-consistent subgraph of $FPFMAP(P)$ that has a nonempty edge set. Then

    $$\Phi(x) := \{u : (x, u) \text{ is not isolated in } G\}.$$

    is a fixed point free isotone relation. What additional properties does Φ have if G is obtained via a $(2, 1)$-consistency algorithm or via stronger consistency enforcing algorithms? Can these properties be used to characterize those ordered sets P which have no fixed point free order-preserving map and for which a $(2, 1)$-consistency algorithm applied to G erases all edges?

12. In [149], it is shown that any constraint satisfaction problem corresponds to a pair of relational structures in such a way that the solutions correspond to the homomorphisms between the structures. As we have seen, existence of specific homomorphisms and the number of homomorphisms are studied extensively in order theory. Can advances in this direction be translated into results on general constraint satisfaction problems? (Also see Remark 15.)

13. Any algorithm that finds all solutions to a constraint satisfaction problem can be used to enumerate all endomorphisms of an ordered set. This can then be used to compute, for example, the spectrum (see [111]) and other data on the ordered set in question. Because the number of endomorphisms grows rapidly, this is only feasible for sets of small size. Still, experiments could point the way towards progress on the automorphism conjecture or towards finding the ordered set with the fewest endomorphisms for a given size (also see [280]).

14. In Exercise 7-20 we give a condition that ensures that convex subsets of ordered sets intersect. The fact that any set of pairwise intersecting linear intervals has a nonempty intersection has been used in [315]. It is shown there in Theorem 3.2 that, for so-called row-convex constraints, $(2, 1)$-consistency implies consistency. (We must be careful, though. Both conditions must be satisfied. There are row-convex CSPs that lose row-convexity when $(2, 1)$-consistency is enforced.)

   Is there an adaptation of the above result (together with Exercise 7-20) that can be used to prove polynomial decidability of the fixed point property in sets of small width? This would require a way to address the problem outlined in and after Theorem 5.6 in [315].

15. Algorithms for constraint satisfaction problems can be used to enumerate the order-preserving self-maps of an ordered set and to obtain bounds on the quotient in Open Question 2.14. For larger sets, estimation algorithms such as proposed in [45, 119, 168, 231] might at least give an estimate of the number of solutions. The algorithms in [45, 168, 231] perform a randomized partial search. Their primary purpose is to estimate the run time. However, the estimate for the number of consistent nodes at maximum depth could be used as an estimate for the number of solutions. This estimate through randomized algorithms is normally bad in the region of the phase transition (for phase transitions, see [142]). When counting order-preserving maps, solutions abound and the problem is well away from the phase transition region. Thus there is hope that estimates may be reasonable, especially for ordered sets of small height. (Chains and tall sets can be problematic for these algorithms.)

   The idea for [119] is probabilistic. Consider a constraint satisfaction problem with n variables and value domains of size m. For each constraint $C_{i,j}$, the number of consistencies divided by the total number of entries in the matrix representation of $C_{i,j}$ gives a probability $p_{i,j}$ for this constraint to be satisfied. The expected number of solutions then is $m^n \prod_{i=1}^{n} \prod_{j=i+1}^{n} p_{ij}$. Unfortunately, this model only applies to randomly seeded problems where every entry of $C_{i,j}$ is equally likely to be a 1. Even refinements so far have only been useful in randomly seeded problems of different types. How promising is the refinement below in terms of making the probabilistic approach more widely applicable?

   To refine the model, one can add all matrices $C_{i,j}$ with $i < j$ to obtain the sum C^* (recall that all $C_{i,j}$ are $m \times m$ matrices). Rescale C^* to obtain the matrix C by dividing C^* by its largest entry. C gives a measure of how much the consistencies in the problem are clustered. (In the uniform probability model, all entries of C would be close to 1.) One can then replace the number of entries of $C_{i,j}$ in the computation of $p_{i,j}$ with the sum of the entries of C. The factor m^n in the computation needs to be replaced with an estimate on the number of solutions of a constraint satisfaction problem "where all constraints are equal to C." Because C has rational numbers as entries, this can be done with a probabilistic variation on the methods of [168]. Small entries in C give small probability to there being a consistency. Experiments in prediction and analysis of randomly seeded constraint satisfaction problems beyond the simple uniform model were encouraging. Indeed, the probabilistic estimates give a certain likelihood whether the problem is solvable. This likelihood remains a fairly good predictor even in some regions where a randomized search algorithm fails.

The main place for improvement of the probabilistic approach currently seems to be the computation of the analogue of the factor m^n through C. This is a chance for feedback from ordered sets. If an ordered set is labeled in such a way that $i < j$ implies $x_i \not> x_j$, then all $C_{i,j}$ with $i < j$ have the same matrix representation. Thus these problems are among the simplest prototypes for trying to estimate the analogue of the factor m^n through C. Indeed, this factor is nothing but the total number of solutions. Therefore, conversely to the idea at the beginning of this remark, any advance on estimating the number of endomorphisms of an ordered set could help in estimating the number of solutions of a constraint satisfaction problem in the above fashion. The first step consists of problems where all constraints have the same matrix (maps from chains to ordered sets if there are no trivial constraints). Subsequently the method would need to be adapted to allow for rational entries using probabilistic modifications.

16. With the help of several students, I have implemented a multitude of reduction and search algorithms in C++ with a WINDOWS front end, see [286].
17. Limitations to enforcing consistency: It is NP-hard to find a solution in an expanded constraint network, even if every edge is part of a solution (see [110]).

Chapter 6
Graphs and Homomorphisms

Aside from ordered sets, the fixed point property has been investigated in other settings. On one hand, the fixed point property is most likely originated in topology. (See Exercise 6-1 for the topological fixed point property.) On the other hand, in any branch of mathematics in which the underlying structures have a natural type of morphism, we can define a fixed point property as "every endomorphism has a fixed point." Hence, graphs are natural discrete structures for which to investigate the fixed point property. However, we must be careful: For graphs, the natural analogues of order-preserving maps would be the simplicial homomorphisms from Definition 6.1. Nonetheless, graph theorists prefer the notion of a (graph) homomorphism from Definition 6.10. It would be fruitless to argue whether simplicial homomorphisms or graph homomorphisms are "more natural." Each notion is natural in certain settings: For example, simplicial homomorphisms are in one-to-one correspondence with the simplicial maps of the clique complex of a graph, as we will see in Exercise 9-19 when we investigate simplicial complexes, whereas we will see in this chapter that graph homomorphisms are a good framework to investigate natural graph theoretical topics, such as, for example, colorings.

Because simplicial homomorphisms are more closely connected to the topic of this text, we start with simplicial homomorphisms. Remember that graphs were already introduced in Definition 1.4.

6.1 A Fixed Edge Theorem for Simplicial Homomorphisms

Similar to order-preserving maps, simplicial homomorphisms preserve edges or map both endvertices of an edge to a single vertex. Definition 9.16 and Exercise 9-19 will show the motivation for the terminology "*simplicial* homomorphism." This introductory section will solely focus on fixed vertices and fixed edges.

© Springer International Publishing 2016

B. Schröder, *Ordered Sets*, DOI 10.1007/978-3-319-29788-0_6

Fig. 6.1 A 5-cycle

Definition 6.1. *Let $G = (V, E)$ and $H = (W, F)$ be graphs. A function $f : V \to W$ is called a **simplicial homomorphism** iff, for all vertices x, y with $\{x, y\} \in E$, we have $f(x) = f(y)$ or $\{f(x), f(y)\} \in F$. A simplicial homomorphism from a graph to itself is called a **simplicial endomorphism**.*

Unfortunately, a fixed point property for graphs would be quite trivial. No graph that has an edge has the fixed point property for simplicial endomorphisms: Indeed, if $\{a, b\}$ is an edge in a graph G, then the map that maps all vertices not equal to a to a and that maps a to b is a fixed point free simplicial endomorphism. Theorem 6.3 below characterizes the graphs for which every simplicial endomorphism fixes a vertex or an edge. Section 6.2 will consider the property that every simplicial endomorphism fixes a clique.

Crowns in ordered sets are close relatives to cycles in graphs, with one main difference being that cycles can have an odd number of vertices (see Figure 6.1).

Definition 6.2. *Let $G = (V, E)$ be a graph. Then G has a **cycle** iff there is a subset $\{c_1, \dots, c_n\} \subseteq V$ with $n \geq 3$ such that, for $k = 1, \dots, n - 1$, the vertex c_k is adjacent to c_{k+1}, and the vertex c_n is adjacent to c_1. Cycles with n vertices (and no edges beyond the ones specified in the preceding sentence) are also called n-**cycles**.*

Clearly, every n-cycle has a simplicial endomorphism that does not fix any edges or vertices (see Exercise 6-2a). Note however, that, unlike in ordered sets, a cycle in a graph could have more edges than just the ones prescribed by Definition 6.2. If we want to assure that a graph that is contained in another graph has *exactly* the edges inherited from the surrounding graph, we will speak of an *induced* subgraph (see Definition 6.14).

Theorem 6.3 (See [218]). *Let $G = (V, E)$ be a finite graph. Then every simplicial endomorphism of G has a fixed edge or a fixed vertex iff G is connected and has no cycles with ≥ 3 vertices.*

Proof. Let $G = (V, E)$ be a graph. We define an ordered set of height 1 with underlying set $S_G := \{\{v\} : v \in V\} \cup E$ and with the order being containment of sets. (This set is sometimes called the **split** of the graph.) Let $f : V \to V$ be a simplicial endomorphism of G. Then

$$F(x) := \begin{cases} \{f(v)\}; & \text{if } x = \{v\}, \\ \{f(v), f(w)\} & \text{if } x = \{v, w\} \end{cases}$$

is an order-preserving map of S_G. Moreover, if F has a fixed point $\{v\}$ that is minimal in S_G, then f fixes v. If F fixes no singleton, but has a fixed point $\{v, w\}$ that is maximal in S_G, then f fixes $\{v, w\}$. Thus, if S_G has the fixed point property, then every endomorphism of G fixes an edge or a vertex.

Conversely, let $h : S_G \to S_G$ be an order-preserving map of S_G that maps the minimal elements to minimal elements. If, for $x \in V$, we let $H(x)$ be the unique element of the singleton set $h(x)$, then H is a simplicial endomorphism of G. Moreover, if H has a fixed vertex v, then clearly h has a fixed point $\{v\}$, and if H has a fixed edge $\{v, w\}$, then $h[\{\{v\}, \{w\}\}] = \{\{v\}, \{w\}\}$ and hence $h(\{v, w\}) = \{v, w\}$. Thus if every simplicial endomorphism of G has a fixed vertex or a fixed edge, then every order-preserving map of S_G that maps minimal elements to minimal elements has a fixed point. Consequently, by Proposition 3.1, in this case S_G has the fixed point property.

Hence S_G has the fixed point property iff every simplicial endomorphism of G has a fixed edge or a fixed vertex. By Theorem 4.37, S_G has the fixed point property iff S_G is connected and has no crowns. This is the case iff G is connected and has no cycles with ≥ 3 elements.   ■

Exercises

6-1. A **topology** is a set τ of subsets of a set X, such that

- $\emptyset, X \in \tau$.
- If $A \subseteq \tau$, then $\bigcup A \in \tau$.
- If $A_1, \ldots, A_n \in \tau$, then $\bigcap_{i=1}^{n} A_i \in \tau$.

The pair (X, τ) is called a **topological space**. A function f from a topological space (X, τ) to a topological space (X', τ') is called **continuous** iff, for every $U \in \tau'$, we have that $f^{-1}[U] \in \tau$. A topological space (X, τ) has the **(topological) fixed point property** iff every continuous function $f : X \to X$ has a fixed point $x = f(x)$.

A topology τ is called T_0 iff, for all distinct points $x, y \in X$, there is an $A \in \tau$ such that $x \in A, y \notin A$ or $y \in A, x \notin A$.

a. Prove that, if a topological space has the fixed point property, then its topology must be T_0. (See [304].)
b. Let X be a finite set.

   i. Prove that the number of topologies on X is equal to the number of reflexive and transitive relations (preorders) on X.
   ii. Prove that the number of T_0-topologies on X is equal to the number of orders on X. (See [301].)

6-2. Let C_n be an n-cycle.

a. Prove that C_n has a simplicial endomorphism that does not fix any edges or vertices.
b. Prove that every non-surjective simplicial endomorphism of C_n fixes a vertex or an edge.

6-3. Let $G = (V, E)$ be an infinite graph. An **infinite path** is a sequence $\{p_n\}_{n=1}^{\infty}$ of vertices so that $\{p_i, p_{i+1}\} \in E$ for all $i \in \mathbb{N}$. Prove that every simplicial endomorphism of G has a fixed edge or a fixed vertex iff G is connected and has no cycles with ≥ 3 vertices and no infinite paths.

Hint. Exercise 4-39.

6.2   The Fixed Clique Property for Simplicial Endomorphisms

We have seen in Section 6.1 that the fixed point property for simplicial endomorphisms of graphs is not very interesting. Theorem 6.3 showed that, if we consider fixed vertices and fixed edges, then the theory becomes a little more interesting. Vertices and edges are the smallest representatives of the class of complete graphs. So, fixing complete graphs is a natural next step.

Definition 6.4 (See [16], p. 10; for earlier results, see [48], Section 1, [218, 223, 233]; for more recent contributions, see [200, 222]).   *Let $G = (V, E)$ be a graph without infinite cliques. We will say that G has the **fixed clique property** iff, for each simplicial endomorphism $f : V \to V$, there is a clique K of G with $f[K] = K$.*

*We will say that an arbitrary graph G has the **invariant clique property** iff, for each simplicial endomorphism $f : V \to V$, there is a clique C such that $f[C] \subseteq C$.*

The order-theoretical analogue of a clique is a chain. By the Abian–Brown Theorem, if an order-preserving function maps a complete chain to itself, then it must have a fixed point. Therefore, in order theory, a "fixed chain property" is not investigated: In the most interesting cases, it is equivalent to the fixed point property, which is easier to formulate. (Also see Exercise 6-4.) For graphs, the situation is quite different.

Example 6.5.

1. Any finite complete graph has the fixed clique property.
2. Any complete graph has the invariant clique property, but infinite complete graphs do not have the fixed clique property.
3. By Theorem 6.3, any finite connected graph that has no cycles has the fixed clique property.   □

The fixed clique property also is interesting for certain graphs related to ordered sets.

Definition 6.6.   *Let P be an ordered set. Let $G = (V, E)$ be the graph with vertices $V = P$ and edges $E := \{\{p, q\} : p \neq q, p \sim q\}$. Then G is called the **comparability graph** $G_C(P)$ of P.*

The question which graphs are comparability graphs is answered in each of the papers [101, 105, 108]. The simplest characterization appears to be the characterization in [108]. It says that a graph G is a comparability graph iff, for every $2k + 1$-cycle $c_0 \sim c_1 \sim c_2 \sim \cdots \sim c_{2k-1} \sim c_{2k} \sim c_0$ in G, there is a j so that $c_j \sim c_{j+2}$ (index arithmetic modulo $2k + 1$). The "$\Rightarrow$"-direction can be proved in Exercise 6-5. The "$\Leftarrow$"-direction is a bit longer, see [108]. For an in-depth treatment of comparability graphs, consider the survey [154].

As long as the Abian–Brown Theorem is available, the fixed clique property for the comparability graph implies the fixed point property for the ordered set.

Proposition 6.7. *Let P be a chain-complete ordered set. If $G_C(P)$ has the invariant clique property, then P has the fixed point property.*

Proof. Let $f : P \to P$ be an order-preserving map. Then f is a simplicial endomorphism of the comparability graph $G_C(P)$ (see Exercise 6-12). Therefore there is a clique C in $G_C(P)$ so that $f[C] \subseteq C$. In particular, this means that there is a $p \in P$ such that $f(p) \sim p$. By the Abian–Brown Theorem, f must have a fixed point.   ∎

The question beckons if the fixed clique property for the comparability graph is equivalent to the fixed point property for the ordered set. Unfortunately, this is not the case.

Example 6.8. *Even for finite ordered sets, the fixed clique property for the comparability graph is not equivalent to the fixed point property for the ordered set.* Consider the ordered set in part b) of Figure 1.1. The map f such that $a \mapsto i \mapsto b \mapsto k \mapsto a$, $c \leftrightarrow h$, and $d \mapsto g \mapsto e \mapsto f \mapsto d$ is an endomorphism of the comparability graph that does not fix any cliques.   □

If this map seems "strange" to you, consider the following. Visual work with the fixed clique property is made challenging by the fact that there is no notion of "up" or "down" in a graph. Thus, if you are used to working with fixed points in ordered sets, the usual reference framework is gone. However, it is tempting to try to maintain the up-down direction of the ordered set, although it has no meaning in the comparability graph. Moreover, especially when going from an ordered set to its comparability graph, it is tempting to forget the comparabilities induced by transitivity, which are omitted in the Hasse diagram.

Ultimately, simplicial homomorphisms of graphs can be viewed as simplicial maps (see Definition 9.16) between special simplicial complexes (see Exercise 9-19). Hence the discussion of the fixed clique property is best continued within the realm of simplicial complexes and the fixed simplex property. This will be done starting in Section 9.4. As we end our brief excursion into the fixed clique property, a word of warning is in order.

Example 6.9. The **Abian–Brown Theorem fails** for the fixed clique property. Indeed, on a 4-cycle $c_1 \sim c_2 \sim c_3 \sim c_4 \sim c_1$ with no further adjacencies, for the function $c_1 \mapsto c_2 \mapsto c_3 \mapsto c_4 \mapsto c_1$, every vertex is adjacent to its image and there is no fixed clique. Exercise 6-7 shows that the failure is even more spectacular.   □

Despite weaker versions being available (see Exercise 6-8), this fact very much complicates the investigation of the fixed clique and fixed simplex properties.

Fig. 6.2 A graph for which
every simplicial
endomorphism f has a vertex
with $x \sim f(x)$ or $x = f(x)$ and
which does not have the fixed
clique property

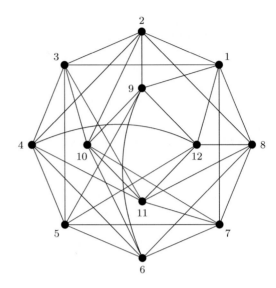

Exercises

6-4. Let P be an ordered set. Say that P has the fixed chain property iff, for every order-preserving
self map $f : P \rightarrow P$, there is a chain $C \subseteq P$ so that $f[C] \subseteq C$.

    a. Prove that, for chain-complete ordered sets, the fixed chain property is equivalent to the
fixed point property.

    b. Prove that, for ordered sets that are not chain-complete, the fixed chain property is
equivalent to the comparable point property.

6-5. Let G be a comparability graph and let $c_0 \sim c_1 \sim c_2 \sim \cdots \sim c_{2k-1} \sim c_{2k} \sim c_0$ be a
$2k+1$-cycle in G. Prove that there is a j so that $c_j \sim c_{j+2}$ (index arithmetic modulo $2k+1$).

6-6. Let $G = (V, E)$ be a graph. Functions $r : V \rightarrow V$ so that $r^2 = r$ are called **simplicial
retractions**. Let $r : V \rightarrow V$ be a simplicial retraction.

    a. Prove that, if G has the fixed clique property, then the induced subgraph $G[r[V]]$, which
consists of the vertices in $r[V]$ and all edges in E between vertices in $r[V]$, has the fixed
clique property, too.

    b. Prove that no graph that contains an infinite clique has the fixed clique property.

6-7. Failure of the Abian–Brown Theorem for the fixed clique property.

    a. Prove that the graph in Figure 6.2 does not have the fixed clique property.

    b. Prove that every simplicial endomorphism f of the graph in Figure 6.2 has a vertex with
$x \sim f(x)$ or $x = f(x)$.

        Note. A verification with a computer program such as [286] should be acceptable
here. A written proof is likely to be very tedious and I have never succeeded in finding a
simple one.

6-8. (Abian–Brown type theorems for the fixed clique property.) Let $G = (V, E)$ be a graph and
let $f : V \rightarrow V$ be a graph endomorphism.

    a. Prove that, if there is a finite clique C with $f[C] \subseteq C$, then there is a clique $K \subseteq C$ with
$f[K] = K$.

    b. Define $N[v] := \{v\} \cup \{w \in V : \{v, w\} \in E\}$. Let $v \in V$ be such that $f[N[v]] \subseteq N[v]$.
Prove that there is a clique $K \subseteq V$ such that $f[K] \subseteq K$.

6-9. Let $G = (V, E)$ be a finite graph. Then a vertex $a \in V$ is called **dominated** by $b \in V \setminus \{a\}$ iff $b \sim a$ and, for all $v \in V \setminus \{a, b\}$, $v \sim a$ implies $v \sim b$. Prove that, if a is dominated by b, then G has the fixed clique property iff $G[V \setminus \{a\}]$ has the fixed simplex property.
   Hint. Use Exercise 6-8b.

6-10. A family $\mathcal{S}$ of sets is said to have the **Helly property** iff, for all $\mathcal{S}' \subseteq \mathcal{S}$, the following holds: If, for all $A, B \in \mathcal{S}'$, we have that $A \cap B \neq \emptyset$, then $\bigcap \mathcal{S}' \neq \emptyset$. For two distinct vertices v, w in a graph $G = (V, E)$, we define $d(v, w)$ to be the smallest $n \in \mathbb{N}$ so that there are $p_0, \ldots, p_n$ so that $v = p_0 \sim p_1 \sim \cdots \sim p_n = w$. We define $d(v, v) := 0$. For a vertex v in a graph $G = (V, E)$ and $r \in \mathbb{N}$, we define $B(v, r) := \{w \in V : d(v, w) \leq r\}$. A graph $G = (V, E)$ is called a **Helly graph** iff the family $\{B(x, r) : x \in V, r \in \mathbb{N} \cup \{0\}\}$ has the Helly property.

   a. Prove that every Helly graph has a dominated vertex.
       Hint. Let $v_0 \in V$ and let $r \in \mathbb{N}$ be so that $B(v_0, r) = V$ and $B(v_0, r - 1) \neq V$. Prove that every $w_0 \in B(v_0, r) \setminus B(v_0, r - 1)$ is dominated by proving that the intersection $\bigcap_{w \sim w_0} B(w, 1) \cap B(w_0, 1) \cap B(v_0, r - 1)$ is not empty.
   b. Prove that every Helly graph has the fixed clique property.
       Hint. Prove that removing a dominated vertex from a Helly graph produces a Helly graph and use Exercise 6-9.

6-11. Let G be a finite graph in which the longest cycle (induced or otherwise) has 3 elements. Prove that G has the fixed clique property.

6-12. Let P, Q be ordered sets and let $f : P \to Q$ be order-preserving. Prove that f is a simplicial homomorphism from $G_C(P)$ to $G_C(Q)$.

6-13. Let G be a k-colorable graph, that is, a graph so that there is a simplicial homomorphism from G to a k-clique so that no two adjacent vertices are mapped to the same image.

   a. Prove that G is contained in the comparability graph of an ordered set of height $k - 1$.
   b. Prove that G contains the covering graph (see Remark 3 at the end of Chapter 7) of an ordered set of height $k - 1$.

6-14. The Dushnik–Miller–Erdös Theorem and Corollary 2.29.

   a. Use the Dushnik–Miller–Erdös Theorem to prove Corollary 2.29.
   b. Let P be an ordered set without infinite antichains. Use the Dushnik–Miller–Erdös Theorem to prove that P contains a chain of cardinality $|P|$.
   c. Let P be an ordered set without infinite antichains. Prove, without using the Dushnik–Miller–Erdös Theorem, that P contains a chain of cardinality $|P|$.
       Hint. Mimic the argument in the proof of the Dushnik–Miller–Erdös Theorem.

6-15. Prove that, for graphs that contain a clique of size $\frac{n}{2}$, there is a fixed $c > 0$ so that the quotient of the number of automorphisms and the number of simplicial endomorphisms are $\leq c 2^{-\frac{7}{80} n}$.

6.3 (Graph) Homomorphisms

We now move on to the notion of (graph) homomorphisms, which is used more often in graph theory than that of simplicial homomorphisms. For a thorough presentation of (graph) homomorphisms, the text [135] is strongly recommended. Here, we will only touch upon the ideas that connect most closely to one of the main drivers of this text, namely, the fixed point property.

Definition 6.10. *Let $G = (V, E)$ and $H = (W, F)$ be graphs. Then the function $f : V \to W$ is called a* **homomorphism** *iff, for all $v, w \in V$, if $\{v, w\} \in E$, then $\{f(v), f(w)\} \in F$. Consistent with standard terminology, an* **endomorphism** *is a homomorphism from G to G, an* **isomorphism** *is a bijective homomorphism whose inverse is a homomorphism, too, and an* **automorphism** *is an isomorphism from G to G.*

From the point-of-view of ordered sets, we must be careful when working with graph homomorphisms, because they do not allow for two vertices that are joined by an edge to be mapped to the same image. Therefore, when we need to emphasize that edges are not allowed to be "collapsed," we will also refer to the functions from Definition 6.10 as **graph homomorphisms**.

Example 6.11. Let C_{2n+1} be a cycle with an odd number of vertices. Then all graph endomorphisms of C_{2n+1} are automorphisms (see Exercise 6-16).   ☐

The fact that edges cannot be "collapsed" leads to a situation that is impossible for ordered sets: There are pairs of graphs G and H so that there is no homomorphism from G to H. Because this is possible, we have the following definition.

Definition 6.12. *Let G, H be graphs. Then G is called* **homomorphic to H** *iff there is a homomorphism from G to H.*

We will now prove a necessary condition for a graph to be homomorphic to another. This will also expose us to some further ideas from graph theory.

Definition 6.13. *Let $G = (V, E)$ be a graph. Then we define the following.*

1. *$\omega(G)$ denotes the size of the largest clique in G. It is called the* **clique number** *of G.*
2. *$\alpha(G)$ denotes the size of the largest set of vertices in G so that no two vertices in the set are joined by an edge. It is called the* **independence number** *of G.*
3. *$\chi(G)$ is the smallest number so that the vertices of G can be* **colored** *with $\chi(G)$ colors so that no two adjacent vertices receive the same color. It is called the* **chromatic number** *of G.*

In the context of homomorphisms, $\chi(G)$ is the smallest $n \in \mathbb{N}$ so that there is a homomorphism from G to the complete graph K_n with n vertices. More importantly, each of these parameters leads to inequalities that must be satisfied if there is a homomorphism from one graph to another. If the inequality is not satisfied, then there is no homomorphism from the first graph to the second. The inequalities for the clique number and for the chromatic number are quite simple and they are given in Exercises 6-17a and 6-17b. For the independence number, the situation is a bit more complicated.

Definition 6.14. *Let $G = (V, E)$ be a graph. For a set of vertices $S \subseteq V$, the* **induced subgraph** *on the vertices of S is the graph $G[S] := (S, E_S)$ with edge set $E_S := \{\{v, w\} \in E : v, w \in S\}$.*

Induced subgraphs, like ordered subsets, are formed by focusing on a subset of the vertices and considering that subset with *all* the edges between vertices in the subset. Such subgraphs are specifically called *induced* subgraphs, because, for a graph $G = (V, E)$, any $H := (W, F)$ with $W \subseteq V$ and $F \subseteq E_W$ is called a **subgraph**. This notion is very sensible in graph theory, because, unlike comparabilities in an ordered set, the edges in a graph do not need to satisfy any additional criteria. In ordered sets, arbitrary comparabilities cannot be erased without running the risk of losing transitivity (see Exercise 6-20).

Definition 6.15. *Let G, H be graphs. We define $n(G)$ to be the number of vertices in G and we define $n(G, H)$ to be the maximum number of vertices in induced subgraphs of G that are homomorphic to H.*

Proposition 6.16. *Let G be a graph and let K_1 be the graph with a single vertex. Then $n(G, K_1) = \alpha(G)$.* ∎

Definition 6.17. *A graph $H = (W, F)$ is called **vertex-transitive** iff, for all vertices $v, w \in V$, there is an automorphism that maps v to w.*

Example 6.18. Cycles are vertex-transitive. □

Proposition 6.19 (See Proposition 1.22 in [135]). *Let G, H, K be graphs so that H is vertex-transitive. If G is homomorphic to H, then*

$$\frac{n(G, K)}{n(G)} \geq \frac{n(H, K)}{n(H)}.$$

Proof. Let $H_1, \ldots, H_q$ be the subgraphs of H with $n(H, K)$ vertices that are homomorphic to K. Because H is vertex-transitive, every vertex of H is contained in the same number, say, p, of the graphs H_i. Hence $qn(H, K) = pn(H)$.

Now let f be a homomorphism from G to H and, for $i = 1, \ldots, q$, let G_i be the preimage of H_i under f. Then G_i is homomorphic to K, which means that G_i has at most $n(G, K)$ vertices. Moreover, every vertex of G is in at least p of the graphs G_i. Hence

$$qn(G, K) \geq \sum_{i=1}^{q} n(G_i) \geq pn(G),$$

and then

$$\frac{n(G, K)}{n(G)} \geq \frac{p}{q} = \frac{n(H, K)}{n(H)}.$$

∎

The No-Homomorphism Lemma, which is often used to prove that there is no homomorphism between two graphs, is now an easy consequence

Corollary 6.20 (No-Homomorphism Lemma). *Let G be a graph and let H be a vertex-transitive graph. If G is homomorphic to H, then*

$$\frac{\alpha(G)}{n(G)} \geq \frac{\alpha(H)}{n(H)}.$$

■

Example 6.21. An odd cycle C_{2k+1} is homomorphic to another odd cycle $C_{2k'+1}$ iff $k \geq k'$ (see Exercise 6-18a). □

Exercises

6-16. Prove that all endomorphisms of an odd cycle are automorphisms.

6-17. Let G and H be graphs.

    a. Prove that, if there is a homomorphism from G to H, then $\omega(G) \leq \omega(H)$.

    b. Prove that, if there is a homomorphism from G to H, then $\chi(G) \leq \chi(H)$.

    c. Describe the properties that a pair of graphs G and H could have if we want that there is no homomorphism from G to H and no homomorphism from H to G.

6-18. Homomorphisms of cycles.

    a. Prove the claim in Example 6.21.

    b. Prove that even cycles are not homomorphic to odd cycles.

    c. Prove that even cycles are homomorphic to all cycles with ≥ 3 vertices.

6-19. Let $G = (V, E)$ be a graph and let $v \in V$. Then the **neighborhood** of v is the set of vertices $N(v) := \{w \in V : \{v, w\} \in E\}$.

    Let $G = (V, E)$ and $H = (W, F)$ be graphs, let f be a homomorphism from G to H, and let $v \in V$ be a vertex of G. Prove that $\chi(G[N(v)]) \leq \chi(H[N(f(v))])$ holds for the induced subgraphs on the neighborhoods.

6-20. Let P be an ordered set and let $x, u \in P$ be so that $x < u$. Prove that erasing the comparability (x, u) from the order relation $\leq$ produces another order relation iff u is an upper cover of x.

6.4   The Fixed Vertex Property for Graph Endomorphisms

At first, the situation for a fixed vertex property for graphs may look bleak. Although the construction described after Definition 6.1 does not work for graph endomorphisms in general, there are other negative indicators. For example, in ordered sets, finite chains have the fixed point property. However, complete graphs K_n with $n \geq 2$ vertices have endomorphisms that do not fix a single vertex: Any fixed vertex free permutation of the vertices of K_n will do! On the other hand, there are positive indicators, too. A graph is called **rigid** iff the identity is the graph's only endomorphism. Asymptotically, almost every graph is rigid (see, for example, [135], Theorem 4.7). Therefore, asymptotically, for almost every graph, every endomorphism has a fixed vertex. So there is reason to investigate a fixed vertex property for graph endomorphisms. We will see that it has good potential for further investigation.

Definition 6.22. *Let* $G = (V, E)$ *be a graph. Then* G *is said to have the* ***fixed vertex property*** *iff every endomorphism* f *of* G *has a fixed vertex* $v = f(v)$.

As is common for fixed point properties for endomorphisms of any kind, the fixed vertex property is inherited by retracts.

Definition 6.23. *Let* $G = (V, E)$ *be a graph. An endomorphism* r *from* G *to* G *that satisfies* $r^2 = r$ *is called a* ***retraction***. *For a retraction* r, *the induced subgraph* $G[r[V]]$ *is also called a* ***retract*** *of* G.

Proposition 6.24. *Let* $G = (V, E)$ *be a graph with the fixed vertex property and let* r *be a retraction from* G *to* G. *Then* $G[r[V]]$ *has the fixed vertex property.*

Proof. See Exercise 6-21.   ■

By Proposition 6.24, finite comparability graphs do not have the fixed vertex property: We can pick a chain C of maximum length and retract each rank to the element of the chain that has the same rank. Although this is disappointing, we will see in Section 6.5 that the whole theory of ordered sets and order-preserving maps can be embedded into work with graphs and graph homomorphisms.

With the, from the order-theoretical viewpoint, most obvious candidates for the fixed vertex property ruled out, we should ask for small examples of graphs with the fixed vertex property. Unlike for ordered sets, there are *no* graphs with $2, 3, 4$, or 5 vertices that have the fixed vertex property. The first nontrivial graph with the fixed vertex property is the 5-wheel (see Figure 6.3). Proving that the 5-wheel has the fixed vertex property is a good exercise in learning to work with graph endomorphisms (see Exercise 6-22a). Note that we cannot simply say that the 5-wheel has the fixed vertex property because there is one vertex that is adjacent to all the other vertices: In a clique, all vertices are adjacent to all other vertices and no nontrivial clique has the fixed vertex property. Proving that the 5-wheel is the smallest nontrivial graph with the fixed vertex property is nowadays best left to a computer: Enumerate all graphs up to a certain size and have another program, such as, for example, [286], check each graph for the fixed vertex property. Nonetheless, proving this fact by hand is a good, if lengthy, exercise (see Exercise 6-25).

For what follows, we will need the notion of the core of a graph.

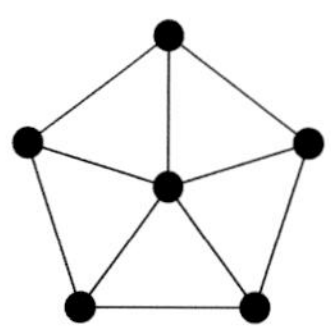

Fig. 6.3 The 5-wheel is the smallest nontrivial graph with the fixed vertex property

Definition 6.25. *A graph C is called a **core** iff all endomorphisms of C are bijective.*

For example, the 5-wheel is a core (see Exercise 6-22c). For a graph $G = (V, E)$, if $W \subseteq V$ is so that $G[W]$ is a core, and if there is an endomorphism from G to $G[W]$, then, for any endomorphism f of G so that $G[f[V]]$ is a core, we must have that $G[W]$ is isomorphic to $G[f[V]]$ and the isomorphism is $f|_W$. Hence, we define the following.

Definition 6.26. *Let G be a graph and let $G[W]$ be a core so that there is an endomorphism from G to $G[W]$. Then we call $G[W]$ the **core** of G.*

Exercises

6-21. Prove Proposition 6.24
6-22. An *n*-**wheel** is a graph that consists of an *n*-cycle with an additional vertex that is adjacent to all other vertices.

    a. Prove that, for odd *n*, the *n*-wheel has the fixed vertex property.
        Hint. Prove that the center vertex must be mapped to itself.
    b. Prove that, for even *n*, the *n*-wheel does not have the fixed vertex property.
        Note. This result shows that there is no analogue of the Abian–Brown Theorem for the fixed vertex property, even when the vertex that is adjacent to all other vertices is unique.
    c. Prove that, for odd *n*, the *n*-wheel is a core.

6-23. Let $G = (V, E)$ be a graph that is not isomorphic to $K_{|V|}$, but which contains a graph H that is isomorphic to $K_{|V|-1}$. Prove that H is a retract of G.
6-24. Let $G = (V, E)$ be a graph that does not contain an isomorphic copy of $K_{|V|-1}$, but which contains a graph H that is isomorphic to $K_{|V|-2}$. Prove that H is a retract of G.
6-25. Prove that the 5-wheel is the smallest nontrivial graph with the fixed vertex property as follows.

    a. Prove that there is no graph with $2, 3, 4$, or 5 vertices that has the fixed vertex property.
        Hint. Use Exercises 6-23 and 6-24.
    b. Prove that, if a graph $G = (V, E)$ has 6 vertices and the fixed vertex property, then G must be isomorphic to the 5-wheel.
        Hint. Use Exercises 6-19, 6-23, and 6-24.

6-26. Let $G = (V, E)$ be a disconnected graph. (Connectedness for graphs is defined similar to connectedness for ordered sets.) Prove that G has the fixed vertex property iff G has a component which has the fixed vertex property and which does not have a homomorphism into any of the other components.
6-27. (Compare with Theorem 4.12.) Let $G = (V, E)$ be a graph so that, for some $a \in V$, $G[V \setminus \{a\}]$ is a retract of G.

    a. Prove that, if $G[V \setminus \{a\}]$ and $G[N(a)]$ have the fixed vertex property, then G has the fixed vertex property.
    b. Prove that, if G is connected with at least 3 vertices and $a \in V$ is a **pendant vertex**, that is, a vertex with exactly one neighbor, then G has the fixed vertex property iff $G[V \setminus \{a\}]$ has the fixed vertex property.

c. Prove that the graph obtained from the 5-wheel by attaching a vertex that is adjacent to
the center and to one vertex on the rim has the fixed vertex property.
Note. This shows that, unlike Theorem 4.12, Exercise 6-27a cannot be turned into an
equivalence, because, in this example, $G[N(a)]$ does not have the fixed vertex property.

6-28. Let $G = (V, E)$ be a graph. A vertex $a \in V$ is called **dominated** by the vertex $w \in V \setminus \{a\}$
iff $\{a, w\} \in E$ and, for all $x \in V$ with $\{a, x\} \in E$, we have $\{w, x\} \in E$.

a. Use the 5-wheel to prove that removing a dominated vertex can turn a graph with the
fixed vertex property into a graph that does not have the fixed vertex property.
b. Use the 5-wheel to prove that adding a dominated vertex can turn a graph with the fixed
vertex property into a graph that does not have the fixed vertex property.
c. With dominated vertices being the graph-theoretical analogue of irreducible points, the
first two parts of this exercise show that there is no analogue of Scholium 4.13 for the
fixed vertex property. Explain why this is not a contradiction to Exercise 6-27a.

6.5   Embedding the Fixed Point Property into the Fixed Vertex Property

We will now prove the surprising fact that the fixed point theory for ordered sets can
be embedded into the fixed vertex theory for graphs.

Definition 6.27. *A **directed graph** (with loops) is a pair (V, A) of a set V of vertices
and a set $A \subseteq V \times V$. The set A is called the set of **arcs** of the directed graph. Visually,
we say that there is an arrow from $v \in V$ to $w \in V$ when $(v, w) \in A$. Note that we
can have $(v, v) \in A$ (a loop), that $(v, w) \neq (w, v)$ and that we can have $(v, w) \in A$
and $(w, v) \in A$ (arrows in each direction).*

Directed graphs ("digraphs") are usually translated into graphs with the "**arrow
construction**" or "**replacement operation**" (see Section 4.4 of [135]). Ordered
sets are directed graphs with loops if we interpret each $p_1 \leq p_2$ as a directed
edge (p_1, p_2). Reflexivity makes the requisite replacement graph (also called a
"**gadget**") not entirely obvious. We will provide the details to assure that the gadget
we will use (see Remark 6.30) has the right properties. The construction is a bit
more complex than needed for this section, but it will also allow the translation
of Open Question 12.12 for products into the analogous question for graphs (see
Exercise 12-17). .

Definition 6.28. *Let $G = (V, E)$ be a graph. A **triangle path** from $x \in V$ to $y \in V$ is
a set of distinct vertices $p_1, \ldots, p_n$ so that $x = p_1 \sim p_2 \sim \cdots \sim p_n = y$ and so that,
for all $i = 1, \ldots, n-2$, we have $p_i \sim p_{i+2}$. A graph is called **triangle connected** iff,
for any two vertices $x, y \in V$, there is a triangle path from x to y.*

Definition 6.29. *Let $C = (V_C, E_C)$ be a rigid triangle connected core and let the
vertices $b, t \in V_C$ be so that the graph $C^{b=t}$ that is obtained by identifying b and t
is a rigid triangle connected core, too. Let P be an ordered set, viewed as a directed*

graph $P = (V_P, E_P)$ with loops. The **arrow construction** for P using (C, b, t) yields a graph $P * (C, b, t)$ so that every directed edge of P is replaced by a copy of C. Formally,

$$V(P * (C, b, t)) = V_P \cup (V_C \setminus \{b, t\}) \times E_P)$$

and

$$E(P * (C, b, t)) = \bigcup_{e \in E_P} \{\{(x, e), (y, e)\} : \{x, y\} \in E_C, x, y \notin \{b, t\}\}$$

$$\cup \{\{p_1, (y, e)\} : e = (p_1, p_2) \in E_P, \{b, y\} \in E_C\}$$

$$\cup \{\{(x, e), p_2\} : e = (p_1, p_2) \in E_P, \{x, t\} \in E_C\}.$$

For $p_1, p_2 \in P$ so that $p_1 \le p_2$, we let C_{p_1, p_2} be the induced subgraph of $P * (C, b, t)$ on the vertices $\{p_1\} \cup (V_C \setminus \{b, t\}) \times \{(p_1, p_2)\} \cup \{p_2\}$. Note that, for $p_1, p_2 \in P$ with $p_1 < p_2$, we have that C_{p_1, p_2} is isomorphic to C, and that, for $p \in P$, we have that $C_{p,p}$ is isomorphic to $C^{b=t}$. For $p_1 \le p_2$, we let i_{p_1, p_2} be the unique isomorphism from C or $C^{b=t}$ to C_{p_1, p_2}.

Remark 6.30. Cores as in Definition 6.29 exist. The graphs H_k from Section 4.4 in [135] are rigid and triangle connected. However, these graphs are used in the replacement construction for digraphs *without loops* and identifying their vertices b and t into one does not produce a rigid core. Consider the graph C (see Figure 6.4) that is obtained as follows. Connect disjoint copies of H_1 and H_2 from Section 4.4 of [135] so that the last vertex of H_1 is joined with an edge to the first and second vertices of H_2, so that the second to last vertex of H_1 is joined to the first vertex of H_2 and let b be the first vertex of H_1 and t be the last vertex of H_2. (In Figure 6.4, H_1 is numbered in the opposite direction of the numbering in [135].) Moreover, a copy T_{10}^i of a triangle 10-cycle is attached to the start of H_1 and to the end of H_2 in the same fashion. (A **triangle n-cycle** is a cycle $c_1 \sim c_2 \sim \cdots \sim c_n \sim c_1$ so that, for all $i = 1, \ldots n$, we have $c_i \sim c_{i+2}$ modulo n.)

Although C is rather large, H_1 and H_2 are needed to provide non-adjacent vertices b and t that can be identified in $C^{b=t}$. The two graphs T_{10}^i allow a relatively simple proof that various products are triangle connected, see Exercise 12-17c.

A written proof that C and $C^{b=t}$ are rigid is left as Exercise 6-29. Note that a computer search for endomorphisms (using, for example, [286]), even using a simple backtracking algorithm, reveals in seconds that C as well as $C^{b=t}$ indeed are rigid. Moreover, there is exactly one homomorphism from C to $C^{b=t}$ and, because C is a core, there is no homomorphism from $C^{b=t}$ to C. (Such a homomorphism, composed with the homomorphism from C to $C^{b=t}$ would induce a non-bijective endomorphism of C.) These facts about endomorphisms of and homomorphisms between C and $C^{b=t}$ are crucial for the proof of Theorem 6.31 below.

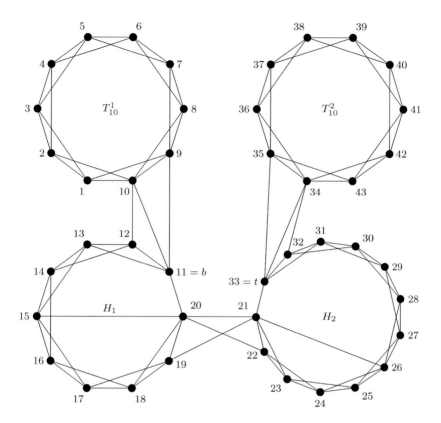

Fig. 6.4 A rigid core C as needed in Definition 6.29

Theorem 6.31. *Let* P, Q *be ordered sets. Then every graph homomorphism* f *from* $P * (C, b, t)$ *to* $Q * (C, b, t)$ *must map* V_P *to* V_Q *in such a way that, if* $p_1 \leq p_2$ *in* P, *then* $f(p_1) \leq f(p_2)$ *in* Q. *Conversely, for every order-preserving map* $F : P \rightarrow Q$, *there is a homomorphism* f *from* $P * (C, b, t)$ *to* $Q * (C, b, t)$ *so that* $f|_{V_P} = F$. *If* $P = Q$, *then* $f : P * (C, b, t) \rightarrow P * (C, b, t)$ *has a fixed vertex iff* $f|_{V_P} : P \rightarrow P$ *has a fixed point.*

Proof (The first paragraph is essentially the argument from [135]). Let f be a homomorphism from $P * (C, b, t)$ to $Q * (C, b, t)$ and let $p_1 \leq p_2$ in P. Note that, for any three elements q_1, q_2, q_3 of Q so that $q_1 \neq q_3$ and so that q_1 is comparable to q_2 and q_2 is comparable to q_3, there is no triangle path from a vertex of $C_{q_1, q_2} - q_2$ to a vertex of $C_{q_2, q_3} - q_2$. Because C_{p_1, p_2} is triangle connected, this means there must be $q_1 \leq q_2 \in Q$ so that $f[C_{p_1, p_2}] \subseteq C_{q_1, q_2}$. Now the rigidity of C and of $C^{b=t}$ assures that $f(p_1) = q_1$ and $f(p_2) = q_2$, which means that $f(p_1) \leq f(p_2)$.

The homomorphism for the converse is the natural extension of F to $P * (C, b, t)$.

Regarding the fixed vertices when $P = Q$, note that, if f has a fixed vertex that is not in V_P, then f has a fixed vertex in some $C_{p_1,p_2} - p_1, p_2$. Hence f maps C_{p_1,p_2} to itself and thus f fixes p_1 and p_2.   ∎

Note that, because C and $C^{b=t}$ are rigid cores, the natural extension of any order-preserving function $F : P \to Q$ to $P * (C, b, t)$ is unique.

Corollary 6.32. *Let P be an ordered set. Then P has the fixed point property iff $P * (C, b, t)$ has the fixed vertex property.*   ∎

Exercises

6-29. Let C be the graph from Figure 6.4 and let $f : C \to C$ be an endomorphism.

    a. Prove that a triangle 10-cycle is a core.
    b. Prove that the only possible f-images of T_{10}^1 and T_{10}^2 are T_{10}^1 and T_{10}^2.
    c. Prove that T_{10}^1 and T_{10}^2 cannot be mapped to the same image set.
        Hint. Use the fact that the shortest triangle path from 10 to 34 is unique, together with the presence of the "extra edges" in $H_1 \cup H_2$, to show that $f[T_{10}^i] = T_{10}^i$ for $i = 1, 2$.
    d. Prove that f must be the identity.
        Hint. Use the fact that the shortest triangle path from 10 to 34 is unique once more.
    e. Prove that $C^{b=t}$ is rigid.

6-30. A homomorphism from the directed graph $D = (V, A)$ to the directed graph $H = (W, B)$ is a function $f : V \to W$ so that, if $(v, w) \in A$, then $(f(v), f(w)) \in B$.

    a. Let $p, q \in \mathbb{N}$ be so that $p \le q$ and p is odd. Let $R'_{p,q}$ be a rigid graph so that the shortest cycle in $R'_{p,q}$ has at least $p + 1$ vertices and so that there are $q - p$ independent vertices $s_1, \ldots, s_{q-p}$ in $R'_{p,q}$. Let $R_{p,q}$ be the directed graph obtained from $R'_{p,q}$ by replacing every edge $\{v, w\}$ with arcs (v, w) and (w, v). Let C_p be a directed p-cycle, that is, C_p has vertices $\{1, \ldots, p\}$ and arcs $(v, v + 1)$ modulo p. Let $C_{p,q} = (V, A)$ be the directed graph obtained as the union of $R_{p,q}$ and C_p with a universal vertex u so that there is an arc (u, v) for all $v \neq u$.

        i. Prove that $C_{p,q}$ has exactly p endomorphisms.
        ii. Prove that all endomorphisms of $C_{p,q}$ are automorphisms.

    b. Define a directed graph with loops $L_{p,q} = (V, A')$ as follows. For any distinct $v, w \in V$, if $v, w \notin \{s_1, \ldots, s_{q-p}\}$, then $(v, w) \in A'$ and $(w, v) \in A'$ iff one of them is in A; if $v = s_i$ and $w \neq s_i$, then $(v, w) \in A'$ iff $(v, w) \in A$ or $(w, v) \in A$; and, for all $j \in \{1, \ldots, q-p\}$, let $(s_j, s_j) \in A'$. That is, the construction of $L_{p,q}$ adds loops at the s_j, replaces the arcs adjacent to s_i with out-arcs, and replaces all other arcs with arcs in both directions.

        i. Prove that every endomorphism of $C_{p,q}$ is an endomorphism of $L_{p,q}$.
        ii. Prove that the endomorphisms of $L_{p,q}$ consist of the endomorphisms of $C_{p,q}$ and of the $q - p$ constant functions that map all $v \in V$ to a vertex s_j.
        iii. Conclude that $|\mathrm{Aut}_{L_{p,q}}^{(}|$ $|\mathrm{End}(L_{p,q})| = \frac{p}{q}$.
            Note. With appropriate replacement graphs, we can prove that the automorphism to endomorphism ratio for graphs can be any rational number that has an odd numerator. By adding cores so that there is no homomorphism to and from $L_{p,q}$, we can show that the limiting set of these ratios as the number of vertices goes to infinity is $[0, 1]$. This effectively settles the version of the **automorphism conjecture** for graph homomorphisms.

Remarks and Open Problems

1. Asymptotically, most graphs may have the fixed vertex property because they are rigid. For ordered sets, the situation is fundamentally different: Most ordered sets do not have the fixed point property, because they are of height 2 (see [165, 166] or Chapter 13) and can be retracted onto a 4-crown tower $\{a, b < c, d < e, f\}$.
2. For more on fixed clique properties in graphs also see [14, 148, 200].
3. For more on the Helly property, see [233–235].

Chapter 7
Lexicographic Sums

In this chapter, we investigate a construction, lexicographic sums, that uses existing ordered sets to build new ordered sets. The pictorial idea is very simple: Take an ordered set T and replace each of its points t with an ordered set P_t. The resulting structure will be a new, larger ordered set. It is then natural to ask how various order-theoretical properties and parameters behave under lexicographic constructions. We will revisit lexicographic sums in later chapters to specifically answer this question in a variety of contexts. At the end of this chapter, in Section 7.6, we will explain the definition of "hard" problems that was mentioned at the start of Section 5.4 (see Definition 7.30). We will then use a construction that is close to the lexicographic sum construction (see Section 7.5), to show that determining if an ordered set has the fixed point property is "hard."

7.1 Definition and Examples

There are two ways to look at lexicographic sums. On one hand, we can consider the construction as a tool to build larger ordered sets from smaller ordered sets. On the other hand, we can analyze a given ordered set and see if we can represent it as a lexicographic sum that is made up of smaller ordered sets. The "building" point-of-view is reflected in the definition below.

Definition 7.1. *Let T be a nonempty ordered set considered as an index set. Let $\{P_t\}_{t \in T}$ be a family of pairwise disjoint nonempty ordered sets that are all disjoint from T. We define the **lexicographic sum** $L\{P_t \mid t \in T\}$ to be the union $\bigcup_{t \in T} P_t$ ordered by $p_1 \leq p_2$ iff*

1. There are distinct $t_1, t_2 \in T$ with $t_1 < t_2$, so that $p_i \in P_{t_i}$, or
2. There is a $t \in T$ so that $p_1, p_2 \in P_t$ and $p_1 \leq_{P_t} p_2$.

© Springer International Publishing 2016
B. Schröder, *Ordered Sets*, DOI 10.1007/978-3-319-29788-0_7

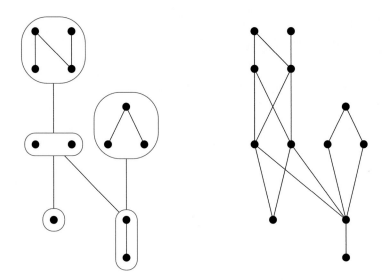

Fig. 7.1 Visualization of a lexicographic sum. To the right, the diagram of the ordered set, to the left, the ordered set with pieces and their comparabilities indicated.

*The ordered sets P_t are the **pieces** of the lexicographic sum and T is the **index set**. For $p \in L\{P_t \mid t \in T\}$, define $I(p)$ to be the unique $t \in T$ such that $p \in P_t$. We will also say that $L\{P_t \mid t \in T\}$ is the **lexicographic sum of the P_t over T**.*

Definition 7.1 is illustrated pictorially in the left ordered set in Figure 7.1. The index set T is the ordered set made up of the large ovals and the connections between the large ovals. The pieces are the ordered sets within the large ovals. The comparabilities that arise are shown in the diagram on the right.

The simplest types of lexicographic sums are the lexicographic sums in which we "stack" ordered sets, and those in which we put the pieces "side by side."

Definition 7.2. *Let $L\{P_t \mid t \in T\}$ be a lexicographic sum.*

1. *If the index set T is a chain, we will call the lexicographic sum **linear**.*
2. *A linear lexicographic sum with finitely many pieces $P_1, \ldots, P_n$ will also be denoted $P_1 \oplus \cdots \oplus P_n$, with the lowest piece at the beginning of the sum.*
3. *A lexicographic sum of $P_1, \ldots, P_n$ where the index set is the n-element antichain will be denoted $P_1 + \cdots + P_n$. It will be called the **disjoint sum** of $P_1, \ldots, P_n$.*
4. *If all pieces are singletons or if there is only one piece, the lexicographic sum will be called **trivial**.*

The point-of-view of breaking down an ordered set into smaller parts that are put together in a lexicographic sum is reflected in the following notation.

Definition 7.3. *Let P be an ordered set.*

1. *If P is isomorphic to a nontrivial lexicographic sum $L\{P_t \mid t \in T\}$, P will be called **decomposable**. In this case, $L\{P_t \mid t \in T\}$ will be called a **lexicographic sum decomposition of** P.*
   *Otherwise P will be called **indecomposable**.*
2. *If P is isomorphic to a nontrivial linear lexicographic sum, P will be called **series decomposable**. The linear lexicographic sum will be called a **series decomposition of** P.*

With this language in mind, consider Figure 7.1 once more. Going from left to right, the set on the right has been built out of five pieces. Formally, these pieces are a one-element antichain, a two-element antichain, a 2-chain, a 3-fence, and a 4-fence. The pieces were put into a lexicographic sum over an index set that is a 4-fence with an additional upper cover attached to the maximal element with two lower covers. Going from right to left, we see that, starting with the set on the right, we can decompose it into the lexicographic sum as indicated above.

Example 7.4.

1. The four crown is a lexicographic sum of two two-element antichains over a 2-chain as the index set.
2. Every disconnected ordered set is decomposable.
3. Every chain with more than two elements is decomposable.
4. The set $\mathcal{P}(X) \setminus \{\emptyset, X\}$ with $\mathcal{P}(X)$ being the power set of the set X is not decomposable.
5. The lexicographic order on $\mathbb{N}^n$ as described in Example 1.2, part 9 can be described inductively as $\mathbb{N}^n = L\{\mathbb{N}_j^{n-1} \mid j \in \mathbb{N}\}$, where all pieces $\mathbb{N}_n^{n-1}$ are isomorphic to $\mathbb{N}^{n-1}$. □

Let us now investigate how lexicographic constructions relate to retractions and to the fixed point property.

Lemma 7.5. *Let $P := L\{P_t \mid t \in T\}$ be a lexicographic sum. Then P has a retract that is isomorphic to T.*

Proof. For each $t \in T$, choose an element $p_t \in P_t$. The map that maps each piece P_t to the chosen element p_t is a retraction of P to an ordered set that is isomorphic to T. ∎

The first step towards the fixed point property is to consider linear lexicographic sums.

Lemma 7.6. *Let $Q := L\{P_c \mid c \in C\}$ be a chain-complete lexicographic sum with a chain C as index set. Then P has the fixed point property iff one piece P_c has the fixed point property.*

Proof. If none of the pieces have the fixed point property, it is easy to see that Q does not have the fixed point property.

Conversely, let one P_c have the fixed point property and let $f : Q \rightarrow Q$ be order-preserving. If f maps P_c to itself, then f has a fixed point. Otherwise for some $p_c \in P_c$ we have $f(p_c) \sim p_c$ and hence, by the Abian–Brown Theorem, f must have a fixed point.

Thus chain-complete linear lexicographic sums have the fixed point property iff they have a piece with the fixed point property.   ∎

Proposition 7.7 (Compare with [139–141]). *Let $L\{P_t \mid t \in T\}$ be a chain-complete lexicographic sum. Then $L\{P_t \mid t \in T\}$ has the fixed point property iff*

1. *T has the fixed point property, and*
2. *For all $t_0 \in T$ one of P_{t_0}, $L\{P_t \mid t < t_0\}$ and $L\{P_t \mid t > t_0\}$ has the fixed point property.*

Proof. Let $P := L\{P_t \mid t \in T\}$ be a chain-complete lexicographic sum with arbitrary index set T. First suppose P has the fixed point property. By Lemma 7.5, part 1 must hold. To prove part 2 let $t_0 \in T$ and fix a point $p_{t_0} \in P_{t_0}$. The map that maps all points that are not comparable to any $p \in P_{t_0}$ to p_{t_0} is a retraction of P onto $L\{P_t \mid t < t_0\} \oplus P_{t_0} \oplus L\{P_t \mid t > t_0\}$. Thus $L\{P_t \mid t < t_0\} \oplus P_{t_0} \oplus L\{P_t \mid t > t_0\}$ must have the fixed point property. By Lemma 7.6, this is the case iff one of its three pieces has the fixed point property. Thus 2 must hold.

Conversely, suppose $P := L\{P_t \mid t \in T\}$ is a chain-complete lexicographic sum that satisfies conditions 1 and 2. Let $f : P \rightarrow P$ be an order-preserving map. Define $F : T \rightarrow \mathcal{P}(T) \setminus \{\emptyset\}$ by $F(t) := I[f[P_t]]$. Then, for some $t \in T$, we must have $F(t) = \{t\}$: Indeed, otherwise, for each $t \in T$, choose $g(t) \in F(t) \setminus \{t\}$. This map would be an order-preserving fixed point free self-map of T, a contradiction to 1. Find a $t_0 \in T$ with $F(t_0) = \{t_0\}$. Then f maps $L\{P_t \mid t < t_0\} \oplus P_{t_0} \oplus L\{P_t \mid t > t_0\}$ to itself. Because, by 2, one of the three pieces must have the fixed point property, we conclude, via Lemma 7.6, that f has a fixed point.   ∎

Note that a characterization of exactly which pieces in a lexicographic sum could be fixed point free ordered sets is unattainable. The reason is that, in a lexicographic sum over a two-point chain that has the fixed point property, either piece could have the fixed point property. This means there can be pairs of points $\{s, t\} \subseteq T$ such that either one could be assigned a fixed point free piece without the sum losing the fixed point property. However if both are assigned a fixed point free piece, the lexicographic sum becomes fixed point free, too. This situation occurs in other index sets, too (see Exercise 7-1).

Exercises

7-1. Find an index set T that is not a chain and that contains elements $s, s' \in T$ such that

- Any finite lexicographic sum $L\{P_t \mid t \in T\}$ with $|P_t| = 1$ for all $t \in P \setminus \{s\}$ or with $|P_t| = 1$ for all $t \in P \setminus \{s'\}$ has the fixed point property, and
- Any finite lexicographic sum $L\{P_t \mid t \in T\}$ with P_s and $P_{s'}$ being two-element antichains does not have the fixed point property.

7-2. A finite ordered set P is called **series parallel** iff P is a singleton, or there are series parallel ordered sets P_1 and P_2 so that $P = P_1 + P_2$ or $P = P_1 \oplus P_2$. A finite ordered set is called **N-free** iff, for all $a \prec b \succ c \prec d$, we have that $a \leq d$. Prove that every series parallel ordered set is N-free.

7-3. Prove that an ordered set is series parallel iff it does not contain a 4-fence.

7-4. Let $P_1, \ldots, P_n$ be ordered sets. Formulate the definition of $P_1 \odot \cdots \odot P_n$ (see Exercise 1-14) in terms of lexicographic sums. Use Example 7.4, part 5 as a model.

7-5. Prove a result analogous to Proposition 7.7 for the connected relational fixed point property for finite ordered sets.

   Hint. Use retractable sets.

   Conclude that there are ordered sets with the connected relational fixed point property that are not connectedly collapsible.

7-6. The **ranked sum** $R\{P_t \mid t \in T\}$ of a set of pairwise disjoint finite ordered sets P_t over the finite index set T is the union $\bigcup_{t \in T} P_t$ ordered as follows. For $a \in P_s$, $b \in P_t$, we let $a \leq b$ iff $s = t$ and $a \leq_t b$, or $s < t$ and $\mathrm{rank}_{P_s}(a) \leq \mathrm{rank}_{P_t}(b)$.

   a. Prove that the order relation of the ranked sum $R\{P_t \mid t \in T\}$ is contained in the order relation of the corresponding lexicographic sum $L\{P_t \mid t \in T\}$.

   b. Show that the ranked sum of two ordered sets with the fixed point property over a 2-chain need not have the fixed point property.

   Note. This means that there is no analogue of Lemma 7.6 for ranked sums.

7.2 The Canonical Decomposition

Consider the ordered set in Figure 7.1 once more. There are indeed several ways to write this set as a lexicographic sum. On one hand, we could write the piece that is a 3-fence as a linear lexicographic sum to obtain a "finer" partition of the ordered set. On the other hand, we could merge the 2-antichain and the 4-fence into a larger piece to obtain a lexicographic sum with a smaller index set. Is there a canonical way to write lexicographic sums? For finite ordered sets, there is a natural idea, which is to "make the pieces as large as possible."

Definition 7.8 (See, e.g., [66, 154]). *Let P be an ordered set and let $S \subseteq P$ be nonempty. Then S is called **order-autonomous** or **autonomous** iff, for all $p \in P \setminus S$, we have that*

1. If there is an $s \in S$ with $p \leq s$, then $p \leq S$, and
2. If there is an $s \in S$ with $p \geq s$, then $p \geq S$.

Note that every piece in any lexicographic sum decomposition of an ordered set is order-autonomous. Conversely, any order-autonomous set in P can be a piece in a lexicographic sum decomposition of P. Moreover we have the following.

Lemma 7.9. *Let P be an ordered set and let S_1 and S_2 be order-autonomous subsets with $S_1 \cap S_2 \neq \emptyset$. Then the union $S_1 \cup S_2$ is order-autonomous.*

Proof. Exercise 7-7. ∎

"Taking pieces that are as large as possible" means finding subsets that are order-autonomous and maximal with respect to inclusion. We just have to be careful not to take the whole set, which is vacuously order-autonomous in itself.

Definition 7.10. *An order-autonomous subset S of the ordered set P is called* **maximal** *iff $S \neq P$ and, for all order-autonomous subsets $Q \subseteq P$ with $S \subseteq Q$, we have that $Q \in \{S, P\}$.*

The canonical decomposition for an ordered set is now a decomposition into disjoint maximal order-autonomous subsets. This idea is important for connected, not series decomposable, ordered sets. (Disconnected ordered sets decompose naturally into their components. For series decomposable ordered sets, a similar idea is discussed in Exercise 7-8.)

Proposition 7.11. *The* **canonical decomposition** *of finite ordered sets. Let P be a finite ordered set. Then every $p \in P$ is contained in a maximal order-autonomous subset S of P. Moreover, if P is not series decomposable with more than two summands and not disconnected with more than two components, then the set S is uniquely determined by p.*

Proof. First consider the case in which P is series decomposable. Decompose P into the lexicographic sum $L\{P_c \mid c \in C\}$ with C as large as possible. Then the sets $B := L\{P_c \mid c \in C \setminus \{\bigvee C\}\}$ and $T := L\{P_c \mid c \in C \setminus \{\bigwedge C\}\}$ are both maximal order-autonomous subsets of P and each $p \in P$ is in at least one of these two sets. If $|C| = 2$, then B and T are disjoint and every order-autonomous subset not equal to P is contained in B or T. This proves the result for series decomposable sets. The proof in case P is disconnected is similar, see Exercise 7-10.

This leaves us with the case in which P is connected and not series decomposable. Because P is finite and every singleton subset of an ordered set is order-autonomous, every $p \in P$ is contained in a maximal order-autonomous subset.

To prove uniqueness, let S_1 and S_2 be two distinct maximal order-autonomous subsets of P with $S_1 \cap S_2 \neq \emptyset$. Suppose, for a contradiction, that $S_1 \neq S_2$. Then, by maximality of S_1 and S_2, we have that both $S_1 \setminus S_2$ and $S_2 \setminus S_1$ are not empty. Because P is connected and $S_1 \cup S_2 = P$, there is a point $p \in S_1 \setminus S_2$ such that p is comparable to a point in S_2, say, $p \leq s_2$, for some $s_2 \in S_2$. Then $p \leq S_2$. However, this implies that every element of $S_2 \setminus S_1$ is an upper bound of p and hence of S_1. This contradicts our assumption that P was not series decomposable. Thus $S_1 = S_2$. Therefore any two maximal order-autonomous subsets in P are either disjoint or equal, which finishes the proof that every point is contained in a unique maximal order-autonomous subset.   ∎

The canonical decomposition of a finite ordered set will be useful whenever a "standardization" of a lexicographic sum is necessary. This is the case, for example, when we talk about comparability invariance and about reconstruction of lexicographic sums in the following. Note that the ambiguity observed in the lexicographic decompositions for series decomposable finite ordered sets gets worse for infinite ordered sets (see Exercise 7-9). For more on order-autonomous sets, consider, for example, [154] and [273].

Because the full strength of the canonical decomposition only applies to connected, non-series decomposable ordered sets we introduce the following terminology.

Definition 7.12. *We shall call an ordered set P **co-connected** iff P is not series decomposable.*

The language is motivated by the fact that, if P is not series decomposable, then the complement of the comparability graph of P must be connected.[1]

Definition 7.13. *Let P be a finite, connected, co-connected ordered set. The **canonical decomposition** of P is the unique decomposition $P = L\{P_t \mid t \in T\}$ so that every P_t is a maximal order-autonomous subset of P.*

Proposition 7.14. *Let P be a finite, connected, co-connected ordered set. The decomposition $L\{P_t \mid t \in T\}$ with $|T| \geq 2$ is the canonical decomposition of P iff T is not decomposable.*

Proof. The direction "⇒" is trivial. For "⇐," suppose, for a contradiction, that $r \in T$ is so that P_r is not a maximal order-autonomous subset of P. Let $A \subset P$ be a maximal order-autonomous subset of P so that A strictly contains $P_r \subset A$. If there is a P_s with $P_s \cap A = \emptyset$, then, by Lemma 7.9 and by the maximality of A, for all P_t, we have that $P_t \subset A$ or $P_t \cap A = \emptyset$. In this case, the set $\{t \in T : P_t \subset A\}$ is a nontrivial order-autonomous subset of T, which is not possible. Thus A intersects every P_t. By Lemma 7.9 and by the maximality of A, there is exactly one piece P_s so that P_s is not contained in A. Consider the set $B := P_s \setminus A \neq \emptyset$. If no element of B is comparable to any element of A, then P is disconnected, which cannot be. Without loss of generality, let $b \in B$ be so that there is an $a \in A$ so that $b > a$. Then, by order-autonomy of A, we have $b > A$ and, in particular, $b > A \setminus P_s$. Because $b \in P_s$ and P_s is order-autonomous, we have that $P_s > A \setminus P_s = P \setminus P_s$ and P is series decomposable, a contradiction. ∎

Exercises

7-7. Prove Lemma 7.9.
7-8. Similar to disconnected ordered sets, in series decomposable ordered sets, maximal order-autonomous subsets can overlap. It thus makes sense to look for some type of minimal order-autonomous subset of series decomposable ordered sets that is similar to connected components for disconnected ordered sets.

An order-autonomous proper subset $S \neq P$ of the ordered set P is called

- A **chain-link** iff every element of S is comparable to all elements in $P \setminus S$,
- A **minimal chain-link** iff S is a chain-link and every proper order-autonomous subset of S is not a chain-link.

---

[1] Terminology suggested by J.-X. Rampon.

Prove the following.

    a. P contains a chain-link iff P is series decomposable.
    b. Let P be a series decomposable ordered set. Then every element $p \in P$ is contained in a unique minimal chain-link.
    c. Let $\overline{G}$ be the complement of the comparability graph of the series decomposable ordered set P. Prove that $C \subseteq P$ is a minimal chain-link iff $\overline{G}[C]$ is a component of $\overline{G}$.

7-9. Construct an infinite ordered set P such that:

    a. P is series decomposable.
    b. No element of P is contained in a maximal order-autonomous subset.

7-10. Prove the part of Proposition 7.11 that refers to disconnected ordered sets.
7-11. Prove that a linear lexicographic sum $A \oplus B$ has a fixed point free automorphism iff both A and B have a fixed point free automorphism.
7-12. Let $P = L\{P_t \mid t \in T\}$ be a lexicographic sum that is

    • The decomposition into components if P is disconnected,
    • Any linear decomposition if P is series decomposable,
    • The canonical decomposition if P is connected and co-connected.

Let $\Phi : P \to P$ be an automorphism. For $t \in T$ let $\tau(t)$ be a (fixed) element of P_t. Prove that:

    a. $I \circ \Phi \circ \tau$ is an automorphism of T.
    b. For all $t \in T$, $\Phi|_{P_t}$ is an isomorphism from P_t to $P_{I \circ \Phi \circ \tau(t)}$.

7-13. Let $P = L\{P_t \mid t \in T\}$ be a lexicographic sum as in Exercise 7-12. Prove that we have the inequality $\dfrac{|\mathrm{Aut}(P)|}{|\mathrm{End}(P)|} \leq \prod_{t \in T} \dfrac{|\mathrm{Aut}(P_t)|}{|\mathrm{End}(P_t)|}$.

7.3 Comparability Invariance

With ordered sets being closely related to graphs, for any property of ordered sets, it is natural to ask if it is "of graph-theoretical nature." How would we define "of graph-theoretical nature," though? Usually, a property is considered to be of graph-theoretical nature iff it is a property of the comparability graph. That is, if and only if two ordered sets with isomorphic comparability graphs either both have the property or neither of them does. Definition 7.15 below formalizes this idea by considering parameters of an ordered set. Note that, although quantities like the height of an ordered set are obvious numerical parameters, indicator functions allow us to encode properties as parameters, too. For example, we can define $\alpha_{FPP}(P)$ to be one iff P has the fixed point property and zero otherwise. In this fashion, the parameter α_{FPP} encodes the fixed point property.

Definition 7.15. *Let C be a class of ordered sets and let α be a parameter for ordered sets. Then α is called a C-**comparability invariant** iff, for any two ordered sets $P, Q \in C$, we have that isomorphism of $G_C(P)$ and $G_C(Q)$ implies $\alpha(P) = \alpha(Q)$.*

The definition of comparability invariants with respect to certain classes of ordered sets is important. Indeed, there are properties that are comparability invariants for large classes of ordered sets (say, for the class of finite ordered sets), but which are not comparability invariants in the class of all ordered sets. For example, the fixed point property is not a comparability invariant in general: Indeed, $\mathbb{N} \cup \{\infty\}$ and $\mathbb{N}$ have isomorphic comparability graphs, but the first ordered set has the fixed point property, while the second does not. In the following, we will concentrate on finite ordered sets, where there is a strong connection between comparability invariance and a certain type of theorem for lexicographic sums. In particular, we will see that the fixed point property is a comparability invariant for finite ordered sets.

Note that, if P is an ordered set, then P and P^d, the dual of P, have the same comparability graph. Thus comparability invariants must be invariant under dualization. Theorem 7.18 shows that, for finite ordered sets, invariance under dualization is close to comparability invariance. (For comparability invariance, we must be able to dualize any piece of a lexicographic decomposition without changing the parameter.) We start with a result on the structure of comparability graphs: The following theorem shows that, if P is indecomposable, its comparability graph has exactly two transitive orientations.

Theorem 7.16. *Let $G = (V, E)$ be a comparability graph. Then one of the following holds for G.*

1. *G has exactly two transitive orientations.*
2. *There is a proper subset $S \subset V$ with at least two elements such that S is order-autonomous in every transitive orientation of G.*
3. *The complement graph $\overline{G}$ of G is not connected.*

This result, which has a surprisingly rich proof, was reported in [66], Lemma on p. 270 as a consequence of [154] Theorems 3.1 and 3.4 and Corollary 4.2. A proof can be constructed in the very instructive sequence of Exercises 7-16–7-19. The theorem is used to prove the following crucial lemma.

Lemma 7.17. *Let P and Q be finite ordered sets with isomorphic comparability graphs and let Φ be a graph isomorphism between $G_C(P)$ and $G_C(Q)$. If P has no nontrivial lexicographic sum decomposition, then the function $\Phi : P \to Q$ is also an order-isomorphism between P and Q or between P and Q^d.*

Proof. Because P has no nontrivial lexicographic sum decomposition, by Theorem 7.16, $G_C(P)$ has exactly two transitive orientations. Because $G_C(P)$ is isomorphic to $G_C(Q)$, $G_C(Q)$ has exactly two transitive orientations, too. Hence, both P and Q are among the two transitive orientations of their respective comparability graphs. Because Φ is a graph isomorphism, we can use Φ to map the transitive orientation of $G_C(P)$ that yields P order-isomorphically to a transitive orientation of $G_C(Q)$. Because $G_C(Q)$ has exactly two transitive orientations, Q and Q^d, said orientation yields either Q or Q^d and Φ is an isomorphism as claimed. ∎

Theorem 7.18 (See [66]). *Let α be a parameter of ordered sets. Then α is a comparability invariant in the class of finite ordered sets iff the following holds.*

For all finite lexicographic sums $L\{P_t \mid t \in T\}$ such that, for at most one $t \in T$, the set P_t has more than one element, we have that

$$\alpha(L\{P_t \mid t \in T\}) = \alpha(L\{P_t^d \mid t \in T\}).$$

Proof. The direction "$\Rightarrow$" is trivial, because the two ordered sets involved have the same comparability graph.

To prove "$\Leftarrow$," let P and Q be two ordered sets so that Φ is a graph isomorphism between the comparability graphs $G_C(P)$ and $G_C(Q)$. We will prove that there is a finite sequence $P = P_0, P_1, \ldots, P_N$ of ordered sets such that P_N is order-isomorphic to Q and P_k is obtained from P_{k-1} by dualization of exactly one order-autonomous subset. (Because an ordered set is vacuously order-autonomous in itself, this includes the possibility of dualizing the ordered set itself.) Because, by assumption, the parameter α is not affected by going from P_{k-1} to P_k, we have $\alpha(P) = \alpha(P_N) = \alpha(Q)$, establishing that α is a comparability invariant for finite ordered sets.

The proof is an induction on the size n of the ordered set P. For $n = 1$, there is nothing to prove. For the induction step, let $|P| = n$ and assume that the result has been proved for ordered sets of size $< n$. We will first consider the case in which P is disconnected (which immediately implies that Q is disconnected). The idea for this case will then merely be transferred to the other cases.

If P is disconnected, then the isomorphism Φ between $G_C(P)$ and $G_C(Q)$ is comprised of isomorphisms of the comparability graphs of the components. Let $C^1, \ldots, C^m$ be the components of P and let $K^1, \ldots, K^m$ be the components of Q, enumerated such that $\Phi|_{C^i}$ is an isomorphism from $G_C(P)[C^i]$ to $G_C(Q)[K^i]$. By induction hypothesis, for each pair $(C^i, \Phi[C^i] = K^i)$, with orders induced by P and Q, respectively, there is a finite sequence $C^i = C_0^i, C_1^i, \ldots, C_{N_i}^i$ such that $C_{N_i}^i$ is order-isomorphic to K^i and C_j^i is obtained from C_{j-1}^i by dualization of exactly one order-autonomous subset. The desired sequence from P to Q is now obtained by replacing each C^i in P (starting with C^1 and ending with C^m) successively with the sets in the sequence $C^i = C_0^i, C_1^i, \ldots, C_{N_i}^i$.

If P is series decomposable, then the isomorphism between $G_C(P)$ and $G_C(Q)$ is comprised of isomorphisms of the components of the complements of the comparability graphs. Let $C^1, \ldots, C^m$ be the components of the complement of $G_C(P)$ and let $K^1, \ldots, K^m$ be the components of the complement of $G_C(Q)$ enumerated so that $\Phi|_{C^i}$ is an isomorphism between $G_C(P)[C_i]$ and $G_C(Q)[K_i]$. From now on, we will interpret the C_i and K_i as ordered subsets of P and Q, respectively.

The set P is a linear lexicographic sum of the C^i and Q is a linear lexicographic sum of the K^i. Without loss of generality, we can assume that $Q = K^1 \oplus K^2 \oplus \cdots \oplus K^m$. Then there is a permutation σ on $\{1, \ldots, m\}$ such that the ordered set P satisfies $P = C^{\sigma(1)} \oplus C^{\sigma(2)} \oplus \cdots \oplus C^{\sigma(m)}$. Recall that every permutation can be represented as a composition of permutations that switch two adjacent elements. Successively switch two adjacent pieces C_i and C_{i+1} in P until the set $P' = C^1 \oplus C^2 \oplus \cdots \oplus C^m$ is reached. Then continue with a successive replacement of the pieces as in the proof for disconnected ordered sets.

Now consider the case that P is connected and co-connected. The above two cases have already shown the main ingredients of the proof. Once we have turned P into a set whose index set is order-isomorphic to the index set of Q via Φ, we can use the induction hypothesis and a replacement process as for the components of disconnected ordered sets to finish the proof. Our main tool to achieve this will be the canonical decomposition.

In case P is connected and co-connected, Q is connected and co-connected, too. Let $P = L\{P_t \mid t \in T\}$ and $Q = L\{Q_u \mid u \in U\}$ be the canonical decompositions of P and Q. We claim that, for all $t \in T$, there is a unique $u \in U$ such that $\Phi[P_t] = Q_u$: Indeed, suppose, for a contradiction, there was a $t \in T$ such that there is no $u \in U$ with $\Phi[P_t] \subseteq Q_u$.

First consider a $q \in Q \setminus \Phi[P_t]$ such that $q > b$ for some $b \in \Phi[P_t]$. This means that $\Phi^{-1}(q)$ is related to $\Phi^{-1}(b)$ in P, say, without loss of generality, $\Phi^{-1}(q) > \Phi^{-1}(b)$. Then $\Phi^{-1}(q) > P_t$. This means that q is related to all elements of $\Phi[P_t]$. This fact will be used frequently.

Now let $B \subseteq Q$ be the union of all Q_u that intersect $\Phi[P_t]$. Because Q is connected and co-connected, we have $|U| \geq 4$. Moreover, because we work with the canonical decomposition, no union of two pieces Q_u is order-autonomous in Q. We claim that, for any two distinct $Q_{u_1}, Q_{u_2} \subseteq B$, we must have $Q_{u_1} \leq Q_{u_2}$ or $Q_{u_2} \leq Q_{u_1}$: Indeed, suppose no element of Q_{u_1} is related to any element of Q_{u_2}. If $q \in Q \setminus (Q_{u_1} \cup Q_{u_2})$ satisfies $q > b'$ for some $b' \in Q_{u_1} \cup Q_{u_2}$, then $q > b$ for some $b \in \Phi[P_t]$. The above shows that then q is related to every element of $\Phi[P_t]$. Thus, in particular, q is related to an element of $\Phi[P_t] \cap Q_{u_1}$ and to an element of $\Phi[P_t] \cap Q_{u_2}$. Because we assumed that Q_{u_1} and Q_{u_2} had no related elements, we conclude $q > Q_{u_1} \cup Q_{u_2}$. This and the dual argument show that $Q_{u_1} \cup Q_{u_2}$ would be order-autonomous in Q. Because $Q_{u_1} \cup Q_{u_2} \neq Q$, this is a contradiction. This means that B is a linear lexicographic sum of pieces Q_u.

Now let u_{max} be the largest element of the set $I_Q[B]$ and let u_{min} be the smallest element of the set $I_Q[B]$. Consider the set

$$C := Q_{u_{min}} \oplus L\{Q_u \mid u_{min} < u < u_{max}\} \oplus Q_{u_{max}}.$$

Because C is series decomposable, we have $C \neq Q$. We claim that C is order-autonomous in Q. Indeed, let $q \in Q \setminus C$ be such that $q > c$ for some $c \in C$. Then $q > c'$ for some $c' \in Q_{u_{min}}$. This means that $q > Q_{u_{min}}$. Let $b \in \Phi[P_t] \cap Q_{u_{min}}$. Then $q > b$ and hence q is related to all elements of $\Phi[P_t]$, which implies q is related to all elements of B. Because $q \notin C$, this means that $q > B$. This finally implies $q > C$. Hence the set C is order-autonomous in Q and not equal to Q, a contradiction to the maximality of the Q_u. Thus, for all $t \in T$, there must be a unique $u \in U$ such that $\Phi[P_t] \subseteq Q_u$. By symmetry we infer $\Phi[P_t] = Q_u$.

The above shows that Φ induces a natural isomorphism Ψ between $G_C(T)$ and $G_C(U)$ by making $\Psi(t)$ the unique element $u \in U$ so that $\Phi[P_t] = Q_u$. Moreover $\Phi|_{P_t}$ is an isomorphism between $G_C(P_t)$ and $G_C(\Phi[P_t])$. T and U are both connected and co-connected ordered sets with no nontrivial order-autonomous subsets. Thus, by Lemma 7.17, T is *order*-isomorphic via Ψ to U or to U^d. If T is isomorphic to U, let $P_1 := P$, otherwise let $P_1 := P^d$.

Now continue with a successive replacement of the pieces as in the proof for disconnected ordered sets.   ∎

With this tool in hand, we can easily prove comparability invariance or non-invariance for many parameters, at least for the class of finite ordered sets. It is interesting that, while the fixed point property turns out to be a comparability invariant in the class of finite ordered sets, the related property of not having a fixed point free automorphism is not a comparability invariant. To prove this fact, we introduce here the concept of rigidity. A **rigid** ordered set has only one automorphism, the identity. As an example, consider a fence with an even number of elements. Rigidity essentially says that the ordered set lacks internal symmetry of a certain kind. It is thus somewhat opposite of having a fixed point free automorphism, which essentially says that the ordered set has a certain type of symmetry.

Corollary 7.19. *In the class of finite ordered sets:*

1. *The fixed point property is a comparability invariant.*
2. *"f is rigid," that is, "f has no nontrivial automorphism," is not a comparability invariant.*
3. *"f has a fixed point free automorphism" is not a comparability invariant.*

Proof. To prove part 1, we prove that the fixed point property is a comparability invariant for finite ordered sets by proving that the condition of Theorem 7.18 holds for ordered sets of any size. We proceed by induction on $|P|$. For $|P| = 1$, there is nothing to prove. For the induction step, assume that the condition of Theorem 7.18 holds for all ordered sets Q with $|Q| < n$.

Let $P = L\{P_t \mid t \in T\}$ be an ordered set of size n that has the fixed point property and is such that, only for $t' \in T$, the piece $P_{t'}$ has more than one element. Then, by Proposition 7.7, T has the fixed point property, and, for all $t_0 \in T$, one of P_{t_0}, $L\{P_t \mid t < t_0\}$ and $L\{P_t \mid t > t_0\}$ has the fixed point property. Then, for all $t_0 \in T$ that are incomparable to t', one of $P_{t_0}^d$, $L\{P_t^d \mid t < t_0\}$ and $L\{P_t^d \mid t > t_0\}$ has the fixed point property (all pieces are singletons and thus they are their own duals). Because an ordered set has the fixed point property iff its dual has the fixed point property, by Lemma 7.6 one of $P_{t'}^d$, $L\{P_t^d \mid t < t'\}$ and $L\{P_t^d \mid t > t'\}$ has the fixed point property. For $t_0 < t'$ note that the induction hypothesis applies to $L\{P_t \mid t > t_0\}$ and thus one of $P_{t_0}^d$, $L\{P_t^d \mid t < t_0\}$ and $L\{P_t^d \mid t > t_0\}$ has the fixed point property. The case $t_0 < t'$ is handled similarly. Thus, by Proposition 7.7, the ordered set $L\{P_t^d \mid t \in T\}$ has the fixed point property.

We have proved the condition in Theorem 7.18 for the fixed point property. Thus the fixed point property is a comparability invariant for finite ordered sets.

To prove 2 and 3, consider the following example. Take a non-self-dual rigid ordered set P. For example, we could take a linear lexicographic sum of a singleton and a fence with an even number of elements. Consider the disjoint union $P + P^d$ of P and its dual P^d. This ordered set is rigid. In particular, it has no fixed point free automorphism. Note that $P + P^d$ has the same comparability graph as $P + P$, the disjoint union of the ordered set P with itself. However, $P + P$ has a fixed point free automorphism (and is thus also not rigid either).   ∎

Exercises

7-14. Prove that the following are comparability invariants for finite ordered sets. Are they comparability invariants for all ordered sets?

   a. The width of an ordered set.
   b. The height of an ordered set.
   c. The number of antichains in an ordered set. (Only consider finite sets.)
   d. The number of chains in an ordered set. (Only consider finite sets.)

7-15. (Convex subsets.) Let P be an ordered set and let $C \subseteq P$. Then C is called **convex** iff, for all $x, y \in C$ with $x \le y$, we have that $\{p \in P : x \le p \le y\} \subseteq C$.

   a. Show that every piece in a lexicographic sum decomposition is convex.
   b. Show that the intersection of convex subsets of an ordered set is again convex.
   c. Show that the union of (even intersecting) convex subsets need not be convex.

7-16. Let P be an ordered set and let $K \subseteq P$ be an ordered subset. Define the **convex hull** of K as $\mathrm{con}_P(K) := \bigcap \{C \subseteq P : K \subseteq C, \text{ and } C \text{ is convex}\}$.

   a. Show that the convex hull of an ordered subset K is the smallest convex subset of P that contains K.
   b. Prove that, if P and Q are ordered sets with the same comparability graph and S is order-autonomous in P, then $\mathrm{con}_Q(S)$ is order-autonomous in Q.
   c. We have encountered a convex hull in the proof of Theorem 7.18. Where?

7-17. Prove that, if P is a finite, decomposable, co-connected ordered set, then P has an order-autonomous subset that is order-autonomous in any transitive orientation of the comparability graph $G_C(P)$ of P.
   Hint. Use Exercise 7-16b to show that maximal order-autonomous sets are order-autonomous in any transitive orientation of $G_C(P)$.

7-18. Let P be a finite ordered set. For any two edges $\{x, y\}$ and $\{x, z\}$ of $G_C(P)$ with $x \ne y$ and $x \ne z$ say $\{x, y\} \wedge \{x, z\}$ iff $y \not\sim z$ or $y = z$ (see [101], p. 25; [154], p. 5).

   a. Prove that the transitive closure of the relation $\wedge$ is an equivalence relation on the nontrivial edges of the comparability graph.
      The equivalence classes of $\wedge$ are called **edge classes** (see [101], p 25; [154], p. 5).
   b. Let C be an edge class. Prove that any transitive orientation of $G_C(P)$ induces one of two dual transitive orientations on the edges in C.
   c. For any edge class C let $V(C)$ be the set of points $p \in P$ such that there is an $x \in P$ with $\{p, x\} \in C$. Prove that $V(C)$ is order-autonomous in any transitive orientation of $G_C(P)$.
   d. Let C be an edge class. Prove that the graph $(V(C), C)$ is connected.
   e. (See [101], Hilfssatz 2.3.) Let P be connected and co-connected. Prove that $G_C(P)$ has exactly one edge class C such that $V(C) = P$.

Hint. Let $\underline{L}$ be a smallest possible set such that the graph induced on $P \setminus L$ by the complement $\overline{G_C(P)}$ of the comparability graph is disconnected. Let $K_1, \ldots, K_n$ be the vertex sets of the components of this graph.

- Note that, for all $i \neq j$ and $k_i \in K_i$, $k_j \in K_j$, we have $k_i \sim k_j$.
- Prove that all edges that go from a K_i to a K_j with $i \neq j$ are in the same edge class. Call this edge class D.
- Prove that all edges from L to a K_i are in D.
- Conclude that $V(D) = P$.
- Prove that, for any further edge class D', the set $V(D')$ must be completely contained in a K_i.

7-19. Use Exercises 7-16 through 7-18 (more precisely, Exercises 7-17, 7-18b, 7-18c, and 7-18e) to prove Theorem 7.16.

7-20. (More intersections of convex sets.) This exercise shows how a natural result for intervals on the real line (see part 7-20a) has analogues for convex sets in ordered sets that are not chains.

a. Prove that if $\{C_i\}_{i \in I}$ is a family of convex subsets of a finite chain such that $C_i \cap C_j \neq \emptyset$ for all $i, j \in I$, then $\bigcap_{i \in I} C_i \neq \emptyset$.
b. Prove that if $\{C_i\}_{i \in I}$ is a family of convex subsets of a finite ordered set of width w such that any subfamily of $\leq 2w$ sets has nonempty intersection, then $\bigcap_{i \in I} C_i \neq \emptyset$.
c. For $w > 0$, find an ordered set of width w and a family $\{C_i\}_{i=1,\ldots,2w}$ with empty intersection such that any $(2w - 1)$-element subfamily has nonempty intersection.

7.4 Lexicographic Sums and Reconstruction

Although it is still an open question whether nontrivial lexicographic sums are reconstructible, we will prove in this section that "many" lexicographic sums are reconstructible. To be precise, we will only prove weak reconstructibility of many lexicographic sums. The proof of recognizability of decomposable ordered sets can be found in [145]. By quoting said paper, we can then establish the reconstruction result in Theorem 7.25. Because disconnected ordered sets are reconstructible by Proposition 2.51 and because series decomposable sets are reconstructible by Exercise 7-21, we will be able to concentrate on connected, and co-connected ordered sets. This allows us to make extensive use of the canonical decomposition.

Definition 7.20. *For every connected and co-connected card C^k of an ordered set we let $L\{C_s^k \mid s \in S^k\}$ stand for the canonical decomposition of C^k.*

Lemma 7.21. *Let P be a finite, decomposable, connected and co-connected ordered set. Let $L\{P_t \mid t \in T\}$ be the canonical decomposition of P and let $x \in P$.*

1. *If $A \subseteq P$ is order-autonomous in P and $x \in P$, then $A \setminus \{x\}$ is order-autonomous in $P \setminus \{x\}$ or empty.*
2. *For every connected and co-connected card C^k, the index set S^k of the canonical decomposition $L\{C_s^k \mid s \in S^k\}$ of C^k has at most as many elements as T. Let*

$$m := \max \left\{ |S^k| : C^k \text{ is a connected and co-connected card} \right\}.$$

3. *There is a connected and co-connected card C^k of P such that the index set S^k of the canonical decomposition $L\{C_s^k \mid s \in S^k\}$ of C^k is isomorphic to T. (Hence $m = |T|$.)*
4. *We will call connected and co-connected cards with $|S^k| = m$ the **crucial cards**. Let $N \subseteq P$ be the set of all points that are in a nontrivial maximal order-autonomous subset of P. The crucial cards are exactly the cards of P that are obtained by removal of an element of N.*
5. *Let C^k be a connected and co-connected card of P. If $|S^k| = m$, then S^k is isomorphic to T.*

Proof. To prove 1, let $A \subseteq P$ be order-autonomous in P and let $x \in P$. If $A \setminus \{x\}$ is empty, there is nothing to prove. If $A \setminus \{x\} \neq \emptyset$, let $p \in P \setminus (A \cup \{x\})$ be comparable to $y \in A \setminus \{x\}$, say $p \geq y$. Then $p \geq A$ and hence we have $p \geq A \setminus \{x\}$. The other comparability is handled dually. This proves that $A \setminus \{x\}$ is order-autonomous in $P \setminus \{x\}$.

For 2, let $C^k := P \setminus \{x\}$ be a connected and co-connected card. Then, for each maximal order-autonomous set $P_t \subseteq P$, the set $P_t \setminus \{x\}$ is order-autonomous in C^k or empty. If $P_t \setminus \{x\}$ is nonempty, it is contained in a unique maximal order-autonomous subset of C^k. Thus C^k has at most as many maximal order-autonomous subsets as P and hence $|S^k| \leq |T|$.

For the proof of 3, let $P_r \subseteq P$ be a maximal order-autonomous subset with $|P_r| \geq 2$. Let $x \in P_r$. Then all $Q_t := P_t \setminus \{x\}$ are nonempty order-autonomous subsets of $P \setminus \{x\}$ and $P \setminus \{x\} = L\{Q_t \mid t \in T\}$. Because T is neither disconnected nor series decomposable, $P \setminus \{x\}$ is connected and co-connected. Because $L\{P_t \mid t \in T\}$ is the canonical decomposition of P, by Proposition 7.14, T is not decomposable. Hence, again by Proposition 7.14, $P \setminus \{x\} = L\{Q_t \mid t \in T\}$ is the canonical decomposition of $P \setminus \{x\}$, and the index set is isomorphic to T.

For the proof of 4, note that, by the proof of 3, if C^k is obtained from P by removing a point out of a maximal order-autonomous set with ≥ 2 elements, then C^k is crucial. On the other hand, if C^k is obtained from P by removing a point p such that $\{p\}$ is maximal order-autonomous, then, by 1, all $P_t \setminus \{p\}$ are order-autonomous in $P \setminus \{p\}$ or empty. There are $m - 1$ nonempty sets $P_t \setminus \{p\}$. Thus, if $P \setminus \{p\}$ is connected and co-connected, then $|S^k| \leq m - 1 < m = |T|$ and C^k is not crucial.

For part 5, note that the crucial cards are exactly those cards that were obtained by removing an element of a nontrivial autonomous subset. The proof of 3 shows that the index set of the canonical decomposition of a crucial card must be isomorphic to T. ∎

Lemma 7.22. *The number m from Lemma 7.21, the crucial cards, and hence T are weakly reconstructible. That is, they can be identified from $\mathcal{D}_P$ if it is known that P is decomposable, connected, and co-connected.*

Proof. The number m can be determined from the formula in part 2 of Lemma 7.21, because the right side of this formula can be determined from the deck. Once we know m, we can determine the crucial cards as those cards for which the index set in the canonical decomposition has m elements. Finally we obtain (the isomorphism type of) T from the crucial cards via part 5 of Lemma 7.21. ∎

Proposition 7.23. *Let P be a finite, decomposable, connected, and co-connected ordered set with at least four elements. Then the maximal order-autonomous subsets of the canonical decomposition of P are weakly reconstructible. That is, if Q is another finite, decomposable, connected, and co-connected ordered set with $\mathcal{D}_P = \mathcal{D}_Q$, then there is a bijection between the maximal order-autonomous subsets of P and Q such that each set is isomorphic to its image.*

Proof. Because the number of maximal order-autonomous subsets of P is known to be m, which is weakly reconstructible from the deck, we only need to reconstruct the non-singleton maximal order-autonomous subsets. Let $C^1, \ldots, C^l$ be the crucial cards. For each crucial card C^k, let g_k be the number of elements of C^k that are contained in non-singleton maximal order-autonomous subsets of C^k. Let g be the number of elements of P that are contained in non-singleton maximal order-autonomous subsets of P. By Lemma 7.21, part 4 we know that $g = l$, so g can be determined. A crucial card C^k is obtained by removing an element from a maximal order-autonomous subset with ≥ 3 elements iff $g_k = g - 1$. C^k is a crucial card obtained by removing an element from a maximal order-autonomous subset with 2 elements iff $g_k = g - 2$.

For each crucial card C^k of P, create a new card D^k as follows. If $g - g_k = 1$, let D^k be the disjoint sum of the non-singleton maximal order-autonomous subsets of C^k. If $g - g_k = 2$, let D^k be the disjoint sum of the non-singleton maximal order-autonomous subsets of C^k and a singleton. The $D^1, \ldots, D^l$ are the deck of an ordered set D. This ordered set D is the disjoint sum of the non-singleton maximal order-autonomous subsets of P.

If D is disconnected with ≥ 3 elements (a fact that can be recognized from $D^1, \ldots, D^l$), we reconstruct D from $D^1, \ldots, D^l$. Because P could have disconnected maximal order-autonomous subsets, not every component of D is necessarily maximal order-autonomous in P. We proceed as follows. There is a $k_0 \in \{1, \ldots, l\}$ such that either no other crucial card C^k has fewer non-singleton maximal order-autonomous subsets than C^{k_0} or no other crucial card C^k has a smaller non-singleton maximal order-autonomous subset. The (not necessarily unique) card C^{k_0} is obtained by removing an element of a smallest possible non-singleton maximal order-autonomous subset of P. If $g - g_{k_0} = 1$, then all non-singleton maximal order-autonomous subsets of C^{k_0} except for the smallest one are in fact non-singleton maximal order-autonomous subsets of P. If $g - g_{k_0} = 2$, then all non-singleton maximal order-autonomous subsets of C^{k_0} are non-singleton maximal order-autonomous subsets of P, and P has exactly one more two-element maximal order-autonomous subset. In either case we obtain the number j of non-singleton maximal order-autonomous subsets of P. If $j = 1$, then the disconnected ordered set D is the only non-singleton maximal order-autonomous subset of P. If $j \geq 2$, then, in both cases above, all but one non-singleton maximal order-autonomous subset of P have been found. Call these sets $Q_1, \ldots, Q_{j-1}$. There is a unique (up to isomorphism) ordered set Q_j such that D is the disjoint sum of the sets $Q_1, \ldots, Q_j$. Q_j is the last non-singleton maximal order-autonomous subset of P.

This leaves the cases in which $D^1, \ldots, D^l$ is the deck of a connected ordered set D with ≥ 3 elements or of a 2-element set (which are not necessarily known to be reconstructible from the deck). In this case, P has exactly one non-singleton maximal order-autonomous subset with $|D|$ elements, namely, the set D.

Let C^1 be a crucial card. If $|D| = 2$, then the ordered subset D is a 2-antichain iff $e_H(C^1) < e_H(P) - 1$ or C^1 has one less minimal or one less maximal element than P. This leaves the case $|D| > 2$. In this case, there is a unique maximal order-autonomous subset D' on C^1 that is not a singleton. Naturally D' was obtained from D by removal of one element. Let U be the set of upper bounds of D' on C^1 and let L be the set of lower bounds of D' on C^1. Then U is also the set of upper bounds of D in P and L is also the set of lower bounds of D in P. P was not series decomposable. Thus, for each ordered set of the form $L \oplus \overline{D} \oplus U$ with $|\overline{D}| = |D|$ we can use Kelly's Lemma (see Proposition 1.40) to find out how many copies of $L \oplus \overline{D} \oplus U$ are contained in P. D will be the unique ordered set for which C^1 contains one less copy of $L \oplus D \oplus U$ than P.   ∎

To state the main result of this section in full generality, we need the recognizability of decomposable ordered sets. For a proof of this result, see [145]. Thus conscientious readers may want to replace the word "reconstructible" in Theorem 7.25 with "weakly reconstructible" (also see Definition 11.15) until after reading the paper [145].

Lemma 7.24 (For the proof, see [145], Theorem 2 and use [144] for $4 \leq n \leq 11$).
If P is a finite ordered set with $|P| \geq 4$, then P is recognizable as decomposable or indecomposable from the deck.   ∎

Theorem 7.25. *Let P be a finite, decomposable ordered set with ≥ 12 elements such that there is a piece P_i of P such that a card C_i of P_i is not isomorphic to any pieces of P. Then P is reconstructible.*

Proof. Find a crucial card that contains a maximal order-autonomous set that is isomorphic to C_i. Replace C_i with P_i to obtain an ordered set isomorphic to P.   ∎

Exercises

7-21. (See [174], Theorem 4.7.) Let P be a series decomposable ordered set. Prove that P is reconstructible.

7-22. Construct an indecomposable ordered set P such that all minimal cards of P are connected and decomposable.

7-23. Let P be a finite ordered set with ≥ 12 elements such that there is a $b > 0$ such that

- There is a maximal order-autonomous subset of P with $> b$ elements, and
- No maximal order-autonomous subset of P is of size b.

Prove that P is reconstructible.

7-24. Let P be a finite ordered set with ≥ 12 elements such that every maximal element of P is contained in a nontrivial order-autonomous subset of P. Prove that P is reconstructible.

7-25. Let P be a finite ordered set with ≥ 12 elements that has only one non-singleton maximal order-autonomous subset. Assume that this subset has at least three elements. Prove that P is reconstructible.

7.5   An Almost Lexicographic Construction

From the proof of Lemma 7.6, it is easy to see that fixed point free order-preserving self-maps of series decomposable finite ordered sets must have a special structure: They must map all pieces to themselves. The following construction exploits a similar idea to generate a class of ordered sets whose fixed point free order-preserving self-maps must have a certain structure. Except for part 5 in Proposition 7.26, the construction is essentially lexicographic, which is why this section fits into this chapter. The structure in Proposition 7.26 will allow us to embed the problem 3SAT (see Example 7.28) into the problem whether a given finite ordered set has a fixed point free order-preserving self-map (see Theorem 7.32).

Proposition 7.26 (See [68], Section 3). *Let X, Y, and Z be finite ordered sets. Let P be an ordered set with the following properties (also see Figure 7.2).*

1. The underlying set for P is the union of the six crown $\{a, b, c, a', b', c'\}$, the ordered sets X, Y, Z, and the $2n$-crown C_{2n}.

2. $\{a, b, c, a', b', c'\}$ is ordered by $a < b', c'$; $b < a', c'$; and $c < a', b'$.

3. X, Y, Z, and C_{2n} carry their original orders.

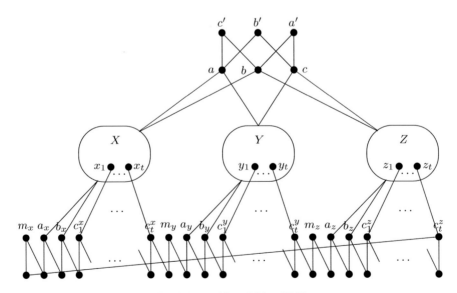

Fig. 7.2 The ordered sets described in Propositions 7.26 and 7.27.

4. $X \le \{a, b\}$, $Y \le \{a, c\}$, $Z \le \{b, c\}$, and $C_{2n} \le \{a, b, c\}$.
5. *Each element of X, Y, Z is above at least two maximal elements of C_{2n}, and no element of X, Y, Z is an upper bound of C_{2n}. (This does not completely specify the comparabilities involving X, Y, and Z, which is intentional.)*
6. *The only further comparabilities are those added to the above by transitivity.*

   If P has a fixed point free order-preserving self-map $f : P \to P$, then the following hold.

1. *$f|_{\{a,b,c,a',b',c'\}}$ and $f|_{C_{2n}}$ are automorphisms.*
2. *$f[X] \subseteq Y, f[Y] \subseteq Z, f[Z] \subseteq X$, or $f[X] \subseteq Z, f[Z] \subseteq Y, f[Y] \subseteq X$.*

Proof. Let $f : P \to P$ be a fixed point free order-preserving self-map. Suppose, for a contradiction, that $f[C_{2n}] \not\subseteq C_{2n}$. Then, without loss of generality, there is a $p \in C_{2n}$ with $f(p) \in Y$. Now, the containments $f[\{a, c, a', b', c'\}] \subseteq\uparrow f(p)$ and $f[\{a, c, a', b', c'\}] \cap Y = \emptyset$ imply that $f[\{a, c, a', b', c'\}] \subseteq \{a, c, a', b', c'\}$, which means that f has a fixed point, a contradiction.

Therefore $f[C_{2n}] \subseteq C_{2n}$, which means, by Exercise 2-28, that $f|_{C_{2n}}$ is an automorphism. However, this implies that $f[\{a, b, c, a', b', c'\}] \subseteq \{a, b, c, a', b', c'\}$, and, using Exercise 2-28 once more, we conclude that 1 must be true.

To prove part 2, first note that, because every element of $X \cup Y \cup Z$ is below two distinct minimal elements of $\{a, b, c, a', b', c'\}$ and above two distinct maximal elements of C_{2n}, we have that $f[X \cup Y \cup Z] \subseteq X \cup Y \cup Z$. Assume without loss of generality that $f(a) = b$, $f(b) = c$, and $f(c) = a$ (the other permutation is handled similarly). Then $f(X) \le \{f(a), f(b)\} = \{b, c\}$, which forces $f[X] \subseteq Z$. The containments $f[Z] \subseteq Y$ and $f[Y] \subseteq X$ are proved similarly.   ∎

Proposition 7.27 (See [68], Section 5). *With assumptions as in Proposition 7.26, let $t > 1$, and let $\{x_1, \ldots, x_t\} \subseteq X$, $\{y_1, \ldots, y_t\} \subseteq Y$, and $\{z_1, \ldots, z_t\} \subseteq Z$ be antichains. Let $n = 3t + 9$ and let the maximal elements of C_{2n}, in cyclic order, be*

$$m_x, a_x, b_x, c_1^x, \ldots, c_t^x, m_y, a_y, b_y, c_1^y, \ldots, c_t^y, m_z, a_z, b_z, c_1^z, \ldots, c_t^z.$$

(See Figure 7.2.) To specify the ordered set P, we only need to specify comparabilities in part 5 of Proposition 7.26.

1. *Let $X > \{a_x, b_x\}$, $Y > \{a_y, b_y\}$, and $Z > \{a_z, b_z\}$.*
2. *For $i \in \{1, \ldots, t\}$ let $x_i > c_i^x$, $y_i > c_i^y$, and $z_i > c_i^z$.*
3. *No further comparabilities beyond those dictated by transitivity.*

   If the thus defined ordered set P has a fixed point free order-preserving map $f : P \to P$, then, for all $i \in \{1, \ldots, t\}$, we have $f(x_i) \ge y_i, f(y_i) \ge z_i$, and $f(z_i) \ge x_i$, or, for all $i \in \{1, \ldots, t\}$, we have $f(x_i) \ge z_i, f(z_i) \ge y_i$, and $f(y_i) \ge x_i$.

Proof. Let $f : P \to P$ be a fixed point free order-preserving map. By Proposition 7.26, we can assume, without loss of generality, that $f[X] \subseteq Y$, $f[Y] \subseteq Z$, and $f[Z] \subseteq X$. Because $f|_{C_{2n}}$ is an automorphism, we have $f[\{a_x, b_x\}] = \{a_y, b_y\}$, $f[\{a_y, b_y\}] = \{a_z, b_z\}$, and $f[\{a_z, b_z\}] = \{a_x, b_x\}$ with a's mapped to a's and b's mapped to b's. Thus, again because $f|_{C_{2n}}$ is an automorphism, for all $i \in \{1, \ldots, t\}$, we have $f(c_i^x) = c_i^y, f(c_i^y) = c_i^z$, and $f(c_i^z) = c_i^x$.

Now, because $f(x_i) \geq f[\{a_x, b_x, c_i^x\}] = \{a_y, b_y, c_i^y\}$ and because all elements in Y that are above this set are also above y_i, we must have $f(x_i) \geq y_i$. Similarly we prove $f(y_i) \geq z_i$ and $f(z_i) \geq x_i$.   ∎

We are now ready to formally prove that determining whether an ordered set has the fixed point property is a "hard" problem. Although the next section only connects to lexicographic sums because it uses Proposition 7.27, we might as well take care of this leftover detail from Chapter 5 right now.

7.6   NP-Completeness of *FPF(P)*

NP-completeness is a formalized way of saying/proving that a problem is "hard." This section will define what exactly this statement means. To discuss the concept of NP-completeness, we need to define a problem in formal logic which can be viewed as another prototype of NP problems. Recall that a logical variable (or literal) is a variable that can be assigned the values TRUE or FALSE and that "∨" denotes the logical OR operation.

Example 7.28 (The problem 3SAT). (See [103].) Given n logical variables ("literals") and m clauses $c_j = q_j \vee r_j \vee s_j$, where each q_j, r_j, s_j can be a literal or its negation, decide if there is a truth value assignment to the literals such that all clauses are true.

□

3SAT is a problem in NP: Because the number of distinct clauses that can be formed with 3 out of n literals or their negations is polynomial in n, for any truth value assignment to the n literals, it will take polynomial time $p(n)$ to check if all clauses are satisfied. The importance of 3SAT stems from a result by Cook, which shows that 3SAT is of a universal nature.

Theorem 7.29 (Cook's theorem, see [49] or [103], Theorem 2.1). *Suppose there is an algorithm that solves 3SAT in polynomial time. Then, for every NP problem, there is a polynomial time solution algorithm.*

The proof of Theorem 7.29 does not fit our scope. It can be found in [49] and in [103], Theorem 2.1, where the result is proved for SAT. Theorem 7.29 shows a fact that the translations shown in Chapter 5 may have already suggested: There are certain "universal problems" into which all NP problems can be translated with a translation that takes polynomial time. So, if one can solve these problems efficiently, then all NP problems can be solved efficiently. These universal problems are called NP-complete.

Definition 7.30. *An NP problem X is said to be **NP-complete** iff an algorithm that solves X in polynomial time would induce an algorithm that solves 3SAT with n variables in polynomial time p(n).*[2]

For problems for which no polynomial algorithm is known yet, it is common to attempt to prove that the problem is NP-complete. Such a proof, if it is possible, normally involves an embedding of 3SAT, or of another NP-complete problem, into the problem in question. In the following, we will exhibit Duffus and Goddard's proof that checking for the existence of a fixed point free order-preserving map is an NP-complete problem. We start by showing that a related problem is NP-complete.

Lemma 7.31 (See [68], Theorem 4.1). *The following problem is NP-complete.*
    Given. *Two ordered sets P and Q and $p_1, \ldots, p_t \in P$, $q_1, \ldots, q_t \in Q$.*
    Question. *Is there an order-preserving map $f : P \to Q$ with $f(p_i) \geq q_i$?*

Proof. We will first embed 3SAT into a similar problem for graphs. Let an instance of 3SAT with clauses $C_1, \ldots, C_m$ and literals $x_1, \ldots, x_n$ be given. Order the literals in each clause in a fixed fashion. (For example, sort them such that the indices increase.) Define graphs $G = (V_G, E_G)$ and $H = (V_H, E_H)$ as follows.

The vertices of G are called $C_1^*, \ldots, C_m^*$ and $x_1^*, \ldots, x_n^*$. There is an edge in G between C_i^* and x_j^* iff the literal x_j or its negation occurs in C_i and there are no other edges.

In H, we have vertices $C_1^1, C_1^2, C_1^3, C_2^1, C_2^2, C_2^3, \ldots, C_m^1, C_m^2, C_m^3$ and vertices $x_1^T, x_1^F, x_2^T, x_2^F, \ldots, x_n^T, x_n^F$. Again, edges only exist between C's and x's. If x_j or its negation occurs in C_i, but not in the l-th place, put an edge between C_i^l and x_j^T and an edge between C_i^l and x_j^F. If x_j occurs in the l-th place of C_i, put an edge between C_i^l and x_j^T. If $\overline{x_j}$ occurs in the l-th place of C_i, put an edge between C_i^l and x_j^F. These are all the edges.

We claim that the instance of 3SAT has a satisfying assignment iff there is a graph homomorphism f from G to H such that $f(C_i^*) \in \{C_i^1, C_i^2, C_i^3\}$ and $f(x_j^*) \in \{x_j^T, x_j^F\}$. Indeed, if there is such a map, then the assignment of "true" to x_j iff $f(x_j^*) = x_j^T$ gives a satisfying assignment. For the converse, fix a satisfying assignment. Then, for each C_i, there is a literal or a negation thereof in C_i that must evaluate to being true, say, one is in the l_i-th place. Map C_i^* to $C_i^{l_i}$, map x_j^* to x_j^T iff x_j is assigned the value "true" and map x_j^* to x_j^F otherwise. This map is as desired.

Now we must add a structure to change the conditions $f(C_i^*) \in \{C_i^1, C_i^2, C_i^3\}$ and $f(x_j^*) \in \{x_j^T, x_j^F\}$ into conditions about adjacencies. To do this, we obtain graphs $K_P = (V_P, E_P)$ from G and $K_Q = (V_Q, E_Q)$ from H as follows. To construct K_P from G, add vertices C_i^0, $i = 1, \ldots, m$ and x_j^a, x_j^b, $j = 1, \ldots, n$ to G. Each C_i^0 is adjacent to C_i^*, while x_j^a and x_j^b are both adjacent to x_j^*. These are all the additions to G.

---

[2]Generally this is done by mapping 3SAT in polynomial fashion into specific instances of X. This definition has be considered somewhat "informal."

To construct K_Q from H, add vertices D_i^0, $i = 1, \ldots, m$ and $x_j^c, x_j^d, j = 1, \ldots, n$ to H. Each D_i^0 is adjacent to C_i^1, C_i^2, C_i^3, while x_j^c and x_j^d are both adjacent to x_j^T and x_j^F. These are all the additions to H.

It is easy to see that there is a graph homomorphism from G to H as described above iff there is a graph homomorphism g from K_P to K_Q such that $g(C_i^0) = D_i^0$, $g(x_j^a) = x_j^c$, and $g(x_j^b) = x_j^d$. Now let P be the dual of the split of K_P and let Q be the dual of the split of K_Q. Then there is a homomorphism g from K_P to K_Q such that $g(C_i^0) = D_i^0$, $g(x_j^a) = x_j^c$, and $g(x_j^b) = x_j^d$ iff there is an order-preserving map $h : P \to Q$ such that $h(C_i^0) \geq D_i^0$, $h(x_j^a) \geq x_j^c$, and $h(x_j^b) \geq x_j^d$. (Note that, because the elements in question are maximal, all the inequalities, in fact, imply equality). ∎

The proof of NP-completeness of checking for the existence of a fixed point free order-preserving map[3] is now accomplished by embedding the above problem into a fixed point setting via the (almost) lexicographic construction in Proposition 7.27.

Theorem 7.32 (See [68], Theorem 1.1). *The following decision problem is NP-complete.*

> **Given.** *A finite ordered set P.*
> **Question.** *Is there a fixed point free order-preserving map $f : P \to P$?*

Proof. First note that, for the sets P and Q in the proof of Lemma 7.31, it is easy to find maps $h' : Q \to P$ and $h'' : P \to P$ such that $h'(D_i^0) \geq C_i^0$, $h'(x_i^c) \geq x_i^a$, $h'(x_i^d) \geq x_i^b$, and $h''(C_i^0) \geq C_i^0, h''(x_i^a) \geq x_i^a, h''(x_i^b) \geq x_i^b$, respectively.

Now, in the construction of Proposition 7.27, let X and Y be copies of P and let Z be a copy of Q. Call the resulting ordered set W. Let $t = 2n + m$, let $x_1, \ldots, x_t$ and $y_1, \ldots, y_t$ each be $C_1^0, \ldots, C_m^0, x_1^a, \ldots, x_n^a, x_1^b, \ldots, x_n^b$ in this order, and let $z_1, \ldots, z_t$ be $D_1^0, \ldots, D_m^0, x_1^c, \ldots, x_n^c, x_1^d, \ldots, x_n^d$ in this order. Now, if the thus constructed ordered set has a fixed point free order-preserving map, then, by Proposition 7.27, without loss of generality there is an order-preserving map $f : Y \to Z$ such that $f(y_i) \geq z_i$. This implies that there is an order-preserving map $h : P \to Q$ such that $h(C_i^0) \geq D_i^0$, $h(x_i^a) \geq x_i^c$, and $h(x_i^b) \geq x_i^d$. Conversely, if there is such a map, it is simple to construct a fixed point free order-preserving map $f : W \to W$.

Thus, if there was a polynomial algorithm that decides if a finite ordered set has the fixed point property, then there would be a polynomial algorithm that decides the problem in Lemma 7.31, which is NP-complete. ∎

We could now continue and show that there are other NP problems that are interesting in order theory and which are NP-complete. In particular, we mention the following.

Corollary 7.33. *The question if there is a solution to a given binary constraint satisfaction problem is NP-complete.*

---

[3]This is equivalent to saying that determining if a given finite ordered set has the fixed point property is co-NP-complete.

Proof. It was already remarked that the constraint satisfaction problem is NP. Suppose we had a polynomial algorithm that decides if a given binary CSP has a solution. Then, by Example 5.13, we would have an algorithm that decides in polynomial time if a given ordered set has a fixed point free order-preserving self-map. By Theorem 7.32, this would lead to a polynomial algorithm for 3SAT. ∎

Instead of proceeding further in this direction, some NP problems are to be shown to be NP-complete in Exercise 7-26 and the complexity status of further problems is discussed in the remarks.

Exercises

7-26. Prove that the following problems are NP-complete.

  a. **Given.** A finite ordered set P.
     Question. Is there a fixed point free automorphism of P?
     Hint. As a lemma, you may use that the following problem is NP-complete (see [196]).
     Given. A finite graph G.
     Question. Is there a fixed point free automorphism of G?
  b. (More on the weak fixed point property.)
     Given. A finite ordered set P.
     Question. Is there a fixed point free order-preserving, rank-preserving map $f : P \to P$?
     Hint. Use the construction in Theorem 7.32.
  c. **Given.** A finite graph $G = (V, E)$.
     Question. Does G have a fixed vertex free endomorphism?
     Hint. Corollary 6.32.

7-27. Technically, Theorem 7.32 shows that it is NP-complete to decide whether a given ordered set of height 5 has a fixed point free order-preserving self-map. Prove that it is NP-complete to decide whether a given ordered set of height 4 has a fixed point free order-preserving self-map.

  Hint. Modify the ordered set in Proposition 7.26 as follows: Keep the fact that C_{2n} is below every element of $\{a, b, c, a', b', c'\}$, but set up the comparabilities of elements of X, Y, Z with minimal elements of C_{2n} instead of maximal elements of C_{2n}.

Remarks and Open Problems

  1. Is it possible to prove an analogue of Proposition 7.7 without the assumption of chain-completeness?
  2. How much does comparability invariance say about a property for ordered sets? True, many properties only depend on the comparability graph, which may say that they are "graph-theoretical in nature." However, as Theorem 7.18 shows, in the setting of finite ordered sets, comparability invariance is equivalent to a simple result about lexicographic sums.

A natural follow-up question is if there are infinitary analogues for Theorem 7.18. A step in this direction is made in [273].

3. Another graph that is associated with ordered sets is the covering graph. The **covering graph** has the points of the ordered set as vertices and there is an edge between p and q iff $p \prec q$ or $q \prec p$. All questions that one can ask about comparability graphs can also be asked for covering graphs. However, the characterization of covering graphs is problematic.

> **Open Question 7.34 (See [249]).** *Characterize the undirected graphs that are covering graphs of ordered sets.*

Certainly, not every graph is a covering graph. The simplest example is the triangle K_3. In general, it is difficult, namely, NP-complete (see Definition 7.30), to determine if a given graph is a covering graph (see [34, 215]). This remains so even for lattices (see [259]), but can become "easy" (that is, polynomially solvable, see Definition 5.4) for some classes of ordered sets (such as distributive and modular lattices, see [6]).

NP-completeness means that a computational characterization is considered "hard." However, this would still allow for structural characterizations, which in turn are computationally "hard," too, but which could lead to deeper understanding. The only **covering graph invariants** found so far seem to be the genus of an ordered set (see [83]) and isometric order embeddability into a Boolean lattice (see [323]). The **genus** of an ordered set is the smallest possible genus of a surface on which the covering graph can be drawn as a planar graph such that a given "up"-direction on the surface makes the covering graph a diagram.

4. The proof of Lemma 7.24 in [145] is quite technical. Is there a short proof of Lemma 7.24?

5. Is the class of decomposable ordered sets reconstructible? Considering Theorem 7.25, one would conjecture that the answer is "yes" and that reconstruction should be possible with a reasonably short argument. However, a proof is not available so far. The main problem seems to be with lexicographic sums that have only one nontrivial piece of size 2.

6. N-free ordered sets have been referred to in a few exercises throughout (the latest being Exercise 7-2). Using results from [174] (or [275]), Lemma 7.22 and Proposition 7.23, it was proved in [276] that N-free ordered sets are reconstructible.

7. Are there reconstruction results for ranked sums (see Exercise 7-6) that are analogous to those for lexicographic sums?

8. Regarding isomorphism, it is not known if checking isomorphisms for graphs is polynomial, NP-complete, or a separate complexity class altogether. For more information on isomorphism problems and their complexity theory, consider [169]. For the special case of graphs with bounded valence, see [197].

9. For further examples of NP-complete problems, consider [103, 252].

10. The Clay Mathematics Institute has chosen the question whether P equals NP, that is, whether every nondeterministic polynomial problem has a polynomial solution algorithm, as one of the seven leading questions in mathematics for the future, see [46].

11. An affirmative answer to the automorphism problem for series parallel ordered sets is presented in [280].

12. To date, the smallest ordered set with the fixed point property for which applying a $(2, 1)$-consistency algorithm does not return an empty graph is constructed through the NP-completeness proof in Theorem 7.32. Take all eight possible clauses on three literals and construct the ordered set as in the proof of Theorem 7.32. Enforcing $(2, 1)$-consistency does not return an empty graph. Yet the ordered set has the fixed point property, because, naturally, the eight clauses cannot all be satisfied simultaneously. This set has over 400 elements. Are there smaller such examples that are more "natural" in the sense that they are not derived from the NP-completeness proof? Insights into this question would shed some further light on how difficult the fixed point property is to decide.

Chapter 8
Lattices

Lattices are (after chains) the most common ordered structures in mathematics. The reason probably is that the union and intersection of sets are the lattice operations "supremum" and "infimum" in the power set ordered by inclusion (see Example 3.21, part 5) and that many function spaces can be viewed as lattices (see Example 3.21, parts 6 and 7). Lattice theory is a well developed branch of mathematics. There are many excellent texts on lattice theory (see, e.g., [21, 54, 56, 98, 112]), so we will concentrate here only on some core topics and on the aspects that relate to unsolved problems and work presented in this text.

8.1 Definition and Examples

What sets lattices apart from other ordered sets is an abundance of suprema and infima.

Definition 8.1. *Let L be an ordered set. Then L is called a **lattice** iff any two elements of L have a supremum and an infimum. L is called a **complete lattice** iff any subset of L has a supremum and an infimum.*

Clearly, complete lattices are stronger structures than (incomplete) lattices. In this chapter, we will first focus on issues related to completeness. General lattice ideas come to the forefront starting in Section 8.4.

Example 8.2.

1. Every chain is a lattice, but not every chain is a complete lattice (consider $\mathbb{N}$).
2. If X is a set, then the power set $\mathcal{P}(X)$ ordered by set inclusion is a complete lattice (see Example 3.21, part 5).
3. The space $C([0, 2], \mathbb{R})$ is a lattice that is not complete (see Example 3.21, parts 6 and 8).

© Springer International Publishing 2016
B. Schröder, *Ordered Sets*, DOI 10.1007/978-3-319-29788-0_8

4. Let $p \geq 1$. The space $L^p(\Omega, \mathbb{R})$ is a lattice that is not complete. (See Exercise 8-11.)
5. Every set of sets that is closed under unions and intersections is a lattice.
6. The closed subspaces of a Hilbert space form a lattice when ordered by inclusion. The supremum of two subspaces X and Y is their direct sum $X \oplus Y$ and their infimum is (of course) $X \cap Y$. $\square$

The existence of suprema of doubleton sets easily implies the existence of suprema of finite nonempty sets. Hence finite lattices are complete.

Proposition 8.3. *Let L be a finite nonempty lattice. Then L is complete.*

Proof. Let L be a finite nonempty lattice. We will prove that every nonempty subset of L has a supremum. If $A \neq \emptyset$ is a subset of L with $|A| \in \{1, 2\}$, then there is nothing to prove. Proceeding by induction, we assume that every subset A of L with $|A| < n$ has a supremum. Let $B \subseteq L$ with $|B| = n$. Let $b \in B$ and let $B' := B \setminus \{b\}$. Then B' has $n - 1$ elements and thus it has a supremum $\bigvee B'$. Now, by Proposition 3.20, $b \vee \bigvee B'$ is the supremum of B.

By duality, we obtain that every nonempty subset of L has an infimum. The supremum and infimum of the empty set are now obtained via $\bigvee \emptyset = \bigwedge L$ and $\bigwedge \emptyset = \bigvee L$. $\square$

Remark 8.4. It can be checked in $O(|P|^4)$ steps if a given finite ordered set is a lattice: There are $O(|P|^2)$ pairs of points, $O(|P|)$ candidates for their supremum and their infimum and it takes $O(|P|)$ steps to check if a given point is the supremum or infimum of two others. This process not only checks if an ordered set is a lattice. It also computes all suprema and infima. $\square$

Proposition 8.5. *Every complete lattice L has a largest and a smallest element. The largest element will be denoted $\mathbf{1}$ and the smallest element will be denoted $\mathbf{0}$.*

Proof. The largest element is $\mathbf{1} := \bigvee L = \bigwedge \emptyset$ and the smallest element is $\mathbf{0} := \bigwedge L = \bigvee \emptyset$.   ∎

Corollary 8.6. *Every finite lattice is reconstructible.*

Proof. Easy consequence of Propositions 8.5 and 1.37.   ∎

The existence of infima and suprema in the definition of complete lattices is somewhat redundant and can be replaced solely by the existence of suprema. This fact is helpful when proving that an ordered set is a complete lattice, because it allows us to focus exclusively on suprema.

Proposition 8.7. *Let P be an ordered set. If every subset of P has a supremum, then P is a complete lattice.*

Proof. We must prove that every $A \subseteq L$ has an infimum. First note that L has a smallest element $\bigvee \emptyset$. Thus, for all $A \subseteq L$, we have that $\downarrow A \neq \emptyset$. However then $\bigvee \downarrow A = \bigwedge A$ and we are done.   ∎

The proof of Proposition 8.3 shows that, in any lattice, any nonempty finite set of elements has a supremum and an infimum. It is natural to ask what types of infinite subsets must have suprema and infima to make an infinite lattice complete. The answer is provided in the following.

Proposition 8.8. *Let L be a lattice. Then the following are equivalent.*

1. *L is complete.*
2. *L is chain-complete.*
3. *Every maximal chain of L is a complete lattice.*

Proof. The implication "1⇒2" is obvious and "1⇒3" follows from Exercise 8-5 and part 6 of Example 4.3.

To prove "3⇒2," let L be a lattice such that every maximal chain in L is a complete lattice. Let $C \subseteq L$ be a nonempty chain. Let K be a maximal chain that contains C and let $c := \bigvee_K C$. Then c is an upper bound of C in L. If $d \geq C$ was an upper bound with $d \not\geq c$, then we would have $C \leq d \wedge c < c$, which would imply that K was not maximal, a contradiction. Thus $c = \bigvee_L C$.

To prove "2⇒1," let L be a chain-complete lattice and let $A \subseteq L$ be a nonempty subset. Construct a chain that has the same upper bounds as A as follows. Well-order A to obtain an indexing $A = \{a_\alpha : \alpha < \xi\}$ for some ordinal number ξ. Let $c_0 := a_0$ and, for $0 < \alpha < \xi$, define

$$c_\alpha := \begin{cases} c_{\alpha^-} \vee a_{\alpha^-}; & \text{if } \alpha \text{ has an immediate predecessor } \alpha^-; \\ \bigvee_{\beta < \alpha} c_\beta; & \text{if } \alpha \text{ is a limit ordinal.} \end{cases}$$

Because L was assumed to be chain-complete, c_α is defined for all $\alpha < \xi$. The set $C := \{c_\alpha : \alpha < \xi\}$ is a chain. Moreover C and A have the same upper bounds: Indeed, $p \geq C$ trivially implies $p \geq A$, and if $q \geq A$, we can prove $q \geq C$ inductively. Thus, because L is chain-complete, C has a supremum and hence A has a supremum.

Infima of nonempty subsets are constructed dually. Because all nonempty subsets of L (in particular L itself) have a supremum and an infimum, L must be a complete lattice.   ∎

Proposition 8.9. *If T and all P_t are complete lattices, then the lexicographic sum $L\{P_t \mid t \in T\}$ is a complete lattice. Conversely, if $L\{P_t \mid t \in T\}$ is a (complete) lattice, then T is a (complete) lattice.*

Proof. Let T and all P_t be complete lattices. Let A be a subset of $L\{P_t \mid t \in T\}$. Then $I[A] \subseteq T$ has a supremum $\bigvee_T I[A]$. The supremum of A in P now is

$$\bigvee A = \bigvee_{P_{\bigvee_T I[A]}} (A \cap P_{\bigvee_T I[A]}).$$

The partial converse is an easy consequence of Lemma 7.5 and Exercise 8-5.   ∎

Note that, even if a lexicographic sum is a complete lattice, it is not necessary for all the pieces to be complete lattices. As an example, consider the linear lexicographic sum of a one-point ordered set, a two-point antichain, and another one-point ordered set. This set is a lattice, and yet the middle piece is not a lattice. We further explore this problem in Exercise 8-17.

Moreover, not even in the class of finite ordered sets is being a lattice a comparability invariant: Every finite lattice L is series decomposable into the top element and the rest of the lattice. Obtain P from L as follows. Remove the top element of L and add a new smallest element. Then $G_C(P)$ is isomorphic to $G_C(L)$. If the top element of L has more than one lower cover, then P is not a lattice.

Exercises

8-1. Let P be a three-dimensional polyhedron. The **faces** of P are P, the empty set, the vertices, the edges, the sides, and the whole polyhedron. The faces are ordered by inclusion.

    a. Prove that the faces form a lattice.
    b. Draw the face lattices for the Platonian solids (tetrahedron, octahedron, icosahedron, cube, and dodecahedron).

8-2. Recall the definition of a topology from Exercise 6-1. Clearly every topology is a lattice.

    a. Give an example of a topology that is not a complete lattice.
    b. Prove that the set of topologies on a set X forms a complete lattice.

8-3. Let P be an ordered set and let $C(P)$ be the set of its convex subsets ordered by inclusion. Show that $C(P)$ is a complete lattice.

8-4. Let P be an ordered set. Prove that, if $L \subseteq P$ is a complete lattice, then L is a retract of P.

8-5. Prove that every retract of a lattice is a lattice and that every retract of a complete lattice is a complete lattice.

8-6. Show that being a lattice also is not a comparability invariant for infinite ordered sets.

8-7. More results related to Proposition 8.9.

    a. Show that, even if T is a complete lattice and all P_t are lattices, the lexicographic sum $L\{P_t \mid t \in T\}$ need not be a lattice.
    b. Show that, if T is a lattice and all P_t are lattices with a smallest and a largest element, then $L\{P_t \mid t \in T\}$ is a lattice.

8-8. Let P be a finite ordered set such that P has a largest element, a smallest element and such that P does not contain any covering four crowns. That is, there are no subsets $\{l_1, l_2, u_1, u_2\}$ in P such that the u_i $(i = 1, 2)$ are upper covers of the l_j $(j = 1, 2)$. Give an example that shows that P need not be a lattice. Then show that, if, for each four crown $\{l_1, l_2, u_1, u_2\}$, there is an element m so that $l_1, l_2 \le m \le u_1, u_2$, then P must be a lattice.

8-9. Let L and M be lattices. Then $f : L \to M$ is called a **lattice homomorphism** iff, for all $x, y \in L$, we have $f(x \vee y) = f(x) \vee f(y)$ and $f(x \wedge y) = f(x) \wedge f(y)$.

    a. Show that every lattice homomorphism is order-preserving.
    b. Show that not every order-preserving map between lattices is a lattice homomorphism.
    c. Show that lattice homomorphisms need not preserve infinite suprema and infima.
    d. Show that a bijective lattice homomorphism is an order isomorphism.

8-10. Permutation lattices. Let S_n be the set of permutations σ of $\{1, \ldots, n\}$. The number of descents in σ is the number of elements k such that $\sigma(k) > \sigma(k + 1)$. Define $\sigma \prec \mu$ iff there is a transposition τ of two adjacent elements such that $\sigma \circ \tau = \mu$ and such that μ has more descents than σ.

  a. Prove that the transitive closure of $\prec$ is an order relation on S_n with $\prec$ being its lower cover relation.
  b. Prove that the thus obtained ordered set is a lattice.

8-11. Let (Ω, Σ, μ) be a measure space and let $p \in [1, \infty)$. We only consider real valued functions. Prove that the space $L^p(\Omega, \Sigma, \mu)$ is a lattice, but not a complete lattice.

8-12. Let (Ω, Σ, μ) be a measure space, let $p \in [1, \infty)$, consider only real valued functions and let $F : L^p(\Omega, \Sigma, \mu) \rightarrow L^p(\Omega, \Sigma, \mu)$ be an order-preserving map. If there is an $f \in L^p(\Omega, \Sigma, \mu)$ and an $n \in \mathbb{N}$ such that $F^n(f) \geq f$, then, for $g := \bigvee_{i=0}^{n} F^i(f) \in L^p(\Omega, \Sigma, \mu)$, we have $F(g) \geq g$. Moreover F has a fixed point above f iff F has a fixed point above g.

8-13. Let (Ω, Σ, μ) be a measure space, let $p \in [1, \infty)$, consider only real valued functions and let $F : L^p(\Omega, \Sigma, \mu) \rightarrow L^p(\Omega, \Sigma, \mu)$ be an order-preserving mapping. Prove that, if the set $\{F^n(f) : n \in \mathbb{N}\}$ is an infinite antichain with no upper bounds, then there is no fixed point of F above f.

8.2   Fixed Point Results/The Tarski–Davis Theorem

When we restrict our scope to lattices, the fixed point problem (see Open Question 1.19) has a complete solution, due to A. Tarski and A. Davis. Although the proof presented here is deceptively short, note that it heavily relies on strong results that we built earlier on.

Theorem 8.10 (The Tarski–Davis Theorem, see [58, 307]). *Let L be a lattice. Then L has the fixed point property iff L is a complete lattice. In this situation, for each order-preserving map $f : L \rightarrow L$, the set Fix(f) is a complete lattice.*

Proof. To show the direction "$\Rightarrow$," let L be a lattice with the fixed point property and assume that L is not complete. Then by, Proposition 8.8, there is a maximal chain $B \subseteq L$ that is not a complete lattice.

Thus there is a chain $C' \subseteq B$ that does not have a supremum in B. Define the chain $K := \uparrow_B C'$ and let $C := \downarrow_B K$. Then K has no infimum in B, C has no supremum in B, and $B = C \cup K$. Thus the (C, K)-core (see Definition 4.15) must be empty, because otherwise B would not be maximal. Hence the (C, K)-core does not have the fixed point property. By Theorem 4.17, this means that L does not have the fixed point property, a contradiction. Thus L must be complete.

The direction "$\Leftarrow$" and the fact that the fixed point set Fix(f) has a smallest element are easy consequences of the Abian–Brown Theorem applied to L and the smallest element of L.

To show that Fix(f) is a complete lattice, we must show that every nonempty set $A \subseteq$ Fix(f) has a supremum in Fix(f). Let $p := \bigvee A$ be the supremum of A in L. Then every fixed point of f that is above A is above p, too. Moreover

$$f(p) = f\left[\bigvee A\right] \geq \bigvee f[A] = \bigvee A = p.$$

Again by the Abian–Brown Theorem, there is a smallest fixed point a above p and a must be the supremum of A in $\mathrm{Fix}(f)$.   ∎

We now detour to a nice set-theoretical application of the Tarski–Davis Theorem and to an investigation of the structure of fixed point sets in a more general setting. If you wish to focus on lattices exclusively, it is safe to go on to Section 8.3.

8.2.1 Preorders/The Bernstein–Cantor–Schröder Theorem

Besides containment, a natural order relation for sets is size. For finite sets, we simply count the elements. For infinite sets, we cannot necessarily count the elements, but injective functions can be used for size comparisons: If there is an injective function $f : A \to B$, then we say A is of smaller cardinality than B and denote this by $A \trianglelefteq B$. This relation defines what is called a preorder on any set of sets (and also on classes of sets, if we want to extend our definitions in that direction).

Definition 8.11. *Let P be a set and let $\trianglelefteq$ be a reflexive and transitive relation on P. Then $(P, \trianglelefteq)$ is called a **preordered set** and $\trianglelefteq$ is called a **preorder**.*

Preorders are a weaker notion than orders. One could say that there is a "glitch" when it comes to antisymmetry. We have seen one example of such a glitch earlier.

Example 8.12. The relation defined in Proposition 1.3 is a preorder.   □

We can translate from preorders to orders simply by forcing antisymmetry. This is done by identifying any two points a and b such that $a \trianglelefteq b$ and $b \trianglelefteq a$. The next result shows that this simple idea is formally sound.

Proposition 8.13. *Let $(P, \trianglelefteq)$ be a preordered set. Then the relation $a \equiv b$ iff $a \trianglelefteq b$ and $a \trianglerighteq b$ is an equivalence relation on P. Let $Q := P/\equiv$ be the set of equivalence classes in P with respect to $\equiv$. The relation $\leq$ on Q defined by $[a] \leq [b]$ iff $a \trianglelefteq b$ is an order relation on Q.*

Proof. It is clear that $\equiv$ is reflexive, symmetric, and transitive. To see that $\leq$ is well-defined, let $a, a' \in [a]$ and $b, b' \in [b]$. Then $a \trianglelefteq b$ together with $a' \trianglelefteq a$ and $b \trianglelefteq b'$ implies that $a' \trianglelefteq b'$ and the converse is proved similarly. Thus the definition of $\leq$ is independent of the choice of representatives.

Reflexivity and transitivity of $\leq$ follow from reflexivity and transitivity of $\trianglelefteq$. Finally, if $[a] \leq [b]$ and $[b] \leq [a]$, then $a \trianglelefteq b$ and $b \trianglelefteq a$, which means $a \equiv b$, that is, $[a] = [b]$.   ∎

We can therefore conclude that our preorder relation $\trianglelefteq$ which measures size on any set or class of sets can be turned into an order relation by identifying any sets for which $A \trianglelefteq B$ and $B \trianglelefteq A$. Although this is formally correct, it would be very unsatisfying if two sets such that A is at most as large as B and B is at most as large as A were not of the same size. Equal size for sets of course means that there is a bijective function between them. Thus it is natural to look for a proof that any two sets that are equivalent as above are in fact of equal size.

The idea for proving this result is quite simple, and the implementation is a beautiful application of fixed point theory for ordered sets. If we have injective maps $f : A \to B$ and $g : B \to A$, then what we need to be concerned with is surjectivity. (Unless of course one of f and g already is surjective, in which case there is nothing to prove.) The elements in $B \setminus f[A]$ are the elements that f "misses." However, these are all inverse images of elements of A under g^{-1}. Thus we should be able to use g^{-1} to "fill in" the holes that f leaves in B. To do this, we partition A into two subsets, X and $A \setminus X$. The set X will be mapped into B with f, while $A \setminus X$ will be mapped with g^{-1}. The trick is how to choose X.

If g^{-1} has to fill up $B \setminus f[X]$, then we must have that $A \setminus X = g[B \setminus f[X]]$, or $X = A \setminus g[B \setminus f[X]]$. Finding X is now the task of finding a fixed point in a complete lattice. Subsequently we must make sure that the map we had in mind truly is the desired bijection. As it turns out, all the needed pieces fall into place.

Theorem 8.14 (The Bernstein–Cantor–Schröder[1] Theorem).   *Let A and B be sets such that there is an injective map $f : A \to B$ and an injective map $g : B \to A$. Then there is a bijective map $h : A \to B$.*

Proof (Guided by Exercise 1J in [325], which is in turn inspired by [167]). Define $F : \mathcal{P}(A) \to \mathcal{P}(A)$ by $F(C) := A \setminus g[B \setminus f[C]]$. To see that F is order-preserving, let $C \subseteq D$. Then $f[C] \subseteq f[D]$, $B \setminus f[C] \supseteq B \setminus f[D]$, $g[B \setminus f[C]] \supseteq g[B \setminus f[D]]$, and finally $F(C) = A \setminus g[B \setminus f[C]] \subseteq A \setminus g[B \setminus f[D]] = F(D)$, which means that F is order-preserving.

Now, because $\mathcal{P}(A)$ is a complete lattice, by the Tarski–Davis Theorem, F has a fixed point

$$E = F(E) = A \setminus g[B \setminus f[E]].$$

Define $h : A \to B$ by

$$h(x) := \begin{cases} f(x); & \text{if } x \in E, \\ g^{-1}(x); & \text{if } x \in A \setminus E. \end{cases}$$

First of all, h is totally defined, because $A \setminus E = g[B \setminus f[E]]$. To see that h is injective, let x and y be two distinct elements of A. Because f and g^{-1} are both injective, the only case we need to consider is $x \in E$ and $y \in A \setminus E$. Note that $y \in A \setminus E = g[B \setminus f[E]]$ means that there is a $z \in B \setminus f[E]$ such that $y = g(z)$. This

---

[1]Not a relative of the author.

in turn implies $h(y) = g^{-1}(y) = z \notin f[E]$ and because $h(x) = f(x) \in f[E]$, we have
that $h(x) \neq h(y)$.

To show that h is surjective, let $b \in B$. If $b \in f[E]$, there is nothing to prove. If
$b \in B \setminus f[E]$, then $g(b) \in g[B \setminus f[E]] = A \setminus E$. But then $b = g^{-1}(g(b)) = h(g(b))$
and h is surjective.   ■

8.2.2   Other Results on the Structure of Fixed Point Sets

Establishing the fixed point property for an ordered set means showing that, for
all order-preserving maps $f : P \to P$, we have that the set $\mathrm{Fix}(f)$ of fixed points
is not empty. It would now also be nice to know more about $\mathrm{Fix}(f)$. As seen in
Theorem 8.10, a sufficient condition for the fixed point property may be strong
enough to imply more properties of $\mathrm{Fix}(f)$.

We start here by first weakening the hypotheses of the Tarski–Davis Theorem.
Then we consider other properties such as dismantlability and connected collapsibility. Although the latter two results do not connect with lattices, it is natural to
present them here, as the investigation of the structure of the fixed point sets clearly
is motivated by the Tarski–Davis Theorem.

Theorem 8.15. *Let P be an ordered set with the fixed point property. Then, for
every order-preserving map $f : P \to P$, we have that every maximal chain in $\mathrm{Fix}(f)$
is a complete lattice.*

Proof. Let $f : P \to P$ be an order-preserving function and let $C \subseteq \mathrm{Fix}(f)$ be a
maximal chain. Suppose that $B \subseteq C$ has no supremum in C. Then $U := \uparrow_C B$ has
no infimum in C and $L := \downarrow_C U$ has no supremum in C. Clearly $C = L \cup U$.
Moreover, because C is maximal in $\mathrm{Fix}(f)$, the restriction of f to the (L, U)-core
$\{x \in P : L \leq x \leq U\}$ does not have any fixed points. However, if P has the
fixed point property, then, by Theorem 4.17, the (L, U)-core also has the fixed point
property, which is a contradiction. Thus every maximal chain in $\mathrm{Fix}(P)$ must be a
complete lattice.   ■

A natural weakening of the definition of a complete lattice is to take away the
top and bottom elements. We will discuss these structures and the related truncated
lattices in more detail in Chapter 9.

Definition 8.16. *An ordered set L is called **conditionally complete** iff any subset
$S \subseteq L$ that has an upper bound and is not empty has a supremum.*

Theorem 8.17. *Let P be a conditionally complete ordered set with the fixed point
property. Then, for each order-preserving map $f : P \to P$, the set $\mathrm{Fix}(P)$ is
conditionally complete.*

Proof. Let $f : P \to P$ be order-preserving and let $A \subseteq \mathrm{Fix}(f)$ be such that
there is a $p \in P$ with $p \geq A$. Then f maps $\uparrow (\bigvee A)$ to itself. Because P is

conditionally complete and has the fixed point property, P must be chain-complete. Thus $\uparrow (\bigvee A)$ is chain-complete. Therefore the Abian–Brown Theorem implies that f has a smallest fixed point in $\uparrow (\bigvee A)$. Hence A has a supremum in $\mathrm{Fix}(f)$. ∎

Theorem 8.18 (See [71], Theorem 3). *Let P be a finite ordered set that is $\mathcal{I}$-dismantlable to a singleton. Then, for each order-preserving map $f : P \to P$, the set $\mathrm{Fix}(f)$ is $\mathcal{I}$-dismantlable to a singleton.*

Proof. The proof is an induction on $n := |P|$. There is nothing to prove for $|P| = 1$. For the induction step, let P be a finite ordered set, $\mathcal{I}$-dismantlable to a singleton, and assume that the result has already been proved for ordered sets with fewer elements than P. Let $f : P \to P$ be order-preserving. Then, by Proposition 4.4, the map $f^{n!}$ is a retraction on P. Note that $f|_{f^{n!}[P]}$ is order-preserving and that, by Proposition 4.43 and Theorem 4.47 (or by Exercise 4-23), $f^{n!}[P]$ is $\mathcal{I}$-dismantlable to a singleton. Thus because $\mathrm{Fix}(f) = \mathrm{Fix}(f|_{f^{n!}[P]})$, we are done by induction hypothesis unless $f^{n!}[P] = P$.

In this case, we have that $f : P \to P$ is an automorphism. First suppose that there is an irreducible point $p \in P$ that is not in $\mathrm{Fix}(f)$. Then, for all $i \in \mathbb{N}$, we have $f^i(p) \notin \mathrm{Fix}(f)$. Because f is an automorphism, $\{f^i(p) : i \in \mathbb{N}\}$ is an antichain and all $f^i(p)$ are irreducible. Thus $P' := P \setminus \{f^i(p) : i \in \mathbb{N}\}$ is $\mathcal{I}$-dismantlable to a singleton and f maps P' to itself. Because $\mathrm{Fix}(f) = \mathrm{Fix}(f|_{P'})$, we are done by induction hypothesis.

This leaves the case in which f is an automorphism and all irreducible points of P are in $\mathrm{Fix}(f)$. Without loss of generality, let $p \in P$ be a point with a unique lower cover l. Because f is an automorphism, we must have that $f(l)$ also is a lower cover of p. Thus $f(l) = l$. This means that $f|_{P \setminus \{p\}}$ is an order-preserving map such that $\mathrm{Fix}(f|_{P \setminus \{p\}})$ is $\mathcal{I}$-dismantlable and $l \in \mathrm{Fix}(f|_{P \setminus \{p\}})$. Because we have $\mathrm{Fix}(f) = \mathrm{Fix}(f|_{P \setminus \{p\}}) \cup \{p\}$, the set $\mathrm{Fix}(f)$ is $\mathcal{I}$-dismantlable to a singleton. ∎

For the weaker condition of connected collapsibility, there is no analogous result (see Exercise 8-16). However, we can at least prove that fixed point sets will be connected.

Theorem 8.19. *Let P be a finite connectedly collapsible ordered set. Then, for each order-preserving map $f : P \to P$, the set $\mathrm{Fix}(f)$ is nonempty and connected.*

Proof. Nonemptyness of $\mathrm{Fix}(F)$ is proved in Theorem 4.30. The proof of connectedness is an induction on $n := |P|$. For $n = 1$, there is nothing to prove. For the induction step $\{1, \ldots, n\} \to (n + 1)$, let P be an $(n + 1)$-element connectedly collapsible ordered set and let $x \in P$ be as in the definition of connected collapsibility. Let $r : P \to P \setminus \{x\}$ be a retraction and let $b := r(x)$. By definition, $P \setminus \{x\}$ and $\updownarrow x \setminus \{x\}$ are connectedly collapsible and clearly both sets have $\leq n$ elements. Thus $H := \mathrm{Fix}(r \circ f|_{P \setminus \{x\}})$ is connected. Clearly $H \setminus \{b\} \subseteq \mathrm{Fix}(f)$, and thus $\mathrm{Fix}(f)$ must be one of the following four sets: H, $H \setminus \{b\}$, $(H \cup \{x\}) \setminus \{b\}$, $H \cup \{x\}$. The case $\mathrm{Fix}(F) = H$ is trivial. In all the other cases we must show that any two elements of $\mathrm{Fix}(f)$ are joined by a fence.

In case $\mathrm{Fix}(f) = H \setminus \{b\}$, we are trivially done if $b \notin H$, so we will assume that $b \in H$. Because $r(f(b)) = b$ and $f(b) \neq b$, we infer that $f(b) = x$. Moreover our assumption on $\mathrm{Fix}(f)$ means that $f(x) \neq x$. Let $p, q \in H \setminus \{b\}$ and let the set $F := \{p = y_0, \ldots, y_k = q\}$ be a fence in H. If $b \notin F$, then $F \subseteq H \setminus \{b\}$ and we are done. If $b \in F$, we can assume without loss of generality that there is exactly one index m such that $b = y_m$ and we can assume that $b < y_{m-1}, y_{m+1}$. By $f(b) = x$ we infer that $x \leq y_{m-1}, y_{m+1}$. If $f(x)$ is not related to x, then, because $f(b) = x$, f maps $\updownarrow x \setminus \{x\}$ to itself and thus, because $\updownarrow x \setminus \{x\}$ is connectedly collapsible with $\leq n$ elements, $\mathrm{Fix}(f) \cap [\updownarrow x \setminus \{x\}]$ is connected. Hence there is a fence from y_{m-1} to y_{m+1} that lies entirely in $\mathrm{Fix}(f)$ and thus there is a fence from p to q in $\mathrm{Fix}(f)$. If $f(x)$ is related to x, there is a smallest fixed point of f that is above x or a largest fixed point of f that is below x. Call this fixed point c. Then $y_{m-1} > c < y_{m+1}$ and p and q are joined by a fence in $\mathrm{Fix}(f)$.

In case $\mathrm{Fix}(f) = (H \cup \{x\}) \setminus \{b\}$, we will again assume that $b \in H$ (the case $b \notin H$ is treated when $\mathrm{Fix}(f) = H \cup \{x\}$). Again we infer $f(b) = x$. If $H = \{b\}$, then we must have $\mathrm{Fix}(f) = \{x\}$ and we are done. Otherwise there is a fixed point $d \in H$ of f that is related to b and not equal to b or x. Because $f(b) = x$, we have that d is related to x, too. Let $p, q \in \mathrm{Fix}(f)$. Because d is related to x we can assume that $p, q \neq x$. Let $F := \{p = y_0, \ldots, y_k = q\}$ be a fence in H. Again we are done unless $b = y_m$ for exactly one m and (without loss of generality) $b < y_{m-1}, y_{m+1}$. Because $f(b) = x$, we have $x \leq y_{m-1}, y_{m+1}$ and p and q are joined by a fence in $\mathrm{Fix}(f)$.

Finally, in case $\mathrm{Fix}(F) = H \cup \{x\}$, let us first assume that $b \in H$. In this case, f maps $\updownarrow x \setminus \{x\}$ to itself, so x and b have a common upper or lower bound in $\mathrm{Fix}(f)$ and thus $\mathrm{Fix}(f)$ is connected. In case $b \notin H$, we have that $f(b) \neq x$. If $f(b)$ is related to x (and thus to b), then there is a point in H that is above x or below x and thus $H \cup \{x\}$ is connected. If $f(b)$ is not related to x, then f maps $\updownarrow \setminus\{x\}$ to itself. Hence again there is a point in H that is above x or below x and thus $H \cup \{x\}$ is connected. ∎

Exercises

8-14. (Alternative proof for one direction of the Tarski–Davis Theorem.) Use Exercise 3-29, Example 4.3, part 6 and Proposition 8.8 to prove that an incomplete lattice does not have the fixed point property.

8-15. Let us analyze the Bernstein–Cantor–Schröder Theorem in the context of order theory, where the natural mappings should always be order-preserving.

a. Show that, for finite ordered sets, there is a Bernstein–Cantor–Schröder Theorem. That is, if A and B are finite ordered sets, such that there are injective order-preserving functions $f : A \to B$ and $g : B \to A$, then A and B are order-isomorphic.

b. Show that there is no Bernstein–Cantor–Schröder Theorem for ordered sets in general. That is, if A and B are (necessarily infinite) ordered sets, such that there are injective order-preserving functions $f : A \to B$ and $g : B \to A$, then A and B need not necessarily be order-isomorphic.

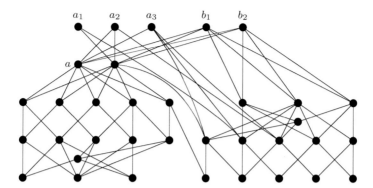

Fig. 8.1 A connectedly collapsible ordered set with an order-preserving function whose fixed point set is not connectedly collapsible.

    c. Let P and Q be ordered sets so that there are order-preserving *surjective* (rather than injective) functions $f : P \to Q$ and $g : Q \to P$. Show that P and Q need not be isomorphic.

        Hint. (J. Hughes) Obtain two infinite ordered sets P and Q from the one-way infinite fence $F = \{f_0 < f_1 > f_2 < f_3 \cdots\}$ as follows. To obtain P, we attach, to each f_{2k}, $2k$ more upper covers $u_1^{2k}, \ldots, u_{2k}^{2k}$ such that f_{2k} is their unique lower cover. To obtain Q, we attach, to each f_{2k}, $2k - 1$ more upper covers $u_1^{2k}, \ldots, u_{2k-1}^{2k}$ such that f_{2k} is their unique lower cover.

    d. Prove that, if P and Q are ordered sets so that there are order-preserving surjective functions $f : P \to Q$ and $g : Q \to P$, then P and Q have equal width.

8-16. Consider the ordered set P in Figure 8.1.

    a. Prove that $P \setminus \{a_1, a_2, a_3, b_1, b_2\}$ is connectedly collapsible.
    b. Prove that P is connectedly collapsible.
    c. Prove that the function f that maps b_1 to b_2 and vice versa and that leaves all other points fixed is order-preserving and that $\mathrm{Fix}(f)$ is not connectedly collapsible.

8-17. Let the lexicographic sum $L\{P_t \mid t \in T\}$ be a complete lattice. Prove each of the following.

    a. Every piece P_t is conditionally complete.
    b. If $t_0 = \bigvee_T \{t \in T : t < t_0\} = \bigwedge_T \{t \in T : t > t_0\}$, then P_{t_0} must be a complete lattice.
    c. State and prove similar results for lattices.

8-18. Let (Ω, Σ, μ) be a measure space and let $p \in [1, \infty)$. We only consider real valued functions. Prove that the space $L^p(\Omega, \Sigma, \mu)$ is conditionally complete.
    Hint. Exercises 8-11 and 2-50 and the Monotone Convergence Theorem.

8.3   Embeddings/The Dedekind–MacNeille Completion

Because complete lattices have many nice properties, it can be helpful to embed a given ordered set in a complete lattice.

Definition 8.20 (See [56], Definition 2.29). *Let P be an ordered set and let L be a complete lattice. If there is an embedding $\phi : P \to L$, then L is called a **completion** of P.*

In Proposition 3.24, we have seen that completions are possible. Every ordered set can be embedded not just into a complete lattice, but into a power set. However, in some ways, such an embedding is quite unsatisfactory. The power set of an ordered set is a lot bigger than the set itself. Thus even sets that are "close" to being complete lattices are embedded into a much bigger structure. For example, consider the natural numbers $\mathbb{N}$ with their canonical order. The easiest way to complete $\mathbb{N}$ is by attaching a top element ∞ to it. Completing via Proposition 3.24 leads to an embedding of $\mathbb{N}$ into $\mathcal{P}(\mathbb{N})$. The power set of $\mathbb{N}$ does not have the same cardinality as $\mathbb{N}$ and it is also not a chain.

It is thus natural to ask whether there is a more "economical" way to embed an ordered set into a complete lattice. As we will see, there is a "smallest" completion, the Dedekind–MacNeille completion.

Definition 8.21. *Let P be an ordered set. We define the Dedekind–MacNeille completion of P to be*

$$DM(P) := \{A \subseteq P : A = \downarrow\uparrow A\}$$

ordered by inclusion. This ordered set is also called the MacNeille completion or the completion by cuts.

Example 8.22.

1. The Dedekind–MacNeille completion of $\mathbb{N}$ is isomorphic to $\mathbb{N} \cup \{\infty\}$.
2. The Dedekind–MacNeille completion of $\mathbb{Q}$ is $\mathbb{R} \cup \{\pm\infty\}$. □

Proof. For part 1, note that, for all $A \subseteq \mathbb{N}$, we have $\downarrow\uparrow A = \{1,\ldots,\bigvee A\}$ if A has a largest element and $\downarrow\uparrow A = \mathbb{N}$ if A is nonempty and unbounded. Because $\downarrow\uparrow \emptyset = \{1\}$, we have that $DM(\mathbb{N}) = \{\{1,\ldots,n\} : n \in \mathbb{N}\} \cup \{\mathbb{N}\}$, which is isomorphic to $\mathbb{N} \cup \{\infty\}$.

For part 2, suffice it to say that one way to construct $\mathbb{R}$ from $\mathbb{Q}$ in set theory is to perform a completion by cuts construction that introduces all points except the largest and the smallest element. Details can be found, for example, in Section 5.5 of [283]. ■

Theorem 8.23 (See [56], Theorem 2.33). *Let P be an ordered set. Then $DM(P)$ is a complete lattice. Moreover, $\phi_{DM} : P \to DM(P)$, defined by $\phi_{DM}(p) := \downarrow p$, is an embedding that preserves all suprema and infima that exist in P.*

Proof. The smallest element of $DM(P)$ is $\downarrow\uparrow \emptyset$. Thus, by Proposition 8.7, to prove that $DM(P)$ is complete, we only need to show that all nonempty subsets of $DM(P)$ have a supremum. Let $\{A_i\}_{i\in I}$ be a nonempty family of subsets of P with $A_i = \downarrow\uparrow A_i$. Then $\downarrow\uparrow \bigcup_{i\in I} A_i$ is an upper bound of $\{A_i\}_{i\in I}$ in $DM(P)$. (It is an element of $DM(P)$ by Exercise 3-19.) Now suppose $B = \downarrow\uparrow B \in DM(P)$ is another upper bound of

$\{A_i\}_{i \in I}$ in $DM(P)$. Then $\bigcup_{i \in I} A_i \subseteq B$, which means that $\uparrow \bigcup_{i \in I} A_i \supseteq \uparrow B$ and then $\downarrow\uparrow \bigcup_{i \in I} A_i \subseteq \downarrow\uparrow B$. This proves that $DM(P)$ is a complete lattice.

It is easy to see that ϕ_{DM} is an embedding. Now let $A \subseteq P$ be a set that has a supremum $\bigvee A$ in P. Then, by part 3 of Lemma 3.23, we have $\downarrow\uparrow A = \downarrow \bigvee A$. This implies that

$$ \bigvee_{DM(P)} \phi_{DM}[A] = \downarrow\uparrow \bigcup_{a \in A} \downarrow a = \downarrow\uparrow A = \downarrow \bigvee A = \phi_{DM}\left(\bigvee A\right). $$

Thus ϕ_{DM} preserves suprema. Preservation of infima is proved similarly, see Exercise 8-20.   ■

The above immediately shows that the Dedekind–MacNeille completion is very economical when we start with a complete lattice: The Dedekind–MacNeille completion leaves complete lattices essentially unchanged.

Proposition 8.24. *Let L be a complete lattice. Then $DM(L)$ is isomorphic to L via the function ϕ_{DM} from Theorem 8.23.*

Proof. If L is a complete lattice, then, by part 3 of Lemma 3.23, for every $A \subseteq L$ we have that $\downarrow\uparrow A = \downarrow \bigvee A$. Thus $\phi_{DM} : L \to DM(L)$ is surjective and hence it is an isomorphism.   ■

In a smallest completion L of the ordered set P, all elements of $L \setminus \phi[P]$ should be "packed tightly" around $\phi[P]$. The following notion formalizes this idea.

Definition 8.25. *Let L be an ordered set and let $P \subseteq L$. Then P is called **join-dense** in L iff, for each $l \in L$, there is a set $A \subseteq P$ such that $l = \bigvee_L A$. The dual notion is called **meet-dense**.*

Proposition 8.26. *Let P be an ordered set. Then $\phi_{DM}[P]$ is join-dense and meet-dense in $DM(P)$.*

Proof. Every element of $DM(P) \setminus \{\mathbf{0}\}$ is of the form $\downarrow\uparrow A$ for some nonempty $A \subseteq P$. Because $\bigvee_{DM(P)} \phi_{DM}[A] = \downarrow\uparrow \bigcup_{a \in A} \downarrow a = \downarrow\uparrow A$, this means that $\phi_{DM}[P]$ is join-dense in $DM(P)$. ($\mathbf{0}$ is of course the supremum of the empty set.)

It would save work if we could infer meet-density by duality. However this is not possible, because we have not proved that the definition of the MacNeille completion is self-dual. Thus we continue as follows.

First we find a representation of infima in the Dedekind–MacNeille completion. Let $\{A_i\}_{i \in I}$ be a nonempty family of subsets of P with $A_i = \downarrow\uparrow A_i$. Then $\downarrow\uparrow \bigcap_{i \in I} A_i$ is a lower bound of $\{A_i\}_{i \in I}$ in $DM(P)$. Let $B = \downarrow\uparrow B \in DM(P)$ be another lower bound of $\{A_i\}_{i \in I}$ in $DM(P)$. Then $\bigcap_{i \in I} A_i \supseteq B$, which means $\uparrow \bigcap_{i \in I} A_i \subseteq \uparrow B$ and then $\downarrow\uparrow \bigcap_{i \in I} A_i \supseteq \downarrow\uparrow B$. Therefore

$$ \downarrow\uparrow \bigcap_{i \in I} A_i = \bigwedge_{DM(P)} \{A_i : i \in I\}. $$

Now let $A = \mathord{\downarrow}\mathord{\uparrow} A$ be an element of $DM(P)$. Then

$$A = \mathord{\downarrow}\mathord{\uparrow}(\mathord{\downarrow}\mathord{\uparrow} A) = \mathord{\downarrow}\mathord{\uparrow} \bigcap_{p \in \mathord{\uparrow} A} \mathord{\downarrow} p = \bigwedge_{DM(P)} \{\mathord{\downarrow} p : p \in \mathord{\uparrow} A\}.$$

Thus $\phi_{DM}[P]$ is also meet-dense in $DM(P)$.   ∎

The following theorem shows that every completion of an ordered set will contain a copy of its Dedekind–MacNeille completion. This is why we say the Dedekind–MacNeille completion is the smallest possible completion of an ordered set.

Theorem 8.27. *Let P be an ordered set and let L be a completion of P. Let the map $\Phi : P \to L$ be an embedding. Then there is an embedding $\Phi' : DM(P) \to L$ such that $\Phi = \Phi' \circ \phi_{DM}$.*

Proof. For $A = \mathord{\downarrow}_P\mathord{\uparrow}_P A$, define $\Phi'(A) := \bigvee_L \mathord{\downarrow}_L\mathord{\uparrow}_L \Phi[A]$. Clearly Φ' is an order-preserving function from $DM(P)$ to L and $\Phi = \Phi' \circ \phi_{DM}$. To show that Φ' is an embedding, let $\Phi'(A) \leq_L \Phi'(B)$.

Suppose $A \not\subseteq B$. Then there is an $a \in A \setminus B$. In particular, there is an upper bound u of $B = \mathord{\downarrow}_P\mathord{\uparrow}_P B$ such that $a \not\leq u$. However, by our assumption, we have

$$\Phi(a) \leq \bigvee_L \mathord{\downarrow}_L\mathord{\uparrow}_L \Phi[A] = \Phi'(A) \leq \Phi'(B) = \bigvee_L \mathord{\downarrow}_L\mathord{\uparrow}_L \Phi[B] \leq \Phi(u),$$

a contradiction. Thus $A \subseteq B$ and Φ' is an embedding.   ∎

Exercises

8-19. Find the Dedekind–MacNeille completion of

   a. A three-element fence,
   b. A four crown,
   c. An antichain,
   d. The ordered set in part e) of Figure 1.4.

8-20. Prove that ϕ_{DM} in Theorem 8.23 preserves infima.

8-21. For every $n \in \mathbb{N}$, find an ordered set P of width n for which the width of $DM(P)$ is larger than $2n$.

8-22. a. For every $n \in \mathbb{N}$, find a finite ordered set P_n of height 1 such that $DM(P_n)$ has height $\geq n$.
   b. Find an ordered set of height 1 whose Dedekind–MacNeille completion has infinite height.

8-23. Let P and Q be ordered sets and let $f : P \to Q$ be order-preserving.

   a. Prove that there is an order-preserving map $f_{DM} : DM(P) \to DM(Q)$ such that the equality $f_{DM} \circ \phi_{DM(P)} = \phi_{DM(Q)} \circ f$ holds.
   b. Give an example that shows that f_{DM} is not unique.

8.4   Irreducible Points in Lattices

With suprema/joins and infima/meets abounding in lattices, it is natural to ask which points are absolutely needed to know as much as possible about the lattice. One possible viewpoint is that points that can be reconstructed as joins or meets of other elements could be considered less essential than those that cannot.

Definition 8.28. *Let P be an ordered set. If $x \in P$ is such that, for all sets $X \subseteq P$, the equality $\bigvee X = x$ implies $x \in X$, then x will be called **join-irreducible**. If $x \in P$ is such that, for all sets $X \subseteq P$, the equality $\bigwedge X = x$ implies $x \in X$, then x will be called **meet-irreducible**. $J(P)$ and $M(P)$ will denote the sets of join- and meet-irreducible elements, respectively. If $x \in P$ is either join- or meet-irreducible, we will call x **irreducible**.*

To show consistency between this definition and our earlier notion of irreducibility, we note the following.

Proposition 8.29. *Let L be a finite lattice and let $x \in L$. Then x is join-irreducible iff x has a unique lower cover.*

Proof. To prove "$\Leftarrow$," let l be the unique lower cover of x and let $X \subseteq L$ be a set such that $\bigvee X = x$. Then $X \subseteq\downarrow x$. Suppose $x \notin X$. Then $X \subseteq\downarrow l$ and $\bigvee X \leq l$, a contradiction. Thus we must have $x \in X$ and x is join-irreducible.

For "$\Rightarrow$," we will prove the contrapositive, that is, we will prove that every element that has no or more than one lower cover is not join-irreducible. Every element with more than one lower cover is the supremum of its lower covers, and $\mathbf{0}$, the only element without lower covers, is the supremum of the empty set. ∎

The above explains the choice of language for irreducible points in the first place: Because lattice theory has attracted attention earlier than order theory in general, some notation was adopted from lattice theory. However, the original motivation behind the notation can be mysterious when the subject is not approached through lattice theory. Irreducible points are a prime example.

As it turns out, not only can irreducible elements not be broken down, they can also be used to represent all elements of a finite lattice.

Proposition 8.30. *Let L be a finite lattice. Then, for every element x of $L \setminus \{\mathbf{0}\}$, we have that $x = \bigvee[(\downarrow x) \cap J(L)]$.*

Proof. For a contradiction, assume the contrary. Then there is a point $x \in L \setminus \{\mathbf{0}\}$ such that $x \neq \bigvee[(\downarrow x) \cap J(L)]$ and such that, for all $y < x$ that are not equal to $\mathbf{0}$, we have $y = \bigvee[(\downarrow y) \cap J(L)]$. In particular, $x \notin J(L)$, so x has more than one lower cover. Let $l_1, \ldots, l_k$ be the lower covers of x. Then $x = \bigvee_{i=1}^{k} l_i$ and, for each l_i, we have $l_i = \bigvee[(\downarrow l_i) \cap J(L)]$. This implies

$$x = \bigvee_{i=1}^{k} l_i = \bigvee_{i=1}^{k} \bigvee[(\downarrow l_i) \cap J(L)] = \bigvee[(\downarrow x) \cap J(L)],$$

a contradiction. ∎

A first consequence of the join-density of the join-irreducible elements is that, on finite lattices, automorphisms are completely determined by their restriction to $J(L)$. This is a key to solving the automorphism problem for lattices.

Proposition 8.31. *Let L be a finite lattice. If f and g are automorphisms of L and $f|_{J(L)} = g|_{J(L)}$, then $f = g$.*

Proof. First note that, for any lattice automorphism Φ, and all subsets $A \subseteq L$ we have $\Phi\left(\bigvee A\right) = \bigvee \Phi[A]$ and $\Phi(0) = 0$. Thus $f(0) = 0$ and, for all $x \in L \setminus \{0\}$, we have

$$f(x) = f\left(\bigvee [(\downarrow x) \cap J(L)]\right) = \bigvee f[(\downarrow x) \cap J(L)]$$

$$= \bigvee g[(\downarrow x) \cap J(L)] = g\left(\bigvee [(\downarrow x) \cap J(L)]\right) = g(x).$$

∎

Lemma 8.32. *For all $a \in \mathbb{N}$, we have $(a + 1)^a \geq 2^a a!$.*

Proof. This is a proof by induction on a. For $a \in \{1, 2, 3, 4\}$, the assertion is easily verified directly.

For the induction step $a \to (a + 1)$, assume that the assertion is true for an $a \geq 4$. We will need to prove it for $a + 1$. Note that

$$((a + 1) + 1)^{(a+1)} = \left(\frac{a + 2}{a + 1}\right)^a (a + 2)(a + 1)^a \geq \left(\frac{a + 2}{a + 1}\right)^a (a + 2)2^a a!$$

$$\geq \left(\left(1 + \frac{1}{a + 1}\right)^{a+1}\right)^{\frac{a}{a+1}} 2^a (a + 1)!$$

From analysis, we know that the terms $\left(1 + \frac{1}{a + 1}\right)^{a+1}$ form an *increasing* sequence with limit e. Thus $\left(\left(1 + \frac{1}{a + 1}\right)^{a+1}\right)^{\frac{a}{a+1}}$ is increasing, too. For $a = 4$, we have

$$\left(\left(1 + \frac{1}{a + 1}\right)^{a+1}\right)^{\frac{a}{a+1}} = \left(1 + \frac{1}{a + 1}\right)^a = \left(\frac{6}{5}\right)^4 = \frac{1296}{625} > 2.$$

Hence this term exceeds 2 for all values $a \geq 4$. This finishes the proof of the inequality $((a + 1) + 1)^{(a+1)} \geq 2^{a+1}(a + 1)!$. ∎

Theorem 8.33 (See [192], Theorem 1). *For every finite ordered set P, we have*

$$\frac{|\mathrm{Aut}(P)|}{|\mathrm{End}(P)|} \leq 2^{-\sqrt{\max\{|J(P)|,|M(P)|\}}}.$$

Proof. Without loss of generality, let $\max\{|J(P)|, |M(P)|\} = |J(P)| =: \alpha$. Then there are either $m \geq \sqrt{\alpha}$ join-irreducible elements with the same lower cover or there are $m \geq \sqrt{\alpha}$ join-irreducible elements such that no two have the same lower cover. In either case, denote these m join-irreducible elements by $v_1, \ldots, v_m$. For every automorphism f, we define the sets

$$A(f) := \{g \in \mathrm{Aut}(P) : f|_{P \setminus \{v_1,\ldots,v_m\}} = g|_{P \setminus \{v_1,\ldots,v_m\}}\}, \qquad \text{and}$$

$$H(f) := \{h \in \mathrm{End}(P) : f|_{P \setminus \{v_1,\ldots,v_m\}} = h|_{P \setminus \{v_1,\ldots,v_m\}}\}.$$

The sets $A(f)$ form a partition of $\mathrm{Aut}(P)$. Moreover $A(f) \cap A(g) = \emptyset$ is equivalent to $H(f) \cap H(g) = \emptyset$. Thus, if we can prove $\dfrac{|A(f)|}{|H(f)|} \leq 2^{-m}$, then we have

$$|\mathrm{Aut}(P)| = \left| \bigcup_{f \in \mathrm{Aut}(P)} A(f) \right| \leq 2^{-m} \left| \bigcup_{f \in \mathrm{Aut}(P)} H(f) \right| \leq 2^{-m} |\mathrm{End}(P)|,$$

which was to be proved.

In case all v_i have distinct lower covers, we have $|A(f)| = 1$ and $|H(f)| = 2^m$ for all $f \in \mathrm{Aut}(P)$. This leaves the case that all v_i have the same lower cover v.

Note that $h \mapsto f^{-1}h$ is a bijection between $A(f)$ and $A(\mathrm{id}_P)$ as well as between $H(f)$ and $H(\mathrm{id}_P)$. Thus we can concentrate on $A(\mathrm{id}_P)$. Let $S_1, \ldots, S_k$ be the partition of $\{v_1, \ldots, v_m\}$ that is obtained from the equivalence relation $v_i \equiv v_j$ iff $\uparrow v_i \setminus \{v_i\} = \uparrow v_j \setminus \{v_j\}$. Then every map in $A(\mathrm{id}_P)$ must map the S_i to themselves and we obtain $|A(\mathrm{id}_P)| = \prod_{i=1}^{k} |S_i|!$. On the other hand, every map in $H(\mathrm{id}_P)$ can map every element of S_i to any element of $S_i \cup \{v\}$. Thus, by Lemma 8.32, we obtain the following.

$$|H(\mathrm{id}_P)| = \prod_{i=1}^{k} (|S_i| + 1)^{|S_i|} \geq \prod_{i=1}^{k} 2^{|S_i|} (|S_i|!) = 2^m \prod_{i=1}^{k} |S_i|! = 2^m |A(\mathrm{id}_P)|.$$

This finishes the proof. ∎

We can now solve the automorphism problem for lattices.

Theorem 8.34 (See [192], Theorem 2).   *Let L be an n-element lattice. Then*

$$\frac{|\mathrm{Aut}(L)|}{|\mathrm{End}(L)|} \leq C 2^{-\sqrt{\log_2(n)}},$$

for some $C > 0$.

Proof. If $|J(L)| \geq \log_2(n)$, we are done by Theorem 8.33. If $|J(L)| \leq \log_2(n)$, we argue as follows. By Proposition 8.31, two automorphisms of L are equal iff they are equal on the join-irreducible elements of L. Thus L has at most $|J(L)|!$ automorphisms. By Theorem 2.12, we obtain the following.

$$\frac{|\mathrm{Aut}(L)|}{|\mathrm{End}(L)|} \leq \frac{\log_2(n)!}{2^{\frac{n}{2}}} \leq 2^{-\frac{n}{2}+\log_2(n)\log_2(\log_2(n))} \leq C2^{-\sqrt{\log_2(n)}}.$$

∎

Exercises

8-24. Join and meet irreducibility in infinite lattices.

  a. Prove that there are infinite complete lattices that do not have any join-irreducible elements. In particular this means that there is no analogue for Propositions 8.30 and 8.31 in general infinite lattices.
  b. Let L be a lattice with a smallest element **0**. Then $a \in L$ is called an **atom** iff a is an upper cover of the smallest element **0**. Prove that, in a finite lattice, every element except **0** is above an atom. Give an example that not every complete lattice has an atom.
  c. Let S be an infinite set.

    i. Prove that the set of infinite-coinfinite subsets of S, that is, the set of infinite subsets $A \subset S$ such that $S \setminus A$ is also infinite, is a complete lattice.
    ii. A lattice L is called **atomic** iff every element of L is a supremum of atoms. Prove that the set of finite or cofinite subsets of S, that is, the set of subsets $A \subset S$ such that A or $S \setminus A$ is finite, is an atomic lattice.

  d. Recall that a complete lattice satisfies the **descending chain condition** iff it contains no copies of the dual of $\mathbb{N}$. Prove that, if $a \not\leq b$ in a complete lattice with the descending chain condition, then there is a join-irreducible element below a that is not below b.
  e. Prove analogues of Propositions 8.30 and 8.31 in complete lattices with the descending chain condition.

8-25. Let P be a finite ordered set and let $\Phi : P \to P$ and $\Psi : P \to P$ be two automorphisms.

  a. Let $a \in P$ be minimal in the set $\{p \in P : \Phi(p) \neq p\}$. Prove that there is a point $b \neq a$ such that a and b have the same lower covers.
  b. Conclude that $\Phi = \Psi$ iff $\Phi(p) = \Psi(p)$ for all $p \in P$ such that there is a $q \in P$ that has the same lower covers as p.

8-26. Prove that, for any finite ordered set P, we have $\dfrac{\mathrm{Aut}(P)}{\mathrm{End}(P)} \leq 2^{-\frac{\sqrt{|S(P)|}}{2}}$, where $S(P)$ is the set of all elements a that are retractable to an element $b \not\geq a$.

8.5 Finite Ordered Sets vs. Distributive Lattices

A nice property of union and intersection in set systems is that they distribute over each other. Because the lattice operations $\vee$ and $\wedge$ are modeled after union and intersection, it is natural to give special consideration to lattices for which $\vee$ and $\wedge$ satisfy a distributive law. We will only discuss some basic ideas about distributive lattices here. For further information on this topic, consider [21, 54, 56, 98, 112]. (Lattices as algebraic objects is a vast topic.)

Definition 8.35. *A lattice L is called **distributive** iff, for all $x, y, z \in L$, we have*

- $x \vee (y \wedge z) = (x \vee y) \wedge (x \vee z)$, *and*
- $x \wedge (y \vee z) = (x \wedge y) \vee (x \wedge z)$.

Example 8.36.

1. Every power set is a distributive lattice.
2. Figure 8.2 shows some examples of distributive and non-distributive lattices. In the non-distributive lattices, three points for which distributivity fails are marked. Note that, in each row, we take a distributive lattice and, through adding of points, we first obtain a non-distributive lattice and then a distributive lattice again. □

Join-irreducible elements have another characterization in distributive lattices. This characterization will be important for us when we prove Theorem 8.40, the characterization theorem for finite distributive lattices.

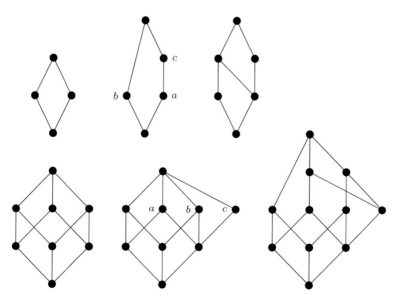

Fig. 8.2 Some examples of distributive and non-distributive lattices.

Lemma 8.37. *Let L be a finite distributive lattice. Then $x \in L$ is join-irreducible iff, for all $a, b \in L$, we have that $x \leq a \vee b$ implies $x \leq a$ or $x \leq b$.*

Proof. To prove "$\Leftarrow$," let $x \in L$ be such that, for all $a, b \in L$, we have that $x \leq a \vee b$ implies $x \leq a$ or $x \leq b$. Then, by induction, for all $Y \subseteq L$ we have that $x \leq \bigvee Y$ implies there is a $y \in Y$ with $x \leq y$. However, then, for $x = \bigvee X$, there must be a $y \in X$ such that $x \leq y$. Because x cannot be strictly less than y we must have $x = y \in X$ and x is join-irreducible.

To prove "$\Rightarrow$," let $x \in L$ be join-irreducible and let $a, b \in L$ with $x \leq a \vee b$. Then

$$x = x \wedge (a \vee b) = (x \wedge a) \vee (x \wedge b).$$

Because x is join-irreducible, we must have $x \in \{x \wedge a, x \wedge b\}$, say, $x = x \wedge a$. Then $x \leq a$ and we are done. ∎

Distributive lattices arise naturally when considering subsets of ordered sets. Before we can state the next theorem, we need to generalize our notion of a down-set.

Definition 8.38. *Let P be an ordered set. Then $A \subseteq P$ is called a **down-set** iff, for all $a \in A$, we have $\downarrow a \subseteq A$. An **up-set** is defined dually.*

Proposition 8.39. *Let P be a finite ordered set and let $D(P)$ be the set of down-sets of P ordered by inclusion. Then*

1. *$D(P)$ is a distributive lattice.*
2. *The join-irreducible elements of $D(P)$ are exactly the ideals $\downarrow p$ with $p \in P$.*
3. *The map $\phi : p \mapsto \downarrow p$ is an isomorphism between P and $J(D(P))$.*

Proof. Because the intersection and the union of two down-sets are again a down-set (see Exercise 8-28), $D(P)$ is a lattice. Union and intersection are the join and meet operations on $D(P)$. Because union and intersection are distributive, $D(P)$ is a distributive lattice. This proves 1.

To prove 2, let $p \in P$ and consider $\downarrow p$. The set $(\downarrow p) \setminus \{p\}$ is the unique lower cover of $\downarrow p$ in $D(P)$. Conversely, let $D \in D(P)$ be join-irreducible. If D had two maximal elements m_1 and m_2, then $D \setminus \{m_1\}$ and $D \setminus \{m_2\}$ would both be lower covers of D in $D(P)$, which is not possible. Thus D has exactly one maximal element m, which means $D = \downarrow m$.

Finally, to prove 3, note that ϕ trivially is injective and order-preserving both ways and, by 2, ϕ is surjective. ∎

Distributive lattices and lattices that arise as above are closely related. The next theorem shows that every finite distributive lattice arises as in Proposition 8.39.

Theorem 8.40 (Birkhoff's characterization theorem for finite distributive lattices). *Let L be a finite distributive lattice. Then L and $D(J(L))$ are isomorphic via the map $\eta : x \to (\downarrow x) \cap J(L)$. Here $D(\cdot)$ is as in Proposition 8.39 and $J(L)$ is the ordered subset of join-irreducible elements.*

Proof. By Proposition 8.30, η is injective (note that $\eta(\mathbf{0}) = \emptyset$). Moreover, because $x \leq y$ implies $\downarrow x \subseteq \downarrow y$, we have that η is order-preserving. To show that η is an embedding, note that $\eta(a) \subseteq \eta(b)$ implies $a = \bigvee \eta(a) \leq \bigvee \eta(b) = b$. (Note that we did not use the distributivity of L yet. This means that η is an embedding for all finite lattices.)

To prove surjectivity of η, we proceed as follows. Let $A \in D(J(L))$ and let $a := \bigvee A$. We want to show that $\eta(a) = A$. To do so, first note that we have $A \subseteq (\downarrow a) \cap J(L) = \eta(a)$. For the reverse inclusion, let $x \in \eta(a)$. Then $x \leq \bigvee A$ is join-irreducible. Thus, by Lemma 8.37 (extended inductively to sets of arbitrary size, see Exercise 8-29), there is a $y \in A$ with $x \leq y$. Because A is a down-set, this means that $x \in A$. Thus $A \supseteq (\downarrow a) \cap J(L) = \eta(a)$ and η is surjective. ∎

The above means that there is a one-to-one correspondence between isomorphism classes of finite ordered sets and the isomorphism classes of finite distributive lattices. Proposition 8.39 shows that $[P] \mapsto [D(P)]$ is injective and Theorem 8.40 shows that it is surjective with inverse $[L] \mapsto [J(L)]$.

This means that any problem for ordered sets can also be formulated as a problem for distributive lattices and vice versa. Some of these translations will appear awkward and may not have much impact. (For example, try to translate the statement "The ordered set P has the fixed point property" into its analogue for $D(P)$.) Other translations lead to nice problems and results for distributive lattices and for ordered sets. For example, see Exercise 8-32 and Remark 5 at the end of this chapter.

Exercises

8-27. Show that Lemma 8.37 does not hold in arbitrary lattices.

8-28. Let P be an ordered set and let $\{D_i\}_{i \in I}$ be a family of down-sets in P. Prove that $\bigcup_{i \in I} D_i$ and $\bigcap_{i \in I} D_i$ are down-sets.

8-29. Let L be an arbitrary lattice and let $x \in L$. Prove that the following are equivalent.

   a. For all $a, b \in L$ we have that $x \leq a \vee b$ implies $x \leq a$ or $x \leq b$.
   b. For all finite sets $Y \subseteq L$ we have that $x \leq \bigvee Y$ implies there is a $y \in Y$ with $x \leq y$.

8-30. Let P be a finite ordered set. Prove that $h(D(P)) = |P|$.

8-31. Let P be a finite ordered set and let $A \subseteq P$ be an antichain. Prove that $w(D(P)) \geq \binom{|A|}{\lfloor \frac{|A|}{2} \rfloor}$.

8-32. In [29], an efficient method for checking if two distributive lattices were isomorphic was suggested. We investigate some of the details here and discuss consequences in Remark 5 at the end of this chapter.

   a. Let D be a distributive lattice and let $m \in J(D)$ be minimal in $J(D)$. Prove that $a \in D$ is join-irreducible in D iff $m \vee a$ is join-irreducible in $\uparrow m$.
   b. Let P be a finite ordered set and let $m \in P$ be a minimal element. Prove that $D(P \setminus \{m\})$ is isomorphic to $\uparrow_{D(P)} \{m\}$.

8.6   More on Distributive Lattices

Distributivity and similar algebraic properties have received much attention in lattice theory. In this section, we will give a pictorial characterization of distributive lattices. This characterization is useful in determining from the diagram if a given lattice is distributive or not. Rather than doing algebraic computations, we only need to search for certain substructures. We start by proving some equations that are true in arbitrary lattices. This will show more clearly what distinguishes distributive lattices from other lattices. We also take this opportunity to state some other (obvious) algebraic properties of lattices.

Lemma 8.41. *Let L be a lattice and let $a, b, c \in L$. Then the following hold.*

1. $a \wedge (b \vee c) \geq (a \wedge b) \vee (a \wedge c)$.
2. $a \vee (b \wedge c) \leq (a \vee b) \wedge (a \vee c)$.
3. $a \vee b = b \vee a$ and $a \wedge b = b \wedge a$ *(commutativity of $\vee$ and $\wedge$)*.
4. $(a \vee b) \vee c = a \vee (b \vee c)$ and $(a \wedge b) \wedge c = a \wedge (b \wedge c)$ *(associativity of $\vee$ and $\wedge$)*.
5. $a = a \vee b$ *iff* $a \geq b$ and $a = a \wedge b$ *iff* $a \leq b$.

Proof. Left as Exercise 8-33.   ∎

This means that demanding distributivity amounts to demanding that two inequalities that hold in general become equalities. Moreover, the equalities in the definition of distributivity actually are equivalent.

Proposition 8.42. *Let L be a lattice. The following are equivalent.*

1. *For all $a, b, c \in L$ we have $a \wedge (b \vee c) = (a \wedge b) \vee (a \wedge c)$.*
2. *For all $a, b, c \in L$ we have $a \vee (b \wedge c) = (a \vee b) \wedge (a \vee c)$.*

Proof. We will only need to prove the direction "1⇒2," because the other direction follows by duality. Assume that, for all $a, b, c \in L$, we have

$$a \wedge (b \vee c) = (a \wedge b) \vee (a \wedge c).$$

Then

$$(a \vee b) \wedge (a \vee c) = [(a \vee b) \wedge a] \vee [(a \vee b) \wedge c]$$
$$= a \vee [(a \vee b) \wedge c]$$
$$= a \vee [c \wedge (a \vee b)]$$
$$= a \vee [(c \wedge a) \vee (c \wedge b)]$$
$$= [a \vee (c \wedge a)] \vee (c \wedge b)$$
$$= a \vee (b \wedge c).$$ ∎

Fig. 8.3 The sublattices that
are forbidden in distributive
lattices.

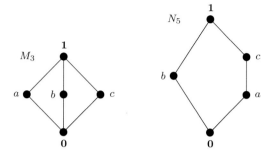

This means that, to prove distributivity, we only need to prove that one of the two inequalities in Lemma 8.41, parts 1 and 2 actually is an equality. Next, we need the notion of a sublattice.

Definition 8.43. *Let L be a lattice and let $S \subseteq L$. Then S is called a **sublattice** iff, for all $x, y \in S$, we have that $x \vee y \in S$ and $x \wedge y \in S$.*

Clearly every sublattice is an ordered subset of L and a lattice. Note however that an ordered subset S that is a lattice need not be a sublattice. This is because the supremum of two elements in S could be different from the supremum of the same two elements in L, see Exercise 8-35.

Using sublattices, we can now give a pictorial characterization of distributive lattices in terms of sublattices. This is one of the nice features of order and lattice theory. Algebraic concepts often can be translated into natural pictures.

Theorem 8.44. *Let L be a lattice. Then L is a distributive lattice iff L does not have any sublattices isomorphic to either of the two forbidden lattices M_3 and N_5 in Figure 8.3.*

Proof. To prove "$\Rightarrow$," note that every sublattice of a distributive lattice must again be distributive. Yet we have the following.

In the left lattice (M_3) $a \vee (b \wedge c) = a \neq 1 = (a \vee b) \wedge (a \vee c)$.

In the right lattice (N_5) $a \vee (b \wedge c) = a \neq c = (a \vee b) \wedge (a \vee c)$.

For the direction "$\Leftarrow$," we prove the contrapositive. (This proof is essentially the proof of [56], 6.10.) Let L be a non-distributive lattice. By Lemma 8.41 part 1 and by Proposition 8.42, there must be $a, b, c \in L$ such that

$$a \wedge (b \vee c) > (a \wedge b) \vee (a \wedge c).$$

Equality of any two of a, b, c, or comparability of b and c, or $a \leq b$, or $a \leq c$, or $a \geq b, c$ implies equality above. Thus we must have the following.

1. The three points a, b, c are distinct.
2. The points b and c are not comparable.
3. a is comparable to at most one of b and c, and, if it is comparable to one of them, then a is larger than that point.

First we show that, if $a > c$, then L contains a sublattice isomorphic to N_5. By the above, we have that $b \not> a, c$. Thus $a \wedge b < b < b \vee c$. Moreover

$$b \vee c > a \wedge (b \vee c) > (a \wedge b) \vee (a \wedge c) = (a \wedge b) \vee c > a \wedge b,$$

where the middle inequality is by assumption and the outer inequalities are via the incomparability of b with a and c. Moreover, the above and the incomparability of b with a and c imply that b is not comparable to $a \wedge (b \vee c)$ or to $(a \wedge b) \vee c$. Therefore

$$B := \{a \wedge b, b, b \vee c, a \wedge (b \vee c), (a \wedge b) \vee c\}$$

is isomorphic to the lattice N_5. To prove that B is a sublattice, we need to show that $b \vee [(a \wedge b) \vee c] = b \vee c$ and $b \wedge [a \wedge (b \vee c)] = a \wedge b$. These equalities are easily proved:

$$b \vee [(a \wedge b) \vee c] = [b \vee (a \wedge b)] \vee c = b \vee c$$
$$b \wedge [a \wedge (b \vee c)] = [b \wedge (b \vee c)] \wedge a = a \wedge b.$$

This leaves the case in which none of a, b, c are comparable to each other. Moreover, we can assume that, whenever $x, y, z \in L$ are so that two of them are comparable, then

$$x \wedge (y \vee z) = (x \wedge y) \vee (x \wedge z).$$

Note that our situation is not symmetric in a, b, and c. Thus it is unlikely that we will find a copy of M_3, which is a very symmetric structure, as directly as we found a copy of N_5. We first generate a situation that is symmetric in a, b, and c. Consider the points

$$s := (a \wedge b) \vee (a \wedge c) \vee (b \wedge c),$$
$$l := (a \vee b) \wedge (a \vee c) \wedge (b \vee c).$$

It is easy to see that $l \geq s$. Now note that $a \wedge l = a \wedge (b \vee c)$, while

$$\begin{aligned} a \wedge s &= a \wedge [[(a \wedge b) \vee (a \wedge c)] \vee (b \wedge c)] \\ &= [a \wedge [(a \wedge b) \vee (a \wedge c)]] \vee [a \wedge (b \wedge c)] \\ &= [(a \wedge b) \vee (a \wedge c)] \vee [a \wedge b \wedge c] \\ &= (a \wedge b) \vee (a \wedge c). \end{aligned}$$

This means that $s = l$ is not possible, so $l > s$. We now consider the three symmetrically defined points

$$(a \wedge l) \vee s, \qquad (b \wedge l) \vee s, \qquad (c \wedge l) \vee s.$$

Clearly all three points are less than or equal to l and greater than or equal to s. Now note that

$$[(a \wedge l) \vee s] \wedge [(b \wedge l) \vee s]$$
$$= [[(a \wedge l) \vee s] \wedge (b \wedge l)] \vee [[(a \wedge l) \vee s] \wedge s]$$
$$= [[(l \wedge a) \vee (l \wedge s)] \wedge (b \wedge l)] \vee s$$
$$= [l \wedge (a \vee s) \wedge (b \wedge l)] \vee s$$
$$= [l \wedge (b \wedge l) \wedge (a \vee s)] \vee s$$
$$= [(b \wedge l) \wedge (a \vee s)] \vee s$$
$$= [b \wedge (a \vee b) \wedge (b \vee c) \wedge (a \vee c) \wedge (a \vee (a \wedge b) \vee (a \wedge c) \vee (b \wedge c))] \vee s$$
$$= [b \wedge ((a \vee c) \wedge (a \vee (b \wedge c)))] \vee s$$
$$= [b \wedge [((a \vee c) \wedge a) \vee ((a \vee c) \wedge (b \wedge c))]] \vee s$$
$$= [b \wedge [a \vee (b \wedge c)]] \vee s$$
$$= [(b \wedge a) \vee (b \wedge (b \wedge c))] \vee s$$
$$= s.$$

Moreover, by duality, we can assume that, whenever any two of x, y, z are comparable, then $x \vee (y \wedge z) = (x \vee y) \wedge (x \vee z)$. (Indeed, otherwise L contains a sublattice isomorphic to N_5 and we are done.) This implies

$$[(a \wedge l) \vee s] \vee [(b \wedge l) \vee s]$$
$$= s \vee (a \wedge l) \vee (b \wedge l)$$
$$= (a \wedge b) \vee (a \wedge c) \vee (b \wedge c) \vee (a \wedge (b \vee c)) \vee (b \wedge (a \vee c))$$
$$= [(a \wedge b) \vee (a \wedge c) \vee (a \wedge (b \vee c))] \vee [(b \wedge c) \vee (b \wedge (a \vee c))]$$
$$= (a \wedge (b \vee c)) \vee (b \wedge (a \vee c))$$
$$= (b \wedge (a \vee c)) \vee (a \wedge (b \vee c))$$
$$= [(b \wedge (a \vee c)) \vee a] \wedge [(b \wedge (a \vee c)) \vee (b \vee c)]$$
$$= [a \vee (b \wedge (a \vee c))] \wedge (b \vee c)$$
$$= [(a \vee b) \wedge (a \vee (a \vee c))] \wedge (b \vee c)$$
$$= l.$$

Symmetric arguments show that the other infima are equal to s and that the other suprema are equal to l. Now suppose for a contradiction that any two of $(a \wedge l) \vee s$, $(b \wedge l) \vee s$, and $(c \wedge l) \vee s$ are equal. Then their supremum and infimum would be equal. Because $s \neq l$, this cannot be and hence our three points must be distinct. Finally, if any one of them was equal to s or l, then the other two would have to be equal to l or s, respectively, which cannot be. Thus $\{l, s, (a \wedge l) \vee s, (b \wedge l) \vee s, (c \wedge l) \vee s\}$ is a sublattice of L that is isomorphic to M_3.   ∎

Remark 8.45. By Theorem 8.44, it can be checked in $O(|P|^3)$ steps if a given lattice is distributive, provided that suprema and infima can be computed in a constant number of steps.   □

Exercises

8-33. Prove Lemma 8.41.

8-34. (Lattices as algebraic objects.) Let L be a set with the binary operations $\vee : L \times L \to L$ and $\wedge : L \times L \to L$ so that, for all $a, b, c \in L$, we have the following.

- $a \vee a = a, a \wedge a = a$ (idempotency).
- $a \vee b = b \vee a, a \wedge b = b \wedge a$ (commutativity).
- $a \vee (b \vee c) = (a \vee b) \vee c, a \wedge (b \wedge c) = (a \wedge b) \wedge c$ (associativity).
- $a \vee (b \wedge a) = a, a \wedge (b \vee a) = a$ (absorption laws).

  a. Prove that, in a lattice P as in Definition 8.1, the above holds for $\vee$ denoting the supremum and $\wedge$ denoting the infimum.
  b. Prove that, in a structure as defined above, $a \vee b = b$ iff $a \wedge b = a$.
  c. Prove that, if, for a structure as defined above, we define $a \leq b$ iff $a \vee b = b$, then L equipped with this order is a lattice as in Definition 8.1 and $a \vee b$ is the supremum of a and b and $a \wedge b$ is the infimum of a and b.

8-35. Give an example of a lattice L and an ordered subset $S \subseteq L$ such that S is a lattice, but not a sublattice of L.

8-36. In the lattice N_5 as drawn in Figure 8.3, we have $a \wedge (b \vee c) = (a \wedge b) \vee (a \wedge c)$. Why is this not a contradiction to Proposition 8.42?

8-37. Let L be the set of infinite subsets S of $\mathbb{N}$ such that the complement $\mathbb{N} \setminus S$ also is infinite. Prove that $L \cup \{\emptyset, \mathbb{N}\}$ is a non-distributive lattice.

8-38. Let L be a distributive lattice. Prove that if $a \geq c$, then $a \wedge (b \vee c) = (a \wedge b) \vee c$.

8-39. Let L be a lattice and let $a, b, c \in L$. Prove that if $b \leq c$ and $a \vee b \geq c$, then $a \vee b = a \vee c$.

8-40. The implication $[a \geq c] \Rightarrow [a \wedge (b \vee c) = (a \wedge b) \vee c]$ in Exercise 8-38 is called the **modular law** and lattices satisfying this law are called **modular lattices**.

  a. Prove that a lattice is modular iff it does not contain a sublattice isomorphic to the lattice N_5 on the right in Figure 8.3.
  b. Prove that, if L is a modular lattice and $a, b \in L$ are such that $a \wedge b$ is a lower cover of a and b, then $a \vee b$ is an upper cover of a and b.
  c. A lattice L is called **upper semi-modular** iff, for all $a, b \in L$, we have that, if $a \wedge b$ is a lower cover of a and b, then $a \vee b$ is an upper cover of a and b. **Lower semi-modular** lattices are defined dually. Prove that a lattice is modular iff it is upper semi-modular and lower semi-modular.
  d. A lattice that is lower semi-modular or upper semi-modular is called **semi-modular**. Give an example of a semi-modular lattice that is not modular.

e. Show that the closed subspaces of a normed vector space ordered by inclusion form a modular lattice that is not distributive. Use the intersection as the infimum and the direct sum as the supremum of two spaces.

8-41. Prove that, in a finite semi-modular lattice, all maximal chains have the same length.

Note. This is called the **Jordan–Dedekind chain condition**. Lattices that satisfy this condition are also called **graded**, which, in general, is a weaker concept. The connection between being graded and the Jordan–Dedekind chain condition is examined further in Exercise 13-3.

8-42. A lattice with the largest element **1** and the smallest element **0** is called **complemented** iff there is a map $x \mapsto x'$ such that $x \leq y$ implies $x' \geq y'$ and such that $x \vee x' = \mathbf{1}$. A **Boolean lattice** is a complemented distributive lattice. (When considered as algebraic objects, they are also called **Boolean algebras**.)

a. Prove that, in a Boolean lattice, for every element x, there is exactly one element c so that $x \vee c = \mathbf{1}$ and $x \wedge c = \mathbf{0}$.

b. Prove that, if B is a Boolean lattice and $x \in B$ is join-irreducible, then $\mathbf{0} \prec x$.

c. Prove that every finite Boolean lattice is isomorphic to a power set of a finite set.

d. Give an example of a complete Boolean lattice that is not isomorphic to a power set.

e. Give an example of an atomic Boolean lattice that is not isomorphic to a power set.

f. Prove that every complete atomic Boolean lattice is isomorphic to a power set.

Remarks and Open Problems

1. For more on lattices see, for example, [21, 54, 56, 98, 112].

2. A variation on the lattice theme is the idea of a directed ordered set. Here we only look at upper bounds and we do not demand existence of a least upper bound, but only the existence of some upper bound. For more on directed sets see [56].

3. Is there an analogue of Theorem 8.18 for infinite dismantlability?

4. Let P be an ordered set without a largest or a smallest element. If P has the fixed point property, must $DM(P) \setminus \{\mathbf{0}, \mathbf{1}\}$ have the fixed point property?

   The other direction is false, as will be shown in Exercise 9-23a.

5. Consider the results of Exercise 8-32. Recall that any finite distributive lattice D is the lattice $D(P)$ for some finite ordered set P and that $D(P)$ is isomorphic to $D(Q)$ iff P is isomorphic to Q. Exercise 8-32 shows that the decks of minimal ideals of $D(P)$ and $D(Q)$ are isomorphic iff the minimal decks of P and Q are isomorphic. Thus, if ordered sets were reconstructible from their minimal decks, isomorphism of finite distributive lattices could be checked through isomorphism of their minimal ideal decks. In [29] this (false) conjecture is translated into a fast (wrong) algorithm to check isomorphism of finite distributive lattices.

   Although the results of [277, 278] apparently show that the approach in [29] is not salvageable, even with stronger hypotheses, it still is remarkable that reconstruction results can be translated into results and algorithms regarding isomorphism. This possibility should merit future investigation.

6. If P and Q are ordered sets and there are order-preserving *bijective* (rather than injective or surjective alone) functions $f : P \to Q$ and $g : Q \to P$, must P and Q be isomorphic?

7. Are Theorem 8.33 and Exercise 8-26 a starting point for solving the automorphism problem for collapsible ordered sets?
8. Another automorphism problem. For lattices, there is the notion of a lattice
   endomorphism, which is an endomorphism that preserves suprema and infima.
   Let the set of lattice endomorphisms of L be $\text{End}_l(L)$. The quotient $\frac{|\text{Aut}(L)|}{|\text{End}_l(L)|}$ can
   remain stationary at 1 (antichain with a top and a bottom attached) or converge
   to zero (chains). What is the limiting set of the quotients $\frac{|\text{Aut}(L)|}{|\text{End}_s(L)|}$ as $|L| \to \infty$?
   What is the class of lattices for which the limiting set is $\{0\}$? How about $\{1\}$?

Chapter 9
Truncated Lattices

How much does a (finite) lattice change when we remove the (trivially always present) elements **0** and **1**? With only the top and bottom elements gone, the picture does not change much at all. However, in terms of order-theoretical properties there is a significant change. Note that both the proof of reconstructibility of finite lattices (Corollary 8.6) as well as the characterization of the fixed point property for lattices (Theorem 8.10) heavily relied on the existence of the smallest (or the largest) element. The smallest element was important in Theorem 8.34, which settles the automorphism conjecture for finite lattices, too (see Exercise 9-8). Thus, in terms of three of our main open questions, the loss of **0** and **1** is significant. The question arises what "intrinsic" parts of the lattice structure can be used to tackle problems such as reconstruction or the fixed point property. To this end, in this chapter we investigate lattices from which top and bottom element have been removed.

9.1 Definition and Examples

The basics of truncated lattices and the related idea of conditional completeness (see Definition 8.16) are, unsurprisingly, quite uncomplicated. This section provides a few facts that are easily translated from lattices.

Definition 9.1. *An ordered set T is called a **truncated lattice** iff any two elements of T that have a common upper bound have a supremum and any two elements of T that have a common lower bound have an infimum.*

Scholium 9.2. *Let P be a finite ordered set. Then P is a truncated lattice iff each nonempty subset $A \subseteq P$ that has an upper bound has a lowest upper bound $\bigvee A$.*

Proof. See Exercise 9-1 ∎

© Springer International Publishing 2016
B. Schröder, *Ordered Sets*, DOI 10.1007/978-3-319-29788-0_9

Remark 9.3. It can be checked in polynomial time if a given finite ordered set is a truncated lattice: For each pair of elements, simply add a check for boundedness in the approach from Remark 8.4.   □

The corresponding proof for the concept of conditional completeness is similar.

Scholium 9.4. *The ordered set P is conditionally complete, iff any nonempty subset* $S \subseteq P$ *that has a lower bound has an infimum.*

Proof. See Exercise 9-2.   ∎

Example 9.5.

1. Any lattice is a truncated lattice, too.
2. Any complete lattice is conditionally complete, too.
3. For any finite lattice L, the set $T = L \setminus \{\mathbf{0}, \mathbf{1}\}$ is a truncated lattice.
4. A finite ordered set P is a truncated lattice iff it is conditionally complete.
5. The set of continuous real valued functions on $[0, 1]$ is a truncated lattice, but it is not conditionally complete. Indeed, the pointwise supremum of two continuous functions is continuous, too. However, (see part 8 of Example 3.21) the family $\{f_n\}_{n\in\mathbb{N}}$ of continuous functions

$$f_n(x) := \begin{cases} 1; & \text{for } x \in \left[0, 1 - \frac{1}{n}\right), \\ -nx + n; & \text{for } x \in \left[1 - \frac{1}{n}, 1\right], \\ 0; & \text{for } x \in (1, 2]. \end{cases}$$

   has an upper bound, but no supremum in $C([0, 1], \mathbb{R})$.
6. The possibly most important example of a conditionally complete ordered set is given in Exercise 8-18.   □

Exercises

9-1. Prove Scholium 9.2.
    Hint. Induction on $n = |A|$ as in Proposition 8.3.
9-2. Prove Scholium 9.4.
    Hint. Similar to the proof of Proposition 8.7.
9-3. Prove that every retract of a truncated lattice is again a truncated lattice. Prove the same result for conditionally complete ordered sets.
9-4. Prove that a finite ordered set L is a truncated lattice iff, for any two antichains $\{a, b\}$ and $\{c, d\}$ with $\{a, b\} \leq \{c, d\}$, there is an element $m \in L$ that is above $\{a, b\}$ and below $\{c, d\}$.
9-5. Let T be a finite truncated lattice and let $f, g : T \rightarrow T$ be automorphisms. Prove that if $f|_{J(T) \cup \text{Min}(T)} = g|_{J(T) \cup \text{Min}(T)}$, then $f = g$.
9-6. Let P be a conditionally complete ordered set that has no largest and no smallest element. Prove that the Dedekind–MacNeille completion of P is the set P with a largest element $\mathbf{1}$ and a smallest element $\mathbf{0}$ attached.
9-7. Let k be a fixed integer. Prove the automorphism conjecture for truncated lattices of width $\leq k$.
    Hint. Exercise 8-25.
9-8. Prove that Theorem 8.34 also is valid for lattices from which top *or* bottom element have been removed. Why does this proof not work for truncated lattices in general?

9.2 Recognizability and More

Considering the proof of the reconstructibility of finite lattices (see Corollary 8.6), we realize that, for reconstruction of finite truncated lattices, we will need new ideas. Fortunately, there is enough structure in finite truncated lattices to allow us to recognize them and to almost reconstruct them. We start by proving some general facts about the deck of a truncated lattice. These facts center around finding out how the property of being a truncated lattice is preserved or lost when going to a one-point deleted subset.

Lemma 9.6. *Let T be a finite truncated lattice, let $x \in T$, and let $C := T \setminus \{x\}$.*

1. *If $A \subseteq C$ and $\bigvee_T A \neq x$, then $\bigvee_C A = \bigvee_T A$.*
2. *If x has a unique upper cover y, then C is a truncated lattice.*
3. *If $A \subseteq C$ is such that $\bigvee_T A = x$ and $y := \bigvee_C A$ exists, then y is the unique upper cover of x and C is a truncated lattice.*
4. *If $\bigvee_T(\downarrow x) \setminus \{x\} \neq x$, then x has a unique lower cover and C is a truncated lattice.*
5. *If x is maximal, then C is a truncated lattice.*
6. *If x is not maximal, minimal, or irreducible, then C is not a truncated lattice.*

Proof. To prove 1, note that $\bigvee_T A \in C$ is an upper bound of A in C. Now each upper bound b of A in C is also in T and hence $b \geq \bigvee_T A$.

For 2, assume that $\emptyset \neq A \subseteq C$ has an upper bound in C. Then A has an upper bound in T and $\bigvee_T A$ exists. If $\bigvee_T A \neq x$, then, by 1, A has a supremum in C. If $\bigvee_T A = x$, then $y \in C$ is an upper bound of A and if $b \in C$ is an upper bound of A, then $b > x$ and hence $b \geq y$. Hence, in this case, $y = \bigvee_C A$. This means that C is a truncated lattice.

To prove 3, let $A \subseteq C$ be such that $\bigvee_T A = x$ and $y := \bigvee_C A \neq x$ exists. Because $y \in T$, clearly $y > x$. Now let $b > x$ be an upper bound of x in T. Then $b \in C$ and b is an upper bound of A. Hence $b \geq \bigvee_C A = y$ and thus y is the unique upper cover of x. Now by 2, C is a truncated lattice.

For 4, $\bigvee_T(\downarrow x) \setminus \{x\} \neq x$ means that x is join-irreducible. By Proposition 8.29, x has a unique lower cover. By the dual of 2, we conclude C is a truncated lattice.

To prove 5, let $x \in T$ be a maximal element and let $\emptyset \neq A \subseteq C$. If A has an upper bound b in C, then $b \neq x$. Hence $\bigvee_T A \neq x$ and thus, by 1, A has a supremum in C. Therefore C is a truncated lattice.

Finally, to prove 6, let $x \in T$ be non-maximal, non-minimal, and non-irreducible. Let U be the set of upper covers of x and let D be the set of lower covers of x. Then $|U|, |D| \geq 2$, U and D are antichains, all elements of U are upper bounds of D, and there is no point in C that is an upper bound of D and a lower bound of U. Thus C is not a truncated lattice, because D has upper bounds and no supremum in C. ∎

In particular, we obtain the following characterization when being a finite truncated lattice is hereditary.

Proposition 9.7. *Let T be a finite truncated lattice and let $x \in T$. Then the card $C = T \setminus \{x\}$ is a truncated lattice iff x is maximal, minimal, or irreducible.*   ∎

Proposition 9.8. *Let P be a finite ordered set that is not a truncated lattice. Then P has at most four cards that are truncated lattices.*

Proof. Because P is not a truncated lattice, there are two elements $a, b \in P$ that have an upper bound but not a supremum. Let $c, d \in P$ be two minimal upper bounds of a and b. Then $\{a, b, c, d\}$ is a four crown, which is contained in all cards $C_{x \notin \{a,b,c,d\}} := P \setminus \{x\}$ with $x \notin \{a, b, c, d\}$. There is no element in P that is between $\{a, b\}$ and $\{c, d\}$. Thus on all the cards $C_{x \notin \{a,b,c,d\}}$, a and b have upper bounds, but not a supremum. Therefore none of the cards $C_{x \notin \{a,b,c,d\}}$ is a truncated lattice. That leaves at most four cards, $P \setminus \{a\}$, $P \setminus \{b\}$, $P \setminus \{c\}$, and $P \setminus \{d\}$, that could be truncated lattices.   ∎

Theorem 9.9. *Finite truncated lattices with at least four elements are recognizable.*

Proof. Ordered sets with a largest or smallest element and disconnected ordered sets are reconstructible (see Propositions 1.37 and 2.51). Thus we shall prove recognizability of connected truncated lattices with at least two maximal elements and at least two minimal elements. Any such truncated lattice has at least four cards that are truncated lattices, too. By Proposition 9.8 any deck with ≥ 5 cards that are truncated lattices is the deck of a truncated lattice.

Thus, by Propositions 9.7 and 9.8, the only truncated lattices that are not trivially recognizable from their decks would have to have exactly two minimal elements, exactly two maximal elements, and no non-extremal irreducible elements. Let T be such a truncated lattice. Let $a, b \in T$ be the minimal elements of T. If T has no non-extremal elements, then, because T is connected, T must be the fence with four elements, which is reconstructible from its deck (see Exercise 1-26a).

Finally, in case T has non-extremal elements, recall that T has no irreducible non-extremal elements. Thus any non-extremal element of T must be above both a and b. (Indeed, otherwise by the dual of Exercise 4-5, T would have an irreducible non-extremal point.) This means a and b have a non-extremal supremum $a \vee b$ and all elements of $T \setminus \{a, b\}$ are above $a \vee b$. The upper covers of $a \vee b$ thus have only one lower cover. This means all upper covers of $a \vee b$ must be maximal, which in turn means that T has exactly five elements $a, b \leq a \vee b \leq c, d$. This ordered set is reconstructible from its deck (see Exercise 1-26b).

Thus truncated lattices are recognizable.   ∎

Theorem 9.10. *Truncated lattices with at least four elements and a non-extremal, non-irreducible element are reconstructible.*

Proof. Truncated lattices are recognizable by Theorem 9.9. Truncated lattices with a non-extremal, non-irreducible element are recognizable as exactly those truncated lattices that have a card that is not a truncated lattice.

Let T be a truncated lattice with a non-extremal, non-irreducible element. Find a card $C = T \setminus \{x\}$ that is not a truncated lattice. Then, by Lemma 9.6, part 4, we have that $\bigvee_T (\downarrow x) \setminus \{x\} = x$. By Lemma 9.6, part 5, x cannot be maximal, so

$(\downarrow x) \setminus \{x\}$ has an upper bound in C. Because x is not irreducible, by Lemma 9.6, part 3, $(\downarrow x) \setminus \{x\}$ does not have a supremum in C. Now let $A \subseteq C$ be a subset with an upper bound in C and no supremum. If there was an $a \in A$ with $a \not\leq x$, then $\bigvee_T A \neq x$ and thus by Lemma 9.6, part 1, A would have a supremum in C contrary to the assumption. Thus every $A \subseteq C$ that has an upper bound and no supremum is contained in $(\downarrow x) \setminus \{x\}$. However this means that

$$(\downarrow x) \setminus \{x\} = \bigcup\{A \subseteq C : A \text{ has an upper bound and no supremum in } C\},$$

and the right-hand side can be reconstructed from C. We obtain $(\uparrow x) \setminus \{x\}$ via the dual argument. Hence T is reconstructible. ∎

For an alternative proof of Theorem 9.10, see Exercise 9-9.

Exercises

9-9. Let C be the card used in the proof of Theorem 9.10. Prove that if T has at least two maximal and two minimal elements, then $T = DM(C) \setminus \{\mathbf{0}, \mathbf{1}\}$.
   Hint. Use Theorem 8.27.
9-10. Let L be a finite lattice and let $x \in L$. Prove that $L \setminus \{x\}$ is a lattice iff x is irreducible.
9-11. Let T be a truncated finite atomic lattice. Prove that T has the fixed point property iff every order-preserving map that maps minimal elements to minimal elements and that preserves suprema of sets of minimal elements has a fixed point.
9-12. Prove that any truncated lattice that contains a four crown as an ordered subset is reconstructible.

9.3 Simplicial Complexes

To start the discussion on the fixed point property for truncated lattices, we first consider an analogous property for simplicial complexes. Recall that Section 6.2 discusses the fixed clique property (see Definition 6.4, every simplicial endomorphism of the graph under consideration fixes a clique). The set of cliques in a graph is a special example of a more general structure, called a simplicial complex.

Definition 9.11. *A **simplicial complex** is a pair $K = (V, S)$ of a set V of vertices and a set S of nonempty subsets of V called **simplices** (singular: **simplex**) such that the following hold.*

1. *Every singleton subset of V is a simplex.*
2. *Every nonempty subset of a simplex is a simplex.*

*When ordered by inclusion, we will call the set S of simplices the **truncated face lattice** of K, also denoted **TFL**(K).*

Clearly the truncated face lattice of a simplicial complex is a truncated lattice (see Exercise 9-13). Simplicial complexes originate in geometry, which motivates the following choice of language.

Definition 9.12. *Let $K = (V, \mathcal{S})$ be a simplicial complex. A q-**simplex** of K is a simplex with exactly $q + 1$ elements. We also say that the **dimension** of the simplex S is q iff S is a q-simplex. If the (q-dimensional) simplex S is contained in the simplex T, then we also say that S is a (q-**dimensional**) **face** of T.*

Geometrically, a 0-simplex is a point, a 1-simplex is a line segment between two points, a 2-simplex is a triangle, a 3-simplex is a tetrahedron, etc. Simplicial complexes can be visualized as geometric figures made up of simplices. They are triangulated by their simplices, which makes them easier to analyze than smooth figures. We will first work with simplicial complexes from a purely algebraic point-of-view. For the topological realization of a simplicial complex, see Section 9.5 and also Figure 9.1 which shows a triangulation of the 2-dimensional unit sphere S^2. The truncated face lattice of the complex is shown on the bottom of the figure.

Any truncated face lattice is conditionally complete with the supremum being the union and the infimum being the intersection. Although we made no finiteness assumptions, we speak of truncated face lattices, because our main focus will be on finite simplicial complexes. Moreover, although these structures are often called **face lattices** in algebraic topology, because some authors consider the empty set a simplex, we keep the word "truncated," to emphasize that we do not consider

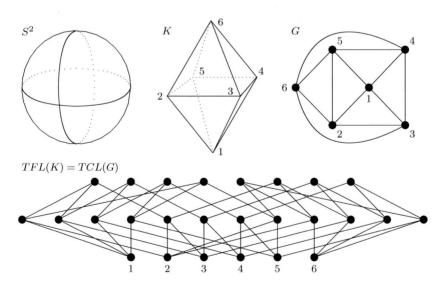

Fig. 9.1 The 2-dimensional unit sphere S^2, a simplicial complex K that triangulates S^2 (the simplices are the points, the edges, and all triangular faces), the unique graph G whose truncated clique lattice equals the truncated face lattice of K (such a graph need not always exist), and the truncated clique/face lattice $TFL(K) = TCL(G)$

the empty set to be a simplex. Note that the minimal elements in a truncated face lattice are the singleton sets and that every element is the supremum of the minimal elements below it. Graphs provide the first large class of examples of simplicial complexes.

Example 9.13. Let $G = (V, E)$ be a graph. Then $K_G = (V, \mathcal{C})$ with $\mathcal{C}$ being the set of cliques (see Definition 5.24) of G is a simplicial complex. This complex is called the **clique complex** of G. The truncated face lattice of the clique complex K_G will also be called the **truncated clique lattice** $TCL(G)$ of G.

For an ordered set P, we can consider the clique complex of its comparability graph (see Definition 6.6). Described directly, the vertices of this simplicial complex are the set P and the simplices are the chains of P. This complex is called the **P-chain complex**[1] of P and the truncated face lattice of the P-chain complex is called the **truncated chain lattice** $TCL(P)$ of P. There is no conflict of notation, because the argument of $TCL(\cdot)$ reveals whether we are looking at a truncated chain lattice or a truncated clique lattice. □

The simplicial complex in Figure 9.1 is indeed the clique complex of the graph that is shown. However, not every simplicial complex is a clique complex: Just consider the complex with vertices $\{a, b, c\}$ and simplices $\{a\}, \{b\}, \{c\}, \{a, b\}, \{a, c\}, \{b, c\}$. The following result characterizes which simplicial complexes arise from graphs and which do not.

Definition 9.14. *Let T be a truncated lattice and let M be the set of minimal elements of T. T is said to satisfy the **clique condition** iff, for all $C \subseteq M$, we have that, if any two elements of C have a supremum in T, then C has a supremum in T. We will say that a simplicial complex satisfies the clique condition iff its truncated face lattice does.*

Proposition 9.15 (See [107], Proposition 4.4, or originally [297]). *Let $\mathcal{S}$ be a truncated face lattice. Then there is a graph G such that $\mathcal{S}$ is the truncated clique lattice of G iff $\mathcal{S}$ satisfies the clique condition.*

Proof. For the "$\Rightarrow$" direction, let $\mathcal{S}$ be the truncated clique lattice of the graph $G = (V, E)$. Let A be a set of minimal elements of $\mathcal{S}$ (that is, A is a set of singleton sets). If any two elements of A have a supremum (=union) in $\mathcal{S}$, then, for any $\{v\}, \{w\} \in A$, we have $v \sim w$ in G. Thus $\bigcup A$ is a clique in G and thus it is an element of $\mathcal{S}$. The set $\bigcup A$ is the supremum of A in $\mathcal{S}$ and hence $\mathcal{S}$ satisfies the clique condition.

For the "$\Leftarrow$" direction, let $\mathcal{S}$ satisfy the clique condition. Let $V := \{x : \{x\} \in \mathcal{S}\}$ and let $E := \{\{x, y\} : \{x, y\} \in \mathcal{S}\}$. Then $G = (V, E)$ is a graph. Clearly every $C \in \mathcal{S}$ is a clique in G. Moreover, by the clique condition, every clique in G is in $\mathcal{S}$ and thus $\mathcal{S}$ is the truncated clique lattice of G. ∎

---

[1] We must be careful here. This natural naming convention is distinct from the idea of a chain complex in algebraic topology as introduced in Definition A.2.

Exercises

9-13. Let $K = (V, S)$ be a simplicial complex. Prove that S is a truncated lattice.
9-14. Prove that the truncated chain lattice of a $2n$-crown is a $4n$-crown.
9-15. Show that the graph G in Figure 9.1 is a comparability graph.
9-16. Compute the truncated chain lattice of each of the following ordered sets.

   a. An n-fence.
   b. An n-chain.
   c. An n-antichain.

9-17. Show that the truncated clique/face lattice in Figure 9.1 is minimal automorphic without referring to topology.
   Hint. I believe that this is very tedious. Hence computer searches should be permissible. For more on algorithmic approaches, see Chapter 5.
9-18. Let P be a finite ordered set so that, for every $p \in P$, the ordered set $\downarrow p \setminus \{p\}$ is isomorphic to a truncated Boolean lattice. Prove that there is a simplicial complex $K = (V, S)$ so that P is isomorphic to S, ordered by inclusion.

9.4   The Fixed Point Property for Truncated Lattices

We can now characterize the fixed point property for finite truncated lattices in terms of simplicial maps on simplicial complexes.

Definition 9.16. *Let $K = (V, S)$ and $H = (W, T)$ be simplicial complexes and let $f : V \to W$ be a function. Then f is called a **simplicial map** iff, for all $S \in S$, we have $f[S] \in T$.*

Just as for graphs and simplicial endomorphisms (see Section 6.1), a fixed point property for simplicial self-maps is not realizable for nontrivial simplicial complexes. It is much more natural to consider the fixed simplex property below. The natural connection between this property and truncated lattices is described in Theorem 9.21. The connection to the fixed clique property is made in Exercise 9-20.

Definition 9.17. *A simplicial complex $K = (V, S)$ has the **fixed simplex property** iff, for every simplicial map $f : V \to V$, there is a simplex S so that $f[S] = S$. A simplicial complex has the **invariant simplex property** iff, for every simplicial map $f : V \to V$, there is a simplex S so that $f[S] \subseteq S$.*

Every order-preserving self-map of an ordered set is also a simplicial endomorphism of the comparability graph. Every simplicial endomorphism is a simplicial map of the clique complex and vice versa (see Exercise 9-19). This explains the terminology in Definition 6.1, and it shows that a graph has the fixed clique property iff its clique complex has the fixed simplex property (see Exercise 9-20). Thus (see Proposition 6.7), for a finite ordered set, the fixed clique property for the comparability graph and the fixed simplex property for the P-chain complex both imply the fixed point property for the ordered set. Moreover, the fixed simplex property is very close to being a property for truncated lattices. After all, it focuses on simplicial maps, which correspond to certain containment preserving maps on truncated face lattices. This is the connection we will investigate now.

As we have seen in Proposition 3.1, it can be helpful to slightly modify a given map when analyzing the existence of fixed points. The more properties a map has, the easier it should be to analyze. A natural property to consider in the context of simplicial complexes is the preservation of unions.

Definition 9.18. *Let X be a set. If $Q \subseteq \mathcal{P}(X)$ is ordered by inclusion, we will call a map $f : Q \to Q$ **union-preserving** iff, for all families $\{A_i\}_{i \in I}$ of sets in Q with $\bigcup_{i \in I} A_i \in Q$, we have that*

$$f\left(\bigcup_{i \in I} A_i\right) = \bigcup_{i \in I} f(A_i) \in Q.$$

Lemma 9.19. *Let $K = (V, \mathcal{S})$ be a simplicial complex.*

1. *If $f : V \to V$ is a simplicial map, then $f^*(S) := f[S]$ defines an order-preserving, union-preserving map on $\mathcal{S}$ that maps minimal elements to minimal elements.*
2. *If $g : \mathcal{S} \to \mathcal{S}$ is an order-preserving, union-preserving map on $\mathcal{S}$ that maps minimal elements to minimal elements, then defining $g_*(v)$ to be the unique element of $g(\{v\})$ defines a simplicial map for K such that $(g_*)^* = g$.*

Proof. To prove part 1, let $f : V \to V$ be a simplicial map of K. Then the function $f^* : \mathcal{S} \to \mathcal{S}$ certainly is well-defined and maps minimal elements to minimal elements. Moreover, if $\{A_i\}_{i \in I}$ is a family of simplices of K with $\bigcup_{i \in I} A_i \in \mathcal{S}$, then $f\left[\bigcup_{i \in I} A_i\right] \in \mathcal{S}$. Hence

$$f^*\left(\bigcup_{i \in I} A_i\right) = f\left[\bigcup_{i \in I} A_i\right] = \bigcup_{i \in I} f[A_i] = \bigcup_{i \in I} f^*(A_i).$$

Thus f^* is union-preserving and a union-preserving map is order-preserving.

For part 2, let $g : \mathcal{S} \to \mathcal{S}$ be an order-preserving, union-preserving map on $\mathcal{S}$ that maps minimal elements to minimal elements, and let $g_*(v)$ be the element of $g(\{v\})$. Let $S \in \mathcal{S}$. To show that g_* is a simplicial map, we must prove that $g_*[S] \in \mathcal{S}$. However, this is trivial, because

$$g_*[S] = \bigcup_{s \in S} \{g_*(s)\} = \bigcup_{s \in S} g(s) = g(S) \in \mathcal{S}.$$

Thus g_* is a simplicial map.

Moreover clearly $(g_*)^*(\{v\}) = g(\{v\})$ for all $v \in V$. Because $(g_*)^*$ and g are both union-preserving and every element of $\mathcal{S}$ is the union of a unique set of minimal elements of $\mathcal{S}$, this means that $g = (g_*)^*$.   ∎

Proposition 9.20. *Let $K = (V, \mathcal{S})$ be a simplicial complex without infinite simplices. Then K has the fixed simplex property iff $\mathcal{S}$ has the fixed point property.*

Proof. First note that, because K has no infinite simplices, S has no infinite chains. Thus if $g : S \to S$ is order-preserving, then existence of an element that is comparable to its image implies existence of a fixed point.

Assume that K has the fixed simplex property. We claim that, for each simplicial map $f : V \to V$, the order-preserving map $f^* : S \to S$ has a fixed point. Indeed, if $S \in S$ is a simplex with $f[S] = S$, then $f^*(S) = S$. Thus, by part 2 of Lemma 9.19, each order-preserving, union-preserving map on S that maps minimal elements to minimal elements has a fixed point. Now let $g : S \to S$ be an order-preserving map. For each $v \in V$, choose $h(\{v\})$ to be a singleton contained in $g(\{v\})$. For each non-singleton $S \in S$, let $h(S) := \bigcup\{h(\{v\}) : v \in S\}$. Let $U \subseteq S$ have a union in S. Then

$$h\left(\bigcup_{S \in U} S\right) = \bigcup \left\{ h(\{v\}) : v \in \bigcup_{S \in U} S \right\} = \bigcup_{S \in U} \bigcup \{h(\{v\}) : v \in S\} = \bigcup_{S \in U} h(S).$$

Thus h is order- and union-preserving and hence $h = (h_*)^*$ has a fixed point. Therefore there is an $S \in S$ such that $g(S) \supseteq h(S) = S$. This implies that g has a fixed point. Because g was arbitrary, S has the fixed point property.

Conversely, suppose that S has the fixed point property. Let $f : V \to V$ be a simplicial map. Then f^* has a fixed point $S \in S$, which is a simplex of K such that $f[S] = f^*(S) = S$. Thus K has the fixed simplex property.   ∎

Proposition 9.20 shows how to encode the fixed simplex property into the fixed point property for a truncated lattice. The interesting fact now is that the fixed point property for a truncated lattice can also be encoded as the fixed simplex property for the appropriate simplicial complex. The idea is simple. Any lattice can be represented in various ways as a set system. Thus the same is true for truncated lattices. The trick is to find the right representation. The set system we will ultimately choose is not necessarily isomorphic to the truncated lattice we start with, but it will have the fixed point property if and only if the original truncated lattice does.

Theorem 9.21. *Let T be a truncated lattice such that the following hold.*

1. *T has no elements with unique lower covers.*
2. *T has no infinite chains.*
3. *No element of T is above infinitely many minimal elements.*

Let $M \subseteq T$ be the set of minimal elements of T. We define B_T to be the ordered set of all nonempty subsets H of M that have an upper bound in T, ordered by set inclusion. Then B_T is a truncated face lattice and the following are equivalent.

1. *T has the fixed point property.*
2. *B_T has the fixed point property.*
3. *The simplicial complex (M, B_T) has the fixed simplex property.*

Moreover, if T satisfies the clique condition, then there is a graph G_T that has the fixed clique property iff T has the fixed point property.

Proof. Clearly B_T is a truncated face lattice. Our first step towards proving the equivalence is to turn the original truncated lattice T into a system of sets. For each $x \in T$ let $M_x := (\downarrow x) \cap M$. Order

$$A_T := \{M_x : x \in T\}$$

by inclusion. Then

$$\Phi : T \to A_T; \qquad x \mapsto M_x$$

is order-preserving and surjective. To see that Φ is an isomorphism, note that, because all non-minimal points of T have at least two lower covers, we have $x = \bigvee[(\downarrow x) \cap M]$ for all $x \in T$. Now $M_x \subseteq M_y$ implies that

$$x = \bigvee (\downarrow x) \cap M = \bigvee M_x \le \bigvee M_y = \bigvee (\downarrow y) \cap M = y.$$

Thus Φ is an isomorphism and, in particular, T has the fixed point property iff A_T does.

For $Y \in B_T$ let $r(Y)$ be the supremum of $\{\{y\} : y \in Y\}$ in A_T. The function r is well-defined, because every $Y \in B_T$ has an upper bound in A_T, no element of A_T is above infinitely many minimal elements, and A_T is a truncated lattice. Clearly r is a retraction and $r(Y) \supseteq Y$ for all Y. Because no element of T is above infinitely many minimal elements, B_T has no infinite chains. Hence, by Theorem 4.11, A_T has the fixed point property iff B_T does. By Proposition 9.20, this is the case iff (M, B_T) has the fixed simplex property. We have thus proved the claimed equivalences.

Finally, if T satisfies the clique condition, then, by Proposition 9.15, the simplicial complex (M, B_T) is induced by a graph G_T. This, together with Exercise 9-20, proves the "Moreover"-part. ∎

Remark 9.22. Proposition 9.15 distinguishes truncated clique lattices from general truncated face lattices. It seems difficult to try to work without the clique condition and obtain fixed simplex theorems. For the fixed simplex property, not even a weak analogue of the Abian–Brown Theorem like Exercise 6-8b is available. To see this consider the simplicial complex with vertices a, b, c and simplices $\{a\}, \{b\}, \{c\}$, $\{a, b\}, \{a, c\}, \{b, c\}$. We have $N(a) = N(b) = N(c) = \{a, b, c\}$ and this simplicial complex does not have the fixed simplex property. (Its truncated face lattice is the six crown). □

The connections between the properties investigated in this chapter are summarized in Figure 9.2.

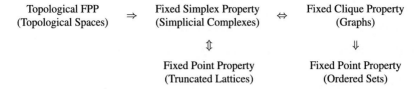

Fig. 9.2 Relation between the properties discussed in Sections 6.2, 9.4, and 9.5. Any of the shown implications is valid whenever the representative of the more general structure was induced by a representative of the more restrictive structure

Exercises

9-19. Let $G = (V, E)$ and $H = (W, F)$ be graphs and let $f : V \to W$ be a function. Prove that f is a simplicial homomorphism from G to H iff f is a simplicial map from the clique complex of G to the clique complex of H.

9-20. Let $G = (V, E)$ be a graph. Prove that G has the fixed clique property iff its clique complex K_G has the fixed simplex property.

9-21. Let $K = (V, S)$ and $H = (W, T)$ be simplicial complexes and let $f : V \to W$ be a simplicial map. Then f is called **dimension-preserving** iff the image of every q-dimensional simplex is again a q-dimensional simplex.

Let $G = (V, E)$ and $H = (W, F)$ be graphs and let $f : V \to W$ be a function. Prove that f is a homomorphism from G to H (in the sense of Definition 6.10) iff f is a dimension-preserving simplicial map from the clique complex of G to the clique complex of H.

9-22. Prove that, if T is a finite truncated lattice, $x \in T$, $f : T \to T$ is order-preserving, $f^n(x) = x$ and $\{x, f(x), \dots, f^n(x) = x\}$ has an upper bound, then f has a fixed point.

9-23. Truncated Dedekind–MacNeille completions and the fixed point property.

a. Find a finite ordered set P that does not have the fixed point property such that the truncated lattice $DM(P) \setminus \{\mathbf{0}, \mathbf{1}\}$ has the fixed point property.

b. Find a finite ordered set P that has the fixed point property such that $DM(P) \setminus \{\mathbf{0}, \mathbf{1}\}$ does not have the fixed point property.

9-24. Let $K = (V, S)$ be a simplicial complex. The vertex $a \in V$ is called **irreducible** iff there is a $b \in V \setminus \{a\}$ such that, for all simplices $\sigma \in \Sigma$ that contain a, the set $\sigma \cup \{b\}$ is a simplex, too.

a. Prove that K has the fixed simplex property iff the full subcomplex (also see Definition 9.33) $K[V \setminus \{a\}] := (V \setminus \{a\}, S \setminus \{\sigma \in S : a \in \sigma\})$ has the fixed simplex property. *Hint.* Use Proposition 9.20 and Theorem 4.11.

b. Define retractable vertices for simplicial complexes.

c. Explain why the idea for part 9-24a does not translate to prove an analogue of the sufficient condition for the fixed point property for P in Theorem 4.12 for simplicial complexes.

Note. I do not know if the analogue mentioned in Exercise 9-24c is even true. Exercise 9-33 shows that the condition is not necessary, and I have some notes that show that, if the condition is not sufficient, the counterexample would satisfy a number of very restrictive conditions. It is shown in [289] that, for graphs and the fixed clique property, there is a full analogue of Theorem 4.12.

9.5   Triangulations of S^n

As noted earlier, originally, simplicial complexes are geometric objects. Indeed, using a sufficiently fine triangulation, geometric objects in any dimension can be approximated through a union of line segments, triangles, tetrahedra, and higher-dimensional simplices. In this section, we have no choice but to assume some background in topology. The main purpose of this section is to present a family of minimal automorphic ordered sets in Theorem 9.28. Hence, this section can be skipped if necessary.

As we have approached simplicial complexes from an algebraic or combinatorial point-of-view, we first define the topological realization of a simplicial complex.

Definition 9.23. *Let $K = (V, S)$ be a finite simplicial complex. We define the* **topological realization** *$|K|$ of K to be the metric space consisting of the set of functions $\alpha : V \to [0, 1]$ such that, for each α, the following hold.*

1. $\{v \in V : \alpha(v) \neq 0\} \in S$.
2. $\sum_{v \in V} \alpha(v) = 1$.

The metric on the topological realization is

$$d(\alpha, \beta) := \sqrt{\sum_{v \in V} (\alpha(v) - \beta(v))^2}.$$

Clearly the topological realization of a finite simplicial complex can be embedded as a subspace in $\mathbb{R}^{|V|}$ with the usual topology. Note that there may be lower-dimensional spaces into which the complex embeds. This is the case for the simplicial complex shown in Figure 9.1, which has six vertices, but which can be realized in $\mathbb{R}^3$.

Simplicial maps between simplicial complexes can be extended to affine maps by affine interpolation. A function $f : S \to T$ from a subset $S \subseteq \mathbb{R}^n$ to a subset $T \subseteq \mathbb{R}^m$ is called an **affine map** iff, for all vectors $x_1, \ldots, x_k \in S$ and all scalars $\alpha_1, \ldots, \alpha_k \in \mathbb{R}$ such that $\sum_{i=1}^k \alpha_i = 1$ and $\sum_{i=1}^k \alpha_i x_i \in S$, we have that $f\left(\sum_{i=1}^k \alpha_i x_i\right) = \sum_{i=1}^k \alpha_i f(x_i)$. Affine interpolation now is the following process. A simplicial map g is defined on the "corners" x_i of the simplices. We can extend g to an affine map g_a by defining $g_a\left(\sum_{i=1}^k \alpha_i x_i\right) := \sum_{i=1}^k \alpha_i g(x_i)$, where we took the liberty of identifying the elements of the simplicial complex with their topological realizations. The fact that g is a simplicial map together with $\sum_{i=1}^k \alpha_i = 1$ assures that the new map g_a does not map any points in the topological realization of the preimage to points outside the topological realization of the image. It is easy to see that a simplicial complex has the fixed simplex property iff every affine map as above fixes a topological realization of a simplex.

Because every affine map is continuous, we can now connect to the fixed point property from topology. A topological space has the **fixed point property** iff each continuous self-map has a fixed point. (Topological) Retractions (idempotent *continuous* self-maps in topology) play a similar role in topology as in ordered sets and the topological fixed point property is, for example, inherited by retracts (see Exercise 9-25).

If the topological realization of a simplicial complex has the topological fixed point property, then the simplicial complex has the fixed simplex property. Indeed, first note that every affine map is continuous. If every continuous map on the topological realization has a fixed point, then every affine map has a fixed point. An affine map induced by a simplicial map can only have a fixed point if it fixes some simplex.

However, the converse of the above is false. If a simplicial complex has the fixed simplex property, then its topological realization can still fail to have the fixed point property. For an example, see Exercise 9-27. This may not be too surprising, because not every continuous function on a given topological realization of a simplicial complex can be approximated closely by affine maps that map vertices to vertices.

For the examples in this section, we consider the n-dimensional unit sphere.

Definition 9.24. *Let S^n denote the n-dimensional unit sphere, that is,*

$$S^n := \left\{ (x_1, \ldots, x_{n+1}) \in \mathbb{R}^{n+1} : x_1^2 + \cdots + x_{n+1}^2 = 1 \right\}.$$

We will always consider S^n with the standard topology inherited from $\mathbb{R}^{n+1}$.

The unit sphere has a property analogous to being minimal automorphic in ordered sets: It does not have the topological fixed point property itself (consider the map that maps every point to its antipode), yet every one of its nontrivial retracts does have the fixed point property (see Lemma 9.25). This should allow us to construct a wealth of examples of minimal automorphic ordered sets. Indeed, because the topological fixed point property is stronger than the combinatorial fixed point/clique/simplex properties, we will be able to obtain simplicial complexes for which every retract has the fixed simplex property. If we then structure the simplicial complex so that we can capture a fixed point free continuous map on it, we are done.

Lemma 9.25. *Let $n \geq 2$. Every retract of S^n that is not equal to S^n has the topological fixed point property.*

Proof. Let $r : S^n \to S^n$ be a (topological) retraction with $r[S^n] \neq S^n$. Then $r[S^n]$ is isomorphic to a retract of the n-dimensional unit ball which, by Brouwer's fixed point theorem, has the topological fixed point property.   ∎

Definition 9.26. *Let K be a finite simplicial complex. Then K is called a **triangulation of** S^n iff the topological realization $|K|$ of K is homeomorphic to S^n.*

An example of a triangulation of S^2 is given in Figure 9.1.

Lemma 9.27. *Every nontrivial retract of the truncated face lattice T_K of a triangulation of S^n with $n \geq 2$ has the (order-theoretical) fixed point property.*

Proof. Let T_K be the truncated face lattice of the triangulation K of S^n. Let $r : T_K \to T_K$ be a nontrivial retraction with $r[T_K] \neq T_K$.

We claim there is a minimal element m of T_K that is not in $r[T_K]$. Suppose otherwise. Then r fixes all minimal elements of T_K, which implies $r(x) \geq x$ for all $x \in T_K$. This implies there is a $y \in T_K$ such that $r(y) > y$ and $r(p) = p$ for all $p \geq y$, which means that y has exactly one upper cover. Therefore there is a maximal element $M \in T_K$ such that the simplex $M \setminus y$ has exactly one point (otherwise y has more than one upper cover). We have established that M is the only upper bound of y.

Now we need to switch our viewpoint to topology. Recall that, for any x and $\varepsilon > 0$, $B_\varepsilon(x)$ denotes the solid ball of radius ε with center x. The topological realization $|y|$ is contained in the boundary of exactly one n-dimensional simplex $|M|$. Therefore, every point x that lies in the interior of the topological realization $|y| \subseteq |K|$ of y is such that, for all $\varepsilon > 0$ small enough, we have that $B_\varepsilon(x) \cap |K|$ is homeomorphic to the upper half space $\{(x_1, \ldots, x_n) : x_j \in \mathbb{R}, x_n \geq 0\}$. Because S^n is an n-dimensional manifold without boundary, this is a contradiction to $|K|$ being homeomorphic to S^n. Thus there must be a minimal element $m \in T_K \setminus r[T_K]$, which proves the claim.

Now let $f : r[T_K] \to r[T_K]$ be an order-preserving map. Let $F := f \circ r$. For each minimal element $p \in T_K$ choose a minimal element $G(p) \in r[T_K]$ such that $G(p) \leq F(p)$. For $q \in T_K$ not minimal, let

$$G(q) := \bigvee \{G(p) : p \in T_K, \ p \leq q \text{ is minimal}\}.$$

Then G is order-preserving on T_K. By Proposition 4.4, $G^{|T_K|!}$ is a retraction. We have shown above that the retract $G^{|T_K|!}[r[T_K]]$ does not contain all minimal elements. Thus $G^{|T_K|!}$ induces a simplicial map on K that is a *nontrivial* retraction. This map can be extended to a continuous retraction $R : |K| \to |K|$. Now, by Lemma 9.25, $R[|K|] = \left| G^{|T_K|!}[K] \right|$ has the topological fixed point property. Thus the continuous map G_c on $R[|K|]$ induced by G has a fixed point p. Let S be the smallest simplex in $G^{|T_K|!}[K]$ such that $p \in G_c[|S|] \subseteq R[|K|]$. Then G maps S to a sub-simplex of S, that is, $G(S) \supseteq S$ and thus G has a fixed point. Thus $F \geq G$ has a fixed point, which must be a fixed point of f.   ∎

Theorem 9.28. *Let K be a triangulation of S^n that has a realization $|K|$ such that the antipodal map $x \mapsto -x$ maps k-simplices of K to k-simplices of K. Then the truncated face lattice T_K of K is minimal automorphic.*

Proof. For each simplex S of K, let $A(S) := -S$. By hypothesis, this function is well-defined. Because no simplex is equal to its antipode, $A : T_K \to T_K$ is a fixed point free order-preserving automorphism of T_K. By Lemma 9.27, all nontrivial retracts of T_K have the (order-theoretical) fixed point property.   ∎

For an example of a triangulation and truncated face lattice as in Theorem 9.28, consider Figure 9.1.

Remark 9.29. It is clear that the above construction can be carried out for other kinds of polyhedral complexes such as buckyballs, etc. Moreover, it is clear that this is a result about the fixed clique and the fixed simplex properties as well as the fixed point property.   □

Exercises

9-25. Let X be a topological space with the (topological) fixed point property and let $r : X \to X$ be a continuous retraction. Prove that $r[X]$ has the (topological) fixed point property.

9-26. Consider the ordered set in Figure 1.3.

    a. Sketch its comparability graph.
    b. Show that its comparability graph is a triangulation of the Möbius strip.

9-27. An example of a simplicial complex that has the fixed simplex property and whose topological realization does not have the topological fixed point property. Consider the ordered set P_{C_8} in Figure 9.3. Note that, with the indicated sets, P_{C_8} is a set of sets ordered by set inclusion. From P_{C_8}, considered as a set of sets, construct an ordered set Q_{C_8}, ordered by set inclusion, by adding to P_{C_8} all subsets of $\{c, d, e, f\}$.

    a. Prove that the ordered set P_{C_8} in Figure 9.3 has the fixed point property. (We do not need that P_{C_8} is a set of sets for this part.)
    b. Prove that Q_{C_8} has the fixed point property.
    c. Prove that the simplicial complex $K_{C_8} = (V_{C_8}, \mathcal{S}_{C_8})$ on the vertex set $V_{C_8} = \{a, c, d, e, f\}$ and with simplex set being $\mathcal{S}_{C_8} = Q_{C_8}$, where we consider Q_{C_8} as a set of sets, has the fixed simplex property.
    d. Prove that the topological realization of K_{C_8} does not have the topological fixed point property.
    Hint. The topological realization of the simplicial complex with vertex set $V = \{c, d, e, f\}$ and whose simplices are all nonempty subsets of V is a tetrahedron in three-dimensional space. The remaining simplices can be realized as triangles that are attached to this tetrahedron.
    e. Prove that P_{C_8} is isomorphic to the lattice of 0-, 1-, and 2-dimensional faces of a pyramid with square base.

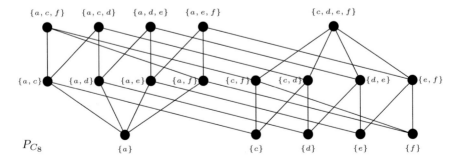

Fig. 9.3 The ordered set P_{C_8}, which has the fixed point property

9.6 Reduction Theorems from Algebraic Topology

We will now prove some results that depend on algebraic topology, namely, homology to be precise. If you are not versed in these fields, you can read Appendix A to gain the necessary preliminaries, or you can simply use Theorems 9.30, 9.34, and 9.35 as axioms. Especially for the first reading, the latter approach may be most effective.

The main idea from algebraic topology that we will need is the notion of acyclicity. Acyclicity is defined through the homology complex, see Definition A.15. In particular, although the word may suggest it, acyclicity does not imply the absence of cycles in the graph or the simplicial complex. We note that acyclicity implies the fixed clique and fixed simplex properties.

Theorem 9.30 (This is exactly Lemma A.16). *Let $K = (V, S)$ be a finite acyclic simplicial complex. Then K has the fixed simplex property.*

The key combinatorial notion will be Constantin and Fournier's notion of a (weakly) escamotable point.

Definition 9.31. *Let $K = (V, S)$ be a simplicial complex. The **link** of a simplex α in K is the set system $Lk(\alpha) := \{\beta \in S : \alpha \cup \beta \in S, \alpha \cap \beta = \emptyset\}$. Moreover, analogous to graphs, for $v \in V$, we can define $N(v) := \{w \in V : \{v, w\} \in S\}$. We will call the simplicial complex $(N(v) \setminus \{v\}, Lk(\{v\}))$ the **link complex** of v.*

Definition 9.32. *Let $K = (V, S)$ be a simplicial complex. Call $v \in V$ **weakly escamotable**[2] iff the link complex $(N(v), Lk(\{v\}))$ is acyclic.*

Escamotable points in simplicial complexes are such that their addition or removal does not affect whether the simplicial complex is acyclic or not. This is an approach we know well from Section 4.2.1. Full subcomplexes of simplicial complexes are defined similar to induced subgraphs (see Definition 6.14).

Definition 9.33. *Let $K = (V, S)$ be a simplicial complex. Then the simplicial complex $K' := K[V'] := (V', S')$ is called a **full subcomplex** of K iff $V' \subseteq V$ and, for all simplices $S \in S$ with $S \subseteq V'$, we have $S \in S'$.*

The difference between the link complex and the full subcomplex $K(N(v) \setminus \{v\})$ is discussed in Exercise 9-28.

Theorem 9.34 (This is exactly Corollary A.19). *Let $K = (V, S)$ be a finite connected simplicial complex and let $v \in V$ be weakly escamotable. Then, for all*

---

[2]The original notion is that of an escamotable *point*, see [48] Définition 3.1. A point p in an ordered set is escamotable iff its center-deleted neighborhood $\updownarrow p \setminus \{p\}$ is contractible. (Contractibility is a topological notion that implies acyclicity.) We use the weaker notion here, as it allows us to avoid the use of topology in our proofs. We also obtain slightly stronger fixed point results. The word "escamotable" is French and means "retractable." To avoid confusion of terms, we use the original French term.

$q \in \mathbb{N} \cup \{0\}$, we have $H_q(K) = H_q(K[V \setminus \{v\}])$. *(These are the homology groups defined in Definition A.9.) In particular, $K[V \setminus \{v\}]$ is acyclic iff K is acyclic.*

Exercise 9-34 shows that, in Theorem 9.34, "is acyclic" cannot be replaced with "has the fixed simplex property." The last result whose proof we relegate to Appendix A is that, unsurprisingly, sufficiently nice simplicial complexes are acyclic.

Theorem 9.35 (Easy consequence of Exercise A-7). *A finite simplicial complex $K = (V, \mathcal{S})$ that has a vertex c so that, for all $\sigma \in \mathcal{S}$, we have that $\sigma \cup \{c\} \in \mathcal{S}$ is acyclic.*

The main theorem that applies algebraic topology to the fixed point theory for ordered sets is now a removal theorem that reduces the analysis of a simplicial complex to the analysis of a certain full subcomplex. Removing excess points from complexes under investigation to ease the analysis appears to be a standard technique in algebraic topology, just as it is for us.

Theorem 9.36 (Compare [48], Théorème 4.1.). *Let $K = (V, \mathcal{S})$ be the clique complex of a finite graph. Let $B \subsetneq V$ be such that, for every nonempty simplex $\sigma \subseteq V \setminus B$, the simplicial complex $K[\{v \in B : \sigma \cup \{v\} \in \mathcal{S}\}]$ is acyclic. Then every $x \in V \setminus B$ is weakly escamotable. Moreover, for all $q \in \mathbb{N} \cup \{0\}$, we have $H_q(K) = H_q(K[B])$, and K is acyclic iff $K[B]$ is acyclic.*

Proof. This is an induction on $n := |V \setminus B|$. For $n = 1$, this is Theorem 9.34. So now let $n > 1$ and suppose the result holds for all $k < n$. Let $x \in V \setminus B$ and let $V' := \{v \in V : \{v, x\} \in \mathcal{S}\} \setminus \{x\}$, $K' := K[V'] = (V', Lk(\{x\}))$ and let $B' := V' \cap B \neq \emptyset$. We claim K' and B' satisfy the hypothesis of the theorem. In fact, if $\sigma \subseteq V' \setminus B'$ is a nonempty simplex in K', then, by the clique condition, $\sigma \cup \{x\}$ is a simplex in K and, again via the clique condition,

$$K'[\{v \in B' : \sigma \cup \{v\} \in \mathcal{S}\} = K[\{v \in B : (\sigma \cup \{x\}) \cup \{v\} \in \mathcal{S}\}],$$

is acyclic. Thus K' and B' satisfy the hypothesis of the theorem and, because we have $|V' \setminus B'| < n$, the induction hypothesis can be applied.

Now note that $K'[B'] = K[\{v \in B : \{v, x\} \in \mathcal{S}\}]$ is acyclic by assumption. Thus, by induction hypothesis, K' is acyclic and thus x is weakly escamotable.

To prove the "moreover" part, we must prove that K is "dismantlable via weakly escamotable points" to $K[B]$. To see this, it is good enough to show that $K[V \setminus \{x\}]$ and B satisfy the assumption of the theorem. Let $\sigma \subseteq V \setminus (\{x\} \cup B)$ be a simplex. Then $K[\{v \in B : \sigma \cup \{v\} \in \mathcal{S}\}]$ is acyclic by assumption. This finishes the proof by induction. ∎

For more on Theorem 9.36, see Remarks 6 and 7 at the end of this chapter. Clearly a set B as above must intersect every maximal simplex, which is exactly the notion of a cutset.

Definition 9.37. *Let $K = (V, S)$ be a finite simplicial complex. Then $A \subseteq V$ is called*

1. *A **cutset** iff A intersects every maximal simplex,*
2. *An **astral subset** iff there is a $k \in V$ such that, for all $a \in A$, we have $\{a, k\} \in S$,*
3. *An **astral subset with a center** c iff*

   a. *A is astral,*
   b. *For all $a \in A$, we have $\{a, c\} \in S$,*
   c. *If $k \in V$ is such that, for all $a \in A$, we have $\{a, k\} \in S$, then $\{k, c\} \in S$,*

4. *A **coherent cutset** iff A is a cutset and every astral subset of A has a center,*
5. ***Semi-bounded by** B iff $A \subseteq B$, every element of B is a center of a nonempty subset of A and, for every astral subset S of A, the set B contains a center of S.*

The name "cutset" is well motivated, as a cutset essentially "cuts into" all parts of a simplicial complex. Astral subsets A can be visualized like stars with rays emanating from k to the elements of A. However note that the element k need not be in A. In ordered sets, the points k often are upper or lower bounds, though they need not be (see Exercise 9-29). The center of an astral subset A is essentially the most natural choice for the heart of the "star" made up by an astral subset. In ordered sets, the centers will normally be a supremum or an infimum, though, again, they need not be. Semi-boundedness assures that everything that should have a center actually has one in B. This is similar to the idea of conditional completeness (where everything that needs a supremum has one). Semi-boundedness allows us to choose a smaller number of all the centers that might be available (normally one for each astral set), which makes combinatorial work easier.

We can now prove that an ordered set with a sufficiently nice cutset has the fixed point property. Where Theorem 9.36 provides a general framework in case one has a good handle on the property of acyclicity, Theorem 9.40 puts us back on more solid combinatorial footing. All hypotheses are stated in combinatorial terms, which makes the visualization of the result slightly easier. The proof relies once again on algebraic topology, however, and the connection to ordered sets needs to be made via the comparability graph. For a special case that may clarify a few things, see Exercise 9-32.

Definition 9.38. *Let $K = (V, S)$ be a simplicial complex, let $\sigma \in S$, and let $A \subseteq V$. We define the following.*

1. *$W_\sigma := \{v \in V \setminus \sigma : \sigma \cup \{v\} \in S\}$.*
2. *$C_A := \{v \in V : v \text{ is the center of a nonempty subset of } A\}$.*

Lemma 9.39 (See [48], Lemme[3] 4.5.). *Let $K = (V, S)$ be the clique complex of a finite graph, let A be a coherent cutset of K that is semi-bounded by B, and let $\sigma \subseteq V \setminus B$ be a nonempty simplex. Then $A' := W_\sigma \cap A$ is a coherent cutset of $K[W_\sigma \cap B]$ and, in $K[W_\sigma \cap B]$, A' is semi-bounded by $B' := C_{A'} \cap B$.*

---

[3]Not a typo. The paper is in French. :)

Proof. To prove that A' is a cutset of $K[W_\sigma \cap B]$, let τ be a maximal simplex in $K[W_\sigma \cap B]$. By the clique condition, $\tau \cup \sigma$ is a simplex in K. Let μ be a maximal simplex in K that contains $\tau \cup \sigma$. Because A is a cutset, there is an $a \in A \cap \mu$. Because $a \notin \sigma$ and $a \in \mu$, we have $a \in W_\sigma$. Together with $a \in A$, we obtain $a \in A'$. Because $a \in \mu$ and $\tau \subseteq \mu$, we obtain that $\{a\} \cup \tau$ is a simplex in K. Now, because $\{a\} \cup \tau \subseteq W_\sigma \cap B$, we obtain that $\{a\} \cup \tau$ is a simplex in $K[W_\sigma \cap B]$. By maximality of τ, $a \in \tau$ and we have proved that A' is a cutset of $K[W_\sigma \cap B]$.

To prove that, in $K[W_\sigma \cap B]$, A' is a coherent cutset that is semi-bounded by $B' := C_{A'} \cap B$, let $S \subseteq A'$ be a nonempty astral subset of A'. Because $S \subseteq A$ and A is semi-bounded by B, there is a $c \in B$ that is a center of S. Because $S \subseteq A'$, we have $c \in C_{A'} \cap B$. ∎

Theorem 9.40 (See [12], Corollary 2.6; [48], Thèorème 4.6; [79], main theorem on p. 117; [262], Theorem 3; [271], Theorem 5.3.). *Let $K = (V, S)$ be the clique complex of a finite graph. Suppose A is a coherent cutset of K that is semi-bounded by B. Then every $x \in V \setminus B$ is weakly escamotable in K. In particular, for all $q \in \mathbb{N} \cup \{0\}$, we have $H_q(K) = H_q(K[B])$, and K is acyclic iff $K[B]$ is acyclic.*

Proof. We will prove by induction on $|V|$ that the simplicial complex and the set B satisfy the hypothesis of Theorem 9.36. The base step $|V| = 1$ is trivial.

For the induction step, let $K = (V, S)$ be a clique complex with $|V| = n$, let A be a coherent cutset of K that is semi-bounded by B, and assume that, for simplicial complexes with fewer than n vertices that satisfy the assumption of this theorem, it has been proved that the hypothesis of Theorem 9.36 is satisfied. Let $\sigma \subseteq V \setminus B$ be a nonempty simplex. Let $A' := W_\sigma \cap A$, $V' := W_\sigma \cap B$, and $B' := C_{A'} \cap B$. By Lemma 9.39, A' is a coherent cutset of $K[V']$ that is, in $K[V']$, semi-bounded by B'. Because no element of σ is in V', we have $|V'| < n$. Hence, by induction hypothesis, $K[V']$ and the set B' satisfy the hypothesis of Theorem 9.36. Hence, by Theorem 9.36, $K[V']$ is acyclic iff $K[B']$ is.

To see that $K[B']$ is acyclic, we will show that there is a $b \in B$ so that, for all simplices τ of $K[B']$, we have that $\{b\} \cup \tau$ is a simplex in $K[B']$. Because $A' \subseteq W_\sigma$, every element of σ is a center for A', that is, A' is astral in K. Because, in K, A is semi-bounded by B, there is a center b of A' in B, and hence b is in B'. Moreover, by definition of centers, for every $v \in C_{A'}$, we have that $\{v, b\}$ is a simplex. By the clique condition, this means that, for every simplex τ of $K[B']$, we have that $\{b\} \cup \tau$ is a simplex in $K[B']$.

Thus, by Theorem 9.35, $K[B']$, and hence $K[V']$, is acyclic. Because σ was arbitrary, K and B satisfy the hypothesis of Theorem 9.36. This concludes the induction, and the conclusion of this theorem now follows from Theorem 9.36. ∎

Exercises

9-28. Let $K = (V, S)$ be a simplicial complex and let $v \in V$ be a vertex.

   a. Give an example of a simplicial complex $K = (V, S)$ and a vertex $v \in V$ so that the full subcomplex $K[N(v) \setminus \{v\}]$ is not equal to the link complex $(N(v), Lk(v))$.

b. Let $G = (V, E)$ be a graph and let $C(G)$ be the clique complex of G. Prove that, in $C(G)$, for all $v \in V$, the full subcomplex $K[N(v) \setminus \{v\}]$ is equal to the link complex $((N(v), Lk(v))$.

9-29. Find an ordered set that has an astral subset A with a center such that the center is not part of A and is not an upper or a lower bound of A.

9-30. Prove that the P-chain complex of an $\mathcal{I}$-dismantlable finite ordered set is acyclic.
    Hint. Theorems 9.34 and 9.35.

9-31. Call a finite ordered set P **escamotably collapsible** iff the points in P can be enumerated as $P = \{a_1, \dots, a_n\}$ so that, for $i = 1, \dots n - 1$, the point a_i is weakly escamotable in $P \setminus \{a_1, \dots, a_{i-1}\}$. Prove that, if P is escamotably collapsible, $a \in P$ is weakly escamotable, and P is of height 2, then $P \setminus \{a\}$ is escamotably collapsible, too.
    Hint. Suppose, for a contradiction, there are $b_1 = a, b_2, \dots, b_k$ so that, for $i = 1, \dots, k$, the point b_i is weakly escamotable in $P \setminus \{b_1, \dots, b_{i-1}\}$, and so that $Q = P \setminus \{b_1, \dots, b_k\}$ has at least two points and no weakly escamotable points. First use Theorem 9.34 and the fact that an ordered set of height 1 is acyclic iff it contains no crowns (see Exercise A-11) to prove that Q must have height 2. Then prove that the point $a_j \in Q$ with the smallest possible j is not weakly escamotable in $P \setminus \{a_1, \dots, a_{i-1}\}$.
    Note. This result seems obvious in light of results such as Theorem 4.31. However, the fact that, in Theorem 4.31 we are working with order-preserving retractions, whereas, for escamotable points, we do not, is crucial: In [17], there is an explicit example of an ordered set that can be escamotably collapsed to a singleton as well as to a non-singleton ordered set that does not have escamotable points. (Use the simplex sets of the simplicial complexes in question.) I conjecture that a result similar to Theorem 1.1 in [305] holds for escamotable collapsibility.

9-32. Let P be a finite ordered set that has a cutset C such that all nonempty subsets of C have a supremum or an infimum.

   a. Prove that P has the fixed point property.
   b. Find a proof that does not rely on algebraic topology.
       Note. Such a proof can be found in [79].
   c. Prove that the $(2, 1)$-core of $FPF(P)^{\mathrm{exp}}$ is empty.
       Note. A proof can be found in [65].

9-33. Consider the simplicial complex $K = (V, S)$ whose simplex set S is depicted in Figure 9.4.

   a. Prove that K has the fixed simplex property.
       Hint. Prove that S has the fixed point property.
   b. Define retractable vertices for simplicial complexes.
   c. Prove that the vertex a is retractable to the vertex b.
   d. Prove that the link complex $(N(a) \setminus \{a\}, Lk(\{a\}))$ of a does not have the fixed simplex property.
       Hint. Prove that its simplex set does not have the fixed point property.

   Note. This example shows that, for simplicial complexes and the fixed simplex property, there cannot be a full analogue of Theorem 4.12. Theorem 4.12 does translate to graphs and the fixed simplex property (see [289]). It is not known whether retractability of a and the fixed simplex property for $(N(a) \setminus \{a\}, Lk(\{a\}))$ and for $K[V \setminus \{a\}]$ imply the fixed simplex property for K. I conjecture this is the case, but, so far, I only have an argument that a counterexample would have very special structure.

9-34. In Theorem 9.34, "is acyclic" cannot be replaced with "has the fixed simplex property." Consider the ordered set P in Figure 9.5.

   a. Prove that this ordered set does not have the fixed point property.
   b. Prove that the chain complex of $P \setminus \{a\}$ has the fixed simplex property.

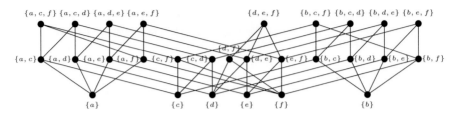

Fig. 9.4 The simplices of a simplicial complex with the fixed simplex property and a retractable vertex a so that the link complex of a does not have the fixed simplex property

Fig. 9.5 An ordered set that shows that, in Theorem 9.34, "is acyclic" cannot be replaced with "has the fixed simplex property." (See Exercise 9-34)

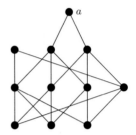

> *Note.* I do not know how hard a written proof would be. Verifying that $TCL(P \setminus \{a\})$ has the fixed point property using, say, [286] works quite well after the requisite input file is generated.

  c. Prove that a, as a vertex of the chain complex, is weakly escamotable.

9.7 Truncated Noncomplemented Lattices

We conclude our presentation of truncated lattices with one of the most intriguing results in the fixed point theory of ordered sets: Baclawski and Björner's theorem that every finite truncated noncomplemented (see Definition 9.41) lattice has the fixed point property. The original proof (see [12], Corollary 3.2) is quite complex (it ultimately goes back to the intricate topological arguments in [22], Theorem 3.2, and to the arguments in [10], Corollary 6.3, respectively) and seems to be not algorithmic. We use the arguments of Section 9.6 to give another proof of this result. Analysis of Section 9.6 and Appendix A reveals that this yields a completely algebraic proof of the result. A combinatorial proof of the result was given in [11]. We provide this argument in Appendix B.

Definition 9.41. *Two points* x, y *in a lattice* L *with largest element* **1** *and smallest element* **0** *are called* **complements** *iff*

$$x \vee y = \mathbf{1} \qquad \text{and} \qquad x \wedge y = \mathbf{0}.$$

A lattice L with largest element **1** *and smallest element* **0** *is called* **noncomplemented** *iff there is an $x \in L$ such that, for all $y \in L$, we have $x \vee y < $* **1** *or $x \wedge y > $* **0**.

The next notion is a somewhat natural generalization of noncomplementedness.

Definition 9.42 (See [48], Définition 5.1 and preceding remarks).   *Let P be a finite ordered set. We define the* **lower distance** *$d(x, y)$ from $x \in P$ to $y \in P$ to be the length n of the shortest fence $x = x_0 \geq x_1 \leq x_2 \geq \cdots x_n = y$ from x to y. We set*

$$x \bigvee y := \{x_{n-1} : x = x_0 \geq x_1 \leq x_2 \geq \cdots x_{n-1} \sim x_n = y \text{ and } n \text{ minimal}\}.$$

$r_x := \max\{n \in \mathbb{N} : d(x, y) = n$ *for some $y \in P\}$ will be called the* **lower x-radius of** *P. P will be called* **weakly noncomplemented from below** *("mal complémenté par le bas" in the original [48]) iff there is an $x \in P$ such that, for all $y \in P \setminus \{x\}$, the set $x \bigvee y$ is acyclic.*

Baclawski and Björner's classical result now easily follows.

Theorem 9.43 (See [48], Théorème 5.2.).   *Every finite ordered set that is weakly noncomplemented from below is acyclic and thus has the fixed point property.*

Proof. The proof is an induction on r_x, the lower x-radius of P, with x as in the definition of weak noncomplementedness from below. For $r_x \in \{0, 1\}$ there is nothing to prove, as P has a largest element. Now assume that $r_x = n + 1$ and the theorem holds for all weakly noncomplemented from below sets P with $r_x \leq n$. Then $B := \{y \in P : d(x, y) \leq n\}$ is weakly noncomplemented from below and thus acyclic. It is thus sufficient to verify that the conditions of Theorem 9.36 are satisfied. We can assume without loss of generality that n is even.

Let C be a chain in $P \setminus B$ and let y be the largest element of C. Then there is a fence $x = x_0 \geq x_1 < x_2 > \cdots < x_n > x_{n+1} = y$ from x to y. We will now show that $\{v \in B : C \cup \{v\}$ is a chain $\} = x \bigvee y$, which is acyclic, thus concluding the proof. Let $v \in B$ be such that $C \cup \{v\}$ is a chain. Then there is a fence $x = v_0 \geq v_1 < v_2 > \cdots \sim v_{k-1} \sim v_k = v$ with k minimal. Moreover, because v is comparable to y, we infer $k = n$ and that $v_{n-2} > v_{n-1} < v_n = v > y$ (otherwise $d(x, y) \leq n$). Thus $v \in x \bigvee y$. Conversely, if $v \in x \bigvee y$, then $d(x, v) \leq n$, so $v \in B$. Moreover we must have $v > y$ (otherwise $d(x, y) \leq n$), which means that $C \cup \{v\}$ is a chain. This proves that $\{v \in B : C \cup \{v\}$ is a chain $\} = x \bigvee y$, which completes the proof.   ∎

Theorem 9.44. *(Compare with [12], Corollary 3.2.) Every finite truncated noncomplemented lattice T is weakly noncomplemented from below. Thus every finite truncated noncomplemented lattice has the fixed point property.*

Proof. Let $x \in T$ be an element that has a supremum or an infimum with any other element of T. For all y with $x \sim y$, we have that $x \bigvee y$ has a largest element, which means $x \bigvee y$ is acyclic. If $x \not\sim y$, then x and y have a supremum or an infimum. If $x \wedge y$ exists, it is the largest element of $x \bigvee y$, which is then acyclic by Theorem 9.35. Otherwise $x \vee y$ exists. Then every element of $x \bigvee y$ has an infimum with $x \vee y$, which means $x \bigvee y$ is C-dismantlable to the singleton $\{x \vee y\}$. Again, this time by Lemma A.20, $x \bigvee y$ is acyclic. This proves that T is weakly noncomplemented from below and thus T has the fixed point property. ∎

There is an interesting fact to be noted here. By Theorem 9.21, there is a direct connection between the fixed point property for finite truncated lattices and the fixed simplex property for simplicial complexes. Yet to prove that finite truncated noncomplemented lattices have the fixed point property we are going "the long route" through the P-chain complex, rather than the direct route through Theorem 9.21. Are there stronger direct applications of algebraic topology via Theorem 9.21?

Exercises

9-35. Prove that every retract of a truncated noncomplemented lattice is a truncated noncomplemented lattice.

9-36. Let L be a lattice with the largest element $\mathbf{1}$ and the smallest element $\mathbf{0}$. Then L is called **strongly complemented** iff, for each $x \in L$, there is a complement $c_1 \in L$ that is a join of atoms and there is a complement $c_2 \in L$ that is a meet of coatoms.

  a. Give an example of a lattice that is not strongly complemented, but in which each element has a complement.
  b. Let L be a finite lattice and let $x \in L \setminus \{\mathbf{0}, \mathbf{1}\}$ be so that x is not the supremum of a set of atoms. Prove that there is a join-irreducible element in $\downarrow x$.
  c. (See [12], Corollary 3.2.) Prove that, if L is a finite lattice that is not strongly complemented, then $L \setminus \{\mathbf{0}, \mathbf{1}\}$ has the fixed point property.

Remarks and Open Problems

1. It has been proved in [146] that, if two ordered sets P, P' with more than 3 elements have the same comparability graph and $P \setminus \{x\}$ is isomorphic to $P' \setminus \{x\}$ for all x, then they must be isomorphic. Thus graph reconstruction *almost* implies order reconstruction. It is not known if order reconstruction is equivalent to graph reconstruction. The following problem is equivalent to graph reconstruction. Is every truncated clique lattice T of a graph reconstructible from its deck of subsets $T \setminus (\uparrow m)$ where m is minimal in T? It is now natural to ask if ordered sets in general are reconstructible from these maximal filter deleted subsets.

   Unfortunately, for ordered sets in general the answer is negative. In [224], it is shown that ternary and higher order relations are not reconstructible from their decks. A counterexample for ternary relations is given in Figure 9.6. The key idea is that a Möbius band untwists when it is cut and that, for ternary and higher order relations, we can model a cut with the removal of a point.

Fig. 9.6 Two nonisomorphic
ternary relations with the
same decks. Every ternary
edge (a, b, c) as indicated
contains the points nearest its
ends and the point nearest its
corner. The point in the
corner is b. The rightmost
point in the figure is c, except
for the edges that wrap
around. For these edges, the
leftmost point in the figure
is c

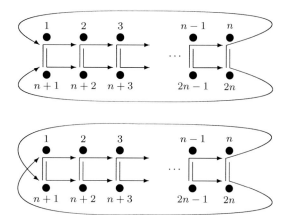

From this example, we can construct nonisomorphic ordered sets whose
maximal filter deleted decks are isomorphic. The points of the respective ordered
sets are the singletons $\{1\}, \ldots, \{2n\}$, for each ternary edge (a, b, c) the triples
$(\{a, b, c\}, 1, 1)$, $(\{a, b, c\}, 2, 1)$, $(\{a, b, c\}, 2, 2)$, $(\{a, b, c\}, 3, 1)$, $(\{a, b, c\}, 3, 2)$,
$(\{a, b, c\}, 3, 3)$, and the set $\{a, b, c\}$. The comparabilities are that each triple
$(\{a, b, c\}, i, j)$ is below $(\{a, b, c\}, i, j')$ iff $j \leq j'$ and each $(\{a, b, c\}, i, j)$ is below
$\{a, b, c\}$; each $\{x\}$ is below $(\{x, y, z\}, i, 1)$ iff there is a ternary edge with elements
x, y, z such that x is the ith element. These, plus the comparabilities dictated by
transitivity, are all comparabilities. The two possible sets constructed as above
from the example in Figure 9.6 are not isomorphic, yet their decks of maximal
filter deleted subsets are.

Note that this is an example involving general ordered sets, not truncated
lattices. The question if truncated lattices or truncated clique lattices are reconstructible from their maximal filter deleted subsets remains thus (of course) open.
Are there any lattice theoretical insights that can help to advance knowledge on
graph reconstruction through this formulation of the problem?

2. For more on algebraic topology, see [124, 295] and Appendix A.
3. There is no known good characterization of which finite truncated lattices have
   the fixed point property. The problem is conjectured to be co-NP-complete, but
   this has not (yet?) been proved. It is not possible to use the construction from
   Proposition 7.26 directly: The supremum of C_{2n}, which would need to be added,
   would lead to a fixed point.
4. Underneath our algebraic/homological investigation of the fixed point property
   lurks the topological fixed point property and a sufficient condition (for the
   P-chain complexes of finite ordered sets) called contractibility. (In general,
   contractibility does not imply the topological fixed point property, see [159].)
   Although I decided to keep the presentation discrete to keep the involved
   quantities computable, the topological approach can yield powerful results.
      Consider an ordered set P such that the topological realization $|P|$ of the P-
   chain complex of P has the topological fixed point property. Then the topological

realization $|TCL(P)|$ of the chain-complex of the truncated chain lattice of P has the topological fixed point property, too, because the $TCL(P)$-chain complex of $TCL(P)$ is a barycentric refinement of the P-chain complex of P. This means that $|P|$ and $|TCL(P)|$ are isomorphic, which implies the result.

Hence, iteration of the $TCL(\cdot)$ operator with the right starting set leads to a sequence of ordered sets that rapidly grows in size and such that every member has the fixed point property. It can be shown that, if we start with a finite ordered set that is $\mathcal{I}$-dismantlable to a singleton, then all sets in this sequence are $\mathcal{I}$-dismantlable to a singleton, too. However, if we start with a finite ordered set that is not $\mathcal{I}$-dismantlable to a singleton, then the size of the $\mathcal{I}$-cores of these sets tends to infinity.

The comparability graph of an ordered set P has the fixed clique property iff $TCL(P)$ has the fixed point property. Hence Example 6.8 shows that, in general, application of the $TCL(\cdot)$ operator need not preserve the fixed point property.

5. Part 4 shows that iterating the $TCL(\cdot)$ operator can yield examples of large nontrivial truncated lattices with the fixed point property. There is another operator whose iteration leads to a sufficient condition for the fixed point property.

The **clique graph** of a graph G has as vertices the maximal (with respect to inclusion) cliques of G and two vertices are connected by an edge iff the intersection of the corresponding cliques is not empty. Clearly, the operation of computing the clique graph can be iterated. There are three possibilities for the sequence of iterated clique graphs: Either the sequence ends with the graph with one vertex (in which case we say that G is K-**null**), or the sequence ends in a cycle of nontrivial graphs (in which case we say that G is K-**bounded**), or the size of the iterated clique graphs goes to infinity (in which case we say that G is K-**divergent**).

In [127], it is shown that, if the comparability graph of a finite ordered set is K-null, then the ordered set has the fixed point property. Interestingly enough, the proof does not establish the fixed clique property for the comparability graph. Instead, it establishes that every simplicial endomorphism has a vertex that is equal or adjacent to its image. Aside from Exercise 6-7, the connection between this property and the fixed clique property is unexplored. It is shown in [227] that, for every $\mathcal{I}$-dismantlable finite ordered set, the comparability graph is K-null. (A shorter proof is given in [100].) Examples in [127] show that there are ordered sets that are not $\mathcal{I}$-dismantlable and which have a K-null comparability graph. It is not known whether connected collapsibility implies that the comparability graph is K-null. In [181], ordered sets are presented which have the fixed point property and whose comparability graph is K-divergent, and, in [127], examples of ordered sets without the fixed point property and K-divergent comparability graphs are given. For the relationship between the fixed point property and K-boundedness of the comparability graph, we can say that crowns do not have the fixed point property and their comparability graphs are K-bounded. The question remains if there is an ordered set with the fixed point property and a K-bounded comparability graph. Corollary 2.4 in [127] provides a way to construct ordered sets with a K-bounded comparability graph, but I conjecture that these ordered sets do not have the fixed point property.

6. Does Theorem 9.36 hold for general simplicial complexes? I have tried to prove Theorem 9.36 for general simplicial complexes and was not successful. In the proof presented here, we have that $K[V'] = (V', Lk(\{x\}))$, which requires the clique condition (see Exercise 9-28). Moreover, the proof of the crucial equality $K'[\{v \in B' : \sigma \cup \{v\} \in \mathcal{S}\} = K[\{v \in B : (\sigma \cup \{x\}) \cup \{v\} \in \mathcal{S}\}]$ relies on the clique condition.

7. Although the "moreover" part of Theorem 9.36 looks like something we should take for granted, and although the proof is quite straightforward, the situation is more subtle than it looks. The remark after Exercise 9-31 shows that, for "dismantling via weakly escamotable points," and unlike for similar notions for finite ordered sets (see Theorem 4.25), cores are not unique. Hence, for this notion, removals "in the wrong order" could destroy proofs.

Chapter 10
Dimension

Dimension theory is a prominent area in ordered sets. In dimension theory, orders are represented using the orders that occur most frequently, namely, total orders.

As for Chapter 8 on lattices, it must be said that this chapter can only provide brief exposure to the basics of dimension theory. For a thorough presentation of this subject, consider [311].

10.1 (Linear) Extensions of Orders

In many situations, a given order must be turned into a linear order. For example, we may have several tasks to be processed on a machine, say, the whole job is to produce a booklet on a sophisticated copier[1]. Some of these tasks depend on the output of other tasks (booklets cannot be stapled before all pages are produced and sorted), but there are also pairs of tasks that could be processed in either order (if some pages are multicolored, any of the colors could be run first). The relation "task y needs the output of task x" defines an order relation on the set of tasks. Tasks that are independent of each other are not comparable. Still, to perform the job on one machine, all tasks must be put in one linear order.

Similarly, in the evaluation of employees, to assign raises, bonuses, etc., a supervisor will consider comparable and incomparable performances. For example, the quality of papers of two mathematicians working in the same research area might be comparable. Yet the same department may also have two members who have published outstanding works in entirely different areas. Say, one person published on nonlinear partial differential equations and the other on algebraic topology. How would we rank these accomplishments? Note that a valuation of one subject over the other often depends on personal preferences. In a "fair" order, which puts greater

---

[1]If this is not interesting enough, consider scheduling computations on a processor.

© Springer International Publishing 2016
B. Schröder, *Ordered Sets*, DOI 10.1007/978-3-319-29788-0_10

accomplishments higher than lesser ones, the two works would be incomparable. However, any such incomparabilities must be resolved when raises are assigned, because any two raises, being numbers, are comparable. Thus here, too, an order is unavoidably turned into a linear order.[2] Ties are possible, but, for our work, we will assume that ties are broken somehow.

Our first task is to investigate how an order can be turned into a linear order. The first step is to show that, to any order, we can add another comparability of previously incomparable elements. For the first time in this text, we must consider the order relation separate from the underlying set, because we must keep track of several different order relations on the same underlying set.

Lemma 10.1. *Let $(P, \leq)$ be an ordered set and let $p, q \in P$ be such that $p \not\sim q$. Then there is an order $\leq'$ on P such that $\leq$ is contained in $\leq'$ and $p \leq' q$.*

Proof. Let $\leq'$ be the transitive closure of $\leq \cup \{(p, q)\}$. Clearly $\leq'$ is reflexive and transitive. Antisymmetry remains to be proved. Let $a \leq' b$ and $b \leq' a$. We must prove that $a = b$ and we will do so by showing $a \leq b$ and $b \leq a$.

First assume, for a contradiction, that $a \not\leq b$. Then $a \leq' b$ implies $a \leq p$ and $q \leq b$. Now, because $p \not\sim q$, we must have $b \not\leq a$. Thus, in turn, $b \leq' a$ implies $b \leq p$ and $q \leq a$. We can now conclude $q \leq b \leq p$, a contradiction. Hence we must have $a \leq b$.

The dual argument shows that $b \leq a$ must hold, too. Now $a \leq b$ and $b \leq a$ naturally imply $a = b$.

Thus $\leq'$ is an order that contains $\leq$ and the new comparability $p <' q$.   ∎

Lemma 10.1 guarantees that, for any order relation that is not a total order, there are order relations that properly contain it. We will now investigate the structure of the set of orders that contain a given order $\leq$. The last part of the following theorem is also known as Szpilrajn's Theorem (see [303]), which says that every order $\leq$ is contained in a linear order.

Theorem 10.2. *Let $(P, \leq)$ be an ordered set and let*

$$\mathcal{E}_\leq := \{\sqsubseteq : \sqsubseteq \text{ is an order on } P \text{ and } \sqsubseteq \text{ contains } \leq\}.$$

*Then $\mathcal{E}_\leq$ is a conditionally complete ordered set when ordered by inclusion. The maximal elements of $\mathcal{E}_\leq$ are linear orders. Finally (**Szpilrajn's Theorem**, see [303]), every element $\sqsubseteq \in \mathcal{E}_\leq$ is below a maximal element.*

Proof. To prove that $\mathcal{E}_\leq$ is conditionally complete, let $\mathcal{A} \subseteq \mathcal{E}_\leq$ be a nonempty subset of $\mathcal{E}_\leq$ with upper bound $\sqsubseteq$. Let $\sqsubseteq_\mathcal{A}$ be the transitive closure of $\bigcup \mathcal{A}$. Then $\leq$ is a subset of $\sqsubseteq_\mathcal{A}$, $\sqsubseteq$ is a superset of $\sqsubseteq_\mathcal{A}$, and $\sqsubseteq_\mathcal{A}$ contains all relations in $\mathcal{A}$. $\sqsubseteq_\mathcal{A}$ is transitive by definition, reflexivity holds because $\sqsubseteq_\mathcal{A}$ contains all orders in

$\mathcal{A}$ and antisymmetry holds because $\sqsubseteq_A$ is contained in $\sqsubseteq$. Because every order that contains all orders in $\mathcal{A}$ must contain the transitive closure of the union of $\mathcal{A}$, $\sqsubseteq_A$ is the supremum of $\mathcal{A}$ in $\mathcal{E}_{\le}$.

By the dual of Scholium 9.4, the above shows that $\mathcal{E}_{\le}$ is conditionally complete. To shed another light on the structure of $\mathcal{E}_{\le}$, it is worth mentioning that infima are obtained via intersections. This is because the intersection of any family of orders is again an order (see Exercise 1-2).

To show that the maximal elements of $\mathcal{E}_{\le}$ are linear orders, note that, if $\sqsubseteq$ is not a linear order, then we can use Lemma 10.1 to find an order $\sqsubseteq'$ in $\mathcal{E}_{\le}$ that strictly contains $\sqsubseteq$.

Finally, to show that every element $\sqsubseteq \in \mathcal{E}_{\le}$ is below a maximal element, note that $\uparrow_{\mathcal{E}_{\le}} \sqsubseteq$ is inductively ordered: Indeed, for every chain $\mathcal{C}$ of orders that contain $\sqsubseteq$, the union $\sqsubseteq_u := \bigcup \mathcal{C}$ is an upper bound of $\mathcal{C}$ in $\mathcal{E}_{\le}$. Thus, by Zorn's Lemma, $\uparrow_{\mathcal{E}_{\le}} \sqsubseteq$ has maximal elements. ∎

Knowing that every order relation can be extended to a linear order, the following definition is sensible.

Definition 10.3. *Let $(P, \le)$ be an ordered set. Then $\sqsubseteq$ is called a **linear extension** of $\le$ iff $\sqsubseteq$ is a total order and it contains $\le$.*

We conclude with a note on the number of linear extensions of a finite ordered set. The proof also gives an idea how to compute all linear extensions of an order.

Proposition 10.4. *For a finite ordered set P, let $L(P)$ be the number of linear extensions of P. Then*

$$L(P) = \sum_{m \text{ minimal}} L(P \setminus \{m\}).$$

Proof. For an ordered set Q, let $\mathcal{L}(Q)$ be the set of all linear extensions of Q. The smallest element in a linear extension of $\le$ must be a minimal element of P. Thus

$$\mathcal{L}(P) = \bigcup_{m \text{ minimal}} \mathcal{L}(P \setminus \{m\}).$$

The result follows because the sets on the right side are pairwise disjoint. ∎

Exercises

10-1. Let P be a finite ordered set of height h and let $r_i(P)$ be the number of elements of rank i in P. Prove that P has at least $\prod_{i=0}^{h}[r_i(P)!]$ linear extensions. Then explain why the number of linear extensions can exceed this lower bound.

10-2. Let $(P, \le)$ be an ordered set and let $\le'$ be a linear extension of $\le$. Prove that the identity from $(P, \le)$ to $(P, \le')$ is an injective, surjective order-preserving map.

10-3. Compute all linear extensions of the following sets.

   a. A four crown.
   b. An antichain.
   c. A chain.
   d. A five fence.

10-4. Prove that the number of linear extensions is reconstructible from the maximal deck $\mathcal{M}_P$.
10-5. Prove that, if $\leq$ and $\sqsubseteq$ are orders on P and $\leq$ is strictly contained in $\sqsubseteq$, then $\sqsubseteq$ has strictly fewer linear extensions than $\leq$.
10-6. Prove that, in the class of finite ordered sets, the number of linear extensions of an ordered set is a comparability invariant.
10-7. An **alternating cycle** in an ordered set $(P, \leq)$ is a set $\{(x_i, y_i) : i = 1, \ldots, k\}$ so that $x_i \not\leq y_i$ and $y_i \leq x_{i+1}$ for $i = 1, \ldots, k$ modulo k. (That is, it's true for $i = 1, \ldots, k-1$ and $y_k \leq x_1$.)

   a. Prove that no order that extends $\leq$ contains an alternating cycle.
   b. A **strict alternating cycle** is an alternating cycle such that the only comparabilities $y_i \leq x_j$ are the ones indicated above. Prove that a transitive relation that contains $\leq$ and fails to be antisymmetric must contain a strict alternating cycle.

10-8. An ordered set is called **weakly ordered** iff it does not contain the disjoint sum of a singleton and a two element chain. Prove that every weakly ordered set is a linear sum of antichains.
10-9. Find the number of linear extensions of a linear lexicographic sum of antichains. (That is, the number of linear extensions of a weak order.)
10-10. We have seen ways in which the adding of a relation leads to new ordered sets. Now consider removal.

   a. Give an example showing that, in general, removal of one comparability from an order relation will lead to a relation that need not be an order.
   b. Give an example that shows that, in general, there is no unique largest order relation that is contained in the relation obtained when deleting one comparability from an order relation.
   c. The edge reconstruction problem in graph theory is the problem of reconstructing a graph from all its one-edge-deleted subgraphs. Prove that ordered sets of height greater than 1 are reconstructible from their one-edge-deleted directed subgraphs

10-11. For $n \in \mathbb{N}$, consider all chains (with respect to inclusion) of order relations $\leq$ on the set $[n] := \{1, \ldots, n\}$. Prove that every such chain is contained in a chain of length $\frac{n}{2}(n-1)$.

10.2 Balancing Pairs

Clearly, if $a \leq b$, then a will be below b in any linear extension of $\leq$. What can be said for incomparable elements? Is there a bound for how many times one element can occur above another in a linear extension? Would such a result have applications in the ordering of tasks?

   Consider the subject of sorting from an order-theoretical point-of-view. Essentially, whenever a set of objects has to be sorted, the task is to find a linear order $\leq_l$ for the objects. This order $\leq_l$ already exists "in the theory" and it can be evaluated for any pair of objects. However, in the form of a list of the objects in that order, $\leq_l$ is not yet known to us. The decision if $a <_l b$ for given elements a, b will require some effort, computational or otherwise. (Consider sorting the words in a dictionary or listing computer files in order of increasing dates.)

The goal of an efficient sorting algorithm is then to find a list of the objects in the order $\leq_l$ by using as few evaluations of $\leq_l$ as possible. (That is, using as little unnecessary effort as possible.) At intermediate stages in the sorting, the prior evaluations of $\leq_l$ induce a (partial) order $\leq$ on the elements.

Clearly $\leq_l$ is a linear extension of $\leq$. Thus, sorting a list can also be seen as the task of finding a particular linear extension of a given order relation. Every evaluation of $\leq_l$, say, for a, b with $a \leq_l b$, rules out all the linear extensions $\leq'$ in which $b \leq' a$. The sorting is most efficient if, at each stage, the evaluation of $\leq_l$ rules out many linear extensions. It is inefficient if an evaluation of $\leq_l$ rules out few linear extensions. Structurally, a sorting algorithm that functions efficiently produces a very short chain in $\mathcal{E}_\leq$ from $\leq$ to $\leq_l$.

How fast could we guarantee such a search algorithm to be in principle? To tackle this problem, we define, for any two incomparable elements, the proportion of linear extensions of $\leq$ in which a is less than b.

Definition 10.5. *Let P be a finite ordered set and let $a \not\sim b$ be elements of P. We define **the proportion of linear extensions of $\leq$ in which a is less than b** as*

$$P(a < b) := \frac{\text{number of all linear extensions } \leq_e \text{ of } \leq \text{ with } a <_e b}{\text{number of all linear extensions of } \leq}.$$

Example 10.6. Clearly, $P(a < b)$ is a rational number. The following shows that every rational number $\frac{n}{d} \in (0, 1)$ occurs as a proportion $P(a < b)$ in some ordered set. Let P_d be an ordered set that consists of a $(d-1)$-chain $1, \ldots, d-1$ with another element e added that is incomparable with any element except itself. Then $P(e > d - n) = \frac{n}{d}$. □

For efficient sorting, we would want to find lots of pairs a, b such that $a <_l b$ and $P(a < b)$ is small. However, the existence of such pairs cannot be guaranteed. Moreover, if $P(a < b)$ is small and $a >_l b$, then the comparison was essentially wasted. Thus, to obtain a performance guarantee for sorting, we need the existence of pairs a, b such that $P(a < b)$ is near $\frac{1}{2}$. Independent of whether a is on the "right" or the "wrong" side of b, the comparison will reduce the remaining effort by a factor near 2. Pairs with $P(a < b)$ near $\frac{1}{2}$ are called **balancing pairs**.

Now assume that there is a $c > 0$ such that, for every ordered set, there exist a, b with $c \leq P(a < b) \leq 1 - c$. Then a bound on the efficiency of a sorting algorithm can be given. If, at every step, we were able to find some points a, b that satisfy $c \leq P(a < b) \leq 1 - c$, then the number of remaining linear extensions is reduced by at least a factor $(1 - c)$ in every step of the search. The search stops when the resulting order is total. If n is the number of steps, then we must have that $(1 - c)^n L(P, \leq) \leq 1$ or

$$n \leq \log_{1-c}\left(\frac{1}{L(P, \leq)}\right) = -\frac{\ln(L(P, \leq))}{\ln(1 - c)}.$$

The closer $1 - c$ is to $\frac{1}{2}$, the smaller this quantity is. It has been proved in [152] that $c = \frac{3}{11}$ can be achieved for the above estimate, though none of the proofs fits our scope. The question that remains is how large a c is possible. In the three-element ordered set $\{a, b, c\}$ with $a < b$ and no further comparabilities except reflexivity, all the proportions $P(a < b)$ are either $\frac{1}{3}$ or $\frac{2}{3}$. The following problem asks whether this is the worst case.

> **Open Question 10.7. The $\frac{1}{3}$–$\frac{2}{3}$ problem** (See [161, 268].) *If P is a finite ordered set that is not a chain, is it true that there is a pair of incomparable elements $a, b \in P$ such that*
> $$\frac{1}{3} \leq P(a < b) \leq \frac{2}{3}?$$
> *The $\frac{1}{3}$–$\frac{2}{3}$ conjecture, see [161], asserts that the answer is positive.*

In [36], Brightwell, Felsner, and Trotter prove that, in any finite ordered set, there are a, b such that

$$\frac{5 - \sqrt{5}}{10} \leq P(a < b) \leq \frac{5 + \sqrt{5}}{10}.$$

The paper [36] also contains a good overview on the history and the motivation for the study of balancing pairs. The result from [36] immediately implies the following.

Proposition 10.8. *Let $(P, \leq)$ be a finite ordered set. Then, for any linear extension $\leq_l$ of $\leq$, there is a chain in $\mathcal{E}_\leq$ from $\leq$ to $\leq_l$ such that the chain has at most*
$$-\frac{\ln(L(P, \leq))}{\ln\left(\frac{5+\sqrt{5}}{10}\right)} + 1 \text{ elements.}$$

The $\frac{5-\sqrt{5}}{10} \leq P(a < b) \leq \frac{5+\sqrt{5}}{10}$ bounds can be seen as optimal in a certain, slightly generalized sense.

Example 10.9 (See [32]). Consider the ordered set Q with a countable ground set $\{x_n : n \in \mathbb{Z}\}$, ordered such that $x_i < x_j$ iff $i + 1 < j$ (see Figure 10.1). Let Q_n be the ordered subset of Q with ground set $\{x_{-2n}, x_{-(2n-1)}, \ldots, x_{(2n-1)}, x_{2n}\}$. We shall prove that

$$\lim_{n \to \infty} P_{Q_n}(x_0 > x_1) = \frac{5 - \sqrt{5}}{10}.$$

Let a_n be the number of linear extensions of the subset of Q_n with ground set $\{x_1, \ldots, x_{2n}\}$ and let b_n be the number of linear extensions of the subset of Q_n with ground set $\{x_2, \ldots, x_{2n}\}$. The point x_0 has three possible positions in a linear

Fig. 10.1 An ordered set Q that shows that existence of points a, b such that
$$\frac{5 - \sqrt{5}}{10} \leq P(a < b) \leq \frac{5 + \sqrt{5}}{10}$$
is best possible when using a certain infinitary generalization of the proportion $P(a < b)$

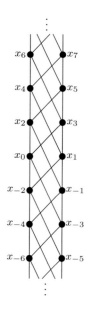

extension of Q_n: Above x_1, below x_{-1}, or between x_{-1} and x_1. Let the numbers of such linear extensions be $L_{x_0 > x_1}$, $L_{x_0 < x_{-1}}$, and $L_{x_{-1} < x_0 < x_1}$, respectively. Then, using symmetry, duality, and the fact that adding or removing a smallest element does not change the number of linear extensions we obtain

$$L_{x_0 > x_1} = a_n b_n, \quad L_{x_0 < x_{-1}} = a_n b_n, \quad L_{x_{-1} < x_0 < x_1} = a_n^2.$$

Consequently we have

$$P_{Q_n}(x_0 > x_1) = \frac{L_{x_0 > x_1}}{L_{x_0 > x_1} + L_{x_0 < x_{-1}} + L_{x_{-1} < x_0 < x_1}}$$
$$= \frac{a_n b_n}{a_n b_n + a_n b_n + a_n^2} = \frac{1}{\frac{a_n}{b_n} + 2}.$$

Moreover, a_n and b_n satisfy the recursion $a_1 = 2$, $b_1 = 1$,

$$a_n = b_n + a_{n-1} \quad \text{and} \quad b_n = a_{n-1} + b_{n-1}.$$

This implies

$$\frac{a_{n-1}}{b_n} = \frac{1}{1 + \frac{b_{n-1}}{a_{n-1}}} \quad \text{and then} \quad \frac{a_n}{b_n} = \frac{1}{1 + \frac{b_{n-1}}{a_{n-1}}} + 1,$$

which implies the recurrence relation

$$\frac{b_n}{a_n} = \frac{1}{1 + \frac{1}{1 + \frac{b_{n-1}}{a_{n-1}}}} \qquad \text{with} \qquad \frac{b_1}{a_1} = \frac{1}{2}.$$

The above implies that $\lim\limits_{n \to \infty} \dfrac{b_n}{a_n} = \dfrac{-1 + \sqrt{5}}{2}$ (see Exercise 10-12).

Moreover, we have

$$P_{Q_n}(x_0 > x_1) = \frac{1}{\frac{a_n}{b_n} + 2} = \frac{1}{\frac{1}{1 + \frac{b_{n-1}}{a_{n-1}}} + 3}.$$

On one hand, this means that $P_{Q_n}(x_0 > x_1)$ can be computed. On the other hand, we have

$$\lim_{n \to \infty} P_{Q_n}(x_0 > x_1) = \frac{1}{\frac{1}{1 + \frac{-1 + \sqrt{5}}{2}} + 3} = \frac{5 - \sqrt{5}}{10}.$$

∎

An infinite ordered set is called **thin** iff there is a $k > 0$ such that each element is incomparable with at most k others. The ordered set in Example 10.9 is thin. In [32], it is shown that, for thin ordered sets, the limit of the $P_n(a < b)$, computed for increasing sequences of subintervals $(\uparrow x_n) \cap (\downarrow y_n)$ such that eventually each point is in an interval (just like we did for one sequence of subintervals in the ordered set in Example 10.9), exists and is independent of the sequence of subintervals. For infinite thin ordered sets, $P(a < b)$ is defined as that limit. Hence, in the ordered set in Example 10.9, for any pair of incomparable elements a, b, we have

$$P(a < b) \in \left\{ \frac{5 - \sqrt{5}}{10}, \frac{5 + \sqrt{5}}{10} \right\}.$$

Thus, in a way, the bounds $\frac{5 \pm \sqrt{5}}{10}$ are the best possible ones.   □

Exercises

10-12. The sequence $\left\{ \dfrac{b_n}{a_n} \right\}_{n=1}^{\infty}$ in Example 10.9.

   a. Prove that $\dfrac{1}{2} \leq \dfrac{b_n}{a_n} < 1$ for all $n \in \mathbb{N}$.

   b. Prove that $\dfrac{b_n}{a_n} \leq \dfrac{b_{n+1}}{a_{n+1}}$ for all $n \in \mathbb{N}$.

c. Prove that $\left\{ \dfrac{b_n}{a_n} \right\}_{n\in\mathbb{N}}$ converges and that $\displaystyle\lim_{n\to\infty} \dfrac{b_n}{a_n} = \dfrac{-1+\sqrt{5}}{2}$.

10-13. An observation by M. Pouzet regarding the $\frac{1}{3}$-$\frac{2}{3}$ conjecture.

  a. Let P be a finite ordered set and let $\varphi : P \to P$ be an order-automorphism.

  i. Prove that, if $x_1 \sqsubset \ldots \sqsubset x_n$ is a linear extension of $\leq$, then so is $\varphi(x_1) \sqsubset \ldots \sqsubset \varphi(x_n)$.
  ii. Prove that, for any incomparable $x, y \in P$, we have that $P(x < y) = P(\varphi(x) < \varphi(y))$.

  b. Now suppose that P is so that, for any incomparable $x, y \in P$, we have that $P(x < y) < \frac{1}{3}$ or $P(x < y) > \frac{2}{3}$.

  i. Prove that the relation $x \ll y$ iff $P(x < y) > \frac{2}{3}$ is a linear order on P.
  ii. Prove that any order-automorphism of P is an order-automorphism with respect to $\ll$.

  c. Conclude that, if there is an ordered set P that violates the $\frac{1}{3}$-$\frac{2}{3}$ conjecture, then P has no nontrivial order-automorphisms.

10.3   Defining the Dimension

By Lemma 10.1, every order is the intersection of all orders that contain it. In dimension theory, we now want to use "nice," that is, linear, orders in the intersection, and we want to use as few linear orders as possible. This is possible because of the following corollary to Lemma 10.1 and Theorem 10.2.

Corollary 10.10. *Every order $\leq$ on a set P such that there are $a, b \in P$ with $a \not\sim b$ has a linear extension $\leq_1$ such that $a \leq_1 b$ and another linear extension $\leq_2$ such that $b \leq_2 a$. Thus, for every order $\leq$, there is a family $\{\leq_\alpha\}_{\alpha\in I}$ of linear orders such that $\leq = \bigcap_{\alpha\in I} \leq_\alpha$.*

Proof. Exercise 10-14.   ∎

The above means that any order can be realized as the intersection of linear orders, which leads to the following definition.

Definition 10.11. *Let $(P, \leq)$ be an ordered set. Then a family $\{\leq_i\}_{i\in I}$ of linear orders $\leq_i$ on P is called a **realizer** of $\leq$ iff $\leq = \bigcap_{i\in I} \leq_i$.*

Example 10.12. Consider the ordered set in Figure 1.4, part e). One realizer for its order is

$$\leq_1 : a <_1 b <_1 c <_1 d <_1 e <_1 g <_1 f <_1 h <_1 k,$$

$$\leq_2 : c <_2 b <_2 a <_2 f <_2 e <_2 k <_2 d <_2 h <_2 g,$$

$$\leq_3 : a <_3 b <_3 c <_3 d <_3 f <_3 h <_3 e <_3 g <_3 k,$$

$$\leq_4 : b <_4 c <_4 f <_4 a <_4 d <_4 e <_4 g <_4 h <_4 k,$$

$$\le_5 : b <_5 a <_5 d <_5 c <_5 e <_5 f <_5 g <_5 h <_5 k.$$

Another realizer is

$$\le_1 : a <_1 b <_1 c <_1 d <_1 e <_1 g <_1 f <_1 h <_1 k,$$
$$\le_2 : c <_2 b <_2 f <_2 a <_2 e <_2 k <_2 d <_2 h <_2 g,$$
$$\le_3 : a <_3 b <_3 d <_3 c <_3 f <_3 h <_3 e <_3 g <_3 k.$$

$\square$

Example 10.12 shows that realizers can have different numbers of orders in them. It is natural to ask what the smallest possible number of orders in a realizer would be. This is the definition of the dimension.

Definition 10.13. *Let* $(P, \le)$ *be an ordered set. Then* $(P, \le)$ *is of* **(linear) dimension** k *and we write* $\dim(P, \le) = k$ *iff* $k \in \mathbb{N}$ *is the smallest natural number such that there is a realizer for* $\le$ *that has* k *orders.*

Example 10.14.

1. A four crown has dimension 2.
2. A six crown has dimension 3.
3. The ordered set in Figure 10.2 is of dimension 2.

$\square$

Proof. We will only prove the claim in part 1. Part 2 will be proved in Example 10.16 and 3 is proved in Figure 10.2 itself.

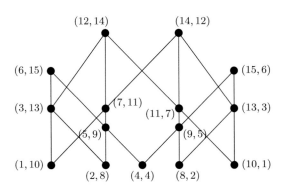

Fig. 10.2 An ordered set of dimension 2. The linear orders whose intersection this order is are indicated in the first and second components of the labels, respectively. Alternatively, the up-set of every element is contained in the $90° + \varepsilon$ cone above the point (standard visualization for two-dimensional sets). The example is due to Rutkowski and it shows that there are ordered sets of dimension 2 that have the fixed point property (see Exercise 10-21) and no irreducible point.

Let $a, b \leq c, d$ be a four crown. Because a four crown is not a chain, it has dimension at least 2 and we are done if we can find a realizer with two orders. This realizer is given as follows. The first linear order is $a <_1 b <_1 c <_1 d$ and the second linear order is $b <_2 a <_2 d <_2 c$. ∎

Example 10.15. Let S be a set of points in $\mathbb{R}^k$, such that no two points are equal in any coordinate. Order S by $(x_1, \ldots, x_k) \leq (y_1, \ldots, y_k)$ iff $x_i \leq y_i$ for all indices $i = 1, \ldots, k$.

The pair $(S, \leq)$ is an ordered set of dimension at most k. A realizer is given by the projections onto the coordinate axes. Indeed, any finite ordered set of dimension k can be embedded into $\mathbb{R}^k$ in this fashion. (This explains the terminology "dimension." For a proof, see Exercise 10-18 or Theorem 12.10.) Note, however, that not every ordered set that arises in this form is of dimension k: The ordered set could be of lower dimension than k. □

The visualization suggested in Example 10.15 is also given in Figure 10.2. The x-axis of $\mathbb{R}^2$ goes up at a $45° - \varepsilon$ angle versus the horizontal, while the y-axis forms a $135° + \varepsilon$ degree angle with the horizontal. The indicated realizer is exactly what is obtained through projections onto the x- and y-axes.

Example 10.16. The **standard example** St_n of an ordered set of dimension n. As was mentioned in Example 10.15, we can geometrically construct ordered sets whose dimension is at most the dimension of the surrounding space. However, these sets could have a dimension that is considerably less. For example, a chain has dimension 1, but it can be embedded into any $\mathbb{R}^n$ in the fashion of Example 10.15. The natural question is "Are there finite ordered sets of arbitrarily high dimension?" The following example answers this question in the affirmative.

Consider the ordered set $\{a_1, \ldots, a_n, b_1, \ldots, b_n\}$ ordered as follows.

1. $\{a_1, \ldots, a_n\}$ and $\{b_1, \ldots, b_n\}$ are both antichains.
2. No b_i is a lower bound of an a_j.
3. $a_i \leq b_j$ iff $i \neq j$.

Call an ordered set as above St_n. For a visualization of these sets, see Figure 10.3. Why is this set at least n-dimensional? The key is in the incomparable pairs $\{a_i, b_i\}$. Every realizer $\leq_1, \ldots, \leq_k$ of St_n must have one order $\leq_l$ in which $a_i >_l b_i$ and another order $\leq_{l'}$ in which $a_i <_{l'} b_i$.

Fig. 10.3 The standard example of a 5-dimensional ordered set

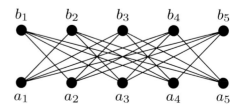

Fix a realizer $\leq_1, \ldots, \leq_k$. For $i \in \{1, \ldots, n\}$ fixed, let $\leq_l$ be an element of the realizer with $a_i >_l b_i$. For $j \neq i$ we have $a_i < b_j$ and $b_i > a_j$. Thus for $j \neq i$ we have that $a_j <_l b_i <_l a_i <_l b_j$.

This means that, for every order $\leq_l$ in the realizer of $\leq$, there is at most one i with $a_i >_l b_i$. Therefore there must be at least n orders in the realizer of $\leq$, one for each $a_i >_l b_i$.

We leave the proof that $\dim(St_n) \leq n$ as Exercise 10-19. $\square$

Unlike other properties that we have investigated recently (lattice and completeness), dimension allows for some type of heredity. Specifically, we have monotonicity.

Proposition 10.17. *Let P be an ordered set and let $Q \subseteq P$ be an ordered subset. Then $\dim(Q) \leq \dim(P)$.*

Proof. Let $\leq_1, \ldots, \leq_k$ be a realizer of $\leq_P$. Then the intersection of the k orders $\leq_1 |_{Q \times Q}, \ldots, \leq_k |_{Q \times Q}$ is $\leq_Q$ and hence $\dim(Q) \leq k = \dim(P)$. $\blacksquare$

As a particular consequence, we can give a lower bound on the dimension of a power set. The fact that the exact dimension of a power set $\mathcal{P}(\{1, \ldots, n\})$ is n is then a consequence of Theorem 12.11.

Corollary 10.18. $\dim(\mathcal{P}(\{1, \ldots, n\})) \geq n$.

Proof. Just note that the subset formed by the singleton sets together with the $(n-1)$-element sets is isomorphic to St_n. $\blacksquare$

Proposition 10.19. *The dimension of a lexicographic sum is*

$$\dim(L\{P_t \mid t \in T\}) = \max\{\dim(T), \dim(P_t) : t \in T\},$$

provided the right side exists as a finite number. If the right side is not finite, then the set $L\{P_t \mid t \in T\}$ is not finite dimensional.

Proof. By Lemma 7.5 and Proposition 10.17, the inequality

$$\dim(L\{P_t \mid t \in T\}) \geq \max\{\dim(T), \dim(P_t) : t \in T\}$$

is trivial. In particular, if the right side is not finite, then the dimension of P is not finite either.

Now assume that the right side is finite, say, k. For each $X \in \{T\} \cup \{P_t : t \in T\}$ denote the order on X by $\leq^X$ and find a realizer $\leq_1^X, \ldots, \leq_{l_X}^X$, with $l_X \leq k$. For $i \in \{l_X + 1, \ldots, k\}$ let $\leq_i^X := \leq_{l_X}^X$. For $i = 1, \ldots, k$, let P_t^i be the set P_t ordered by $\leq_i^{P_t}$ and let T^i be the set T ordered by $\leq_i^T$.

We claim that the orders $\leq_i$ of the lexicographic sums $P_i := L\{P_t^i \mid t \in T^i\}$ have the order $\leq$ of $L\{P_t \mid t \in T\}$ as their intersection. We first note that $\leq_i |_{P_t \times P_t} = \leq_i^{P_t}$. This means that $\bigcap_{i=1}^k \leq_i |_{P_t \times P_t} = \bigcap_{i=1}^k \leq_i^{P_t} = \leq^{P_t}$, so the pieces are given their original order by the intersection of these linear extensions of $\leq$. Therefore, if

$p, q \in P_s \subseteq L\{P_t \mid t \in T\}$, then $p \leq q$ iff $p \leq^{P_s} q$ iff, for all i, we have $p \leq_i q$. Now consider $p \in P_r$ and $q \in P_s$ with $r \neq s$. Then $p < q$ iff $r <^T s$ iff $r <_i^T s$ for all i iff $p <_i q$ for all i.

This proves that $\leq$ is the intersection of the $\leq_i$. Hence

$$\dim(L\{P_t \mid t \in T\}) \leq \max\{\dim(T), \dim(P_t) : t \in T\}$$

and we are done.   ■

It follows directly from Proposition 10.19, Theorem 7.18, and the fact that $\dim(P) = \dim(P^d)$ that, in the class of finite ordered sets, dimension is a comparability invariant. It should be noted that dimension is a comparability invariant for all ordered sets, see [8]. Comparability invariance of the dimension for finite sets was first established, independent of each other, in the papers [7, 114, 312].

10.3.1 A Characterization of Realizers

To obtain an upper bound on the dimension of a specific ordered set, we could specify a realizer. The number of orders in the realizer is then an upper bound on the dimension. To make this process more feasible, it would be good to have a criterion to recognize a realizer that is more efficient than computing the intersection of its orders. The notion of a critical pair is the key to such a criterion.

Definition 10.20. *Let P be an ordered set. Then the pair $(x, y) \in P^2$ is called a* ***critical pair*** *in P iff*

$$x \not\sim y, \ (\downarrow x) \setminus \{x\} \subseteq (\downarrow y) \setminus \{y\}, \text{ and } (\uparrow x) \setminus \{x\} \supseteq (\uparrow y) \setminus \{y\}.$$

Example 10.21. Note that, in the standard example St_n, the critical pairs are exactly the pairs of unrelated minimal and maximal elements.   □

Essentially, in a critical pair (x, y), we have that there is more above x than there is above y and there is less below x than there is below y. Making $x > y$ ("**reversing**" the critical pair) will thus introduce more new comparabilities than making $x < y$. This seems to indicate that $x > y$ is harder to achieve in a given linear extension. Interestingly, once the above is achieved for all critical pairs, we have a realizer.

Proposition 10.22 (Compare with [238]). *Let $(P, \leq)$ be a finite ordered set. Then $\leq_1, \ldots, \leq_d$ is a realizer of $\leq$ iff, for each critical pair (x, y) of P, there is a k such that $x \geq_k y$.*

Proof. Because $x \not\sim y$ for each critical pair (x, y), the part "$\Rightarrow$" is trivial.

To prove "$\Leftarrow$" let $\leq_1, \ldots, \leq_d$ be such that, for each critical pair (x, y) of P, there is a k such that $x \geq_k y$. Let $a, b \in P$ be such that $a \not\sim b$. We must show that there is a $k \in \{1, \ldots, d\}$ such that $a \geq_k b$. By symmetry, this will prove the result.

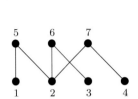 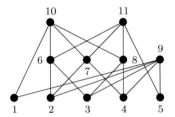

Fig. 10.4 A three-dimensional set (left) and a four-dimensional ordered set with a critical pair whose removal decreases the dimension by 2 (right). The right example first appeared in [243]

The proof will be an induction on $c(a, b) := |(\downarrow a) \setminus (\downarrow b)| + |(\uparrow b) \setminus (\uparrow a)|$. For the base step, note that, if $c(a, b) = 2$, then (a, b) is a critical pair. Thus, by hypothesis, there is a k with $a \geq_k b$.

Proceeding to the induction step, we assume that $c(a, b) = n$ and that the result holds for all pairs a', b' with $a' \not\sim b'$ and $c(a', b') < n$. Suppose without loss of generality that $(\downarrow a) \setminus (\downarrow b) \neq \emptyset$. Pick $a' \in (\downarrow a) \setminus (\downarrow b)$. Then $a' \not\sim b$ and $c(a', b) < c(a, b)$. By induction hypothesis, there is a k such that $a' \geq_k b$. This proves $a \geq_k a' \geq_k b$ and we are done. ∎

Aside from giving a criterion for what a realizer is, Proposition 10.22 can be used to provide lower bounds on the dimension of an ordered set. This should not be surprising. For example, our computation of the dimension of the standard example heavily relied on the existence of critical pairs that could not be "reversed" simultaneously.

Example 10.23. To further demonstrate how Proposition 10.22 can be used to provide lower bounds on the dimension of an ordered set, consider Figure 10.4.

1. Consider the ordered set on the left in Figure 10.4. In Exercise 10-17, you can show that this ordered set has dimension at most 3. We will use Proposition 10.22 to show that the dimension of this ordered set is at least 3.

   To see this, first note that $(1, 6)$, $(2, 1)$, $(1, 7)$, $(4, 5)$, $(2, 4)$, $(4, 6)$, $(3, 7)$, $(2, 3)$, and $(3, 5)$ are all critical pairs. Moreover, no two pairs that are listed consecutively can be reversed by the same linear extension. Thus, if the set had dimension 2, then the first and the last critical pair would have to be reversed by the same linear extension. This, however, is impossible. Thus any realizer of this ordered set must have at least three total orders.

2. Now consider the ordered set on the right in Figure 10.4. In Exercise 10-17 you can show that this ordered set has dimension at most 4. We will show that the dimension of this ordered set is at least 4.

   For a contradiction, assume that the set has dimension 3 and that $\{\leq_1, \leq_2, \leq_3\}$ is a realizer. $(1, 11)$ and $(5, 10)$ are critical pairs that cannot be reversed in the same linear extension. We can assume that $1 >_1 11$ and $5 >_2 10$. Now $(6, 9)$, $(7, 9)$, and $(8, 9)$ are critical pairs that cannot be reversed in any linear extension that reverses $(1, 11)$ or $(5, 10)$. Therefore we must have $6 >_3 9$, $7 >_3 9$, and $8 >_3 9$.

No two of the critical pairs $(4, 6)$, $(3, 7)$, and $(2, 8)$ can be reversed in the same linear extension. Moreover, $(4, 6)$ and $(6, 9)$ cannot be reversed simultaneously and the same is true for $(3, 7)$ and $(7, 9)$ and for $(2, 8)$ and $(8, 9)$. Thus none of $(4, 6)$, $(3, 7)$, and $(2, 8)$ can be reversed in $\leq_3$, contradiction. ∎

Readers versed in graph theory will notice that the above arguments rely on a connection to vertex coloring: Consider the graph whose vertices are the critical pairs so that two pairs are joined by an edge iff they cannot occur in the same linear extension. In each of the above arguments, we showed that this graph cannot be colored with less than a certain number of colors. (Sketching the graphs we argue about and verifying the connection are advisable.) The connection between dimension and coloring will be made more precise in Remark 2 at the end of this chapter. □

Exercises

10-14. Prove Corollary 10.10.
10-15. Compute the dimension of the following.

    a. An antichain.
    b. A chain.
    c. The ordered set in part b) of Figure 1.1.
    d. The ordered set in Figure 1.3.
    e. The ordered set in Figure 1.4, part e).

10-16. (Crowns and fences again)

    a. Prove that every fence (including the one-way and the two-way infinite fence) has dimension 2.
    b. Prove that every $2n$-crown with $n \geq 3$ has dimension 3.

10-17. Consider the ordered sets in Figure 10.4.

    a. Prove that the ordered set on the left in Figure 10.4 has dimension at most 3. (Together with the argument in Example 10.23 this shows that its dimension is exactly 3.)
    b. Prove that the ordered set on the right in Figure 10.4 has dimension at most 4. (Together with the argument in Example 10.23 this shows that its dimension is exactly 4.)
    c. Prove that if the points 8 and 9 are removed from the ordered set on the right in Figure 10.4, then the remaining ordered set has dimension 2.

10-18. Prove that any finite ordered set of dimension $\leq n$ can be embedded into $\mathbb{R}^n$ as given in Example 10.15.
10-19. Prove that the standard example really has dimension at most n. Then prove that removal of any point from the standard example reduces the dimension by 1.
10-20. Compute the Dedekind–MacNeille completion of the standard example.
10-21. Prove that the ordered set in Figure 10.2 has the fixed point property.
10-22. Dedekind's problem for two-dimensional ordered sets. Let P be a finite ordered set of dimension 2.

    a. Prove that there are a bijection $f : \{1, \ldots, n\} \to P$ and a permutation σ of $\{1, \ldots, n\}$ such that $a \not\sim b$ iff $\sigma(f^{-1}(a)) > \sigma(f^{-1}(b))$.

  b. Prove that the number of antichains of P is the number of decreasing subsequences of $(\sigma(1), \ldots, \sigma(n))$ with σ as above.

10-23. (Covers in the set of extensions of an order, also see [39].) Let $(P, \leq)$ be a finite ordered set. Call an order $\sqsubseteq$ that contains $\leq$ an **immediate extension** of $\leq$ iff the only extensions of $\leq$ that are contained in $\sqsubseteq$ are $\leq$ and $\sqsubseteq$.

  a. Prove that all immediate extensions are obtained by adding a (non-reversed) critical pair to the order relation.
  b. Prove that the dimension can increase or decrease when going from an order to an immediate extension.
  c. Let $\leq$ be an order that is not a total order. Is there always an immediate extension $\sqsubseteq$ of $\leq$ such that $\dim(P, \sqsubseteq) \leq \dim(P, \leq)$?

10-24. Prove that the standard example of an n-dimensional set (which has $2n$ elements) has $n!$ automorphisms and at least n^n endomorphisms. Conclude that $\lim\limits_{n \to \infty} \dfrac{|\mathrm{Aut}(St_n)|}{|\mathrm{End}(St_n)|} = 0$.

10.4   Bounds on the Dimension

Although there is an efficient algorithm to determine if a given arbitrary ordered set has dimension ≤ 2 (see Remark 11 at the end of this chapter), for every $t \geq 3$, it is NP-complete to decide if a given finite ordered set has dimension at most t (see Remark 10 at the end of this chapter). Because determining the dimension is hard in general, upper or lower bounds are important tools. In this section, we consider two upper bounds on the dimension in terms of easier order-theoretical parameters. Although the bounds appear weak, they are both sharp.

We start by considering the connection between the dimension and the width.

Theorem 10.24. *Let P be a finite ordered set. Then* $\dim(P) \leq w(P)$.

Proof. (Adapted from [311]; the argument is essentially that for Theorem 3.3 in [138], combined with Dilworth's Theorem. The result appears first in [63], though the connection would have to be made through Theorem 12.11.) We will first show that, if C is a chain in $(P, \leq)$, then there is a linear extension $\sqsubseteq$ of $\leq$ such that, for all $c \in C$ and $p \in P$ such that $c \not< p$, we have $c \sqsupseteq p$.

Enumerate the elements of C as $c_1 < c_2 < \cdots < c_k$. Let A_k be the set of all elements $p \in P \setminus C$ such that c_k is the largest element of C that is below p. Let $A_0 := P \setminus (C \cup A_1 \cup \cdots \cup A_k)$. Note that, if $p \in A_i$, $q \in A_j$, and $i < j$, then $p \not> q$. For each $i \in \{0, \ldots, k\}$ and each of the sets $A_i = \{a_i^1, a_i^2, \ldots, a_i^{n_i}\}$, find a linear extension $\sqsubseteq_i$ of $\leq$ on A_i so that $a_i^1 \sqsubseteq_i a_i^2 \sqsubseteq_i \cdots \sqsubseteq_i a_i^{n_i}$. The desired linear extension of the ordered set P then is, with elements listed in increasing order, $a_0^1, a_0^2, \ldots, a_0^{n_0}$, $c_1, a_1^1, a_1^2, \ldots, a_1^{n_1}, c_2, a_2^1, \ldots, a_{k-1}^{n_{k-1}}, c_k, a_k^1, a_k^2, \ldots, a_k^{n_k}$.

Now let $P = C_1 \cup \cdots \cup C_{w(P)}$ be a decomposition of P into chains as guaranteed by Dilworth's theorem. For each i, let $\leq_i$ be a linear extension of P such that, for all $c \in C_i$ and $p \in P \setminus C_i$ with $c \not< p$, we have $c >_i p$. Let $p, q \in P$ be incomparable elements. Find i, j such that $p \in C_i$ and $q \in C_j$. Then $p >_i q$ and $q >_j p$. This shows that $\leq$ is the intersection of the $\leq_i$.   ∎

The standard example St_n shows that the above inequality cannot be improved in general. To relate the dimension to the size of the ordered set, too, we need the following, somewhat technical, result.

Lemma 10.25 (See [158, 309]). *Let P be an ordered set and let $A \subseteq X$ be an antichain. Then $\dim(P) \leq \max\{2, |P \setminus A|\}$.*

Proof. The result is proved if we can prove it in the case that A is a maximal antichain. Thus, in this proof, we will assume that A is a maximal antichain. Because A is maximal, for each $x \in P \setminus A$, there is an $a \in A$ such that $a \leq x$ or $a \geq x$. Note also that there is no $x \in P \setminus A$ such that there are $a, b \in A$ with $a \leq x \leq b$. We define

$$D := \{x \in P \setminus A : (\exists a \in A) x \leq a\},$$

$$U := \{y \in P \setminus A : (\exists a \in A) y \geq a\}.$$

The proof is an induction on $|P \setminus A|$. The base step covers the cases $|P \setminus A| \leq 2$.

For $|P \setminus A| = 0$, a realizer with two orders consists of an extension $\leq_1$ of A and of its dual $\leq_2 := \leq_1^d$.

For $|P \setminus A| = 1$, assume without loss of generality that the element p of $P \setminus A$ is in D. A realizer with two orders consists of an extension $\leq_1$ of A such that $(A \setminus \uparrow p) \leq_1 p \leq_1 (\uparrow p) \setminus \{p\}$ and of the extension $\leq_2$ for which p is the smallest element and $\leq_2 |_{A \times A}$ is the dual of $\leq_1 |_{A \times A}$.

For $|P \setminus A| = 2$, assume first that the elements p, q of $P \setminus A$ are in D. Also assume that $q \not\geq p$. Find an extension $\leq_1$ of A such that the following holds.

$$(A \setminus (\uparrow p \cup \uparrow q)) \leq_1 q \leq_1 (\uparrow q) \setminus (\{q\} \cup \uparrow p) \leq_1 p$$
$$\leq_1 (\uparrow p \cap \uparrow q) \setminus \{p\} \leq_1 (\uparrow p \setminus (\{p\} \cup \uparrow q)).$$

In case $p \not\geq q$ find an extension $\leq_2$ for which p is the smallest element, such that

$$p \leq_2 (\uparrow p \setminus (\{p\} \cup \uparrow q)) \leq_2 q \leq_2 (\uparrow p \cap \uparrow q) \setminus \{p\}$$
$$\leq_2 (\uparrow q) \setminus (\{q\} \cup \uparrow p) \leq_2 (A \setminus (\uparrow p \cup \uparrow q))$$

and such that $\leq_2 |_{A \times A}$ is the dual of $\leq_1 |_{A \times A}$. In case $q < p$, we must put q below p in $\leq_2$ above.

The case $p, q \in U$ is dual to the above. The case $p \in D$ and $q \in U$ is handled similarly and can be produced in Exercise 10-25.

This leaves the induction step for $|P \setminus A| = k \geq 3$. The induction hypothesis is that the result holds for all ordered sets P and maximal antichains $A \subseteq P$ with $|P \setminus A| = k - 1$. Without loss of generality we can assume that $D \neq \emptyset$. Let $x \in D$ be a minimal element. Apply the induction hypothesis to $P \setminus \{x\}$ and A to obtain a realizer of $\leq |_{P \setminus \{x\} \times P \setminus \{x\}}$ with $k - 1$ linear orders. Turn each of these orders into an extension of $\leq$ by making x the smallest element. Find an extension $\leq_x$ of $\leq$ so that $x \leq_x y$ only if $x \leq y$. The thus obtained k linear orders are a realizer of P. ∎

The above allows us to connect the dimension with the size of the ordered set.

Theorem 10.26. Hiraguchi's Theorem, *see [138]. For* $|P| \geq 4$, *the dimension of P is bounded by half its size. More precisely,* $\dim(P) \leq \left\lfloor \dfrac{|P|}{2} \right\rfloor$.

Proof (Also see [310]). If $w(P) \leq \left\lfloor \dfrac{|P|}{2} \right\rfloor$, this follows from Theorem 10.24. If $w(P) > \left\lfloor \dfrac{|P|}{2} \right\rfloor$, then P has an antichain A with $|P \setminus A| \leq \left\lfloor \dfrac{|P|}{2} \right\rfloor$. Thus, by Lemma 10.25, we have $\dim(P) \leq \max\{2, |P \setminus A|\} \leq \left\lfloor \dfrac{|P|}{2} \right\rfloor$ for $|P| \geq 4$. ∎

Again, the standard example St_n shows that the above inequality cannot be improved in general. The original proof in [138] uses theorems that involve removals of four points which reduce the dimension by at most two each and removals of two points which reduce the dimension by at most 1 each. An open question is if there is a way to prove the result using just removals of 2 points at every stage.

Open Question 10.27 (See [153]). *Does every ordered set P (with* ≥ 3 *elements) contain a pair of elements* a, b *whose removal decreases the dimension by at most 1? That is, are there* $a, b \in P$ *such that we have the inequality* $\dim(P) - \dim(P \setminus \{a, b\}) \leq 1$?

If the answer to the above is affirmative, then Theorem 10.26 would be an easy corollary. It was conjectured for a while, that critical pairs would be pairs as asked for in Open Question 10.27. The ordered set on the right in Figure 10.4, due to Reuter, shows that this is not the case. In part 2 of Example 10.23, we have seen that the ordered set has dimension 4. It is easy to check that $(8, 9)$ is a critical pair. Finally the set obtained by removing 8 and 9 has dimension 2 (see Exercise 10-17c). Thus removal of critical pairs can reduce the dimension by more than 1. Examples of larger size and any dimension > 4 are given in [157].

Exercises

10-25. Finish the proof of Lemma 10.25 by proving that, if A is a maximal antichain in P so that $P \setminus A = \{p, q\}, p \in D$ and $q \in U$, then P is two-dimensional.
    Hint. Distinguish the cases $p < q$ and $p \not< q$. For an alternative proof, see [311], Theorem 11.1.

10-26. Prove that if P can be embedded in Q, then $\dim(P) \leq \dim(Q)$.

10-27. Prove Theorem 10.24 for infinite ordered sets.

10-28. (See [138].) Prove that removal of a single point cannot decrease the dimension of an ordered set by more than 1. That is, prove that, for an ordered set P and any element $x \in P$, we have that $\dim(P \setminus \{x\}) \in \{\dim(P), \dim(P) - 1\}$.

10-29. Consider the ordered set in Figure 10.5. (This set occurred in [243].)

   a. Find the dimension of the ordered set.
   b. Prove that the pair $(8, 11)$ is a critical pair such that its removal reduces the dimension to 2.

Fig. 10.5 The ordered set for
Exercise 10-29

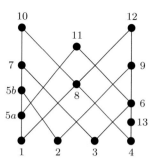

10.5 Ordered Sets of Dimension 2

Ordered sets of dimension 2 are the simplest higher-dimensional ordered sets. In this
section we discuss an interesting structural property of ordered sets of dimension 2
that relates to graph theory. We leave a discussion of the fixed point property for
ordered sets of dimension 2 to Corollary 11.23.

A standard tool in graph theory is the complement of a graph, which is obtained
by erasing all original edges and by connecting any two vertices that were not
originally connected by an edge.

Definition 10.28. *Let* $G = (V, E)$ *be a graph. The **complement (graph)** of* G *is
defined to be the graph with vertex set* V *and edge set*

$$E_C := \{\{v, w\} : v, w \in V, v \neq w, \{v, w\} \notin E\}.$$

Clearly the complement of a graph is a graph, too. It should be equally clear that
the complement of a comparability graph need not be a comparability graph.

Example 10.29. Let P be a six crown. Then the complement graph of the comparability graph of P is itself not a comparability graph.

Proof. Let $(P, \leq)$ be a six crown as shown in Figure 10.6. The complement $\overline{C}$ of
the comparability graph of $(P, \leq)$ is shown to the right of P. Assume that $\overline{C}$ was
the comparability graph of an ordered set $(P, \leq')$. Because $\{a, b, c\}$ is a clique in
$\overline{C}$, $\{a, b, c\}$ must be a chain with respect to $\leq'$. Without loss of generality we can
assume that $a \leq' b \leq' c$. However, then $b \leq' e$ would imply $e \geq' a$, which is not
possible. On the other hand, $b \geq' e$ would imply $e \leq' c$, which is not possible either.
Thus $\overline{C}$ cannot be a comparability graph.   ∎

It is natural to ask for which comparability graphs the complement graph is again
a comparability graph. This is answered in the next theorem.

Theorem 10.30. *Let* P *be an ordered set. Then the complement graph of the
comparability graph* $G_C(P)$ *is a comparability graph iff* P *is of dimension* ≤ 2.

Fig. 10.6 A six crown and the complement of its comparability graph. Note that the left is a diagram of an ordered set, while the right is merely a graph. The graph is drawn to assist in the visualization of the argument in Example 10.29

Proof. To prove "$\Leftarrow$," we let $(P, \leq)$ be an ordered set of dimension 2 and we let $\{\leq_1, \leq_2\}$ be a realizer for $\leq$. (In case P is one-dimensional, that is, a chain, choose $\leq_1=\leq_2=\leq$.) Define $\leq'$ by $p \leq' q$ iff $p \leq_1 q$ and $p \geq_2 q$. Then $\leq'$ is another order of dimension at most 2. We will show that the comparability graph of $(P, \leq')$ is the complement of the comparability graph of $(P, \leq)$.

First suppose $p \sim q$ and $p \neq q$. Then $p <_1 q$ and $p <_2 q$ or $p >_1 q$ and $p >_2 q$ and hence $p \not\sim' q$. Now suppose $p \not\sim q$. Then $p <_1 q$ and $p >_2 q$ or $p >_1 q$ and $p <_2 q$. This means that $p <' q$ or $p >' q$, that is, $p \sim' q$. Thus we have proved $p \not\sim q$ iff $p \sim' q$.

To prove "$\Rightarrow$," let $(P, \leq)$ be an ordered set such that the complement graph H of the comparability graph $G_C(P)$ is a comparability graph, too. Let $\leq'$ be an order relation on P such that $G_C(P, \leq') = H$. Define $p \leq_1 q$ iff $p \leq q$ or $p \leq' q$. To show that $\leq_1$ is a total order on P we proceed as follows. Reflexivity is trivial. Because any two elements of P are comparable via either $\leq$ or $\leq'$, it is also clear that any two elements in P are comparable via $\leq_1$. To prove antisymmetry, let $p \leq_1 q$ and $q \leq_1 p$. Assume without loss of generality that $p \sim q$ and $p \not\sim' q$. Then $p \leq q$ and $q \leq p$, which implies $p = q$. Finally, to prove transitivity, assume that $p <_1 q <_1 r$. In case $p \sim q, q \sim r$ or $p \sim' q, q \sim' r$ there is nothing to prove. Now consider the case $p \sim q$ and $q \sim' r$. This means that $p \leq q$ and $q \leq' r$. Without loss of generality, we can assume that $p \sim r$. Because $q \leq' r$, we have $q \not\sim r$, which means $r \not\leq p$. Thus $p \leq r$, which means $p \leq_1 r$. Thus $\leq_1$ is a total order on P.

We define $p \leq_2 q$ iff $p \leq q$ or $p \geq' q$. Then (with the same argument as above) $\leq_2$ is a linear order on P, too. Now note that $\leq=\leq_1 \cap \leq_2$. ∎

The above is the foundation for the only known reconstruction result that relates to dimension. We will only present the idea of the proof.

Theorem 10.31. *Ordered sets of dimension 2 are recognizable from the deck.*

Idea of the Proof. Recognizability of ordered sets of dimension 2 is a corollary to the recognizability of comparability graphs in [316]. By Theorem 10.30, an ordered set is of dimension 2 iff the complement of its comparability graph is a comparability graph, too. For the deck of P, one can form the deck of the complement of the comparability graph of P by forming the complement of the comparability graph of every card. P is of dimension 2 iff this is the deck of a comparability graph.

The idea for recognizability of comparability graphs in [316] is to prove that the forbidden subgraphs for comparability graphs identified in [101] are all reconstructible. Once this result is established, comparability graphs are recognizable as those graphs that are not equal to any of the forbidden subgraphs and whose cards are all comparability graphs. Because the proof of [101] is too lengthy to be included in this text, we will not elaborate further.

Exercises

10-30. Give examples of two-dimensional ordered sets with the fixed point property so that every orientation of the complement of the comparability graph

  a. Has the fixed point property.
  b. Does not have the fixed point property.

10-31. Prove that every two-dimensional ordered set has a retractable point.

10-32. Show that, for every even $n \in \mathbb{N}$, there is a sequence of n orders $\leq_1 \subseteq \leq_2 \subseteq \cdots \subseteq \leq_n$ on $\{1, \ldots, n\}$ such that the even-indexed orders have the fixed point property and the odd-indexed orders do not. (Also see Open Problem 9 at the end of this chapter.)

10-33. (Private communication with I. Zaguia.) A **weak order** is a union of antichains $A_1, \ldots, A_n$ such that $A_1 \leq A_2 \leq \cdots \leq A_n$. Let $(P, \leq)$ be a finite ordered set and let $f : P \to P$ be an order-preserving map. We will try to construct a weak order $\leq_f$ that extends $\leq$ such that f is still an order-preserving map on $(P, \leq_f)$. So let us assume that such a weak order $\leq_f$ exists.

  a. Prove that, if $a = f^k(a)$ for some $k \geq 1$, then $\{a, \ldots, f^k(a)\}$ must be an antichain in $\leq_f$.
  b. Prove that, if $\leq_f$ exists and $f(a) \neq a = f^k(a)$ for some $k > 1$, then, for all x such that there are i, j with $f^i(x) = f^j(a)$, no $f^u(x)$ is strictly $\leq$-comparable to any $f^v(a)$.
  c. Prove that, for any map as in 10-33b, a weak order as desired exists.

10-34. (Inspired by [81], also see [89].) Call a map $f : P \to Q$ strictly order-preserving iff $x < y$ implies $f(x) < f(y)$.

  a. Prove that, if there is a strictly order-preserving map $f : P \to \mathbb{Z}$ from P to the integers, then the intervals $[a, b] = \uparrow a \cap \downarrow b$ of P must all be of finite height.
  b. Prove that, if P is countable, then the above finiteness condition is sufficient for existence of a strictly order-preserving map into the integers.
  c. Consider the following ordered set P.

   • The minimal elements are the (unordered) natural numbers.
   • The maximal elements are the (unordered) functions $f : \mathbb{N} \to \mathbb{N}$.
   • Between any $n \in \mathbb{N}$ and any $f : \mathbb{N} \to \mathbb{N}$, there is a chain of length $f(n)$ such that n is the bottom element, f is the top, and the chain elements are not comparable to any other natural numbers or maps.

Show that all intervals in P are finite and yet there is no strictly order-preserving map of P into $\mathbb{Z}$.

Hint. Suppose $\alpha : P \to \mathbb{Z}$ is as desired and show that $h(n) := (\alpha(n))^2 \vee 1$ could not be mapped into $\mathbb{Z}$.

d. Show that there is a lattice that satisfies the interval finiteness condition which cannot be mapped into $\mathbb{Z}$ with a strictly order-preserving map.

Remarks and Open Problems

1. For more on dimension theory, see [155, 311]. Major open problems are presented in [311].
2. Given a finite ordered set P, construct the following hypergraph $\mathcal{K}_P$. The vertices of $\mathcal{K}_P$ are the critical pairs of P and the hyperedges of $\mathcal{K}_P$ are those sets $\{(x_i, y_i) : i = 1, \ldots, k\}$ of critical pairs so that $\{(y_i, x_i) : i = 1, \ldots, k\}$ forms an alternating cycle. Using Exercise 10-7, it can be shown that the dimension of P is the chromatic number of $\mathcal{K}_P$. (That is, the smallest number k such that k colors can be used to color the vertices of $\mathcal{K}_P$ so that no hyperedge connects vertices that are all of the same color.)

   For more on the connection between coloring and dimension, see [90].
3. We have seen in Exercise 10-10c that the naive order analogue of the edge reconstruction problem is trivial except in the case of height 1.

   Consider the following enhanced deck $\mathcal{EN}_P$ of an ordered set P defined as follows. For every non-maximal element $x \in P$, let $(P, \leq \setminus U(x))$ be the ordered set obtained by erasing all comparabilities of the form (x, p) from P. Let $\mathcal{EN}_P([C])$ be the number of non-maximal points $x \in P$ such that $(P, \leq \setminus U(x)) \in [C]$. Is P reconstructible from the maximal deck plus $\mathcal{EN}_P$? To get a start at reconstruction "from the other end," that is, to find a set of nontrivial parameters that completely determine the order, one could modify this problem and add any other order parameters that are reconstructible from the deck $\mathcal{D}_P$ to the hypothesis.
4. How many indecomposable ordered sets of a given width, with the same number of comparabilities and with the same number of edges in the diagram, can have the same number of linear extensions? This will likely depend on the number of elements. The idea is of course that (if there is little ambiguity or none, which would mean only the set and its dual fit the description) in this fashion the given (reconstructible) parameters would allow us to (almost) reconstruct the ordered setup to duality. We shall revisit this idea briefly in Remark 10 in Chapter 13.
5. Linear extensions may be most natural in order, but they can also be applied to graphs. One application of such extensions is the notion of the width of a constraint graph of a constraint satisfaction problem (see Exercise 5-45).

6. Let $\leq_l$ be a linear extension of $\leq$. Then the **jump number** of $(\leq, \leq_l)$ is defined to be

$$s_{(\leq, \leq_l)} := \#\{(x, y) : x \prec_l y \text{ and } x \not\geq y\}.$$

The **jump number** $s(\leq)$ of $(P, \leq)$ is the minimum of all $s_{(\leq, \leq_l)}$.

This parameter can be understood as follows. Suppose several jobs need to be scheduled on one machine and there is a (partial) order $\leq$ that must not be violated. One would expect the jobs to run most smoothly if there are few interruptions to the "natural" order $\leq$. This means we would want to minimize the number of jumps. The jump number exhibits to what extent this is possible.

For more on the jump number, see [302] and Chapter 9 in [311].

7. How hard is it to reconstruct ordered sets of dimension 2?

This has not been done yet, but the canonical visualization of ordered sets of dimension 2 suggests a deceptively simple approach. If we just knew the right way to draw the cards, reconstruction should be trivial ... (?)

8. For a connection between dimension and homology, see [242].

9. In Exercise 10-32, we found a chain of orders on n elements such that going from one order to the next one always changes the status of the fixed point property. What is the longest chain of orders $\leq_1 \subseteq \leq_2 \subseteq \cdots \subseteq \leq_k$ on $\{1, \ldots, n\}$ such that the even-indexed orders have the fixed point property and the odd-indexed orders do not?

10. By [329] (also see Chapter 4, Theorem 2.7 in [311]), for any $k \geq 3$, it is NP-complete to decide if a given ordered set has dimension $\leq k$.

11. There is a polynomial algorithm to decide if the dimension of a given ordered set is ≤ 2. Consider [311], Corollary 2.6 (originally [109], Chapter 5) or [296] for the best algorithm.

12. Is it true that, for every indecomposable ordered set of dimension 2, we have that every automorphism Φ satisfies $\Phi^2 \neq \text{id}$? For decomposable ordered sets of dimension 2, this is not true: Consider three disjoint 2-chains; for a connected example, attach a bottom element. Moreover, Exercise 11-14 shows that, if this is true, it specifically is a property for linear dimension. Can this be used to solve the automorphism problem for ordered sets of dimension 2?

Chapter 11
Interval Orders

Scheduling theory is concerned with problems such as scheduling talks at a conference or allocating processor time to several concurrently running programs. The tasks involved each take a certain amount of time. Thus, abstractly, each task can be represented as an interval on the real line. Intervals can be ordered in a natural fashion: An interval I is before another interval I' iff the interval I is completely to the left of I'. This reflects the idea that two tasks can only be related if one task is finished before the other.

Studying interval orders means studying the properties of ordered sets that arise as above. Just as the chapters on lattices and on dimension, this chapter also can only be seen as a brief introduction. For further work on interval orders, consider [91] and papers such as [24].

11.1 Definition and Examples

The real line is not the only structure in which intervals can be defined. Indeed, intervals can be defined in any ordered set. For this chapter, we will focus on intervals in arbitrary chains.

Definition 11.1. *Let P be an ordered set. Then, for $a \leq b$ in P, the **interval** $[a, b]$ is defined to be*

$$[a, b] := (\uparrow a) \cap (\downarrow b).$$

Definition 11.2. *Let $(P, \leq)$ be an ordered set. Then $\leq$ is called an **interval order** iff there is a set J of intervals in a chain C such that $(P, \leq)$ is isomorphic to $(J, \leq_{int})$, where $[a, b] \leq_{int} [c, d]$ iff $b \leq c$ or $[a, b] = [c, d]$ (as in Example 1.2, part 7). The ordered set $(J, \leq_{int})$ is called the **interval representation** of P. We will also call P an **interval ordered set**.*

© Springer International Publishing 2016
B. Schröder, *Ordered Sets*, DOI 10.1007/978-3-319-29788-0_11

The most natural examples of interval ordered sets are of course sets of intervals on the real line. An interesting property of interval ordered sets is that all interval ordered sets with up to n elements are contained in a certain interval ordered set with $\frac{n}{2}(n+1)$ elements. Thus, in a sense, Example 11.3, part 4 gives all finite examples of interval ordered sets.

Example 11.3. 1. Every finite chain carries an interval order.
2. Any set of intervals in $\mathbb{R}$ ordered by $\leq_{\text{int}}$ is an interval ordered set.
3. Every subset of an interval ordered set carries an interval order.
4. Let P_n be the set of all intervals with more than one point and integer endpoints in $\{0, \ldots, n\}$, ordered by $[a, b] \sqsubseteq [c, d]$ iff $b \leq c$ or $[a, b] = [c, d]$. Then every interval ordered set with up to n elements is isomorphic to a subset of P_n in which no two intervals have the same left endpoint.

Proof. All claims except for part 4 are straightforward. To prove part 4, we proceed by induction on n. Because there is nothing to prove for $n = 1$, we proceed to the induction step.

Let $Q := \{[a_1, b_1], \ldots, [a_n, b_n], [a_{n+1}, b_{n+1}]\}$ be a set of $n + 1$ intervals ordered by $\leq_{\text{int}}$. Assume without loss of generality that $[a_{n+1}, b_{n+1}]$ is an interval such that no other interval has a larger left endpoint. Then, for every non-maximal (with respect to the order $\leq_{\text{int}}$) interval $[a_k, b_k]$, we have $b_k \leq a_{n+1}$.

By induction hypothesis, the subset $H := \{[a_1, b_1], \ldots, [a_n, b_n]\}$ of Q is already isomorphic to a subset $H' = \{[a_1', b_1'], \ldots, [a_n', b_n']\}$ (such that no two a_i' are equal) of P_n via an isomorphism Φ that maps $[a_k, b_k]$ to $[a_k', b_k']$. Replace every interval $[a_k', b_k'] \in H'$ that is the Φ-image of a $(Q, \leq_{\text{int}})$-maximal element, with the interval $[a_k'', b_k''] := [a_k', n + 1]$. For all other intervals, let $[a_k'', b_k''] := [a_k', b_k']$. The thus obtained ordered subset $H'' = \{[a_1'', b_1''], \ldots, [a_n'', b_n'']\}$ of P_{n+1} is still isomorphic to H. (Recall that no two intervals in H' have the same left endpoint.) Let Ψ be the isomorphism from H to H''.

We claim that $\Psi^{-1}[H'' \cap P_n]$ is the set of intervals whose right boundary is at most a_{n+1}. To see this, let $[a_k, b_k] \in Q$ be an interval with $k \leq n$ and $b_k \leq a_{n+1}$. Then, by definition, $[a_k'', b_k''] = [a_k', b_k']$ and $b_k' \leq n$. This means that $[a_k, b_k] \in \Psi^{-1}[H'' \cap P_n]$. Conversely, if $[a_k, b_k] \in \Psi^{-1}[H'' \cap P_n]$, then $\Psi([a_k, b_k]) = [a_k', b_k'] = [a_k'', b_k'']$. This means that $[a_k, b_k]$ cannot have been maximal in Q, which means that $b_k \leq a_{n+1}$.

We thus conclude that the ordered subset $H'' \cup \{[n, n + 1]\}$ of P_{n+1} is isomorphic to Q. The isomorphism is the map that, for $k \leq n$, maps $[a_k, b_k]$ to $[a_k'', b_k'']$ and which maps $[a_{n+1}, b_{n+1}]$ to $[n, n + 1]$. ∎

Example 11.3, part 4 essentially provides us with all finite examples of interval ordered sets. It also allows us to prove that, although the standard example St_n is not interval ordered, there are interval ordered sets of any dimension.

Example 11.4. The examples P_n in part 4 in Example 11.3 show that there are interval orders of any dimension.

Suppose, for a contradiction, that the dimension of the P_n is uniformly bounded by d. Then, for all $n \in \mathbb{N}$, there are linear extensions $\leq_1^n, \ldots, \leq_d^n$ whose intersection is the order on P_n. For $n \in \mathbb{N}$, consider P_n. For any three numbers $a < b < c$ in

$\{0, \ldots, n\}$ note that $[a, b]$ and $[b-1, c]$ are incomparable elements in P_n. Thus there must be a $k_n(a, b, c) \in \{1, \ldots, d\}$ such that $[a, b] \geq^n_{k_n(a,b,c)} [b-1, c]$.

It can be concluded from Ramsey's theorem (see [241]) that, for large enough n, there will be a quadruple $a < b < c < d$ such that $k_n(a, b, c) = k_n(b-1, c, d) = k$. (This is because there will be a quintuple $a < b - 1 < b < c < d$ such that k_n is constant on all its sub-triples.) But then $[a, b] \geq^n_k [b-1, c] \geq^n_k [c-1, d]$, which is impossible because $[a, b] \leq [c-1, d]$.

Thus the sequence of the dimensions of the P_n must be unbounded.   ∎

It would be quite difficult to check if a given ordered set with n elements carries an interval order by testing if it is isomorphic to an ordered subset of P_n. We will now prove a powerful graphical characterization for interval orders via forbidden suborders. We have seen a trivial characterization of this kind for chains in Proposition 2.21.

Theorem 11.5. *(Fishburn-Mirkin.) Let $(P, \leq)$ be an ordered set. Then $\leq$ is an interval order iff P does not contain the ordered set $\mathbf{2} + \mathbf{2}$ depicted in Figure 11.1 as an ordered subset.*

Proof (Adapted from the introduction to [208]. A simple inductive proof for finite ordered sets is given in [13]). First assume that P has an interval order and suppose for a contradiction that P contains the ordered set $\mathbf{2} + \mathbf{2}$, with the first 2-chain being $A < C$ and the second being $E < G$. In the interval representation of P, let $A = [a, b]$ and $C = [c, d]$. Then $b \leq c$. Let $E =: [e, f]$ in the interval representation of P. Then, because E is not comparable to either A or C, we have $e < b$ and $f > c$. However, now, for $G = [g, h]$, we must have $g \geq f > c \geq b$, a contradiction to the assumption that G and A were incomparable.

Conversely, assume that P is an ordered set that contains no copy of $\mathbf{2} + \mathbf{2}$. We need to find an interval representation for P. Recall that, when embedding ordered sets into lattices via Proposition 3.24, Theorem 8.23, or Proposition 8.39, the points p generally were mapped to their principal ideals $\downarrow p$. A similar idea will be used here.

Fig. 11.1 The ordered set $\mathbf{2} + \mathbf{2}$ and an illustration why it cannot be contained in an interval ordered set

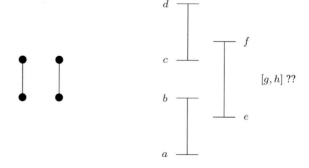

We claim that, for any two elements $x, y \in P$, we have

$$(\downarrow x) \setminus \{x\} \subseteq (\downarrow y) \setminus \{y\} \qquad \text{or} \qquad (\downarrow y) \setminus \{y\} \subseteq (\downarrow x) \setminus \{x\}.$$

For $x \sim y$, this is clear. For $x \nsim y$ suppose, for a contradiction, that the result was not true. Then there would be an $x' < x$ such that $x' \nless y$ and a $y' < y$ such that $y' \nless x$. Then $x' \nsim y'$ and, because $x \nsim y$, we also have $x' \nsim y$ and $y' \nsim x$. However, this means that $\{x, y, x', y'\}$ is isomorphic to $\mathbf{2} + \mathbf{2}$, contradicting our assumption. This means that $\{(\downarrow x) \setminus \{x\} : x \in P\}$ is linearly ordered. This insight is the key to constructing the ground set for the interval representation.

For each $x \in P$, we define

$$L(x) := (\downarrow x) \setminus \{x\} \qquad \text{and} \qquad R(x) := \bigcap \{(\downarrow p) \setminus \{p\} : p \in P, p > x\}.$$

Clearly, $L(x) \subseteq R(x)$. These points will of course become the left and right endpoints in our interval representation. However, first we must prove that the set $C := \{L(x) : x \in P\} \cup \{R(x) : x \in P\}$ is linearly ordered by inclusion. Let a, b be elements of C. If $a = L(x)$ and $b = L(y)$, there is nothing to prove. If $a = R(x)$ and $b = R(y)$, by the dual of the previous claim, we can assume without loss of generality that $(\uparrow x) \setminus \{x\} \subseteq (\uparrow y) \setminus \{y\}$. However then $R(x) \supseteq R(y)$. Finally, in case $a = L(x)$ and $b = R(y)$, there is nothing to prove if $x \sim y$. In case $x \nsim y$, every strict lower bound l of x is below every strict upper bound u of y (otherwise $\{x, y, l, u\}$ is isomorphic to $\mathbf{2} + \mathbf{2}$). However then $a = L(x) \subseteq R(y) = b$.

Now let $\Phi(x) := [L(x), R(x)]$. We claim that $x < y$ iff $R(x) \subseteq L(y)$. First, if $x < y$, then $(\downarrow y) \setminus \{y\}$ is one of the sets over which we take the intersection when computing $R(x)$. Thus $R(x) \subseteq (\downarrow y) \setminus \{y\} = L(y)$. Second, if $R(x) \subseteq L(y)$, then $x \in R(x) \subseteq L(y) = (\downarrow y) \setminus \{y\}$. This means $x < y$. We have proved that P carries an interval order. ∎

Theorem 11.5 allows for a quick visual identification of interval ordered sets, which can be used to prove first results about interval ordered sets. For example, being an interval ordered set is a comparability invariant for all ordered sets, see Exercise 11-1.

Proposition 11.6. *Even if T and all P_t carry interval orders, the lexicographic sum $L\{P_t \mid t \in T\}$ need not have an interval order.*

Proof. A trivial example will do. The two-element chain and the two-element antichain both carry interval orders. Yet the lexicographic sum of two two-element chains over a two-element antichain is the forbidden set $\mathbf{2} + \mathbf{2}$. ∎

The proof of Theorem 11.5 has several consequences. First note that, in the construction of the interval representation, $x \in R(x)$ and $x \notin L(x)$. Therefore

Scholium 11.7. *Every interval ordered set has an interval representation in which none of the intervals is a singleton.* ∎

So, even if we start with an interval order in which some of the intervals are singletons, we can find another representation in which this is not the case. Another important part of the proof of Theorem 11.5 is the insight that, in an interval ordered set, the sets of strict lower bounds (and by duality the sets of strict upper bounds) are linearly ordered. We record this fact for future reference.

Scholium 11.8. *Let P be an ordered set with an interval order and let $x \not\sim y$ in P. Then $(\uparrow x) \setminus \{x\} \subseteq (\uparrow y) \setminus \{y\}$ or $(\uparrow y) \setminus \{y\} \subseteq (\uparrow x) \setminus \{x\}$.*   ∎

Exercises

11-1. Prove that being an interval ordered set is a comparability invariant in the class of all ordered sets.

11-2. Prove that the standard example St_n is not interval ordered.

11-3. Prove that, if we demand strict inequality between endpoints in the definition of interval orders, we arrive at the same class of orders. That is, if $[a, b] \leq_{int} [c, d]$ iff $b < c$ or $[a, b] = [c, d]$, then every interval ordered set in this sense is interval ordered in the sense of Definition 11.2 and vice versa.

Prove that, in addition to the above, and similar to Scholium 11.7, we can demand that all intervals contain more than one point.

11-4. Let P_n be the set in Example 11.3, part 4. Prove that no proper subset of P_n contains all interval ordered sets with up to n elements.

Hint. Stack chains and antichains.

11-5. Let P be an ordered set. Prove that the following are equivalent.

a. P has an interval order.
b. For all $x, y \in P$ we have $(\uparrow x) \setminus \{x\} \subseteq (\uparrow y) \setminus \{y\}$ or $(\uparrow y) \setminus \{y\} \subseteq (\uparrow x) \setminus \{x\}$.
c. For all $x, y \in P$ we have $(\downarrow x) \setminus \{x\} \subseteq (\downarrow y) \setminus \{y\}$ or $(\downarrow y) \setminus \{y\} \subseteq (\downarrow x) \setminus \{x\}$.

11-6. In this exercise, the proofs of Example 11.3, part 4 and Theorem 11.5 are unified and the results are strengthened a bit. This exercise shows very nicely that, with deep enough insight to define the right auxiliary quantities, one can achieve elegant proofs. This result can be found in [91] or in [208].

Let P be a finite ordered set. For $x \in P$ define

$$L(x) := (\downarrow x) \setminus \{x\},$$

$$R(x) := (\uparrow x) \setminus \{x\},$$

$$L^* := \{L(x) : x \in X\},$$

$$R^* := \{R(x) : x \in X\},$$

$$l(x) := |\{L \in L^* : L \subseteq L(x)\}|,$$

$$r(x) := |\{R \in R^* : R \supseteq R(x)\}|.$$

Prove that, for all x, we have $l(x) \leq r(x)$ and that the following are equivalent.

a. P carries an interval order.
b. P carries an interval order with endpoints of comparable intervals being distinct.
c. P does not have a suborder that is isomorphic to $\mathbf{2} + \mathbf{2}$.
d. R^* is linearly ordered by inclusion.

    e. L^* is linearly ordered by inclusion,
    f. $x < y$ is equivalent to $r(x) < l(y)$.

11-7. Prove that the Dedekind–MacNeille completion of an interval ordered set is again interval ordered.

11-8. **Semi-orders.** An ordered set P is called **semi-ordered** iff it is isomorphic to a set J of intervals of unit length on the real line ordered by $[a, b] \le [c, d]$ iff $b \le c$ or $[a, b] = [c, d]$. Such orders are also called **unit interval orders**.

    a. (The Scott–Suppes Theorem, see [291].) Prove that a countable ordered set P is semi-ordered iff P does not contain a subset of the form $\omega \oplus \mathbf{1}$, $(\omega \oplus \mathbf{1})^d$, $\mathbf{2} + \mathbf{2}$, or $\mathbf{3} + \mathbf{1}$, where $\mathbf{n}$ is a chain with n elements.
    b. Verify that the ordered set in Figure 10.1 is semi-ordered.
    c. Prove that every semi-ordered set with up to n elements is isomorphic to an ordered subset of the set P_n^l of all intervals of unit length with rational endpoints in

$$\left\{ \frac{p}{q} : p = 0, \ldots, n^2, q = 0, \ldots, n \right\},$$

ordered by $[a, b] \sqsubseteq [c, d]$ iff $b \le c$ or $[a, b] = [c, d]$.

11.2   The Fixed Point Property for Interval Orders

Scholium 11.8 immediately leads to another structural insight into interval orders, which is in turn related to the fixed point property.

Corollary 11.9. *Finite interval ordered sets are collapsible.*

Proof. This proof is an induction on the size n of the ordered set. For $n = 0, 1$ there is nothing to prove. For the induction step, assume that P is an interval ordered set with $|P| = n > 1$ and that all interval ordered sets with fewer than n elements are collapsible.

If P has a smallest element, then P is $\mathcal{I}$-dismantlable to a singleton, and hence it is collapsible, too. So suppose that P has at least two minimal elements. Let $m_1, \ldots, m_k$ be the minimal elements of P. By Scholium 11.8, we can assume that $(\uparrow m_i) \backslash \{m_i\} \supseteq (\uparrow m_{i+1}) \backslash \{m_{i+1}\}$ for all $i = 1, \ldots, k - 1$. This implies that m_2 is retractable to m_1. Because subsets of interval ordered sets are again interval ordered, the induction hypothesis applies to $P \backslash \{m_2\}$ as well as to $(\uparrow m_2) \backslash \{m_2\}$. Hence both these sets are collapsible, which implies that P is collapsible. ∎

We trivially conclude that a finite interval ordered set has the fixed point property iff it is connectedly collapsible. However, even more is the case.

Proposition 11.10. *A finite interval ordered set has the fixed point property iff it is $\mathcal{I}$-dismantlable to a singleton.*

Proof. All we need to prove is "⇒." Let P be a finite interval ordered set with no irreducible points and the fixed point property. Because a finite ordered set has the fixed point property iff its $\mathcal{I}$-core has the fixed point property, we are done if we can show that P must be a singleton.

For a contradiction, assume that P is not a singleton. Then $P_0 := P$ has at least two minimal elements a_0 and b_0. By Scholium 11.8 we can assume, without loss of generality, that a_0 is retractable to b_0. The set $P_1 := (\uparrow a_0) \setminus \{a_0\}$ is not empty, because, otherwise, P_0 would be disconnected and thus not have the fixed point property. Moreover, P_1 is also interval ordered, has the fixed point property (see Theorem 4.12), and no point in P_1 has a unique upper cover in P_1 (otherwise the unique upper cover in P_1 would also be a unique upper cover in P_0).

We now proceed inductively. Assume

$$a_0, b_0, \ldots, a_n, b_n, \quad \text{and} \quad P_{k+1} := (\uparrow a_k) \setminus \{a_k\} \neq \emptyset \ (k \in \{0, \ldots, n\}),$$

have already been chosen and are such that the following hold.

1. For $k \in \{0, \ldots, n\}$,

    a. a_k, b_k are minimal in P_k.
    b. a_k is retractable to b_k in P_k

2. For $k \in \{0, \ldots, n+1\}$, the set P_k

    a. Has an interval order.
    b. Has the fixed point property.
    c. Has no points with a unique upper cover.

Then a_n has at least two upper covers, which means P_{n+1} has at least two minimal elements. Choose a_{n+1} and b_{n+1} minimal in P_{n+1} such that a_{n+1} is retractable to b_{n+1} in P_{n+1}. Then $P_{n+2} := (\uparrow a_{n+1}) \setminus \{a_{n+1}\} \neq \emptyset$ (otherwise P_{n+1} is disconnected). Moreover P_{n+2} has an interval order, the fixed point property, and no points with a unique upper cover in P_{n+2} (otherwise P_{n+1} has a point with a unique upper cover).

The above inductive process never stops, contradicting the finiteness of P. Thus P must have been a singleton and we are done.   ∎

Exercises

11-9. Interval orders and dismantlability.

    a. Give an example of an interval ordered set that is not dismantlable,
    b. Give an example of a dismantlable ordered set that does not carry an interval order.

11-10. Let P be a finite interval ordered set and let $f : P \to P$ be an automorphism. Prove that, for all $p \in P$, we have $(\uparrow p) \setminus \{p\} = (\uparrow f(p)) \setminus \{f(p)\}$ and $(\downarrow p) \setminus \{p\} = (\downarrow f(p)) \setminus \{f(p)\}$.

11-11. Let P be an n-element interval ordered set. Prove that there is a $C > 0$ such that we have the inequality $\dfrac{|\mathrm{Aut}(P)|}{|\mathrm{End}(P)|} \leq C 2^{-\sqrt{\log_2(n)}}$.

    Hint. Use Exercises 11-10 and 8-26.

11.3   Dedekind's Problem for Interval Orders and Reconstruction

Dedekind's problem of finding the number of antichains in an ordered set has a particularly easy solution for interval orders.

Lemma 11.11. *Let P be a finite interval ordered set. Let m be a minimal element such that* $(\uparrow m) \setminus \{m\} \supseteq (\uparrow m') \setminus \{m'\}$ *for all other minimal elements m' of P. Then* $\#A(P) = \#A(P \setminus \{m\}) + 2^{|\mathrm{Min}(P)|-1}$.

Proof. Every antichain in P either contains m or it does not. The antichains that do not contain m are the antichains of $P \setminus \{m\}$. The antichains that contain m are the antichains of $P \setminus (\uparrow m)$ with m added to each antichain. Therefore we have $\#A(P) = \#A(P\setminus\{m\})+\#A(P\setminus(\uparrow m))$. By choice of m, the set $P\setminus(\uparrow m)$ is the set of all minimal elements of P except m. Because the number of antichains in a $(|\mathrm{Min}(P)| - 1)$-element antichain is $2^{|\mathrm{Min}(P)|-1}$, the result follows.   ∎

The recursive formula is now easily translated into a closed formula.

Theorem 11.12. *Let P be a finite interval ordered set and let* $u_1, \ldots, u_n$ *be the sequence of the filter sizes* $|\uparrow p|$ *of P in nonincreasing order and with multiplicity. Then*

$$\#A(P) = 1 + \sum_{i=1}^{n} 2^{n-u_i-i+1}.$$

Proof. This proof is an induction on n. For the base step $n = 1$, there is nothing to prove. For the induction step, assume that the result holds for interval ordered sets of size $n - 1$ and let P be an interval ordered set with n elements.

Let $x \in P$ be an element with $|\uparrow x| = u_1$. Then x must be a minimal element such that $(\uparrow x) \setminus \{x\} \supseteq (\uparrow m') \setminus \{m'\}$ for all other minimal elements m' of P. By Lemma 11.11 we have

$$\#A(P) = \#A(P \setminus \{x\}) + 2^{|\mathrm{Min}(P)|-1}.$$

Now, with $i = 1$, we have $|\mathrm{Min}(P)| - 1 = n - u_1 = n - u_1 + i - 1$.

By induction hypothesis, we have

$$\#A(P \setminus \{x\}) = 1 + \sum_{i=2}^{n} 2^{(n-1)-u_i-(i-1)+1},$$

which finishes the proof.   ∎

The above indicates that, once we know something about the filters or ideals of the set, we know a lot about an interval ordered set. Indeed, finite interval ordered sets are determined up to isomorphism by their ideal decks. For convenience of notation in the proof below, we introduce the ideal size sequence.

Definition 11.13. *Let P be a finite ordered set. The* **ideal size sequence** *of P is the function $I_P : \mathbb{N} \to \mathbb{N}$ that assigns to each natural number k the number of ideals of size k in P.*

Theorem 11.14. *Let P and Q be finite interval ordered sets. Then P is isomorphic to Q iff $\mathcal{I}_P = \mathcal{I}_Q$, that is, iff the ideal decks are equal.*

Proof. The direction "$\Rightarrow$" is trivial. This proof of the direction "$\Leftarrow$" is by induction on $n := |P|$. For $n = 1$, there is nothing to prove.

For the induction step, assume that the result holds for interval ordered sets of size $n - 1$ and let P, Q be interval ordered sets of size n such that $\mathcal{I}_P = \mathcal{I}_Q$. Let k be the largest number such that P (and hence Q) has an ideal of size k. Because $\mathcal{I}_P = \mathcal{I}_Q$, there are elements $m^P \in P$ and $m^Q \in Q$ with $|\downarrow m^P| = |\downarrow m^Q| = k$ such that $\downarrow m^P$ is isomorphic to $\downarrow m^Q$. Then m^P and m^Q are maximal and each of m^P and m^Q is an upper bound of all non-maximal elements of $P \setminus \{m^P\}$ and $Q \setminus \{m^Q\}$, respectively. Moreover $\mathcal{I}_{P \setminus \{m^P\}} = \mathcal{I}_{Q \setminus \{m^Q\}}$. Therefore, by induction hypothesis, there is an isomorphism $\Phi : P \setminus \{m^P\} \to Q \setminus \{m^Q\}$. If we have $\Phi[(\downarrow m^P) \setminus \{m^P\}] = (\downarrow m^Q) \setminus \{m^Q\}$, then we are done, because Φ can be extended to an isomorphism of P and Q by mapping m^P to m^Q. If this is not the case, the following argument shows how to adjust Φ to obtain an isomorphism from $P \setminus \{m^P\}$ to $Q \setminus \{m^Q\}$ such that strict lower bounds of m^P go to strict lower bounds of m^Q.

Let P' be the set obtained from $P \setminus \{m^P\}$ by removing all its maximal elements and let Q' be the set obtained from $Q \setminus \{m^Q\}$ by removing all its maximal elements. Let $J_P := I_{(\downarrow m^P) \setminus \{m^P\}} - I_{P'}$ and let $J_Q := I_{(\downarrow m^Q) \setminus \{m^Q\}} - I_{Q'}$. Because $\downarrow m^P$ is isomorphic to $\downarrow m^Q$ and P' is isomorphic to Q', we have $J_P = J_Q$. $J_P(i)$ is the number of lower covers of m^P that have i lower bounds and are maximal in $P \setminus \{m^P\}$.

Let $p \in P$ be a lower cover of m^P in P. If $\Phi(p)$ is not a lower bound of m^Q, then $\Phi(p)$ must be maximal in $Q \setminus \{m^Q\}$. Maximal elements in $Q \setminus \{m^Q\}$ that are not below m^Q in Q are also maximal in Q. Thus $\Phi(p)$ is maximal in Q. Moreover p is maximal in $P \setminus \{m^P\}$. With $l := |\downarrow p|$ we must have $J_P(l) > 0$. Then m^Q has $J_Q(l) = J_P(l) > 0$ lower covers with l lower bounds that are maximal in the ordered set $Q \setminus \{m^Q\}$. The set M_l^Q of maximal elements of $Q \setminus \{m^Q\}$ with l lower bounds is thus partitioned into two nonempty sets. These sets are comprised of those elements of M_l^Q that are maximal in Q and those that are not maximal in Q. The same holds for the set M_l^P of maximal elements of $P \setminus \{m^P\}$ with l lower bounds. Corresponding subsets are of equal size because $J_P(l) = J_Q(l)$.

Because Φ maps maximal elements of $P \setminus \{m^P\}$ with l lower bounds to maximal elements of $Q \setminus \{m^Q\}$ with l lower bounds, there must be a maximal element q in $Q \setminus \{m^Q\}$ that is below m^Q in Q, has l lower bounds, and which is not the Φ-image of a lower cover of m^P. Note that $(\downarrow q) \setminus \{q\} = (\downarrow \Phi(p)) \setminus \{\Phi(p)\}$, because both have equally many lower bounds. Define

$$\Phi'(x) := \begin{cases} \Phi(x); & \text{if } x \notin \{\Phi^{-1}(q), p\}, \\ \Phi(p); & \text{if } x = \Phi^{-1}(q), \\ q; & \text{if } x = p. \end{cases}$$

Then Φ' is an isomorphism from $P \setminus \{m^P\}$ to $Q \setminus \{m^Q\}$ and $\Phi'(p)$ is below m^Q.

If $\Phi'[(\downarrow m^P) \setminus \{m^P\}] = (\downarrow m^Q) \setminus \{m^Q\}$, then we are done. Otherwise note that Φ' maps one more strict lower bound of m^P to a point below m^Q than Φ. Thus continuing in this fashion will eventually produce a map with the desired property. This proves that P and Q are isomorphic. ∎

A similar proof can be used in Exercise 11-12 to prove the more common and computationally simpler characterization of isomorphism of interval orders. Aside from being a result about isomorphism, the above is a result on reconstructibility. If we know that a given deck stems from an interval ordered set, then we can reconstruct the ordered set. The typical reconstruction proof for a class of ordered sets consists of proving recognizability and subsequently reconstructibility. If, like here, only the second part of the proof is available, we speak of the following.

Definition 11.15. *A class $\mathcal{O}$ of ordered sets is called **weakly reconstructible** iff for any two P, Q in $\mathcal{O}$ with $\mathcal{D}_P = \mathcal{D}_Q$ we have that P is isomorphic to Q.*

With this definition, we have the following immediate corollary.

Corollary 11.16. *Let P be an interval ordered set with at least four points. Then P is weakly reconstructible from its ideal deck $\mathcal{I}_P$.* ∎

Of course, recognizability of interval ordered sets is quite trivial. Thus we can conclude that interval ordered sets are reconstructible.

Corollary 11.17. *Let P be an interval ordered set with at least four points. Then P is reconstructible.*

Proof. First note that, by Theorem 11.5, interval ordered sets with at least five elements are recognizable from their decks. Because ordered sets with four elements are reconstructible, this means that interval ordered sets are recognizable.

Because the ideal deck is reconstructible by Theorem 3.6, the result follows from Corollary 11.16. ∎

Reconstructibility of interval orders can also trivially be concluded from Exercise 11-12 and Theorem 3.5 in [275]. The presentation here was chosen to provide a self-contained proof.

Exercises

11-12. Let P be a finite ordered set. Let $IF_P(m,n) : \mathbb{N} \times \mathbb{N} \to \mathbb{N}$ be the number of elements $x \in P$ such that $|\uparrow x| = m$ and $|\downarrow x| = n$. Prove that, if P and Q are interval ordered sets, then P is isomorphic to Q iff $IF_P = IF_Q$. Conclude that interval ordered sets are weakly reconstructible from IF_P.

11-13. Let P be a finite interval ordered set. Let $a, b \in P$ be distinct elements such that $|\downarrow a| = |\downarrow b|$. Define the order $\leq'$ on P as follows.

  a. If $p \notin \{a, b\}$ or if q is not a strict upper bound of a or b, then $p \leq' q$ iff $p \leq q$.
  b. For all $q \in P \setminus \{a, b\}$ we set

   - $a \leq' q$ iff $b \leq q$,
   - $b \leq' q$ iff $a \leq q$.

Prove that $(P, \leq)$ is isomorphic to $(P, \leq')$.
  Also give an example that shows that this result is not true when $\leq$ is not an interval order.

11.4 Interval Dimension

The idea of a dimension can be defined using more general classes of orders than just linear orders. Indeed, any class of orders that contains the chains will contain a realizer for every ordered set. However, the availability of more orders might allow for smaller realizers for certain sets. In this section, we will briefly examine how interval orders can be used to define a notion of dimension. Our focus will be on the fixed point property for interval dimension 2.

Definition 11.18. *Let $(P, \leq)$ be an ordered set. Then P is of interval dimension k iff $k \in \mathbb{N}$ is the smallest natural number such that there are k interval orders $\leq_1, \ldots, \leq_k$ on P such that $\leq = \bigcap_{i=1}^{k} \leq_i$.*

Remark 11.19. Clearly the interval dimension of an ordered set is less than or equal to its linear dimension.   □

Example 11.20. Let A be a set of products of intervals $[a_1, b_1] \times \cdots \times [a_n, b_n]$ ordered by $[a_1^1, b_1^1] \times \cdots \times [a_n^1, b_n^1] \leq [a_1^2, b_1^2] \times \cdots \times [a_n^2, b_n^2]$ iff $b_i^1 \leq a_i^2$ for all $i = 1, \ldots, n$ or the two products are equal. Then $(A, \leq)$ is of interval dimension at most n.

  Comparing this example with Example 10.15, we find the ordering that we are familiar with from linear dimension. The change is that, instead of considering sets of points, we now consider n-dimensional boxes.   □

  Although Example 11.20 shows that interval dimension is similar to linear dimension, there are important differences, too. Example 11.4 shows that the interval dimension of a set can be 1 and the linear dimension can be arbitrarily large. Thus there is no companion "converse" inequality for Remark 11.19. For our investigation of the structure of sets of interval dimension 2, we use a feature of interval orders that is not available for chains. Namely, it is possible to erase comparabilities in an interval order and obtain another interval order.

Lemma 11.21. *Let $(P, \leq)$ be a finite ordered set of interval dimension k. Let the orders $\leq_1, \ldots, \leq_k$ be k interval orders such that $\leq = \bigcap_{i=1}^{k} \leq_i$. Then there are interval orders $\leq_1^m, \ldots, \leq_k^m$ with $\leq = \bigcap_{i=1}^{k} \leq_i^m$ that have the same minimal elements as $\leq$.*

Proof. Let M be the set of minimal elements of the ordered set $(P, \leq)$ and, for each $i \in \{1, \ldots, k\}$, let M_i be the set of minimal elements of $(P, \leq_i)$. Clearly, for each i, we have $M_i \subseteq M$. Our proof is an induction on $n := \sum_{i=1}^{k} |M \setminus M_i|$.

The case $n = 0$ is trivial. For the induction step, assume that $n > 0$ and that the result holds for all $j < n$. Let $m \in P$ be $\leq$-minimal and not $\leq_i$-minimal for some i. Define $\leq_i' := \leq_i \setminus \{(x, m) : x <_i m\}$. Clearly $\leq_i'$ is still an order. To see that $\leq_i'$ is still an interval order, assume that there are distinct points $a, b, c, d \in P$ with $a <_i' c, b <_i' d, b \not<_i' c$, and $a \not<_i' d$. Because $\leq_i$ is an interval order that only differs from $\leq_i'$ in the lower bounds of m, we infer $m \in \{a, b, c, d\}$. Because m is minimal in $(P, \leq_i')$ we can assume without loss of generality that $m = a$. But then $a <_i c, b <_i d, b \not<_i c$, and $a \not<_i d$, a contradiction. Thus $\leq_i'$ is still an interval order. Moreover $\{(x, m) : x <_i m\} \cap \leq = \emptyset$, so replacing $\leq_i$ with $\leq_i'$ does not remove any comparabilities in $\leq$. For $j \neq i$ let $\leq_j' := \leq_j$. Then $\leq = \bigcap_{j=1}^{k} \leq_j'$.

Applying the induction hypothesis to the above representation of $\leq$ yields the result.   ∎

We can now use the special form of the above interval realizer to show that ordered sets of interval dimension 2 are collapsible.

Theorem 11.22 (This generalizes [93], Theorems 2 and 6). *Let P be a finite ordered set of interval dimension 2. Then P has a retractable minimal element or an element of rank 1 with a unique lower cover. Consequently, every finite ordered set of interval dimension 2 is collapsible.*

Proof. Let $\leq$ be the order on P and let $\leq_1, \leq_2$ be two interval orders so that we have $\leq_1 \cap \leq_2 = \leq$. Let M_i be the set of minimal elements of $(P, \leq_i)$ and let M be the set of minimal elements of $(P, \leq)$. By Lemma 11.21, we can assume that $M_1 = M_2 = M$.

If $|M| = 1$, then every element of rank 1 has a unique lower cover. Hence we can assume that $|M| \geq 2$. By Scholium 11.8 and finiteness of P, there are elements $a, b \in M$ such that, for all $m \in M$, we have $(\uparrow_1 m) \setminus \{m\} \subseteq (\uparrow_1 b) \setminus \{b\}$ and, for all $m \in M \setminus \{b\}$, we have $(\uparrow_1 m) \setminus \{m\} \subseteq (\uparrow_1 a) \setminus \{a\}$. Assume neither a is retractable to b nor b is retractable to a in P. Then, by definition of interval orders, we must have $(\uparrow_2 b) \setminus \{b\} \subseteq (\uparrow_2 a) \setminus \{a\}$. Moreover there is a $d \in P$ with $d > b$ and $d \not\geq a$. Because $d \geq_2 b \geq_2 a$, we must have $d \not\geq_1 a$. Thus the only minimal $\leq_1$-lower bound of d is b. Hence d is $\leq$-above only one $\leq$-minimal element, namely, b. This implies that $(P, \leq)$ must have an element of rank 1 with a unique lower cover (see Exercise 4-5).

Collapsibility is now proved via an easy induction.   ∎

The above immediately gives us the characterization of the fixed point property for interval dimension 2 in terms of connected collapsibility. Unlike for interval ordered sets, for interval dimension 2, it is not possible to characterize the fixed point property via $\mathcal{I}$-dismantlability. In Figure 10.2, we have an ordered set of linear and interval dimension 2 which has no irreducible point. Yet this ordered set has the fixed point property.

Corollary 11.23. *Let P be a finite ordered set of interval dimension 2. Then P has the fixed point property iff P is connectedly collapsible.*

Proof. By Theorem 11.22, every finite ordered set of interval dimension 2 is collapsible. Thus, by Theorem 4.30, any such ordered set has the fixed point property iff it is connectedly collapsible. ∎

Exercises

11-14. Prove that the ordered set on the left in Figure 10.4 has interval dimension 2. Then show that an indecomposable ordered set of interval dimension 2 can have an automorphism Φ with $\Phi^2 \neq$ id.

11-15. Give an example that shows that the equation in Proposition 10.19 has no analogue for interval dimension.

11.5 Algorithms for Interval Orders

In special cases, there can be a polynomial algorithm for a problem even if no general polynomial algorithm is known. Isomorphism of interval orders is such a special case. Another example is given in Exercise 5-10c.

Proposition 11.24. *Isomorphism of interval orders can be checked in polynomial time. Moreover, if there is an isomorphism, it can be constructed in polynomial time.*

Proof (A more efficient algorithm is to be given in Exercise 11-16). Let P and Q be ordered sets that carry interval orders. We will describe the algorithm that verifies isomorphism in polynomial time. Because sets with different numbers of elements are trivially not isomorphic, we only consider the situation $|P| = |Q|$. (In a program, the following would thus only execute if the check whether $|P|$ equals $|Q|$ succeeds.)

Linearly order the elements of both sets according to the following rules.

1. If $\text{rank}(x) < \text{rank}(y)$, then x comes before y.
2. If $\text{rank}(x) = \text{rank}(y)$ and $|\uparrow x| > |\uparrow y|$, then x comes before y.
3. If $\text{rank}(x) = \text{rank}(y)$, $|\uparrow x| = |\uparrow y|$, and $|\downarrow x| > |\downarrow y|$, then x comes before y.
4. Break the remaining ties arbitrarily.

Because all quantities in question can be computed in polynomial time, the above orderings for the elements of P and Q can be achieved in polynomial time.

Let $p_1, \ldots, p_n$ and $q_1, \ldots, q_n$ be the elements of P and Q, respectively, each listed in one of the orders as above. Isomorphisms preserve the rank and the numbers of upper and lower bounds of an element. Thus, if there is an i such that $\text{rank}_P(p_i) \neq \text{rank}_Q(q_i)$ or $|\uparrow_P p_i| \neq |\uparrow_Q q_i|$ or $|\downarrow_P p_i| \neq |\downarrow_Q q_i|$, then P cannot be isomorphic to Q. Consider the remaining case $\text{rank}_P(p_i) = \text{rank}_Q(q_i)$ and $|\uparrow_P p_i| = |\uparrow_Q q_i|$ and $|\downarrow_P p_i| = |\downarrow_Q q_i|$ for all $i \in \{1, \ldots, n\}$. By Scholium 11.8 and its dual, we have that, for $x \not\sim y$ in an interval order, $|\uparrow x| = |\uparrow y|$ implies $(\uparrow x) \setminus \{x\} = (\uparrow y) \setminus \{y\}$ and $|\downarrow x| = |\downarrow y|$ implies $(\downarrow x)\setminus\{x\} = (\downarrow y)\setminus\{y\}$. Thus, for any p_i, p_j that were tied as in 4, we have $(\uparrow_P p_i) \setminus \{p_i\} = (\uparrow_P p_j) \setminus \{p_j\}$ and $(\downarrow_P p_i) \setminus \{p_i\} = (\downarrow_P p_j) \setminus \{p_j\}$. A similar result holds in Q. Thus, if P is isomorphic to Q, then there must be an isomorphism that, for all $i \in \{1, \ldots, n\}$, maps p_i to q_i.

For our algorithm, this means that we consider the map $\Phi : P \to Q$ that is defined by $\Phi(p_i) := q_i$ and check if it is an isomorphism. If Φ is an isomorphism, then P and Q are isomorphic. If not, then, by the above, P and Q are not isomorphic. Because this last check again only requires a polynomial number of steps, the proposed algorithm is polynomial. Clearly, if there is an isomorphism, the algorithm will produce an isomorphism in a polynomial number of steps.   ∎

Recall that, if an interval order represents the time frames for a given set of tasks, the width gives an upper bound on how many tasks are executed at the same time. This has applications, for example, in the register allocation on a computer CPU. Readers versed in graph theory will recognize greedy coloring in the following proof.

Proposition 11.25. *Let P be a finite interval ordered set and assume that we know an interval representation of P. Assume that the comparability between endpoints of intervals in the representation can be checked in a constant number of steps. Then $w(P)$ can be computed in $O(|P|^2)$ steps.*

Proof (Adapted from [320], Proposition 5.1.11). We will use the notation of the proof of Theorem 11.5 for the interval representation. That is, x is represented by $[L(x), R(x)]$ and the endpoints are ordered by inclusion.

Linearly order the vertices of P by $x \sqsubseteq y$ iff $L(x) \subseteq L(y)$, breaking ties arbitrarily. Recursively define a map $f : P \to \mathbb{N}$ as follows. For the $\sqsubseteq$-smallest element s of P, set $f(s) := 1$. Now suppose $x \in P$ and $f(y)$ has been defined for all $y \in P$ with $y \sqsubseteq x$ and $y \neq x$. In this case, we let $f(x)$ be the smallest number k such that, for all $y \sqsubseteq x$, $y \neq x$, if $L(x) \subsetneq R(y)$ then $f(y) \neq k$.

We claim that the function f can be computed in at most $O(|P|^2)$ steps: Indeed, for each $x \in P$, only the f-values of the elements of P before x in the order $\sqsubseteq$ have to be checked. For each of the less than $|P|$ elements y that is $\sqsubseteq$-before x and satisfies $L(x) \subsetneq R(y)$ (note that we assume that the comparability of endpoints can be checked quickly) we enter $f(y)$ in a list. Once all y's are checked, we find the desired number for $f(x)$ in at most $|P|$ steps.

To finish the proof, we claim that $\max\{f(p) : p \in P\} = w(P)$. The inequality "$\geq$" is clear, because, if $x_1, \ldots, x_j$ is an antichain of P in its original order, then f must assign distinct values to all x_i. For the inequality "$\leq$," assume that $x \in P$ is such that $m := f(x) = \max\{f(p) : p \in P\}$. Then, for each $i \in \{1, \ldots, m-1\}$, there is an $x_i \in P$ with $x_i \sqsubseteq x$ and $f(x_i) = i$. Because $L(x_i) \subseteq L(x) \subsetneq R(x_i)$, x and x_i are not comparable in P. Now let $i \neq j$ be distinct elements of the set $\{1, \ldots, m-1\}$. Assume without loss of generality that we have $L(x_i) \subseteq L(x_j)$. Then $L(x_i) \subseteq L(x_j) \subseteq L(x) \subsetneq R(x_i)$, so x_i and x_j are not comparable in P (in case $L(x_i) = L(x_j)$, we also use $L(x_i) = L(x_j) \subseteq L(x) \subsetneq R(x_j)$). This means $m \leq w(P)$ and we are done.   ∎

The above result naturally leads to the question how hard it is to obtain the interval representation of the ordered set. We can at least get an idea by looking at the proof of Theorem 11.5, where we constructed an interval representation.

In the proof of Theorem 11.5, we constructed an interval representation for an ordered set that did not contain a copy of a set $2 + 2$. We claim this construction takes $O(|P|^3)$ steps if the order relation is known. Indeed, for each $x \in P$, it takes $O(|P|)$ steps to construct $L(x) = (\downarrow x) \setminus \{x\}$ and it takes $O(|P|^2)$ to construct $R(x) := \bigcap\{(\downarrow y) \setminus \{y\} : y \in P, y > x\}$. Thus, in $O(|P|^3)$ steps, the intervals $[L(x), R(x)]$ have been constructed.

Therefore it takes $O(|P|^4)$ steps to check if an ordered set has an interval order (check for subsets $2 + 2$) and, if so, it takes $O(|P|^3)$ steps to construct an interval representation. Note that computing the explicit order relation of the $L(x)$ and $R(x)$ (as needed in Proposition 11.25) takes another $O(|P|^3)$ steps: Indeed, there are up to $2|P|$ endpoints, so there are $O(|P|^2)$ pairs of endpoints and the containment of the sets is checked in $O(|P|)$ steps if the sets are encoded appropriately.

Exercises

11-16. Use Exercise 11-12 to give an algorithm that is more efficient than the algorithm given in Proposition 11.24 to check if two interval orders are isomorphic.

11-17. Revisiting the relation $\sqsubseteq$ from Proposition 1.3, which also features prominently in Section 4.5.

    a. Show that if $\sqsubseteq$ is considered as a relation on a set $\mathcal{J}$ of intervals in an ordered set, then $\sqsubseteq$ is an order relation.

    b. Let $\mathcal{J}$ be a set of intervals on the real line. Show that a finite ordered set P is isomorphic to a set $(\mathcal{J}, \sqsubseteq)$ iff P is two-dimensional.

    c. Let $\mathcal{J}$ be a set of intervals in a totally ordered set. Show that an ordered set P is isomorphic to a set $(\mathcal{J}, \sqsubseteq)$ iff P is two-dimensional.

Remarks and Open Problems

1. For more on interval orders see [24, 91]. For the connection to dimension, see [311], Chapter 8 and [236, 237].

2. Interval ordered sets can be generalized in various ways.

    a. By Theorem 11.5, interval ordered sets are characterized by the forbidden subset $2 + 2$. One could also investigate classes for which other types of subsets are forbidden. The semi-orders in Exercise 11-8 are an example.

    b. The notion of dimension can be generalized by considering a class less restrictive than chains for the realizers. Similarly, interval orders also can be generalized by looking at intervals in less restrictive classes than chains. For some work in this direction, see [209]. One has to be careful with this generalization, though. Every ordered set can be embedded into a lattice. Thus, for example, replacing the totally ordered set in the definition of an interval order with a lattice simply gives us all ordered sets. (In [209], the class of sets in which the intervals are formed has a forbidden subset.)

One can also go in the opposite direction and pose more conditions on the intervals as done in Exercise 11-8.

Parts 2a and 2b are addressed, for example, in [24].

3. Ordered sets can also be defined using shapes in $\mathbb{R}^n$, ordered via geometric containment. If we don't demand that the set is closed under union and intersection, we can indeed represent all ordered sets in such a fashion. For more on these geometric containment orders see [92].

4. Theorem 11.22 is a first step toward a polynomial algorithm that determines if a given ordered set of interval dimension 2 has the fixed point property, see [288].

5. Is Lemma 11.21 extendable? That is, is the interval dimension of an ordered set $(P, \leq)$ the smallest number k for which there are k interval orders (denoted $\leq_1, \ldots, \leq_k$) such that $\leq = \bigcap_{j=1}^{k} \leq_j$ and, for all $p \in P$ and all $j \in \{1, \ldots, k\}$, we have $\text{rank}_{\leq}(p) = \text{rank}_{\leq_j}(p)$?

6. It is proved in [115] that the interval dimension of an ordered set and of its Dedekind–MacNeille completion are equal. The same holds for the linear dimension, see Exercise 12-14.

7. By [329], for any $k \geq 3$, it is NP-complete to decide if a given ordered set has interval dimension $\leq k$.

8. In [179], a polynomial algorithm to decide if the interval dimension of a given ordered set is ≤ 2 is given.

9. (Inspired by Example 11.3, part 4 and Exercise 11-4.) Let a_n be the smallest natural number such that there is an ordered set P of size a_n that contains an isomorphic copy of every ordered set Q of size $\leq n$. Compute a_n or find estimates.

   Clearly, a_n is a finite number, because we can always pick one representative of each of the finitely many isomorphism classes of n-element ordered sets and form an ordered set whose connected components are the thus picked sets. It is equally clear that this construction is much too inefficient to get close to a_n.

   More generally, for any class $\mathcal{C}$ of ordered sets, let $a_n^{\mathcal{C}}$ be defined as the smallest number such that there is an ordered set $P \in \mathcal{C}$ of size $|P| = a_n^{\mathcal{C}}$ that contains isomorphic copies of all ordered sets $Q \in \mathcal{C}$ with $|Q| \leq n$. Compute $a_n^{\mathcal{C}}$ or find estimates.

10. (Variation on 9.) Let b_n be the smallest natural number such that there is an ordered set P of size b_n that contains an isomorphic copy of every ordered set Q of size $\leq n$ as a **covering subset**. That is, Q is an ordered subset and the diagram of Q is contained in the diagram of P. Compute b_n or find estimates. Numbers $b_n^{\mathcal{C}}$ can be defined analogously to numbers $a_n^{\mathcal{C}}$ and the same question can be asked.

11. In [33] it was proved that the $\frac{1}{3}$–$\frac{2}{3}$ conjecture holds for semi-orders. Can it be proved for interval orders?

12. One natural generalization of an interval is the notion of a convex subset of an ordered set. Thus, another possible generalization of interval ordered sets is to represent ordered sets as sets of convex subsets of some type of ordered set. One can order convex sets in various ways. These questions are addressed, for example, in [211]

13. Interval orders have a nice short proof of reconstructibility that uses the fact that isomorphism of interval orders is easily verifiable. Does polynomially verifiable isomorphism imply reconstructibility in general?

Chapter 12
Sets $P^Q = \mathrm{Hom}(Q, P)$ and Products

The order-preserving maps from one ordered set to another form a natural ordered set when given the pointwise order. Products are defined similar to these homomorphism sets. Hence we will investigate homomorphism sets and products of ordered sets in the same chapter. We will introduce some of the salient results on these sets, such as the fixed point theorem for products of two finite ordered sets (see Theorem 12.17), Hashimoto's Refinement Theorem (see Theorem 12.30), and the cancelation property for exponents (see Theorem 12.47). The automorphism conjecture (see Open Question 2.14) as well as the open problems at the end of this chapter show that there are interesting problems related to homomorphism sets and products that remain open.

12.1   Sets $P^Q = \mathrm{Hom}(Q, P)$

We start by defining homomorphism sets and by investigating their relationship to the fixed point property.

Definition 12.1. *Let P, Q be ordered sets. We define the set of **(order) homomorphisms** from Q to P to be*

$$\mathrm{Hom}(Q, P) := \{f : Q \to P | f \text{ is order-preserving }\},$$

*ordered by the pointwise order $f \le g$ iff $f(q) \le g(q)$ for all $q \in Q$. Note that, if $Q = P$, then $\mathrm{Hom}(P, P) = \mathrm{End}(P)$, the set of **(order) endomorphisms** of P.*

Notation 12.2. Because the formal laws associated with forming sets of homomorphisms (see Section 12.5) are similar to exponentiation, we also denote $\mathrm{Hom}(Q, P)$ by P^Q.

© Springer International Publishing 2016
B. Schröder, *Ordered Sets*, DOI 10.1007/978-3-319-29788-0_12

We already have encountered examples of homomorphism sets. They were just represented in a slightly different way.

Example 12.3. Let **2** denote the 2-element chain $\{0, 1\}$ and let n denote the n-element antichain. Then the set $\mathbf{2}^n$ is isomorphic to $\mathcal{P}(n)$, the power set of an n-element set ordered by inclusion. The isomorphism maps each $f : n \to \mathbf{2}$ to the set $\{k : f(k) = 1\}$.   □

There is no guarantee that there is a copy of the domain Q in $\mathrm{Hom}(Q, P)$: Indeed, if P is a singleton, then so is $\mathrm{Hom}(Q, P)$. However, we can find a copy of P.

Lemma 12.4. *Let P be an ordered set. Then P is a retract of $\mathrm{Hom}(Q, P)$.*

Proof. Fix $q_0 \in Q$ and define $R_{q_0} : \mathrm{Hom}(Q, P) \to P$ by $R_{q_0}(f) := f(q_0)$. For any $p \in P$, define f_p to be the constant function that returns p, that is, $f_p(q) := p$ for all $q \in Q$. Let $I : P \to \mathrm{Hom}(Q, P)$ be defined by $I(p) := f_p$. Then R_{q_0} and I are a retraction-coretraction pair as in the alternative definition of retractions given in Exercise 4-1.   ■

Regarding the fixed point property, we can record an extensive characterization. Indeed, especially for endomorphism sets, many notions related to the fixed point property are equivalent.

Theorem 12.5 (See [12], Theorem 4.5, [73], Theorem 6.13, and also [301]).
Let P be a chain-complete ordered set. Then the following are equivalent.

1. *P is C-dismantlable to a singleton.*
2. *For all ordered sets Q, $\mathrm{Hom}(Q, P)$ is C-dismantlable to a singleton.*
3. *For all ordered sets Q, $\mathrm{Hom}(Q, P)$ is connected.*
4. *For all ordered sets Q, $\mathrm{Hom}(Q, P)$ has the fixed point property.*
5. *$\mathrm{End}(P)$ has the fixed point property.*
6. *$\mathrm{End}(P)$ is C-dismantlable to a singleton.*
7. *$\mathrm{End}(P)$ is connected.*

Proof. For the implication "1⇒2," let $n \in \mathbb{N}$ be such that, for $i = 1, \ldots, n$, there are retractions $r_i : P_{i-1} \to P_i$ in $\mathcal{U} \cup \mathcal{D}$ (recall Exercise 4-24) with $P_0 = P$ and P_n a singleton. Define $R_i : \mathrm{Hom}(Q, P_{i-1}) \to \mathrm{Hom}(Q, P_i)$ by $R_i(f) := r_i \circ f$. Clearly the R_i are retractions in $\mathcal{U} \cup \mathcal{D}$ and $\mathrm{Hom}(Q, P_n) \subseteq \mathrm{Hom}(Q, P)$ is a singleton.

For the implication "2⇒4," by Theorem 4.11, we only need to prove that $\mathrm{Hom}(Q, P)$ is chain-complete. To do this, let $C \subseteq \mathrm{Hom}(Q, P)$ be a chain. For each $q \in Q$, the set $\{c(q) : c \in C\}$ is a chain. Define $s(q) := \bigvee \{c(q) : c \in C\}$. Then s is the supremum of C in $\mathrm{Hom}(Q, P)$.

The implications "2⇒3⇒7," "2⇒6⇒7," and the implications "4⇒5⇒7" are trivial.

This leaves us with the implication "7⇒1." To prove that the needed retractions exist, we will proceed by induction on the length n of the fence from id_P to a constant function. In case $n = 0$, there is nothing to prove, because P is a singleton. Now assume that there is a fence of length n from id_P to a constant function and that the result is proved for all sets P' for which there is a shorter fence from $\mathrm{id}_{P'}$ to a constant function. Without loss of generality let $\mathrm{id}_P = F_0 > F_1 < F_2 > \cdots F_n$ be a fence in $\mathrm{End}(P)$ such that F_n is a constant function. By an easy adaptation of Lemma 4.24, there is a retraction $R_1 \le F_1 < F_0$ such that $R_1 = R_1 \circ F_1$. Because $R_1 < \mathrm{id}_P$ we have that $R_1 \in \mathcal{C}$. Moreover

$$\mathrm{id}_{R_1[P]} = R_1|_{R_1[P]} = R_1 \circ F_1|_{R_1[P]} < R_1 \circ F_2|_{R_1[P]} > \cdots R_1 \circ F_n|_{R_1[P]}$$

is a fence in $\mathrm{End}(R_1[P])$ such that $R_1 \circ F_n|_{R_1[P]}$ is a constant function. Thus $R_1[P]$ is $\mathcal{C}$-dismantlable to a singleton by induction hypothesis. This implies that P is $\mathcal{C}$-dismantlable to a singleton. ∎

With the fixed point property for sets $\mathrm{End}(P) = P^P$ characterized in many ways (as long as we assume chain-completeness) one could now consider sets $\mathrm{Hom}(Q, P) = P^Q$. One manifestation of this problem is the famous product problem (see Open Question 12.12), which handles exponents Q that are finite antichains. We will explore this problem in more detail later on. For connected exponents, not much is known, see Remark 4 at the end of this chapter. Note that, for connected exponents, P^Q need not be disconnected, even when both base and exponent are not dismantlable.

Example 12.6. Let P be an arbitrary ordered set. Let Q be an ordered set with n elements and let $nA^2 := L\{A_k \mid k \in \{1, \ldots, n\}\}$, where the set $\{1, \ldots, n\}$ is carrying its natural order and each A_k is a two-element antichain. Then the ordered set $(nA^2 \oplus P)^Q$ is connected.

Proof. Let m_1 and m_2 be the two minimal elements of $nA^2 \oplus P$. We will show that each order-preserving map $f : Q \to nA^2 \oplus P$ is connected to one of the maps $g_i(q) = m_i$ $(i = 1, 2)$. If $|f[Q] \cap \{m_1, m_2\}| \le 1$, then $f \ge g_1$ or $f \ge g_2$, and there is nothing to prove. If $\{m_1, m_2\} \subseteq f[Q]$, then there is a $k \in \{1, \ldots, n\}$ such that $f[Q] \cap A_k = \emptyset$. Fix $a \in A_k$ and define

$$\tilde{f}(q) := \begin{cases} f(q); & \text{if } f(q) > A_k, \\ a; & \text{if } f(q) < A_k. \end{cases}$$

Then $\tilde{f}$ is order-preserving and $f < \tilde{f} > g_1, g_2$. ∎

The above example shows at least that a characterization of the fixed point property for P^Q with both P^Q and Q connected will not be as strong as the characterization for P^P in Theorem 12.5. Indeed, for chain-complete sets, by Lemma 7.6, the ordered set $nA^2 \oplus P$ has the fixed point property iff P has the fixed point property. In particular, $nA^2 \oplus P$ need not be dismantlable. Yet the set $(nA^2 \oplus P)^Q$ is connected for all P.

Exercises

12-1. Prove that, if L is a lattice (complete lattice, distributive lattice), then the ordered set $\text{Hom}(Q, L)$ is a lattice (complete lattice, distributive lattice).

12-2. Let P be a finite ordered set. Prove that, for every fixed point free order-preserving function $f : P \to P$, there is a fixed point free map $h : P \to P$ that maps minimal elements to minimal elements and maximal elements to maximal elements and such that $\text{dist}_{\text{End}(P)}(f, h) \leq 2$.

12-3. Show that, for each $n \in \mathbb{N}$, there are connected ordered sets P and Q such that $h(P) > |Q| + n$ and yet P^Q is not connected.

12-4. Let P be a finite ordered set. Prove that $D(P)$ is isomorphic to $\text{Hom}(P, \mathbf{2})$, where $\mathbf{2}$ is a 2-element chain.

12-5. Let P be a connected finite ordered set that is not $\mathcal{I}$-dismantlable to a singleton. Prove that the component of $\text{End}(P)$ that contains the constant functions is not $\mathcal{I}$-dismantlable to a singleton.

12.2   Finite Products

The possibly simplest sets of homomorphisms $\text{Hom}(Q, P)$ are those in which the domain Q is an antichain. These ordered sets are nothing but the $|Q|$-fold product of an ordered set P with itself. Although we will define products in full generality, this section will mainly focus on finite products. Infinite products will be investigated in Sections 12.3 and 12.4.

Definition 12.7. *Let A be an index set and let $\{(P_\alpha, \leq_\alpha)\}_{\alpha \in A}$ be an indexed family of ordered sets. We define the **product** $\prod_{\alpha \in A} P_\alpha$ **of the ordered sets** P_α to be the (set-theoretical) product*

$$\prod_{\alpha \in A} P_\alpha = \left\{ f : A \to \bigcup_{\alpha \in A} P_\alpha \,\middle|\, f(\alpha) \in P_\alpha \right\}$$

*equipped with the **pointwise order***

$$f \leq g \; :\Leftrightarrow \; (\forall \alpha \in A) f(\alpha) \leq_\alpha g(\alpha).$$

If the index set is finite, say, $\{1, \ldots, n\}$, we denote the product by $P_1 \times \cdots \times P_n$. The elements of a product are often given componentwise as $p = (p_\alpha)_{\alpha \in A}$, or as $p = (p_1, \ldots, p_n)$ for finite index sets.

*The **natural projection** $\pi_{P_\beta} : \prod_{\alpha \in A} P_\alpha \to P_\beta$ is defined to be*

$$\pi_{P_\beta} ((p_\alpha)_{\alpha \in A}) := p_\beta.$$

If there is no confusion possible, then π_{P_β} may also be abbreviated as π_β.

Note that the definition of products requires the existence of selection functions. This means that, for infinite index sets, we need the Axiom of Choice. As previously,

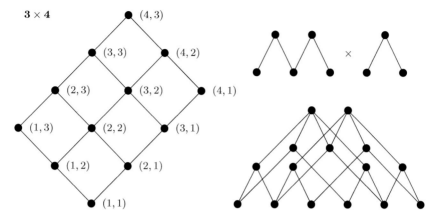

Fig. 12.1 The product of a 3-chain with a 4-chain and the product of a 5-fence with a 3-fence.

we will not hesitate to use the Axiom of Choice. Finite products that are not too large can be visualized as shown in Figure 12.1.

The same task that led us to Lemma 12.4 can be posed for products: Find copies of the building blocks, in this case, the factors, in the whole structure. The examples in Figure 12.1 show that such copies should abound. The next result (the only one on infinite products in this section) verifies this impression. Moreover, all the embedded copies actually are retracts. These embedded copies will become very important when we prove Hashimoto's Refinement Theorem.

Lemma 12.8. *Let* $\prod_{\alpha \in A} X_\alpha$ *be an infinite product of ordered sets* X_α. *For the elements* $s^0 \in \prod_{\alpha \in A} X_\alpha$, $\lambda \in A$, *and* $x_\lambda \in X_\lambda$ *let*

$$(s^0, x_\lambda, \lambda)_\alpha := \begin{cases} s^0_\alpha; & \text{for } \alpha \neq \lambda, \\ x_\lambda; & \text{for } \alpha = \lambda. \end{cases}$$

Let $X^{s^0}_\lambda := \{(s^0, x_\lambda, \lambda) : x_\lambda \in X_\lambda\}$. *Then* $X^{s^0}_\lambda$ *is isomorphic to* X_λ *and* X_λ *is a retract of* $\prod_{\alpha \in A} X_\alpha$ *in the sense of Exercise 4-1.*

Proof. Exercise 12-7.   ∎

Let us now consider how products connect with notions we introduced earlier.

Proposition 12.9. *Let* P, Q *be ordered sets.*

1. *If* P, Q *are (complete) lattices, then* $P \times Q$ *is a (complete) lattice.*
2. *Let* P, Q *be ordered sets with more than one point. Then*

$$\max\{\dim(P), \dim(Q)\} \leq \dim(P \times Q) \leq \dim(P) + \dim(Q).$$

3. *The product of two interval orders need not be an interval order.*

Proof. For part 1, simply note that the needed pointwise suprema and infima will exist and that they will be the suprema and infima in the product.

To prove part 2, first note that the first inequality trivially follows from the fact that $P \times Q$ has retracts isomorphic to P and Q, respectively. To prove the second inequality, let $(P, \leq_P)$ and $(Q, \leq_Q)$ be ordered sets. If one of P, Q is infinite dimensional, then there is nothing to prove. Thus let $\leq_P^1, \ldots, \leq_P^n$ and $\leq_Q^1, \ldots, \leq_Q^m$ be realizers of P and Q, respectively. For $i = 1, \ldots, n$, define $\leq_i$ on $P \times Q$ by $(p, q) \leq_i (p', q')$ iff $p <_P^i p'$ or $p = p'$ and $q \leq_Q^1 q'$. (This is the lexicographic sum of n copies of $(Q, \leq_Q^1)$ over $(P, \leq_i)$.) Similarly, for $j = 1, \ldots, m$, define $\leq^j$ on $P \times Q$ by $(p, q) \leq^j (p', q')$ iff $q <_Q^j q'$ or $q = q'$ and $p \leq_P^1 p'$.

We claim that $\{\leq_1, \ldots, \leq_n, \leq^1, \ldots, \leq^m\}$ is a realizer of the product order $\leq$. Indeed, it is trivial to see that $\leq$ is contained in the intersection of the above orders. Conversely, let $(p, q) \leq_i (p', q')$ for all $i \in \{1, \ldots, n\}$ and also $(p, q) \leq^j (p', q')$ for all $j \in \{1, \ldots, m\}$. If $p \neq p'$, we have $p <_P^i p'$ for all $i \in \{1, \ldots, n\}$ and thus $p <_P p'$. Similarly, if $q \neq q'$, we have $q <_Q^j q'$ for all $j \in \{1, \ldots, m\}$ and thus $q <_Q q'$. This means that $p \leq_P p'$ and $q \leq_Q q'$ which means $(p, q) \leq (p', q')$ in the product order.

Finally, for part 3, note that the product of the two 3-chains $\{1 < 2 < 3\}$ and $\{a < b < c\}$ contains the subset $\{(3, a), (3, b), (1, c), (2, c)\}$. ■

Products can now be used to give a different characterization of linear dimension. (Also see Example 10.15.)

Theorem 12.10 (O. Ore). *The product of m nontrivial chains has dimension m. As a consequence, the linear dimension of an ordered set P is the smallest number k such that P can be embedded as an ordered subset into a product of k chains.*

Proof. Let $C_1, \ldots, C_m$ be m chains, each with at least two elements. Then the product $C_1 \times \cdots \times C_m$ contains a copy of $\mathbf{2}^m$, which (see Example 12.3) is isomorphic to $\mathcal{P}(m)$. By Corollary 10.18 we have $\dim(\mathcal{P}(m)) \geq m$ which means $\dim(C_1 \times \cdots \times C_m) \geq m$. By part 2 of Proposition 12.9 we have the reverse inequality $\dim(C_1 \times \cdots \times C_m) \leq m$. Therefore $\dim(C_1 \times \cdots \times C_m) = m$.

For the second statement, let $(P, \leq)$ be an ordered set of dimension k and let the set of orders $\{\leq_1, \ldots, \leq_k\}$ be a realizer of $\leq$. Let $(H, \sqsubseteq)$ be the product of the ordered sets $(P, \leq_i)$. Then $p \leq p'$ iff $p \leq_i p'$ for all $i \in \{1, \ldots, k\}$, which is the case iff $(p, \ldots, p) \sqsubseteq (p', \ldots, p')$ in H. Thus the map $e : P \to H, p \mapsto (p, \ldots, p)$ is an embedding of P into a product of k chains. Finally P cannot be embedded into a product of fewer than k chains, because this would imply $\dim(P) \leq k$. ■

If we draw products of chains like the left set in Figure 12.1, Theorem 12.10 gives a formal basis for our visualization of ordered sets of a given dimension in Example 10.15. Moreover, Theorem 12.10 gives a strong upper bound on the dimension of finite distributive lattices.

Theorem 12.11 (A product version of Dilworth's Chain Decomposition Theorem, see [63], Theorem 1.2). *Let L be a finite distributive lattice and let k be the largest number of lower covers that any element of L can have. Then L is a sublattice of a product of k chains.*

Proof. By Birkhoff's characterization of finite distributive lattices, see Theorem 8.40, L is isomorphic to $D(J(L))$. We shall first prove that $w(J(L)) = k$.

Let $A = \{a_1, \ldots, a_m\} \subseteq J(L)$ be an antichain. Consider the down-set defined by $A_{J(L)} := (\downarrow_{J(L)} a_1) \cup \cdots \cup (\downarrow_{J(L)} a_m)$. Then the sets $A^i_{J(L)} := A_{J(L)} \setminus \{a_i\}$ are lower covers of $A_{J(L)}$ in $D(J(L))$. Thus $m \le k$, that is, $w(J(L)) \le k$.

Conversely, every $S \in D(J(L))$ is of the form $S = A_{J(L)}$ for some antichain $A = \{a_1, \ldots, a_m\} \subseteq J(L)$. The lower covers of S are exactly the sets $A^i_{J(L)}$ as described above. Thus $k \le w(J(L))$ and hence the two quantities are equal.

By Dilworth's Chain Decomposition Theorem, there are $k = w(J(L))$ chains $C_1, \ldots, C_k$ such that $J(L) = C_1 \cup \ldots \cup C_k$. Add a new smallest element 0 to each chain C_i. For every set $A \in D(J(L))$ (which, as we recall, is a down-set in $J(L)$), let a_i be the supremum of $(C_i \cup \{0\}) \cap A$. The map $\Phi : A \mapsto (a_1, \ldots, a_k)$ embeds $D(J(L))$ into $C_1 \times \cdots \times C_k$. Moreover, it is easy to see that $\Phi[D(J(L))]$ is a sublattice of $C_1 \times \cdots \times C_k$. ∎

12.2.1 Finite Products and the Fixed Point Property

One of the most attractive problems in the theory of ordered sets is the question if the fixed point property is "productive." The problem was open even for the case of two finite factors for over 15 years. M. Roddy solved this (most important) case in [255]. We will present a slight refinement of the argument (due to Roddy) here.

> **Open Question 12.12.** *Let P and Q be ordered sets with the fixed point property. Does $P \times Q$ necessarily have the fixed point property? (The most important special case was solved by Roddy in [255], see Theorem 12.17 here.)*

The key idea is to concentrate on one factor. Factor maps allow us to do so.

Definition 12.13. *Let P, X be ordered sets and let $f : P \times X \to P \times X$ be an order-preserving map. For $x \in X$, we define the **factor map** $f_x : P \to P$ to be $f_x(\cdot) := \pi_P f(\cdot, x)$. For $p \in P$, we define $f_p(\cdot) := \pi_X f(p, \cdot)$.*

As we will be working with factor maps, it will be good for these factor maps to have as simple a structure as possible. Nice maps as defined below fit this description. We will see that, for a finite ordered set, the factor maps can be replaced with nice maps.

Definition 12.14. *An order-preserving map $f : P \to P$ is called **flat** iff, for all $x \in f[P]$, we have that x is comparable to $f(x)$ iff $f(x) = x$. f will be called a **nice map** iff the following hold.*

1. f is flat.
2. f is injective on its range.
3. There is a retraction $r_f : P \to f[P]$ such that $f = f \circ r_f$.

Essentially, nice maps are maps for which the Abian–Brown Theorem is instantaneous and for which, in the finite setting, the range does not change under iterations. As we will see, both properties allow us to eliminate many technical difficulties. We will first show that the fixed point property is productive for ordered sets $P \times X$ and order-preserving maps $f : P \times X \to P \times X$ with sufficiently structured factor maps. The proof of Theorem 12.17 is then a simple application of Proposition 4.4.

Theorem 12.15. *Let P, X be ordered sets such that X has the fixed point property. Assume that P is connected and so that every nonempty chain has an upper bound and a lower bound. Let $f : P \times X \to P \times X$ be an order-preserving map such that the following hold.*

1. The factor maps $f_x : P \to P$ are all nice.
2. Each factor map $f_x : P \to P$ has a fixed point.
3. $x \le y$ implies $r_x := r_{f_x} \le r_{f_y} =: r_y$.

Then f has a fixed point.

The key idea for the proof is the notion of an s-fence defined below: It is easy to find a fixed point p for a factor map f_x. Thereafter, it is equally easy to find a fixed point x' for the factor map f_p. Unfortunately, it can be that $x' \ne x$. Jumping back and forth between P and X in this fashion merely leads to a process in which one hand destroys what the other hand has just repaired. Using an s-fence, a fixed point in X is maintained throughout the argument, which eliminates the problem indicated above. The actual proof is an argument by contradiction. However, the construction that leads to the contradiction can be turned into an algorithm to find the fixed point of a product map.

Definition 12.16. *An $(n + 2)$-tuple $(a_0, \dots, a_n; x)$ is called an **up s-fence** of length n iff the following are satisfied.*

1. $a_0 \le a_1 \ge \cdots a_n$ in P,
2. $x \in X$,
3. $f_{a_0}(x) = x$,
4. $r_x(a_j) = a_j$, for $j > 0$, and
5. $f_x(a_n) = a_n$.

*A **down s-fence** is defined dually and an **s-fence** is one or the other.*

Proof of Theorem 12.15. First we show that P contains an s-fence. Fix $q \in P$ and consider the order-preserving map $x \mapsto f_{r_x(q)}(x)$. Let x_0 be one of its fixed points. Let $a_0 := r_{x_0}(q)$ and find a fixed point p of f_{x_0}. Because P is connected, there is a fence $(a_0, \dots, a_n)$ from a_0 to $a_n := p$ in $r_{x_0}[P]$. Now $(a_0, \dots, a_n; x_0)$ is the desired s-fence.

Let $(a_0, \dots, a_n; x_0)$ be an s-fence of minimum length. If $n = 0$, then we are done, because $f_{x_0}(a_0) = a_0$ and $f_{a_0}(x_0) = x_0$ implies $f(a_0, x_0) = (a_0, x_0)$. We will lead the assumption $n > 0$ to a contradiction. Assume without loss of generality that $a_0 < a_1$. Define sequences $\{x_i\}_{i \in \mathbb{N}}$, $\{a_{0,i}\}_{i \in \mathbb{N}}$ as

$$a_{0,0} = a_0, \; a_{0,1} = a_1, \; x_i = f_{a_{0,i}}(x_{i-1}), \text{ and } a_{0,i+1} = r_{x_i}(a_{0,i}) \; (i \geq 1).$$

Then the sequences $\{x_i\}_{i \in \mathbb{N}}$ and $\{a_{0,i}\}_{i \in \mathbb{N}}$ are both increasing, which can be seen as follows.

$$a_{0,1} = a_1 \geq a_0 = a_{0,0},$$

$$x_1 = f_{a_{0,1}}(x_0) = f_{a_1}(x_0) \geq f_{a_0}(x_0) = x_0,$$

$$a_{0,2} = r_{x_1}(a_{0,1}) \geq r_{x_0}(a_{0,1}) = a_{0,1},$$

$$a_{0,i+1} = r_{x_i}(a_{0,i}) \geq r_{x_{i-1}}(a_{0,i}) \geq r_{x_{i-1}}(a_{0,i-1}) = a_{0,i} \quad (i \geq 2),$$

$$x_{i+1} = f_{a_{0,i+1}}(x_i) \geq f_{a_{0,i}}(x_{i-1}) = x_i \quad (i \geq 1).$$

Let a be any upper bound of the sequence $\{a_{0,i}\}_{i \in \mathbb{N}}$ and let

$$Y := \uparrow \{x_i : i \in \mathbb{N}\}.$$

Define $h : Y \mapsto X$ by $h(y) = f_{r_y(a)}(y)$. We claim that h maps Y to itself. To see this, it is enough to show that $h(y) \geq x_i$ for each $i \in \mathbb{N}$ and $y \in Y$: Indeed, for $y \in Y$ and all $i \in \mathbb{N}$, we have that $r_y(a) \geq r_{x_{i-1}}(a_{0,i-1}) = a_{0,i}$, which gives

$$h(y) \geq f_{a_{0,i+1}}(x_i) = x_{i+1} \geq x_i$$

for all $i \in \mathbb{N}$, and hence $h(y) \in Y$.

By Theorem 4.17, $h : Y \to Y$ has a fixed point. Let y be any fixed point of h and define the $(n + 1)$-tuple $(b_1, \ldots, b_n; y)$ by setting

$$b_1 = r_y(a), \quad b_j = r_y(a_j), \quad 1 < j \leq n.$$

We claim that $(b_1, \ldots, b_n; y)$ is a down s-fence of length $n - 1$, which is the contradiction we seek. To prove this, we must check all properties of an s-fence. Properties 2 and 4 are trivial. For property 1, note that

$$b_1 = r_y(a) \geq r_y(a_2) = b_2,$$

because $a \geq a_{0,1} = a_1 \geq a_2$. The other inequalities of 1 hold because r_y is order-preserving. For 3, note that $f_{b_1}(y) = f_{r_y(a)}(y) = h(y) = y$. Finally, to prove 5, we establish three claims, (A), (B), and (C) below.

(A) $f_y(a_n)$ is a fixed point of f_y. Indeed, $f_y(a_n) \geq f_{x_0}(a_n) = a_n$, and the claim follows from the fact that f_y is nice.

(B) $f_y(a_n) = r_y(a_n)$. To see this, note that, by (A) and the definition of r_y, $f_y(f_y(a_n)) = f_y(a_n) = f_y r_y(a_n)$ and $r_y(a_n) \in f_y[P]$. Because f_y is one-to-one on its range, we have $f_y(a_n) = r_y(a_n)$.

(C) $f_y(b_n) = b_n$ and thus 5 holds.

$$f_y(b_n) = f_y r_y(a_n) \overset{(B)}{=} f_x f_y(a_n) \overset{(A)}{=} f_y(a_n) \overset{(B)}{=} r_y(a_n) = b_n.$$

This completes the proof of Theorem 12.15.

With Theorem 12.15, we can prove that the product of two finite ordered sets with the fixed point property has the fixed point property, too. Below we state the most general result that can be derived from Theorem 12.15. Except for the first factor having width 3 (see [256]), the "remaining cases" to prove that the fixed point property is productive have eluded researchers so far.

Theorem 12.17 (Also see [255], Theorem 1.1). *Let P be a chain-complete ordered set with no infinite antichains and the fixed point property and let Q be an ordered set with the fixed point property. Then $P \times Q$ also has the fixed point property.*

Proof. We first prove the result in the case that P is finite. Let $g : P \times Q \to P \times Q$ be order-preserving. Define $f : P \times Q \to P \times Q$ by defining the factor maps $f_q := g_q^{|P|!+1}$ for $q \in Q$ and $f_p := g_p$ for $p \in P$. Then f is order-preserving. For all $q \in Q$, define $r_q := g_q^{|P|!}$. Then, by Proposition 4.4, r_q is a retraction (onto $f_q[P]$) and $q_1 \leq q_2$ implies $r_{q_1} \leq r_{q_2}$.

Clearly every fixed point of g is also a fixed point of f. Conversely, if (p, q) is a fixed point of f, then $g_p(q) = f_p(q) = q$ and

$$p = f_q(p) = g_q^{|P|!+1}(p) = g_q(r_q(p)) = g_q(r_q(r_q(p))) = r_q(f_q(p)) = f_q(p),$$

which implies $g_q(p) = g_q(r_q(p)) = f_q(p) = p$. Thus f and g have the same fixed points and we are done if we can show that f has a fixed point. To do this, we show that f is a nice map, which will allow us to apply Theorem 12.15.

To see that each f_q is injective on its image, let $a, b \in f_q[P]$. Then $a = g_q^{|P|!+1}(x)$ and $b = g_q^{|P|!+1}(y)$. If $f_q(a) = f_q(b)$, then

$$a = r_q^2(a) = g_q^{3|P|!+1}(x) = g_q^{|P|!-1}f_q(a)$$
$$= g_q^{|P|!-1}f_q(b) = g_q^{3|P|!+1}(y) = r_q^2(b) = b.$$

Thus each f_q is injective on its image. In particular, because P was assumed to be finite, this means that, if $p \in f_q[P]$ is comparable to $f_q(p)$, then $f_q(p) = p$, so f_q is flat. With $r_{f_q} := r_q$ we have via Proposition 4.4 that $f_q = f_q \circ r_{f_q}$. Thus each f_q is nice.

Now f satisfies all hypotheses of Theorem 12.15. This means that f has a fixed point and we are done in the case that P is finite. The rest follows from Li and Milner's Structure Theorem and Exercise 12-15. ∎

Exercises

12-6. Prove that the product of two connected ordered sets is again connected.

12-7. Prove Lemma 12.8, which says that every factor X_λ of a product can be "spliced into any element" s^0 to form a retract.

12-8. Prove that any product of chain-complete ordered sets is chain-complete.

12-9. Give an example of two ordered sets P and Q so that $\dim(P \times Q) = \dim(P) + \dim(Q)$.

12-10. It would be nice if we could extend Dilworth's Theorem to (maybe) say that an ordered set without infinite antichains must be a union of countably many chains. Unfortunately, this is false: Let α be an infinite limit ordinal and consider the product $\alpha \times \alpha$. Prove that:

  a. $\alpha \times \alpha$ has no infinite antichains.

  b. $\alpha \times \alpha$ is not the union of fewer than α chains.

  Note. This particularly simple example was first observed in [220], also see [207].

12-11. Show that the Dedekind–MacNeille completion of a product can be a proper subset of the product of the Dedekind–MacNeille completions of the factors.

12-12. Let P, Q be ordered sets, let $\leq_{P \times Q}$ be the order of the product $P \times Q$, and let $\leq_{L(P,Q)}$ be the order of the lexicographic sum $L\{P_q | q \in Q\}$ with all P_q isomorphic to P. Prove that $\leq_{P \times Q} \subseteq \leq_{L(P,Q)}$.

12-13. Let P be a chain-complete ordered set and let Q be an ordered set (not necessarily chain-complete). Prove that, if $f : P \times Q \to P \times Q$ is order-preserving and $f(p, q) \geq (p, q)$, then f has a fixed point.

12-14. (K. A. Baker, unpublished, see [115, 155].) Prove that forming the Dedekind–MacNeille completion does not change the dimension, that is, $\dim(P) = \dim(DM(P))$.
  Hint. Combine Theorems 8.27 and 12.10.

12-15. Let P be a chain-complete ordered set and let Q be an ordered set with the fixed point property. Prove that if P is C-dismantlable to C and $C \times Q$ has the fixed point property, then $P \times Q$ has the fixed point property.

12-16. Products of power sets.

  a. Prove that $\mathcal{P}(\{1, \ldots, m\}) \times \mathcal{P}(\{1, \ldots, n\})$ is isomorphic to $\mathcal{P}(\{1, \ldots, m + n\})$.

  b. Prove that $\mathcal{P}(\mathbb{N}) \times \mathcal{P}(\mathbb{N})$ is isomorphic to $\mathcal{P}(\mathbb{N})$.

12-17. Let $G = (V, E)$ and $H = (W, F)$ be two graphs. The **direct product** or **categorical product** or **tensor product** of G and H is the graph $G \times H$ whose vertices are the set $V \times W$ and for which there is an edge between (x, u) and (y, v) iff $\{x, y\} \in E$ and $\{u, v\} \in F$.

  a. Prove that any direct product of a finite number of odd wheels has the fixed vertex property.

  b. A triangle walk in a graph is a sequence of vertices $v_0 \sim v_1 \sim \cdots \sim v_k$ so that we have $v_i \sim v_{i+2}$ for $i = 0, \ldots, k-2$. Let X, Y be graphs. Prove that, if $x_0 \sim x_1 \sim \cdots \sim x_k$ and $y_0 \sim y_1 \sim \cdots \sim y_k$ are triangle walks of the same length $k \neq 0$ in X and Y, respectively, so that $(x_i, y_i) = (x_j, y_j)$ implies $i = j$, then $(x_0, y_0) \sim (x_1, y_1) \sim \cdots \sim (x_k, y_k)$ is a triangle path in $X \times Y$.

  c. Prove that the core C in Figure 6.4 is so that $C \times C$, $C \times C^{b=t}$, and $C^{b=t} \times C^{b=t}$ are triangle connected.

      Hint. Extend triangle paths into triangle walks by attaching copies of $x_0 \sim x_1 \sim x_2$. Use Exercise 12-17b and compensate for different lengths of triangle walks modulo 3 by, in one component, going around one of the copies of T_{10} the appropriate number of times.

  d. Let P, Q, R, and S be ordered sets viewed as directed graphs (V_P, E_P), (V_Q, E_Q), (V_R, E_R), and (V_S, E_S) and let C be the graph in Figure 6.4.

    i. Prove that every homomorphism f from the graph $P * (C, b, t) \times Q * (C, b, t)$ to the graph $R * (C, b, t) \times S * (C, b, t)$ must map $V_P \times V_Q$ to $V_R \times V_S$ in such a way that, if $(p_1, q_1) \leq (p_2, q_2)$ in $P \times Q$, then $f(p_1, q_1) \leq f(p_2, q_2)$ in $R \times S$.

       Hint. Use Exercise 12-17c.

    ii. Prove that, conversely, for every order-preserving map $F : P \times Q \to R \times S$, there is a homomorphism f from $P * (C, b, t) \times Q * (C, b, t)$ to $R * (C, b, t) \times S * (C, b, t)$ so that $f|_{V_P \times V_Q} = F$.

    iii. Prove that, if $P \times Q = R \times S$, then a homomorphism f from $P*(C, b, t) \times Q*(C, b, t)$ to $R * (C, b, t) \times S * (C, b, t)$ has a fixed vertex iff $f|_{V_P \times V_Q}$ has a fixed point.

    Hint. Mimic the proof of Theorem 6.31.

  e. (See [287].) Let P and Q be ordered sets. Prove that $P * (C, b, t) \times Q * (C, b, t)$ has the fixed vertex property iff $P \times Q$ has the fixed point property.

  f. Explain why the proof of Theorem 12.17 cannot be translated into a proof that the product of two graphs with the fixed vertex property has the fixed vertex property, too.

12-18. (Compare with Théorème 4.8 in [16].) Let $G = (V, E)$ and $G' = (V', E')$ be finite graphs with the fixed clique property. Prove that $G \times G'$ has the fixed clique property.

    Hint. Find a comparative retraction from $TCL(G \times G')$ to an embedded copy of the product $TCL(G) \times TCL(G')$.

12-19. (More on f-chains. This problem is a precursor to Exercise 12-20.) Let P be a chain-complete ordered set and let $f : P \to P$ be order-preserving. For $x \in P$, if $f(x) \geq x$ ($f(x) \leq x$), let C_x^f be the (dual) maximal f-chain that starts at x.

  a. Prove that, for $x \in P$ and $n \in \mathbb{N}$, if $f(x) \geq x$, then we have $\bigvee C_x^f = \bigvee C_{f^n(x)}^f$.

    Hint. Lemma 3.31.

  b. Let $g : P \to P$ be order-preserving with $f \leq g$ and let $p \leq q$. Prove the following statements.

    1) If $f(p) \geq p$ and $f(q) \geq q$, then $\bigvee C_p^f \leq \bigvee C_q^f$.
    2) If $f(p) \geq p$ and $f(q) \leq q$, then $\bigvee C_p^f \leq \bigwedge C_q^f$.
    3) If $f(p) \leq p$ and $f(q) \geq q$, then $\bigwedge C_p^f \leq \bigvee C_q^f$.
    4) If $f(p) \geq p$ and $g(p) \geq p$, then $\bigvee C_p^f \leq \bigvee C_p^g$.
    5) If $f(p) \leq p$ and $g(p) \geq p$, then $\bigwedge C_p^f \leq \bigvee C_p^g$.

    Hint. Lemma 3.31.

  c. Prove that, if $f^k(p) \geq p$ and $l \in \mathbb{N}$, then $f^{lk}(p) \geq p$ and $\bigvee C_p^{f^k} = \bigvee C_p^{f^{lk}}$.

12-20. Let P be an ordered set. We define $\text{Retr}(P)$ to be the ordered subset of $\text{End}(P)$ that consists of all the retractions on P.

  a. Let P be a chain-complete ordered set that does not contain any infinite antichains. Prove that there is a retraction $R : \text{End}(P) \to \text{Retr}(P)$ such that, for each $f \in \text{End}(P)$, we have $R(f)[P] = \{p \in P : (\exists n \in \mathbb{N}) f^n(p) = p\}$.

    Hint. For each p, there are $k, m \in \mathbb{N}$ with $f^k(p) \sim f^{k+m}(p)$; also use Exercise 12-19.

  b. Let L be a complete lattice. Prove that $\text{Retr}(L)$ is a retract of $\text{End}(L)$.

12-21. Use the proof for Exercise 12-20 to show that, for chain-complete ordered sets without infinite antichains, we can replace order-preserving factor maps with maps that satisfy the hypotheses of Theorem 12.15. This provides an alternative proof for Theorem 12.17 that does not rely on the Li–Milner Structure Theorem.

12-22. Prove that, if the answer to Open Question 12.12 is negative and P, X have the fixed point property and $f : P \times X \to P \times X$ is order-preserving with no fixed points, then there must be an $x \in X$ so that the factor map f_x has an infinite orbit $\{f_x^k(p)\}_{n \in \mathbb{N}}$.

    Hint. If all f_x have finite orbits, use as r_{f_x} the "limit" of $\{f_x^{n!}\}_{n \in \mathbb{N}}$, defined similar to Exercise 4-36. Apply Theorem 12.15.

12-23. (Motivated by a question in [53]. Also see [51, 52] or Exercise 12-20b for the proof that for complete lattices L the set Retr(L) is a complete lattice and [183] for an example that the retraction set of a lattice need not form a lattice.) Let L be a lattice. Prove that, if Retr(L) is a lattice, then L must be a complete lattice.

    Hint. By Proposition 8.8, it is enough to prove that L is chain-complete (why?). Suppose L is not chain-complete. Find, without loss of generality, a well-ordered chain C without a supremum. Partition it into two cofinal subsets C_1, C_2 so that the C-successor of each element in C_i is in the respective other set. Retract onto $C_1 \cup \uparrow C_1$ via

$$r_1(x) := \begin{cases} x; & \text{if } x \geq C_1 \text{ or } x \in C_1, \\ \min_{C_1}\{c \in C_1 : c \not< x\}; & \text{if } x > c \text{ for some } c \in C_1, \\ \min(C_1); & \text{otherwise,} \end{cases}$$

and similarly onto $C_2 \cup \uparrow C_2$. These two retractions do not have a supremum, contradiction.

12.3 Infinite Products

A natural generalization of Open Question 12.12 is to ask if products of infinite families of ordered sets with the fixed point property have the fixed point property, too. At first glance, this question does not appear promising. Even products of nice families of ordered sets with the fixed point property need not be connected (see Lemma 12.20). However, taking trivially disconnected infinite products out of consideration, nothing much is known about the fixed point property for infinite products beyond the (simple) results of this section and the results in [257] and [281]. We will thus be led to Open Question 12.24. Progress on the fixed point property for infinite products would have at least two consequences: Better understanding of order-preserving maps on products and a better understanding of the fixed point property for infinite ordered sets.

    We start with connectivity as a warm-up to infinite products. Although the main scope of the following results are infinite products, they are of course equally valid for any finite products that fall into their scope.

Lemma 12.18. *Let $\{P_\alpha\}_{\alpha \in A}$ be a family of ordered sets. For any two elements $(p_\alpha)_{\alpha \in A}, (q_\alpha)_{\alpha \in A} \in \prod_{\alpha \in A} P_\alpha$ we have*

$$\sup_{\alpha \in A} \operatorname{dist}(p_\alpha, q_\alpha) \leq \operatorname{dist}((p_\alpha)_{\alpha \in A}, (q_\alpha)_{\alpha \in A}) \leq \sup_{\alpha \in A} \operatorname{dist}(p_\alpha, q_\alpha) + 1.$$

Proof. The first inequality follows because every P_α is a retract of $\prod_{\alpha \in A} P_\alpha$. To see the other inequality, let $n := \sup_{\alpha \in A} \operatorname{dist}(p_\alpha, q_\alpha)$. If $n = \infty$, then there is nothing to prove. If n is finite, then, for each α, one can find $x_\alpha^0, \ldots, x_\alpha^{n+1}$ such that $p_\alpha = x_\alpha^0 \geq x_\alpha^1 \leq \cdots x_\alpha^{n+1} = q_\alpha$. (To see this, first find a fence that connects p_α and q_α. Then duplicate elements of that fence as necessary to get the above.) Then $(x_\alpha^0)_{\alpha \in A} \geq (x_\alpha^1)_{\alpha \in A} \leq \cdots (x_\alpha^{n+1})_{\alpha \in A}$ connects the two points in the product. ∎

Proposition 12.19. *Let $\{P_\alpha\}_{\alpha \in A}$ be a family of ordered sets. The following are equivalent.*

1. *$\prod_{\alpha \in A} P_\alpha$ is connected and of finite diameter.*
2. *All P_α are connected and there is an $n \in \mathbb{N}$ such that all P_α are of diameter $\leq n$.*

Proof. Easy consequence of Lemma 12.18. (See Exercise 12-24.) ∎

Proposition 12.19 almost characterizes connected products, but easy examples show that ordered sets can be connected and of unbounded diameter. These must be considered separately to completely characterize connectivity for products.

Lemma 12.20 (Also see [73]). *Let $\{P_\alpha\}_{\alpha \in A}$ be a family of ordered sets. Assume that, for each $n \in \mathbb{N}$, there is an α_n such that $n \neq m$ implies $\alpha_n \neq \alpha_m$ and such that the diameter of P_{α_n} is larger than n. Then $\prod_{\alpha \in A} P_\alpha$ is not connected.*

Proof. For each $n \in \mathbb{N}$, find $p_{\alpha_n}, q_{\alpha_n} \in P_{\alpha_n}$ whose distance is larger than n. For $\alpha \in A \setminus \{\alpha_n : n \in \mathbb{N}\}$ let $p_\alpha, q_\alpha \in P_\alpha$ be arbitrary. Then the distance $\mathrm{dist}((p_\alpha)_{\alpha \in A}, (q_\alpha)_{\alpha \in A})$ cannot be finite, because, for all $n \in \mathbb{N}$, we have the inequality $n < \mathrm{dist}(p_{\alpha_n}, q_{\alpha_n}) \leq \mathrm{dist}((p_\alpha)_{\alpha \in A}, (q_\alpha)_{\alpha \in A})$. ∎

Theorem 12.21. *Let $\{P_\alpha\}_{\alpha \in A}$ be a family of ordered sets. The following are equivalent.*

1. *$\prod_{\alpha \in A} P_\alpha$ is connected.*
2. *All P_α are connected and there is an $n \in \mathbb{N}$ such that all but finitely many P_α are of diameter $\leq n$.*

Proof. "2⇒1": The product of the factors whose diameter is bounded by n is connected by Proposition 12.19. The product of the factors with diameter $> n$ is a finite product of connected sets, hence it is connected (see Exercise 12-6). Thus the product of all factors is connected.

"1⇒2": If 2 was false, then, by Lemma 12.20, the product $\prod_{\alpha \in A} P_\alpha$ would be disconnected. ∎

Whereas connectivity is a simple necessary condition for the fixed point property, C-dismantlability is a strong sufficient condition. Moreover we have seen in Theorem 12.5 that C-dismantlability to a singleton is closely related to connectivity of $\mathrm{End}(P)$. These ideas are again present in the next result.

Proposition 12.22. *Let $\{P_\alpha\}_{\alpha \in A}$ be a family of ordered sets. The following are equivalent.*

1. *All P_α are chain-complete and there is an $n \in \mathbb{N}$ such that, for all $\alpha \in A$, the diameter of $\mathrm{End}(P_\alpha)$ is $\leq n$.*
2. *$\prod_{\alpha \in A} P_\alpha$ is C-dismantlable to a singleton and chain-complete.*

Proof. "1⇒2": For each α, select a $p_\alpha \in P_\alpha$. Then, for each α, there are order-preserving maps $r_\alpha^0, r_\alpha^1, \ldots, r_\alpha^{n+1} \in \mathrm{End}(P_\alpha)$ with $r_\alpha^0 = \mathrm{id}_{P_\alpha}$, $r_\alpha^{n+1}(x) = p_\alpha$ for all $x \in P_\alpha$ and such that $r_\alpha^0 \leq_\alpha r_\alpha^1 \geq_\alpha \cdots r_\alpha^{n+1}$. For $j = 0, \ldots, n+1$, define

$R^j((q_\alpha)_{\alpha \in A}) := (r^j_\alpha(q_\alpha))_{\alpha \in A}$. Then, for every mapping $f \in \text{End}\left(\prod_{\alpha \in A} P_\alpha\right)$, the set $\{R^0 \circ f, R^1 \circ f, \dots, R^{n+1} \circ f = R^{n+1}\}$ contains a fence from f to R^{n+1}. Thus $\text{End}\left(\prod_{\alpha \in A} P_\alpha\right)$ is connected. Chain-completeness is proved in Exercise 12-8. The result follows from Theorem 12.5.

"2⇒1": Because $\prod_{\alpha \in A} P_\alpha$ is $\mathcal{C}$-dismantlable to a singleton and chain-complete, $\text{End}(\prod_{\alpha \in A} P_\alpha)$ is connected. The diameter of $\text{End}(\prod_{\alpha \in A} P_\alpha)$ is now bounded by twice the distance from $\text{id}_{\prod_{\alpha \in A} P_\alpha}$ to any constant map.

Now fix $(p_\alpha)_{\alpha \in A} \in \prod_{\alpha \in A} P_\alpha$. Every set of endomorphisms $\text{End}(P_\gamma)$ can be embedded into $\text{End}\left(\prod_{\alpha \in A} P_\alpha\right)$ via $f \mapsto F_f$, where

$$(F_f((q_\alpha)_{\alpha \in A}))_\beta := \begin{cases} f(q_\gamma); & \text{if } \beta = \gamma, \\ p_\beta; & \text{otherwise.} \end{cases}$$

Moreover, for fixed $(p_\alpha)_{\alpha \in A} \in \prod_{\alpha \in A} P_\alpha$, the map $m(F) := f_F$, where (with notation as in Lemma 12.8)

$$(f_F((q_\alpha)_{\alpha \in A}))_\beta := \begin{cases} \pi_\gamma(F(((p_\alpha)_{\alpha \in A}, q_\gamma, \gamma))); & \text{if } \beta = \gamma, \\ p_\beta; & \text{otherwise,} \end{cases}$$

maps $\text{End}\left(\prod_{\alpha \in A} P_\alpha\right)$ onto the isomorphic copy of $\text{End}(P_\gamma)$. Hence the diameter of each $\text{End}(P_\alpha)$ is bounded by the diameter of $\text{End}\left(\prod_{\alpha \in A} P_\alpha\right)$. Because every factor of a product is a retract of the product, all factors are chain-complete. ∎

Remark 12.23. The following example[1] shows that the technical condition 1 in Proposition 12.22 cannot be replaced with the condition "all P_α have uniformly bounded diameter." Let **2** be the two-element antichain and let F_{2n} be a fence of length $2n$. Let P_n be the lexicographic sum $\mathbf{2} \oplus F_{2n}$. Then $\text{diam}(P_n) = 2$ and $\text{diam}(\text{End}(P_n)) \geq n$. Hence $\prod_{n \in \mathbb{N}} P_n$ is connected, but not dismantlable. Even more is true: It is shown in [257] that this ordered set does not have the fixed point property, as long as we assume the existence of free ultrafilters. □

We have thus completely characterized the relationship between products and the conditions most easily linked to the fixed point property, connectivity, and $\mathcal{C}$-dismantlability. The open question that beckons is the following.

> **Open Question 12.24.** Let $\{P_\alpha\}_{\alpha \in A}$ be a family of ordered sets with the fixed point property such that there is an $n \in \mathbb{N}$ such that $\text{diam}(P_\alpha) \leq n$ for all $\alpha \in A$. Does $\prod_{\alpha \in A} P_\alpha$ have the fixed point property?

Aside from [257] and [281], no progress has been made on this problem. We conclude this section showing that a natural avenue to disprove the fixed point

---

[1]First shown to the author by A. Rutkowski in a different context.

property, minimal finite automorphic retracts, is closed for infinite products. Indeed, by Proposition 12.25 and Theorem 12.17, every finite retract of an infinite product of finitely many nonisomorphic finite ordered sets with the fixed point property must have the fixed point property, too.

Proposition 12.25 (See [73], Lemma 3.2). *Let $S_1, \ldots, S_N$ be pairwise nonisomorphic finite ordered sets and let $\{Q_\alpha\}_{\alpha \in A}$ be a family of ordered sets such that each Q_α is isomorphic to some set S_{k_α}. Let P be a finite retract of the product $\prod_{\alpha \in A} Q_\alpha$. Then there is a finite set $\{\alpha_1, \ldots, \alpha_n\} \subseteq A$ such that P is isomorphic to a retract of $\prod_{m=1}^{n} Q_{\alpha_m}$.*

Proof. Let $r : \prod_{\alpha \in A} Q_\alpha \to P$ be a retraction. For each α, let the function $\phi_\alpha : Q_\alpha \to S_{k_\alpha}$ be an isomorphism. Let $\alpha \equiv \beta$ iff $\phi_\alpha \circ \pi_\alpha|_P = \phi_\beta \circ \pi_\beta|_P$. Clearly $\equiv$ is an equivalence relation. Moreover, because P and S_k are finite and because there are only finitely many sets S_k, there are only finitely many possibilities for maps from P to S_k with arbitrary k. Thus the equivalence relation $\equiv$ partitions the set A into finitely many equivalence classes. Assuming there are n equivalence classes, let $\alpha_1, \ldots, \alpha_n$ be a system of representatives for the equivalence classes. For $q \in \prod_{m=1}^{n} Q_{\alpha_m}$, define the point $q' \in \prod_{\alpha \in A} Q_\alpha$ componentwise as follows. For $\alpha \in \{\alpha_1, \ldots, \alpha_n\}$, let $q'_\alpha := q_\alpha$. For $\alpha \notin \{\alpha_1, \ldots, \alpha_n\}$, let α_m be the unique element of $\{\alpha_1, \ldots, \alpha_n\}$ such that $\alpha \equiv \alpha_m$ and let $q'_\alpha := \phi_\alpha^{-1}(\phi_{\alpha_m}(q_{\alpha_m}))$. Note that $q \mapsto q'$ is an embedding of $\prod_{m=1}^{n} Q_{\alpha_m}$ into $\prod_{\alpha \in A} Q_\alpha$.

Now define $r' : \prod_{m=1}^{n} Q_{\alpha_m} \to P$ by $r'(q) := r(q')$. Because $q \mapsto q'$ is an embedding, r' is order-preserving. Define $c : P \to \prod_{m=1}^{n} Q_{\alpha_m}$ to be the map that maps $(p_\alpha)_{\alpha \in A}$ to $(p_{\alpha_m})_{m=1}^{n}$. Clearly c is order-preserving, too. Finally note that $r' \circ c = \mathrm{id}_P$. This means that P is a retract of $\prod_{m=1}^{n} Q_{\alpha_m}$ in the sense of the alternative definition of retracts given in Exercise 4-1. ∎

Remark 12.26. The condition in Proposition 12.25 that only finitely many factors are nonisomorphic is needed: In Exercise 4-23, it is shown that any retract of a finite $\mathcal{I}$-dismantlable ordered set is again $\mathcal{I}$-dismantlable, and Proposition 12.22 shows that any finite product of finite $\mathcal{I}$-dismantlable ordered sets is again $\mathcal{I}$-dismantlable. However, M. Roddy showed in [257] that $\prod_{n \in \mathbb{N}} P_n$ as in Remark 12.23 can be retracted onto a four crown. Because the four crown is not $\mathcal{I}$-dismantlable, the condition of finitely many nonisomorphic factors cannot be dropped.

Moreover, the set $\prod_{n \in \mathbb{N}} P_n$ is a retract of an infinite power of linear sums of 2-antichains and the "spider" in Figure 1.1 f). Thus the assumption in Proposition 12.25 that the factors are all finite is needed, too.   □

Exercises

12-24. Prove Proposition 12.19.

12.4 Hashimoto's Theorem and Automorphisms of Products

In the representation theory for ordered sets, it is natural to ask about the relation between two product representations of an ordered set. Clearly, their factors need not be isomorphic. Simply consider $P \times (P \times Q)$ and $(P \times P) \times Q$ as products of two factors each. For this example, it is also quite obvious that we get isomorphisms between the factors if we factor into three factors. Interestingly, by Hashimoto's Theorem 12.30, this type of refinement will always work for *connected* ordered sets, while, for disconnected ordered sets, it may not work (see Exercise 12-30).

Hashimoto's Theorem will allow us to show, in Theorem 12.38 and Exercise 12-32, that the property that all automorphisms of an ordered set have a fixed point is "infinitely productive." To this end, we need a stronger result than Hashimoto's original refinement theorem. This requires a more formal framework for products.

Definition 12.27. *A **(product) decomposition** of an ordered set P is an isomorphism* $\Psi : P \to \prod_{\alpha \in A} X_\alpha$ *from P to a product* $\prod_{\alpha \in A} X_\alpha$ *of ordered sets* X_α.

This definition of a product decomposition clearly separates different representations from each other. As we find common refinements of product decompositions, such as in the $P \times (P \times Q)$ versus $(P \times P) \times Q$ example, we will find product decompositions of the factors which we will need to turn into decompositions of the original ordered set. To do this, we need to turn decompositions of the factors into maps from the original product to the refined product. This is easily done using the idea of a product map.

Definition 12.28. *Let* $\prod_{\alpha \in A} X_\alpha$ *and* $\prod_{\alpha \in A} Y_\alpha$ *be two products with the same index set A. Let* $\{f_\alpha : X_\alpha \to Y_\alpha\}_{\alpha \in A}$ *be a family of order-preserving maps. Then the **product map** $f : \prod_{\alpha \in A} X_\alpha \to \prod_{\alpha \in A} Y_\alpha$ is defined componentwise as the unique map such that* $\pi_\beta f((x_\alpha)_{\alpha \in A}) = f_\beta(x_\beta)$.

The strict refinement property now says that any two product representations have refinements that differ from each other only in the way the factors are grouped. Hashimoto's Refinement Theorem then says that this property applies for all connected ordered sets.

Definition 12.29. *An ordered set P has the **strict refinement property** iff, for any two decompositions* $\Psi_A : X \to \prod_{\alpha \in A} X_\alpha$ *and* $\Psi_B : X \to \prod_{\beta \in B} Y_\beta$, *there are decompositions* $a_\alpha : X_\alpha \to \prod_{\beta \in B} Z_{\alpha,\beta}$ *and* $b_\beta : Y_\beta \to \prod_{\alpha \in A} Z_{\alpha,\beta}$ *such that, with a and b being the product maps of the* a_α *and* b_β, *respectively, and* γ *being the natural map that maps* $((z_{\alpha,\beta})_{\beta \in B})_{\alpha \in A}$ *to* $((z_{\alpha,\beta})_{\alpha \in A})_{\beta \in B}$, *we have* $\gamma \circ a \circ \Psi_A = b \circ \Psi_B$.

For an illustration of the strict refinement property and its proof through an extension of Hashimoto's Refinement Theorem, see Figure 12.2.

Theorem 12.30. A strengthening of **Hashimoto's Refinement Theorem**, see [43], [121] and [122]. *Every connected ordered set has the strict refinement property.*

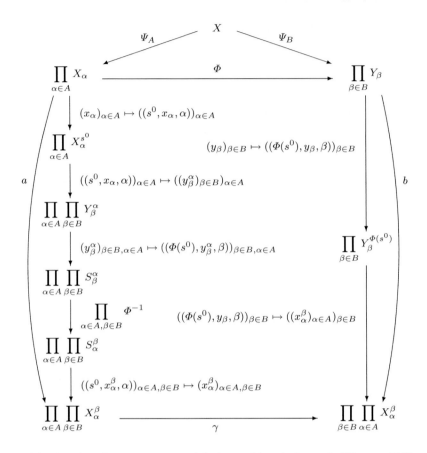

Fig. 12.2 The strict refinement property and the isomorphisms in the proof of Theorem 12.30.

Note that, in the definition of the strict refinement property, we can eliminate the "middleman" X and simply talk about two infinite products that are isomorphic via $\Phi : \prod_{\alpha \in A} X_\alpha \to \prod_{\beta \in B} Y_\beta$. We will do so throughout the proof of Hashimoto's Refinement Theorem. We can (and do) also assume that all factors are connected. The proof of Theorem 12.30 starts with a series of lemmas. These lemmas show that product decompositions have a certain resilience under isomorphisms.

Consider the first lemma. It focuses on elements that have the same first component in one decomposition of an ordered set. The lemma says that the first component in this decomposition is invariant under finite interchanges of components of these elements in any other decomposition. The key to applying this lemma later on is to build the element whose inverse image is of interest from two elements whose preimages have the same component in the right place.

Lemma 12.31 (See [122], Lemma 3). *Let P, Q, U, V be connected ordered sets such that $\Phi : P \times Q \to U \times V$ is an isomorphism. If*

$$\Phi(p, q) = (u, v), \quad \Phi(p, q') = (u', v'), \quad \text{and} \quad \Phi^{-1}(u, v') = (p^*, q^*),$$

then $p^* = p$.

Proof. First assume that q, q' have a common upper bound $\hat{q}$. Let $(\hat{u}, \hat{v}) := \Phi(p, \hat{q})$. Note that $(u, v) \leq (u, \hat{v}) \leq (\hat{u}, \hat{v})$ and $(u', v') \leq (\hat{u}, v') \leq (\hat{u}, \hat{v})$ implies that

$$p = \pi_P \Phi^{-1}(u, v) = \pi_P \Phi^{-1}(u', v')$$
$$\leq \{\pi_P \Phi^{-1}(u, \hat{v}), \pi_P \Phi^{-1}(\hat{u}, v')\} \leq \pi_P \Phi^{-1}(\hat{u}, \hat{v}) = p.$$

Thus there are $\widehat{q_1}, \widehat{q_2}$ such that

$$\Phi^{-1}(u, \hat{v}) = (p, \widehat{q_1}) \quad \text{and} \quad \Phi^{-1}(\hat{u}, v') = (p, \widehat{q_2}).$$

Now $(u, v') \leq \{(\hat{u}, v'), (u, \hat{v})\}$ implies $(p^*, q^*) \leq \{(p, \widehat{q_1}), (p, \widehat{q_2})\}$. In particular, we have $p^* \leq p$.

Moreover, because $q^* \leq \{\widehat{q_1}, \widehat{q_2}\}$, we have $(p, q^*) \leq \{(p, \widehat{q_1}), (p, \widehat{q_2})\}$. This implies $\Phi(p, q^*) \leq \{(u, \hat{v}), (\hat{u}, v')\}$, so, arguing componentwise, $\Phi(p, q^*) \leq (u, v')$. But then $(p, q^*) \leq (p^*, q^*)$ and $p \leq p^*$, which finishes the proof that $p = p^*$.

Now let $q, q' \in Q$ be arbitrary. Because Q is connected, it contains a fence $q = q_0 \leq \widehat{q_1} \geq q_1 \leq \widehat{q_2} \geq q_2 \leq \cdots \leq \widehat{q_n} \geq q_n = q'$. We proceed by induction on the parameter n. For $n = 1$, the result is proved above. For the induction step, let $n > 1$ and assume that the result has been proved for all pairs (q, q') for which a fence as above, but with fewer elements, exists. With

$$\Phi(p, q_i) = (u_i, v_i) \quad \text{and} \quad \Phi(p, \widehat{q_i}) = (\widehat{u_i}, \widehat{v_i})$$

we obtain, by induction hypothesis, that there are $q_{0,n-1} \in Q$ and $q_{1,n} \in Q$ such that $\Phi^{-1}(u_0, v_{n-1}) = (p, q_{0,n-1})$ and $\Phi^{-1}(u_1, v_n) = (p, q_{1,n})$. By duality, we also obtain $\Phi^{-1}(\widehat{u_1}, \widehat{v_n}) = (p, \widehat{q_{1,n}})$ for some $\widehat{q_{1,n}} \in Q$. Now the inequality $\{(u_0, v_{n-1}), (u_1, v_n)\} \leq (\widehat{u_1}, \widehat{v_n})$, implies $\{q_{0,n-1}, q_{1,n}\} \leq \widehat{q_{1,n}}$.

Now apply what we proved first to $\{(p, q_{0,n-1}), (p, q_{1,n})\}$ to obtain that we have $\pi_P \Phi^{-1}(u, v') = \pi_P \Phi^{-1}(u_0, v_n) = p$, as was to be proved. ∎

Lemma 12.31 can now be turned into a result for arbitrary products.

Lemma 12.32 (See [122], Lemma 2). *Let $\Phi : \prod_{\alpha \in A} X_\alpha \rightarrow \prod_{\beta \in B} Y_\beta$ be an isomorphism between two connected products. For $i = 1, 2$, let $x^i = (x^i_\alpha)_{\alpha \in A}$ be elements of $\prod_{\alpha \in A} X_\alpha$ such that $\Phi((x^i_\alpha)_{\alpha \in A}) = (y^i_\beta)_{\beta \in B}$. Fix $\lambda \in A$. If $x^1_\lambda = x^2_\lambda = x_\lambda$ and $y^3 \in \prod_{\beta \in B} Y_\beta$ is such that, for all $\beta \in B$, we have $y^3_\beta \in \{y^1_\beta, y^2_\beta\}$, then $\pi_\lambda \Phi^{-1}(y^3) = x_\lambda$.*

Proof. This follows directly from Lemma 12.31. Simply choose the following, with the obvious identifications made between sets.

$$P := X_\lambda, \quad Q := \prod_{\alpha \in A \setminus \{\lambda\}} X_\alpha,$$

$$U := \prod_{\beta \in \{\mu \in B: y_\mu^3 = y_\mu^1\}} Y_\beta, \quad V := \prod_{\beta \in \{\mu \in B: y_\mu^3 \neq y_\mu^1\}} Y_\beta,$$

$$(p, q) := x^1, \quad (p, q') := x^2, \quad (u, v') := y^3. \qquad \blacksquare$$

The key idea to obtain the common refinements now is to consider the, via Lemma 12.8, embedded factors in one decomposition and examine their images in the other decomposition. Interestingly, as we are about to show now, the image itself is a natural product. This observation provides the desired decomposition of the factors.

Lemma 12.33 (See [122], Lemma 4). *Let* $\Phi : \prod_{\alpha \in A} X_\alpha \to \prod_{\beta \in B} Y_\beta$ *be an isomorphism between two connected products. Fix* $s^0 \in \prod_{\alpha \in A} X_\alpha$ *and* $\lambda \in A$ *and let* $X_\lambda^{s^0}$ *be as in Lemma 12.8. For* $\beta \in B$, *let*

$$Y_\beta^\lambda := \pi_\beta \Phi[X_\lambda^{s^0}].$$

Then $\Phi|_{X_\lambda^{s^0}} : X_\lambda^{s^0} \to \prod_{\beta \in B} Y_\beta^\lambda$ *is an isomorphism.*

Proof. Clearly the map $\Phi|_{X_\lambda^{s^0}}$ is an isomorphism onto its image and it maps $X_\lambda^{s^0}$ into $\prod_{\beta \in B} Y_\beta^\lambda$. To show that $\Phi|_{X_\lambda^{s^0}}$ is surjective, note that $\Phi|_{X_\lambda^{s^0}}[X_\lambda^{s^0}]$ is a nonempty subset of the connected set $\prod_{\beta \in B} Y_\beta^\lambda$. Thus, by Exercise 12-29, we are done if we can show that $\Phi|_{X_\lambda^{s^0}}[X_\lambda^{s^0}]$ is an up-set and a down-set in $\prod_{\beta \in B} Y_\beta^\lambda$. By duality, it suffices to show that it is an up-set.

For $i = 1, 2$ pick elements $y^i = \Phi(x^i) = \Phi((x_\alpha^i)_{\alpha \in A}) = (y_\beta^i)_{\beta \in B}$ and assume that $y^1 \in \Phi|_{X_\lambda^{s^0}}[X_\lambda^{s^0}]$, $y^2 \in \prod_{\beta \in B} Y_\beta^\lambda$, and $y^1 \leq y^2$. We will now first find an element $x^* \in X_\lambda^{s^0}$ such that $\Phi|_{X_\lambda^{s^0}}(x^*) \geq y^2$.

Let $\mu \in B$. By definition of the product $\prod_{\beta \in B} Y_\beta^\lambda$, $y_\mu^2 = \pi_\mu \Phi((s^0, x_\lambda, \lambda))$ for some $x_\lambda \in X_\lambda$. Let $y^\mu := (y^1, y_\mu^2, \mu)$. Then, by Lemma 12.32, $y^\mu \in \Phi|_{X_\lambda^{s^0}}[X_\lambda^{s^0}]$. Moreover, since $y^1 \leq y^\mu \leq y^2$, we have $\Phi^{-1}(y^1) \leq \Phi^{-1}(y^\mu) \leq \Phi^{-1}(y^2)$. This means $\Phi^{-1}(y^\mu) = (x^1, \tilde{x}, \lambda)$ for some $\tilde{x} \in X_\lambda$ with $x_\lambda^1 \leq \tilde{x} \leq x_\lambda^2$.

Let $x^* := (x^1, x_\lambda^2, \lambda)$. Then $y^\mu \leq \Phi(x^*)$. Because μ was arbitrary, this means that $\pi_\mu \Phi(x^*) \geq y_\mu^\mu = y_\mu^2$ for all μ. Therefore (after componentwise comparison) $\Phi(x^*) \geq y^2 \geq y^1$.

However, this implies that $(x^1, x_\lambda^2, \lambda) \geq \Phi^{-1}(y^2) \geq \Phi^{-1}(y^1)$, which means that all components of $\Phi^{-1}(y^2)$, except possibly the λth component, are equal to the corresponding component of s^0. Thus $\Phi^{-1}(y^2) \in X_\lambda^{s^0}$ and we are done. $\blacksquare$

Proof of Theorem 12.30, first stage. Without loss of generality, assume that $A \cap B = \emptyset$. With s^0 and Y_β^λ chosen as in Lemma 12.33, there is an isomorphism $a_\lambda : X_\lambda \to \prod_{\beta \in B} Y_\beta^\lambda$, namely, the composition of $\Phi|_{X_\lambda^{s^0}} : X_\lambda^{s^0} \to \prod_{\beta \in B} Y_\beta^\lambda$ and the isomorphism between X_λ and $X_\lambda^{s^0}$. Similarly, using $\Phi(s^0)$ as the fixed element in $\prod_{\beta \in B} Y_\beta$, we can define isomorphisms $b_\mu : Y_\mu \to \prod_{\alpha \in A} X_\alpha^\mu$: Each b_μ is the composition of the inverse $\Phi^{-1}|_{Y_\mu^{\Phi(s^0)}} : Y_\mu^{\Phi(s^0)} \to \prod_{\alpha \in A} X_\alpha^\mu$ and the isomorphism between Y_μ and $Y_\mu^{\Phi(s^0)}$.

We can show now that X_λ^μ is isomorphic to Y_μ^λ. Let

$$S_\mu^\lambda := \left\{ (\Phi(s^0), y_\mu, \mu) : y_\mu \in Y_\mu^\lambda \right\}, \quad \text{and} \quad S_\lambda^\mu := \left\{ (s^0, x_\lambda, \lambda) : x_\lambda \in X_\lambda^\mu \right\}.$$

Clearly S_μ^λ is isomorphic to Y_μ^λ and S_λ^μ is isomorphic to X_λ^μ. Let $x \in S_\lambda^\mu$. Then $x \in X_\lambda^{s^0}$. Thus, by Lemma 12.33, $\Phi(x) \in \prod_{\beta \in B} Y_\beta^\lambda$, especially $\pi_\mu \Phi(x) \in Y_\mu^\lambda$. Moreover, $s^0 = \Phi^{-1}(\Phi(s^0))$ is in $\prod_{\alpha \in A} X_\alpha^\mu$ and $x_\lambda \in X_\lambda^\mu$ by definition of S_λ^μ. Thus $x \in \prod_{\alpha \in A} X_\alpha^\mu$, which means $x = \Phi^{-1}(y^*)$ for some $y^* \in Y_\mu^{\Phi(s_0)}$, or, equivalently, $\Phi(x) = (\Phi(s^0), y_\mu, \mu)$ for some $y_\mu \in Y_\mu$. Because $\pi_\mu \Phi(x) \in Y_\mu^\lambda$, this means that $\Phi(x) \in S_\mu^\lambda$. Because $x \in S_\lambda^\mu$ was arbitrary, we conclude that $\Phi[S_\lambda^\mu] \subseteq S_\mu^\lambda$. The symmetric argument can be applied to $\Phi^{-1}|_{S_\mu^\lambda}$, showing that $\Phi^{-1}[S_\mu^\lambda] \subseteq S_\lambda^\mu$.

Thus $\Phi|_{S_\lambda^\mu}$ is an isomorphism between S_λ^μ and S_μ^λ, which proves that X_λ^μ is isomorphic to Y_μ^λ.

The above first stage is proved in [122]. Simply put, any two connected products have refinements with isomorphic factors, which is also called the **refinement property**. However, the isomorphism between the refinements has not been put into relation with the isomorphism between the products. For a quick overview of the structures that are isomorphic to $\prod_{\alpha \in A} X_\alpha$ and the isomorphisms between them, consider Figure 12.2. To finish the proof of Theorem 12.30, we will prove that the diagram in Figure 12.2 commutes. That is, independent of which path of arrows we follow, we obtain the same map. This fact was observed, for example, in [151].

Proof of Theorem 12.30, taking stock. We have so far that $\prod_{\alpha \in A} \prod_{\beta \in B} Y_\beta^\alpha$ is isomorphic to $\prod_{\alpha \in A} \prod_{\beta \in B} X_\alpha^\beta$. With all the other isomorphisms, this gives us two distinct isomorphisms between $\prod_{\alpha \in A} X_\alpha$ and $\prod_{\alpha \in A} \prod_{\beta \in B} X_\alpha^\beta$.

For the first isomorphism, we map $(x_\lambda)_{\lambda \in A} \in \prod_{\lambda \in A} X_\lambda$ to

$$\Phi((x_\lambda)_{\lambda \in A}) \in \prod_{\beta \in B} Y_\beta.$$

This in turn is mapped to

$$((\Phi(s^0), \Phi((x_\lambda)_{\lambda \in A})_\beta, \beta))_{\beta \in B} \in \prod_{\beta \in B} Y_\beta^{\Phi(s^0)}.$$

Via the B-fold product of Φ^{-1}, we reach

$$\Psi_1((x_\lambda)_{\lambda \in A}) := (\Phi^{-1}(\Phi(s^0), \Phi((x_\lambda)_{\lambda \in A})_\beta, \beta))_{\beta \in B} \in \prod_{\beta \in B} \prod_{\alpha \in A} X_\alpha^\beta,$$

and $\prod_{\beta \in B} \prod_{\alpha \in A} X_\alpha^\beta$ is isomorphic to $\prod_{\alpha \in A} \prod_{\beta \in B} X_\beta^\alpha$ via the natural map γ^{-1}. On the other hand, we can also map $(x_\alpha)_{\alpha \in A} \in \prod_{\alpha \in A} X_\alpha$ to

$$((s^0, x_\alpha, \alpha))_{\alpha \in A} \in \prod_{\alpha \in A} X_\alpha^{s^0}.$$

The A-fold product of Φ maps this to

$$(\Phi(s^0, x_\alpha, \alpha))_{\alpha \in A} = ((\Phi(s^0, x_\alpha, \alpha)_\beta)_{\beta \in B})_{\alpha \in A} \in \prod_{\alpha \in A} \prod_{\beta \in B} Y_\beta^\alpha.$$

From here, we can map to

$$((\Phi(s^0), \Phi(s^0, x_\alpha, \alpha)_\beta, \beta)_{\beta \in B})_{\alpha \in A} \in \prod_{\alpha \in A} \prod_{\beta \in B} S_\beta^\alpha.$$

The AB-fold product of Φ^{-1} maps this to

$$((\Phi^{-1}(\Phi(s^0), \Phi(s^0, x_\alpha, \alpha)_\beta, \beta))_{\beta \in B})_{\alpha \in A} \in \prod_{\alpha \in A} \prod_{\beta \in B} S_\alpha^\beta.$$

Finally, each element of S_α^β is of the form (s^0, z, α) for some $z \in X_\alpha^\beta$. So, to get to $\prod_{\alpha \in A} \prod_{\beta \in B} X_\alpha^\beta$, we need to take the αth projection inside and map to

$$\Psi_2((x_\lambda)_{\lambda \in A}) := ((\Phi^{-1}(\Phi(s^0), \Phi(s^0, x_\alpha, \alpha)_\beta, \beta)_\alpha)_{\beta \in B})_{\alpha \in A} \in \prod_{\alpha \in A} \prod_{\beta \in B} X_\alpha^\beta.$$

Because a and b in Figure 12.2 are product maps, we are done with the proof of Theorem 12.30 if we can now show that these two isomorphisms are equal, that is, $\gamma^{-1} \circ \Psi_1 = \Psi_2$.

To do this, we must show that, for all $(x_\lambda)_{\lambda \in A} \in \prod_{\lambda \in A} X_\lambda$ and all $\alpha \in A$ and $\beta \in B$, we have that

$$\Phi^{-1}(\Phi(s^0), \Phi((x_\lambda)_{\lambda \in A})_\beta, \beta)_\alpha = \Phi^{-1}(\Phi(s^0), \Phi(s^0, x_\alpha, \alpha)_\beta, \beta)_\alpha.$$

First, fix $\mu \in B$ and consider

$$(x_\lambda)_{\lambda \in A} \in \prod_{\lambda \in A} X_\lambda^\mu = \{\Phi^{-1}(\Phi(s^0), y_\mu, \mu) : y_\mu \in Y_\mu\}.$$

For $\beta \neq \mu$, both sides above return the αth component of s^0 (on the right side, apply Lemma 12.31 to Φ^{-1} to obtain $\Phi(s^0, x_\alpha, \alpha)_\beta = \Phi(s^0)_\beta$). For $\beta = \mu$, both sides above return x_α (on the right side, apply Lemma 12.31 to Φ). Therefore we have $\gamma^{-1} \circ \Psi_1 = \Psi_2$ on $\prod_{\lambda \in A} X_\lambda^\mu$ for arbitrary $\mu \in B$. To finish the proof, we must extend this equality to the whole product. To do this, we need a lemma that is similar to Lemma 12.31. Lemma 12.34 below shows that sometimes components in different product decompositions "stick together." That is, equality of components in one product decomposition can, under the right circumstances, force the equality of corresponding components in the other.

Lemma 12.34. *Let $P, Q, U,$ and V be connected ordered sets such that the function $\Phi : P \times Q \to U \times V$ is an isomorphism. If $\Phi(p, q) = (u, v)$ and $\Phi(p, q') = (u, v')$, then, for each $p' \in P$, there is a $w_{p'} \in U$ such that $\Phi(p', q) = (w_{p'}, z)$ and $\Phi(p', q') = (w_{p'}, z')$ for some $z, z' \in V$.*

Proof. Because P is connected, it suffices to prove the result for $p < p'$. The proof is an induction on $\text{dist}(q, q')$. The base step consists of the cases $q' < q$ and $q' > q$. Let $(w, z) := \Phi(p', q)$ and $(w', z') := \Phi(p', q')$.

First consider the case $q' < q$. In this case, $w' \leq w$. Moreover, the supremum of (p', q') and (p, q) is (p', q). This means that the supremum of (u, v) and (w', z') is (w, z). Because $(p, q') \leq (p', q')$, we have $(u, v') \leq (w', z')$, which means that $u \leq w'$. Therefore (w', z) is an upper bound of (u, v) and of (w', z'), which means $(w', z) \geq (w, z)$. Hence $w' \geq w$, which shows $w' = w$, proving our claim.

For the case $q' > q$, simply interchange the roles of q and q' above.

For the induction step, let $\text{dist}(q, q') = n$ and assume that the statement is true for all pairs q, q' whose distance is less than n. Let $q = q_0, q_1, \ldots, q_n = q'$ be a fence that connects q and q'. For $i = 0, \ldots, n$, let $(u_i, v_i) := f(p, q_i)$. Note that $u_0 = u = u_n$. By Lemma 12.31, there is an $r \in Q$ such that $f(p, r) = (u_0, v_1)$. Now $\text{dist}_{U \times V}((u_0, v_1), (u_n = u_0, v_n)) = \text{dist}_V(v_1, v_n) < n$, which means that

$$n > \text{dist}_{P \times Q}((p, r), (p, q')) = \text{dist}_Q(r, q').$$

The induction hypothesis implies $\pi_U(\Phi(p', r)) = \pi_U(\Phi(p', q'))$. Because $r \sim q$, the base case shows that $\pi_U(\Phi(p', r)) = \pi_U(\Phi(p', q))$. This proves the claim. ∎

Proof of Theorem 12.30, final stage. Keep $\mu \in B$ fixed and fix $\lambda_0 \in A$. Let

$$(x_{\lambda_0}, (x_\lambda)_{\lambda \in A \setminus \{\lambda_0\}}) \in X_{\lambda_0} \times \prod_{\lambda \in A \setminus \{\lambda_0\}} X_\lambda =: P \times Q.$$

We will denote the elements of $X_{\lambda_0} \times \prod_{\lambda \in A \setminus \{\lambda_0\}} X_\lambda^\mu$ by $(x_{\lambda_0}, (x_\lambda^\mu)_{\lambda \in A \setminus \{\lambda_0\}})$. Decompose the range $\prod_{\alpha \in A} \prod_{\beta \in B} X_\beta^\alpha$ as

$$\prod_{\beta \in B} X_\beta^{\lambda_0} \times \prod_{\alpha \in A \setminus \{\lambda_0\}} \prod_{\beta \in B} X_\beta^\alpha =: U \times V.$$

Let $\Psi : X_{\lambda_0} \times \prod_{\lambda \in A \setminus \{\lambda_0\}} X_\lambda \to \prod_{\beta \in B} X_\beta^{\lambda_0} \times \prod_{\alpha \in A \setminus \{\lambda_0\}} \prod_{\beta \in B} X_\beta^\alpha$ be any isomorphism such that, for all $(x_\lambda^\mu)_{\lambda \in A} \in \prod_{\lambda \in A} X_\lambda^\mu$, we have that

$$(\Psi((x_\lambda^\mu)_{\lambda \in A})_\beta)_\alpha = s_\alpha^0 \text{ if } \beta \neq \mu \quad \text{and} \quad (\Psi((x_\lambda^\mu)_{\lambda \in A})_\mu)_\alpha = x_\alpha^\mu.$$

(Ψ will serve as a stand-in for $\gamma^{-1} \circ \Psi_1$ and Ψ_2.) By Lemma 12.34, we know that the $X_\beta^{\lambda_0}$-components of the Ψ-image of any point $(x_{\lambda_0}, (x_\lambda^\mu)_{\lambda \in A \setminus \{\lambda_0\}})$ with $(x_\lambda^\mu)_{\lambda \in A \setminus \{\lambda_0\}} \in \prod_{\lambda \in A \setminus \{\lambda_0\}} X_\lambda^\mu$ solely depend on x_{λ_0}. Thus the U-component of the Ψ-image of $(x_{\lambda_0}, (x_\lambda^\mu)_{\lambda \in A \setminus \{\lambda_0\}})$ is exactly that of $\Psi((s^0, x_{\lambda_0}, \lambda_0))$; call it $\Psi_{\lambda_0}(x_{\lambda_0})$.

Now consider the V-components of the Ψ-image of $(x_{\lambda_0}, (x_\lambda^\mu)_{\lambda \in A \setminus \{\lambda_0\}})$. We have $\Psi(x_{\lambda_0}, (x_\lambda^\mu)_{\lambda \in A \setminus \{\lambda_0\}}) = (u, v)$ and, for a fixed $x_{\lambda_0}^\mu \in X_{\lambda_0}^\mu$, we have

$$\Psi(x_{\lambda_0}^\mu, (x_\lambda^\mu)_{\lambda \in A \setminus \{\lambda_0\}}) = (u', v').$$

By Lemma 12.31, this means that $\Psi^{-1}(u, v') = (\tilde{x}_{\lambda_0}, (x_\lambda^\mu)_{\lambda \in A \setminus \{\lambda_0\}})$. Because the U-component of a Ψ-image of a point in $X_{\lambda_0} \times \prod_{\lambda \in A \setminus \{\lambda_0\}} X_\lambda^\mu$ only depends on the X_{λ_0}-component, we have that $\tilde{x}_{\lambda_0} = x_{\lambda_0}$ and hence $v = v'$. Thus, via the properties of Ψ, $\Psi(x_{\lambda_0}, (x_\lambda^\mu)_{\lambda \in A \setminus \{\lambda_0\}}) = (\Psi_{\lambda_0}(x_{\lambda_0}), v)$ and, for $\beta \neq \mu$, the αth component of v is s_α^0, while, for $\beta = \mu$, the αth component of v is x_α^μ. Define $v =: \Psi_{\neq \lambda_0}((x_\lambda^\mu)_{\lambda \in A \setminus \{\lambda_0\}})$.

Now we essentially repeat the argument only with the factors reversed. We know that, for $(x_{\lambda_0}, (x_\lambda^\mu)_{\lambda \in A \setminus \{\lambda_0\}}) \in X_{\lambda_0} \times \prod_{\lambda \in A \setminus \{\lambda_0\}} X_\lambda^\mu$, we have

$$\Psi(x_{\lambda_0}, (x_\lambda^\mu)_{\lambda \in A \setminus \{\lambda_0\}}) = (\Psi_{\lambda_0}(x_{\lambda_0}), \Psi_{\neq \lambda_0}((x_\lambda^\mu)_{\lambda \in A \setminus \{\lambda_0\}})).$$

By Lemma 12.34, the V-component of any $\Psi(x_{\lambda_0}, (x_\lambda)_{\lambda \in A \setminus \{\lambda_0\}})$ only depends on the components $(x_\lambda)_{\lambda \in A \setminus \{\lambda_0\}}$. Call this V-component $\Psi_{\neq \lambda_0}((x_\lambda)_{\lambda \in A \setminus \{\lambda_0\}})$. We have

$$\Psi(x_{\lambda_0}, (x_\lambda^\mu)_{\lambda \in A \setminus \{\lambda_0\}}) = (\Psi_{\lambda_0}(x_{\lambda_0}), \Psi_{\neq \lambda_0}((x_\lambda^\mu)_{\lambda \in A \setminus \{\lambda_0\}}))$$

and

$$\Psi(x_{\lambda_0}, (x_\lambda)_{\lambda \in A \setminus \{\lambda_0\}}) = (u, \Psi_{\neq \lambda_0}((x_\lambda)_{\lambda \in A \setminus \{\lambda_0\}})).$$

Thus, by Lemma 12.31, we have

$$\Psi^{-1}(\Psi_{\lambda_0}(x_{\lambda_0}), \Psi_{\neq \lambda_0}((x_\lambda)_{\lambda \in A \setminus \{\lambda_0\}})) = (x_{\lambda_0}, (\tilde{x}_\lambda)_{\lambda \in A \setminus \{\lambda_0\}}).$$

Because the V-component of any $\Psi(x_{\lambda_0}, (x_\lambda)_{\lambda \in A \setminus \{\lambda_0\}})$ only depends on the components $(x_\lambda)_{\lambda \in A \setminus \{\lambda_0\}}$, we have $(\tilde{x}_\lambda)_{\lambda \in A \setminus \{\lambda_0\}} = (x_\lambda)_{\lambda \in A \setminus \{\lambda_0\}}$. We have thus proved that, for all $(x_{\lambda_0}, (x_\lambda)_{\lambda \in A \setminus \{\lambda_0\}}) \in X_{\lambda_0} \times \prod_{\lambda \in A \setminus \{\lambda_0\}} X_\lambda$, we have

$$\Psi(x_{\lambda_0}, (x_\lambda)_{\lambda \in A \setminus \{\lambda_0\}}) = (\Psi_{\lambda_0}(x_{\lambda_0}), \Psi_{\neq \lambda_0}((x_\lambda)_{\lambda \in A \setminus \{\lambda_0\}})).$$

This in turn implies that $\Psi((x_\lambda)_{\lambda \in A}) = (\Psi_\lambda(x_\lambda))_{\lambda \in A}$. Note that the above argument applies to both isomorphisms $\gamma^{-1} \circ \Psi_1$ and Ψ_2. Because the λ_0-components of $\gamma^{-1} \circ \Psi_1((s^0, x_{\lambda_0}, \lambda_0))$ equal those of $\Psi_2((s^0, x_{\lambda_0}, \lambda_0))$, all the factor maps for $\gamma^{-1} \circ \Psi_1$ and Ψ_2 are equal. Thus we have proved that $\gamma^{-1} \circ \Psi_1 = \Psi_2$, which completes the proof of Theorem 12.30. ∎

12.4.1   Fixed Points of Automorphisms of Products

Now that Hashimoto's Theorem is available, we turn our attention to fixed points of automorphisms of products. Unlike for the fixed point property, there are disconnected ordered sets for which every automorphism has a fixed point. The role of connectivity for the property that every automorphism has a fixed point is investigated in Exercise 12-25. We will exclusively focus on connected sets here.

The first step is to show that automorphisms of products are essentially made up of automorphisms of the factors. To prove this fact, we must demand that our sets are decomposable into product indecomposable ordered sets defined as follows.

Definition 12.35. *Call an ordered set P **product indecomposable** iff, for every isomorphism $\Phi : P \to A \times B$, we have that either A or B must be a singleton.*

This condition may look technical, but it is needed. Unfortunately, although any two product decompositions of an ordered set have a common refinement, it is not guaranteed that (in infinite ordered sets) the refinement process ever stops. (See Remark 15 in this chapter.)

Corollary 12.36 (See [67], Lemma 2).

1. *Let $P = \prod_{\alpha \in A} X_\alpha$ be a connected ordered set such that no two distinct X_α have a common factor. Then, for any automorphism $\Phi : P \to P$, there are automorphisms $\Phi_\alpha : X_\alpha \to X_\alpha$ such that $\Phi((x_\alpha)_{\alpha \in A}) = (\Phi_\alpha(x_\alpha))_{\alpha \in A}$.*
2. *Let $P = \prod_{\alpha \in A} X_\alpha$ be a connected ordered set such that all $X_\alpha = X$ and X is product indecomposable. Then, for any automorphism $\Phi : P \to P$, there is a permutation σ of A and there are automorphisms $\Phi_\alpha : X \to X$ such that $\Phi((x_\alpha)_{\alpha \in A}) = (\Phi_\alpha(x_{\sigma(\alpha)}))_{\alpha \in A}$.*

Proof. In both cases, we apply the strict refinement property to the decompositions $\Psi_A := \mathrm{id} : \prod_{\alpha \in A} X_\alpha \to \prod_{\alpha \in A} X_\alpha$ and $\Psi_B := \Phi : \prod_{\alpha \in A} X_\alpha \to \prod_{\beta \in A} X_\beta$. By Hashimoto's Theorem, there are decompositions $a_\alpha : X_\alpha \to \prod_{\beta \in A} Z_{\alpha,\beta}$ and $b_\beta : X_\beta \to \prod_{\alpha \in A} Z_{\alpha,\beta}$, such that $\gamma \circ a \circ \mathrm{id} = b \circ \Phi$.

To prove 1, note that X_α and X_β have no common factors unless $\alpha = \beta$. Thus we have that $Z_{\alpha,\beta}$ is trivial unless $\alpha = \beta$. Hence $\prod_{\beta \in A} Z_{\alpha,\beta}$ is isomorphic to $Z_{\alpha,\alpha}$ which is isomorphic to $\prod_{\delta \in A} Z_{\delta,\alpha}$. The isomorphism in each case is the projection on the αth component. Let $\eta_\alpha : \prod_{\beta \in A} Z_{\alpha,\beta} \to \prod_{\delta \in A} Z_{\delta,\alpha}$ be the isomorphism that fixes the αth component. Choose $\Phi_\alpha := b_\alpha^{-1} \circ \eta_\alpha \circ a_\alpha$. Then $\Phi_\alpha : X_\alpha \to X_\alpha$ is an isomorphism. Moreover $\Phi = b^{-1} \circ \gamma \circ a$, which means Φ is the product of the Φ_α.

To prove 2, note that X is product indecomposable. Thus there must be a permutation σ of A such that, for each α, only $Z_{\sigma(\alpha),\alpha}$ is nontrivial. Hence $\prod_{\beta \in A} Z_{\sigma(\alpha),\beta}$ is isomorphic to $Z_{\sigma(\alpha),\alpha}$ which is in turn isomorphic to $\prod_{\delta \in A} Z_{\delta,\alpha}$. The isomorphism in each case is the projection on the αth or $\sigma(\alpha)^{\mathrm{th}}$ component, respectively. Let $\eta'_\alpha : \prod_{\beta \in A} Z_{\sigma(\alpha),\beta} \to \prod_{\delta \in A} Z_{\delta,\alpha}$ be the isomorphism that maps the αth component to the $\sigma(\alpha)^{\mathrm{th}}$ component. Choose $\Phi_\alpha := b_\alpha^{-1} \circ \eta'_\alpha \circ a_{\sigma(\alpha)}$. ∎

Fixed points of automorphisms of ordered sets as in part 1 of Corollary 12.36 are easy to construct. For ordered sets as in part 2, the following lemma shows that fixed points can be constructed, too.

Lemma 12.37 (See [67], Lemma 3). *Let $P = \prod_{\alpha \in A} X_\alpha$ be a connected ordered set such that all $X_\alpha = X$ and X is product indecomposable. If each automorphism of X has a fixed point, then each automorphism of P has a fixed point.*

Proof. Let Φ be an automorphism of P. As in part 2 of Corollary 12.36, let σ be a permutation of A and let $\Phi_\alpha : X \to X$ be automorphisms of X such that $\Phi((x_\alpha)_{\alpha \in A}) = (\Phi_\alpha(x_{\sigma(\alpha)}))_{\alpha \in A}$. We construct a fixed point of Φ componentwise.

For $\alpha \in A$, call $\{\sigma^k(\alpha) : k \in \mathbb{Z}\}$ the *orbit* of α. A has two different types of elements, those with finite orbits and those with infinite orbits. Constructing a fixed point must take into account that, with the representation as above, Φ maps every component x_α of $(x_\alpha)_{\alpha \in A}$ to the $\sigma^{-1}(\alpha)^{\mathrm{th}}$ component.

If the orbit of $\alpha \in A$ has size n, let $y_\alpha \in X_\alpha$ be a fixed point of the composition $\Phi_\alpha \circ \Phi_{\sigma(\alpha)} \circ \Phi_{\sigma^2(\alpha)} \circ \cdots \circ \Phi_{\sigma^{n-1}(\alpha)}$. Let $y_{\sigma^k(\alpha)} := \Phi_{\sigma^k(\alpha)} \circ \cdots \circ \Phi_{\sigma^{n-1}(\alpha)}(y_\alpha)$. Then, for all $p \in P$ with $p_{\sigma^k(\alpha)} = y_{\sigma^k(\alpha)}$ for all $k \in \{0,\dots,n-1\}$, we have the following for all $k \in \{0,\dots,n-1\}$.

$$\pi_{\sigma^k(\alpha)}\Phi(p) = \Phi_{\sigma^k(\alpha)}(\pi_{\sigma^{k+1}(\alpha)}(p)) = \Phi_{\sigma^k(\alpha)}(y_{\sigma^{k+1}(\alpha)})$$
$$= \Phi_{\sigma^k(\alpha)}(\Phi_{\sigma^{k+1}(\alpha)} \circ \cdots \circ \Phi_{\sigma^{n-1}(\alpha)}(y_\alpha)) = y_{\sigma^k(\alpha)} = \pi_{\sigma^k(\alpha)}(p)$$

If the orbit of α is infinite, we argue as follows. Fix $y_\alpha \in X_\alpha$. Inductively (in both directions) we define $y_{\sigma^k(\alpha)} := \begin{cases} \Phi_{\sigma^{k-1}(\alpha)}^{-1}(y_{\sigma^{k-1}(\alpha)}) & \text{if } k > 0, \\ \Phi_{\sigma^k(\alpha)}(y_{\sigma^{k+1}(\alpha)}) & \text{if } k < 0. \end{cases}$ Then, for all $p \in P$ with $p_{\sigma^k(\alpha)} = y_{\sigma^k(\alpha)}$ for all $k \in \mathbb{Z}$, we have the following for all $k \in \mathbb{Z}$.

$$\pi_{\sigma^k(\alpha)}\Phi(p) = \Phi_{\sigma^k(\alpha)}(\pi_{\sigma^{k+1}(\alpha)}(p)) = \Phi_{\sigma^k(\alpha)}(y_{\sigma^{k+1}(\alpha)})$$

$$
= \begin{cases} \Phi_{\sigma^k(\alpha)}(\Phi^{-1}_{\sigma^k(\alpha)}(y_{\sigma^k(\alpha)})) & \text{if } k \geq 0, \\ \Phi_{\sigma^k(\alpha)}(\Phi_{\sigma^{k+1}(\alpha)}(y_{\sigma^{k+2}(\alpha)})) & \text{if } k < 0, \end{cases}
$$

$$
= \begin{cases} y_{\sigma^k(\alpha)} & \text{if } k \geq 0, \\ \Phi_{\sigma^k(\alpha)}(y_{\sigma^{k+1}(\alpha)}) & \text{if } k < 0, \end{cases}
$$

$$
= y_{\sigma^k(\alpha)} = \pi_{\sigma^k(\alpha)}(p).
$$

From the above, it is easy to construct a fixed point. ∎

The above now combines into an analogue of Theorem 12.17 for fixed points of automorphisms.

Theorem 12.38 (See [67], Theorem 4). *Let X and Y be connected ordered sets, each with a decomposition as product indecomposable sets. Then $X \times Y$ has a fixed point free automorphism iff at least one of X and Y has a fixed point free automorphism.*

Proof. The direction "⇐" is trivial. For "⇒," we prove the contrapositive. So assume that all automorphisms of X and Y have a fixed point and let Φ be an automorphism of $X \times Y$. Because X and Y are decomposable into product indecomposable sets, there also is a decomposition $X \times Y$ that is isomorphic to $\prod_{\alpha \in A} Z_\alpha^{k_\alpha}$, with the Z_α being product indecomposable, pairwise nonisomorphic and the k_α being cardinals. By Corollary 12.36, part 1, Φ is the product of automorphisms Φ_α of $Z_\alpha^{k_\alpha}$.

Because all automorphisms of X and Y have a fixed point, all automorphisms of the Z_α have a fixed point. By Lemma 12.37 this means that all automorphisms of the powers $Z_\alpha^{k_\alpha}$ have a fixed point. In particular, each Φ_α has a fixed point, which means that Φ has a fixed point. ∎

Exercises

12-25. (See [67], introduction.) To investigate Open Question 1.24 for products of disconnected ordered sets, let $n \cdot X$ be an ordered set with n components isomorphic to X.

   a. Prove that $n_1 \cdot X_1 + n_2 \cdot X_2 + \cdots + n_k \cdot X_k$ with $X_1, \ldots, X_k$ pairwise nonisomorphic has a fixed point free automorphism iff, for all i with $n_i = 1$, the set X_i has a fixed point free automorphism.

   b. Find two ordered sets P, Q such that all automorphisms of P and Q have a fixed point, but $P \times Q$ has a fixed point free automorphism.

   c. Find an ordered set P such that all automorphisms of P have a fixed point, but $P \times P$ has a fixed point free automorphism.

   d. Prove that, if P is finite with at most four components and all automorphisms of P have a fixed point, then all automorphisms of P^2 have a fixed point.

   e. Can 12-25d be improved?

12-26. Prove that if P is finite, connected and all automorphisms of P have a fixed point, then, for all finite Q, such that each automorphism of Q has a fixed point, every automorphism of $P \times Q$ has a fixed point.

12-27. Prove that, if P is finite with two components, neither of which has a fixed point free automorphism, then there is a finite Q such that each automorphism of Q has a fixed point, and $P \times Q$ has a fixed point free automorphism.

12-28. What is the smallest total number of components such that there are finite P, Q such that each automorphism of P and Q has a fixed point, and $P \times Q$ has a fixed point free automorphism?

12-29. Let C be a connected ordered set and let $\emptyset \neq A \subseteq C$. Prove that, if A is an up-set and a down-set, then $A = C$.

12-30. There is no Hashimoto theorem for disconnected ordered sets, see [123, 151].

　a. Find two factorizations of $P = 1 + 2 + 2^2 + 2^3 + 2^4 + 2^5$ such that no two factors are isomorphic and no factor can be represented nontrivially as a product.

　b. Find a characterization when a finite ordered set has a unique product decomposition.
　　　Hint. Identify each indecomposable factor of a component with a variable in a multivariate polynomial. Consider the factorizations of the polynomial.

12-31. Let $\Phi : P \times Q \to U \times V$, $\Phi_P : P \to U$, and $\Phi_Q : Q \to V$ be isomorphisms such that, for a fixed $(x, y) \in P \times Q$, the following is true.

　a. For all $p \in P$ we have $\Phi(p, y) = (\Phi_P(p), \Phi_Q(y))$.
　b. For all $q \in Q$ we have $\Phi(x, q) = (\Phi_P(x), \Phi_Q(q))$.

　　Prove that $\Phi(p, q) = (\Phi_P(p), \Phi_Q(q))$ for all $(p, q) \in P \times Q$.
　　Hint. Lemmas 12.31 and 12.34.

12-32. State and prove a version of Theorem 12.38 for infinite products.

12-33. More on the tensor product $\times$ of graphs.

　a. The product of connected graphs need not be connected: Show that the product of a 6-cycle C_6 with a complete graph K_2 on 2 vertices is a union of two 6-cycles.

　b. Find another factorization for a union of two 6-cycles, thus showing that there is no fully general Hashimoto theorem for the tensor product of graphs.
　　　Note. There is such a theorem for non-bipartite graphs, though, see [194, 195].

12.5　Arithmetic of Ordered Sets

By now, we have seen several binary operations that take two ordered sets and return a new set. Among them are the disjoint sum $P + Q$, the product $P \times Q$, and the exponentiation $P^Q = \text{Hom}(Q, P)$. Interestingly enough, there are many analogies with regular algebraic operations. Consider that the n-fold sum of P with itself is isomorphic to the product of P with an antichain with n elements. This means that the product of ordered sets really is an abbreviation of repeated summation of ordered sets. Similarly the exponentiation operation is an abbreviation of repeated multiplication: Indeed, the n-fold product of P with itself is isomorphic to the set of order-preserving maps from an antichain with n elements to P. Thus, with isomorphism taking the place of equality, we can define an arithmetic on ordered sets.

In this section, we will investigate how far the arithmetic properties of these operations on ordered sets can go. For the purposes of this arithmetic, we will not distinguish between isomorphic sets here. Nonetheless, it seemed worth pointing out which laws are equalities and which are isomorphisms. Our starting point are the "usual laws of algebra" for the operations mentioned above.

Proposition 12.39 (Arithmetic of ordered sets, see [21], p. 55, ff). *With $\approx$ denoting isomorphism of ordered sets, the following are true.*

$$X + Y = Y + X \qquad (X + Y) + Z = X + (Y + Z)$$

$$X \times Y \approx Y \times X \qquad (X \times Y) \times Z \approx X \times (Y \times Z)$$

$$X \times (Y + Z) = (X \times Y) + (X \times Z) \qquad (X + Y) \times Z = (X \times Z) + (Y \times Z)$$

$$X^{Y+Z} \approx X^Y \times X^Z \qquad (X \times Y)^Z \approx X^Z \times Y^Z$$

$$(X^Y)^Z \approx X^{Y \times Z}$$

Proof. Commutativity and associativity of sum and product are trivial.

We will only prove one distributive law. Clearly the underlying sets of the ordered sets $X \times (Y + Z)$ and $(X \times Y) + (X \times Z)$ are equal. Now consider (x_1, a_1) and (x_2, a_2) with $x_i \in X$ and $a_i \in Y \cup Z$. Then $(x_1, a_1) \leq_{X \times (Y+Z)} (x_2, a_2)$ iff $x_1 \leq_X x_2$ and $a_1 \leq_{Y+Z} a_2$ iff $x_1 \leq_X x_2$ and $a_1, a_2 \in A \in \{Y, Z\}$ and $a_1 \leq_A a_2$ iff $a_1, a_2 \in A \in \{Y, Z\}$ and $(x_1, a_1) \leq_{X \times A} (x_2, a_2)$, which is the case iff $(x_1, a_1) \leq_{X \times Y + X \times Z} (x_2, a_2)$. Hence the two underlying orders are equal, too.

The remaining proofs are very similar to the above, so we will only give the isomorphisms, leaving the details to the reader.

For $X^{Y+Z} \approx X^Y \times X^Z$, the isomorphism is $f \mapsto (f|_X, f|_Y)$. Its inverse is $(f, g) \mapsto f \cup g$.

The identity $(X \times Y)^Z \approx X^Z \times Y^Z$ is left as Exercise 12-34.

Finally, for $(X^Y)^Z \approx X^{Y \times Z}$, the isomorphism is $f \mapsto g(y, z) := f(z)(y)$ with inverse $g \mapsto f(z)(y) := g(y, z)$. ∎

The search for neutral elements for our arithmetic operations is simple, too. The empty set is the neutral element for addition and the singleton ordered set is the neutral element for multiplication. The singleton ordered set also has all the properties that the number 1 has for exponentiation.

Whenever we use finite sets with at least two elements, all our operations return larger sets. (Exercise 12-16b showed that this need not be the case for infinite ordered sets.) Thus it is impossible to define a group structure for ordered sets in this fashion. With inverse elements out of the question, the next natural candidates for laws of algebra are cancelation laws. You can prove cancelation laws for sums, including linear lexicographic sums, in Exercise 12-36. For products, we have the following result. For our purposes, its main scope are finite ordered sets, but it shows that some infinite ordered sets satisfy a cancelation law, too. In Exercise 12-38, you can prove the result for finite ordered sets with the connectivity hypotheses dropped.

Theorem 12.40. *Let* P, Q, R *be connected ordered sets, factorable into product indecomposable factors. Also, let* R *be factorable into finitely many indecomposable factors. If* $P \times R$ *is isomorphic to* $Q \times R$*, then* P *is isomorphic to* Q*.*

Proof. Factor both products into product indecomposable ordered sets. Say, the factorization of $P \times R$ is $\prod_{i \in I} X_i$ and the factorization of $Q \times R$ is $\prod_{j \in J} Y_j$. By Hashimoto's Theorem, we can assume that $I = J$ and that X_i is isomorphic to Y_i.

Let $i_1, \ldots, i_n$ be the indices of the factors of R in $\prod_{i \in I} X_i$. Then the product $\prod_{i \in I \setminus \{i_1, \ldots, i_n\}} X_i$ is isomorphic to P. Let $i'_1, \ldots, i'_n$ be the indices of the factors of R in $\prod_{i \in I} Y_i$. Then the product $\prod_{i \in I \setminus \{i'_1, \ldots, i'_n\}} Y_i$ is isomorphic to Q. By Hashimoto's Theorem again, we can assume without loss of generality that X_{i_k} is isomorphic to $Y_{i'_k}$ for $k \in \{1, \ldots, n\}$. Moreover, after further renumbering, we can assume that if $i_j = i'_j$, then $j = j'$.

This means that the map $f \,:\, I \setminus \{i_1, \ldots, i_n\} \;\rightarrow\; I \setminus \{i'_1, \ldots, i'_n\}$ defined by $f(i) := \begin{cases} i; & \text{if } i \notin \{i_1, \ldots, i_n\} \cup \{i'_1, \ldots, i'_n\}, \\ i_k; & \text{if } i = i'_k \notin \{i_1, \ldots, i_n\}, \end{cases}$ is a bijection between the sets $I \setminus \{i_1, \ldots, i_n\}$ and $I \setminus \{i'_1, \ldots, i'_n\}$ such that X_i is isomorphic to $Y_{f(i)}$. This shows that P and Q are isomorphic. ∎

The non-existence of an infinite factorization is essential, as the following example shows. This means that, although our ordered set arithmetic has no problems accommodating sums and products with arbitrarily many factors, it reaches its limitations in the cancelation laws.

Example 12.41. The condition that R is not factorable into infinitely many factors in Theorem 12.40 is needed: Let P, Q be two ordered sets and let A_ω be a countable antichain. Then $(Q \times P) \times P^{A_\omega}$ is isomorphic to $Q \times P^{A_\omega}$, but $Q \times P$ is not isomorphic to Q for any finite P, Q with at least two elements. □

This leaves the question if our exponentiation operation is "injective," leading to a problem by Birkhoff that was open for over 40 years before it was solved in the year 2000 (see [206]).

> **Formerly Open Question 12.42 (Birkhoff's cancelation problem. Solved by McKenzie in [206]).** *Let* P, Q, R *be (finite) ordered sets. Does the fact that* P^Q *is isomorphic to* R^Q *imply that* P *is isomorphic to* R*?*

Although the answer is negative if base or exponent is allowed to be infinite (see Examples 12.43 and 12.44), it was the finite case that baffled researchers for decades. We will present some examples and a partial result here. For the proof for all finite sets, see [206].

Example 12.43 (See [150], Example 4, p. 21). Let C be a four crown and define $A := L\{C_i \mid i \in \mathbb{N}\}$, where all C_i are pairwise disjoint four crowns and $\mathbb{N}$ carries the discrete order, that is, it is an antichain. Let $B := A + S$, where S is a singleton ordered set. Then A^C is isomorphic to B^C and A is not isomorphic to B.

Fig. 12.3 Ordered sets A and B that show that the cancelation problem has a negative solution for infinite exponents, even if the bases are finite.

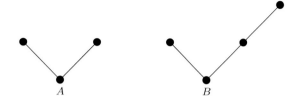

A B

To see this, note that both A^C and B^C have countably many isolated points: Indeed, every order-preserving map that maps C surjectively to any of the C_i is not comparable to any other order-preserving maps. Let H be the set of these maps. Moreover, for B^C, the map f that maps C into S is also not comparable to any other maps. The isomorphism is now made up of the identity between $A^C \setminus H$ and $B^C \setminus H \cup \{f\}$ and any bijection between H and $H \cup \{f\}$. ∎

Example 12.44 (See [150], Example 5, p.21). *Let A, B be the ordered sets in Figure 12.3. Let C be the direct product of infinitely many copies of the two-element chain $\mathbf{2}$. Then A^C is isomorphic to B^C, but A is not isomorphic to B.*

To see this, note that every map from C to A (or B, respectively) must map C into one arm of A (or B, respectively), because the range must have a largest element. Thus A^C is obtained from two copies of $\mathbf{2}^C$ with their bottom elements identified and B^C is obtained from a copy of $\mathbf{2}^C$ and a copy of $\mathbf{3}^C$ with their bottom elements identified ($\mathbf{3}$ naturally denotes the 3-element chain). However, $\mathbf{3}$ is isomorphic to $\mathbf{2}^2$. Thus, by Proposition 12.39, we have that $\mathbf{3}^C$ is isomorphic to $\mathbf{2}^{2 \times C}$, which is isomorphic to $\mathbf{2}^C$. ∎

With infinite bases or exponents not allowing cancelation of exponents, the question that remained was if it was possible to cancel the exponent if all the involved sets were finite. Special cases were proved over the years, such as, for example, that arbitrary linearly ordered exponents could be canceled (see [18]). However, the general proof remained elusive. All partial results required extra hypotheses on the exponent and/or the bases. Consider, for example, the following simple, yet fairly strong cancelation result.

Lemma 12.45. *Let D be an ordered set with an up-retraction r. Then, for all ordered sets P, we have that $r : P^D \to P^D$ defined by $R(f) := f \circ r$ is an up-retraction. Moreover $R[P^D]$ is isomorphic to $P^{r[D]}$.*

Proof. It is clear that R is an up-retraction. To see the "moreover" part, note that $f \mapsto f \circ r$ is a bijective map from $P^{r[D]}$ to $R[P^D]$ that preserves order both ways. Thus, by Proposition 1.15, it is an isomorphism. ∎

Proposition 12.46. *Let P, D, R be finite ordered sets such that the ordered set D is $\mathcal{I}$-dismantlable to a singleton and let P^D be isomorphic to R^D. Then $\mathcal{I} - \mathrm{core}(P)$ is isomorphic to $\mathcal{I} - \mathrm{core}(R)$.*

328 12 Sets $P^Q = \mathrm{Hom}(Q, P)$ and Products

Proof. Recall that, by Proposition 4.21, the $\mathcal{I}$-core and the $\mathcal{C}$-core of a finite ordered set are equal. By Lemma 12.45, we have that P^D can be $\mathcal{C}$-dismantled to a subset that is isomorphic to P. Thus P^D can be $\mathcal{C}$-dismantled to a subset that is isomorphic to $\mathcal{I} - \mathrm{core}(P)$. By the uniqueness of the $\mathcal{I}$-core, we infer that $\mathcal{I} - \mathrm{core}(P^D)$ is isomorphic to $\mathcal{I} - \mathrm{core}(P)$. Similarly $\mathcal{I} - \mathrm{core}(R^D)$ is isomorphic to $\mathcal{I} - \mathrm{core}(R)$, and the isomorphic ordered sets P^D and R^D naturally must have isomorphic $\mathcal{I}$-cores.

∎

Thus the above shows that if the bases have no irreducible points and the exponent is $\mathcal{I}$-dismantlable to a singleton, then the exponent can be canceled. This is a nice application of notions we introduced earlier, but not the solution of the cancelation problem. We will only state the solution here, as the proof is beyond the scope of this text.

Theorem 12.47 (For the proof, see [206]). *Let P, Q, R be finite ordered sets. If $P^R \approx Q^R$, then $P \approx Q$.*

Exercises

12-34. Prove that $(X \times Y)^Z \approx X^Z \times Y^Z$.

12-35. Show that, in general, $X \oplus Y \not\approx Y \oplus X$.

12-36. Let X, Y, Z be ordered sets. Prove each of the following.

    a. If Z has finitely many components and $X + Z \approx Y + Z$, then $X \approx Y$.

    b. If the complement of the comparability graph of Z has finitely many components and we have $X \oplus Z \approx Y \oplus Z$, then $X \approx Y$.

    c. Give examples that show 12-36a and 12-36b are false if the respective finiteness condition is dropped.

    d. Give an example to show that, if $Z \oplus X \approx Y \oplus Z$, then it is not necessary that $X \approx Y$.

12-37. Prove that, if $\mathrm{Hom}(P, P) = \mathrm{Hom}(Q, Q)$, then $P = Q$ or $P = Q^d$.

12-38. (See [150], Theorem 5.4.) Let P, Q, R be finite ordered sets. Prove that, if $P \times R$ is isomorphic to $Q \times R$, then P is isomorphic to Q.

    Hint. If no two connected components of R have the same number of elements, this can be proved via Theorem 12.40. For the general case, identify the indecomposable factors of the components of P, Q, and R with variables. Identify P, Q and R with multivariable polynomials in these variables with integer coefficients and use the cancelation property in the ring of multivariable polynomials.

Remarks and Open Problems

1. For more on the structure theory for ordered sets, see [73].
2. For more on the arithmetic of ordered sets, see [150].
3. Little is known about the reconstruction problems for products and homomorphism sets. It is not even known if products and homomorphism sets are recognizable from the deck.

4. For what finite *connected* ordered sets P, Q does P^Q have the fixed point property? Some first results are in [281].

5. I conjecture that, for a finite ordered set P with the fixed point property and A a countable antichain, the ordered set $P^A = \prod_{n \in \mathbb{N}} P_n$ (with all $P_n = P$) has the fixed point property. Some first results are in [281]. (And Proposition 12.25 shows that all finite retracts of these sets have the fixed point property.)

6. Is there an analogue of Theorem 12.10 for other notions of dimension, such as, say, interval dimension?

7. For refinements of Theorem 12.10, see Chapter 10 in [311].

8. Kukieła and the author have found an example of an infinite ordered set P of height 2 so that a certain 2-antichain is not contained in any retract of P except P itself (see [176]). This is a bit surprising, as, in a chain-complete ordered set, every antichain with 2 elements that consists of maximal or minimal elements is contained in an isometric spanning fence connecting the two elements, and that fence is a retract.

9. Note that the topological fixed point property is not productive (see [38]). Is there a way to get two infinite ordered sets P, Q with the fixed point property such that $P \times Q$ does not have the fixed point property through topology? One possible way could be through infinite truncated lattices that arise via infinite triangulations of the manifolds involved in the topological counterexamples.

10. On the other hand, Roddy proved in [256] that, if P, Q are ordered sets with the fixed point property and P is of width ≤ 3 (no further hypotheses), then $P \times Q$ has the fixed point property. Is this a start towards proving that the fixed point property is "productive"?

11. One crucial challenge in trying to prove that the fixed point property is productive for infinite ordered sets is that we know very little about infinite ordered sets with the fixed point property. Indeed, the ordered set $\prod_{n \in \mathbb{N}} P_n$ in Remark 12.23 as well as the conjecture in 5 in this list were motivated by the desire to find nontrivial infinite ordered sets that have the fixed point property. It appears that all examples of infinite ordered sets with the fixed point property have the fixed point property for some "finitary reason." That is, the proof of the fixed point property focuses on some finitary structure or procedure.

    Thus, finding infinite ordered sets that have the fixed point property for some "non-finitary" reason is an interesting challenge.

12. A property closely related to the fixed point property for products is Duffus and Sauer's **strong fixed point property** (see [77]). The idea for the strong fixed point property was first used at the end of [318] to prove that the product of an ordered set with the relational fixed point property with a set with the fixed point property again has the fixed point property.

    An ordered set has the strong fixed point property iff there is an order-preserving function $\varphi : \mathrm{End}(P) \to P$ such that $f(\varphi(f)) = \varphi(f)$. That is, φ selects a fixed point for each order-preserving self-map and it does so in an order-preserving fashion. In [77], it is proved that the product of two ordered sets with the strong fixed point property again has the strong fixed point property. Moreover the product of an ordered set with the strong fixed point property and an ordered set with the fixed point property again has the fixed point property.

It was hoped that the fixed point property and the strong fixed point property might be equivalent, thus settling the product problem. Unfortunately this is not the case, as is shown in [221].

The above implies that, if there is an infinite counterexample to the product conjecture, then both of the factors must have the fixed point property but not the strong fixed point property. The observation that such sets were hard to find in the finite case is another indication how difficult an infinite counterexample to the product conjecture (if it exists) may be.

In [258], Roddy and the author generalized the strong fixed point property in an attempt to find a stronger "productive" property that may be equivalent to the fixed point property. Although this could be possible, no such property has materialized yet.

13. Another approach towards the product problem was through projectivity as introduced in [50]. An ordered set is called **projective** iff the only order-preserving maps $f : P \times P \to P$ such that $f(x, x) = x$ for all $x \in P$ are the projections. In [50], it is shown that *if* all minimal automorphic finite ordered sets were projective (which is conjectured in [50]), then the product of two finite ordered sets with the fixed point property again has the fixed point property. A conjecture stronger than projectivity of minimal automorphic sets was refuted in [125]. The author does not know of a satisfactory generalization of the approach in [50] to fixed points in products of infinite ordered sets, and Roddy of course solved the product problem in the finite case. Nonetheless, as, for example, [126, 180] show, projectivity is a property of independent interest.

In a way, questions related to projectivity stand representative of a large algebraic part of the theory of ordered sets that is beyond the scope of this text. The main ideas of representation theory go back to [73], where the representation theory is developed and some interesting questions are posed. I cannot comment on which of the problems in [73] are solved and what the exact status of this part of the theory is. For some later developments see, for example, [330, 331].

14. Define the **distortion** of an ordered set P as $D(P) := \max\limits_{f \in \mathrm{End}(P)} \min\limits_{p \in P} \mathrm{dist}(p, f(p))$ (see [217]). Nowakowski conjectures that $D(P \times Q) = \max\{D(P), D(Q)\}$. This conjecture has the product conjecture for the fixed point property as a special case. How much can be proved about the distortion parameter? The only reference I know of is [128].

15. Every finite ordered set is a product of product indecomposable ordered sets. (Simply decompose until this is no longer possible.) However, not every ordered set is a product of product indecomposable ordered sets.

Consider the following counterexample. An atom in a lattice is an upper cover of the smallest element. There is, up to isomorphism, exactly one countable Boolean algebra C that has no atoms. This is actually a pretty deep result. It depends on the Stone representation theorem for Boolean algebras and results leading up to it (for example, see [56], Theorem 10.8 and Propositions 10.6, 10.7), Urysohn's metrization theorem (see [325], Theorem 23.1), and on the fact that every perfect, compact, totally disconnected metric space is homeomorphic to the Cantor set (see [325], Corollary 30.4) together with Example 10.21(3) of [56].

It is easy to show that any factor of C must be atomless, too, and thus, in particular, countable. Moreover, every factor of a Boolean algebra is again a Boolean algebra. Thus every factor of C is a countable atomless Boolean algebra and hence isomorphic to C. Finally C is not product indecomposable, because $C \times C$ is isomorphic to C (because $C \times C$, too, is a countable atomless Boolean algebra).

16. In Exercise 12-20, we have considered two classes of ordered sets (chain-complete without infinite antichains and complete lattices) for which the fixed point property is productive. For both these classes, we proved that the set $\mathrm{Retr}(P)$ is a retract of $\mathrm{End}(P)$. By Exercise 12-21, part of the proof can be used to prove Theorem 12.17 without using the Li–Milner Theorem.

Is there a deeper connection between $\mathrm{Retr}(P)$ being a retract of $\mathrm{End}(P)$ and productivity of the fixed point property? Specifically, if P, Q are ordered sets with the fixed point property and $\mathrm{Retr}(P)$ is a retract of $\mathrm{End}(P)$, does it follow that $P \times Q$ has the fixed point property?

17. For what classes of ordered sets P is $\mathrm{Retr}(P)$ a retract of $\mathrm{End}(P)$?

18. Is there a proof for the cancelation of $\mathcal{I}$-dismantlable exponents that is not too much longer than the proof for Proposition 12.46? By Theorem 12.47, the result is true, but shorter proofs could provide further insights.

19. There also is a cancelation law for bases. It says that, if A, B, C are finite ordered sets and A is not an antichain, then $A^B \approx A^C$ implies $B \approx C$. For a proof see [72]. Clearly it is needed that A is not an antichain, because otherwise A^B is merely an $|A|c(B)$-element antichain, where $c(B)$ denotes the number of components of B.

20. Brouwer's fixed point theorem says that every continuous self-map of $[0,1]^d$ must have a fixed point. Let $n \in \mathbb{N}$. Subdivide the cube $[0,1]^d$ into n^d subcubes $\prod_{k=1}^{d} \left[\frac{x_k-1}{n}, \frac{x_k}{n}\right]$, where $x_k \in \{1, \ldots, n\}$. Consider the cubic complex $C_n^d := \left\{ \prod_{k=1}^{d} \left[\frac{x_k}{n}, \frac{y_k}{n}\right] : y_k \in \{1, \ldots, n\}, x_k \in \{y_k - 1, y_k\} \text{ or } x_k = y_k = 0 \right\}$ as an ordered set ordered by inclusion. Note that $C_n^d = \prod_{k=1}^{d} C_n^1$, where the product is a product of ordered sets. The set C_n^1 is a fence with $n+1$ minimal and n maximal elements, so C_n^d is a product of fences, which is a very nice ordered set.

If $f : [0,1]^d \to [0,1]^d$ is a continuous map, we can define $F_n^f : C_n^d \to \mathcal{P}(C_n^d)$ by $F_n^f(c) := \{k \in C_n^d : f[c] \cap k \neq \emptyset\}$. This is a multivalued function on C_n^d. Moreover $x \subseteq y$ implies that, for each $a \in F_n^f(x)$, we have $a \in F_n^f(y)$, too. Thus F_n^f is almost an isotone relation.

Suppose we could prove that any such function F_n^f (or some similarly defined function) has a fixed point in C_n^d. Then, for each $n \in \mathbb{N}$, there is an $x_n \in [0,1]^d$

with $|f(x_n) - x_n| \leq 2\frac{\sqrt{d}}{n}$. By compactness of $[0, 1]^n$, there is a subsequence $\{x_{n_k}\}_{k \in \mathbb{N}}$ that converges to a point x, which must be a fixed point of f. So this argument *would* prove Brouwer's theorem, if we can prove that every F_n^f has a fixed point.

For $d = 1$, this is indeed possible, but the proof relies more on the linear structure of $[0, 1]$ than on order theory. For higher dimensions, any multivalued functions I could come up with were lacking some feature. Hence, even though C_n^d has very nice order-theoretical structure, a proof of Brouwer's theorem in this fashion was not possible so far.

What kinds of fixed point results for multivalued maps can be proved for products of fences? Could any of them be used to prove Brouwer's theorem?

A combinatorial proof of Brouwer's fixed point theorem that relies on graph theory and Sperner's lemma can be found in [320].

Chapter 13
Enumeration of Ordered Sets

In any discrete setting, one of the most natural questions to ask is "How many of these objects are there?" The automorphism problem (Open Question 2.14) and Dedekind's problem (Open Question 2.30) are questions like that. A counting question can be motivated by pure curiosity or, as the Kelly Lemma (see Proposition 1.40) in reconstruction shows, it can be a useful lemma for proving further results. The two most natural counting questions for ordered sets are still unanswered and it appears that both are quite hard.

> **Open Question 13.1.** *Given an n-element ground set, how many different orders can be imposed on the set?*

Because different orders can be isomorphic, it is natural to ask the following.

> **Open Question 13.2.** *Given an n-element ground set, how many nonisomorphic orders can be imposed on the set?*

In this chapter, we will investigate some techniques that are motivated by the above questions. Our approach towards enumeration will feature two natural steps that can be taken when we cannot solve an enumeration problem exactly: We can enumerate subclasses (see Section 13.2) and we can construct expressions that are asymptotically equal to the unknown numbers (see Section 13.3). Needed tools will be introduced on the spot.

13.1 Graded Ordered Sets

Graded ordered sets can be enumerated exactly as well as asymptotically. They will feature prominently in this chapter. We will introduce them here and prove enumeration results in the following two sections.

© Springer International Publishing 2016
B. Schröder, *Ordered Sets*, DOI 10.1007/978-3-319-29788-0_13

Definition 13.3. *A **graded ordered set** is an ordered set P for which there is an order-preserving function $g_P : P \to \mathbb{N}$ such that, if p is a lower cover of q in P, then $g_P(q) - g_P(p) = 1$. The function g_P is called the **grading function** and the sets $g_P^{-1}(k) \subseteq P$ are the **grade levels** of P.*

To have a common reference point, we will always assume that every connected component of P contains a point b such that $g_P(b) = 1$ (normalization condition).

Note that the rank function is order-preserving from P into $\mathbb{N}$, too. It is tempting to hope that every rank function is a grading function. Figure 2.1b) shows that this is not the case, because this ordered set is not graded. (You can prove this directly, or by first solving Exercise 13-1.) Even knowing this, it is still tempting to hope that, if an ordered set is graded, then the rank function plus 1 will be a grading function. You can show in Exercise 13-3c that even this is not the case. These examples should serve as a caution that it is easy to read too much or too little into an ordered set when it comes to grading functions. One nice result is that, with the normalization as above, if a grading function exists, then it must be unique.

Proposition 13.4. *Every graded ordered set P has at most one grading function.*

Proof. Suppose for a contradiction that P has two functions g_P and g'_P as in Definition 13.3. If $g_P \neq g'_P$, then there is a $p \in P$ such that $g_P(p) \neq g'_P(p)$. By Definition 13.3, p cannot be an isolated point. Let $b \in P$ be such that $g_P(b) = 1$ and such that p and b are in the same connected component of P. By Exercise 13-2, we have that $g'_P(b) = 1$, too. Moreover, there is a sequence $b = p_1, \ldots, p_k = p$ such that, for all $i \in \{1, \ldots, k-1\}$, p_{i+1} is a lower or an upper cover of p_i. Let u be the number of times p_{i+1} is an upper cover of p_i and let l be the number of times p_{i+1} is a lower cover of p_i. By definition of g_P and g'_P we have that

$$u - l = \sum_{i=1}^{k-1}[g_P(p_{i+1}) - g_P(p_i)] = g_P(p_k) - g_P(p_1) = g_P(p) - g_P(b)$$

$$\neq g'_P(p) - g'_P(b) = g'_P(p_k) - g'_P(p_1) = \sum_{i=1}^{k-1}[g'_P(p_{i+1}) - g'_P(p_i)]$$

$$= u - l,$$

a contradiction. Thus we must have that $g_P = g'_P$.   ∎

Exercises

13-1. Prove that a finite ordered set is graded iff, for any sequence of points $x = x_0, x_1, \ldots, x_n = x$ such that, for all $i \in \{1, \ldots, n\}$, we have that x_{i-1} is adjacent to x_i, we have the equality $|\{i : x_{i-1} \prec x_i\}| = |\{i : x_i \prec x_{i-1}\}|$.

   Note. This is a discrete analogue of the analytical result that a vector field is a gradient field iff line integrals over closed curves vanish.

13-2. Let P be a graded ordered set and let $g_P : P \to \mathbb{N}$ be a grading function. Prove that $g_P(x) = 1$ iff there is no sequence $x = x_1, \ldots, x_k$ so that, for all $i \in \{1, \ldots, k-1\}$, x_{i+1} is a lower or an upper cover of x_i and x_{i+1} is more often a lower of x_i than an upper cover of x_i.

13-3. Chain conditions related to graded ordered sets.

   a. Prove that a finite ordered set that has a largest and a smallest element is graded iff any two maximal chains have the same length.
   b. Prove that every ordered set that satisfies the condition in part 13-3a is graded by the function that maps each element to its rank plus 1.
   c. Give an example of a graded ordered set for which the function that maps each element to its rank plus 1 is *not* a grading function.
   d. Give an example of an ordered set that is graded by the function that maps each element to its rank plus 1 and which does not satisfy the chain condition in part 13-3a.

13-4. The ordered set P satisfies the **Jordan–Dedekind chain condition** iff, for all $a < b$ in P, any two maximal chains between a and b have the same length.

   a. Prove that, in finite ordered sets with a largest and a smallest element, the Jordan–Dedekind chain condition is equivalent to the condition in Exercise 13-3a.
   b. Construct a nongraded ordered set that satisfies the Jordan–Dedekind chain condition.

13-5. Recognizability of conditions related to graded sets.

   a. Prove that ordered sets that satisfy the Jordan–Dedekind chain condition are recognizable.
   b. Prove that ordered sets for which the function that maps each element to its rank plus 1 is a grading function are recognizable.
   c. Prove that graded ordered sets are recognizable.

13-6. Let P, Q be ordered sets and let P be graded by g_P. Prove that $f : P \to Q$ is order-preserving iff, for all $k \in g_P[P]$, the restriction $f|_{g_P^{-1}[\{k,k+1\}]}$ of f to "grade levels k and $k+1$" is order-preserving.

13-7. Let P and Q be graded ordered sets. Let $|P| = |Q| = n$ and let the width of P and Q be bounded by w. Prove that isomorphism of P and Q can be checked in $O(w!^4 n)$ time and $O(w!^2 n)$ space.

13-8. Let P be a graded ordered set with $|P| = n$ and let the width of P be bounded by w. Prove that it can be determined in $O(n^{2w+1}w^2)$ time and $O(n^{2w+1})$ space whether P has a fixed point free order-preserving self-map. Proceed as follows.
As stipulated in Definition 13.3, we have $\min g_P[P] = 1$. Let $h := \max g_P[P]$. For all $k \in \{1, \ldots, h\}$, let $\{p_1^k, \ldots, p_{w_k}^k\} = g_P^{-1}(k)$. Define the constraint satisfaction problem $FPFG(P)$ as follows.

- For $k = 1, \ldots, h$, the variable x_k is a stand-in for the w_k-tuple $(p_1^k, \ldots, p_{w_k}^k)$.
- For $k = 1, \ldots, h$, let $D_k := \prod_{i=1}^{w_k} \{y \in P : y \not\geq p_i^k\}$.
- The constraints are the following.

   – No explicit unary constraints. (The constraint $p \not\geq f(p)$ is already built into the way the value domains are set up.)
   – A binary constraint $C_{i(i+1)}$ for every set of variables $\{x_i, x_{i+1}\}$: The pair of vertices $((y_1^i, \ldots, y_{w_i}^i), (y_1^{i+1}, \ldots, y_{w_{i+1}}^{i+1}))$ is in $C_{i(i+1)}$ iff the map

   $$p_1^i \mapsto y_1^i, \ldots, p_{w_i}^i \mapsto y_{w_i}^i, p_1^{i+1} \mapsto y_1^{i+1}, \ldots, p_{w_{i+1}}^{i+1} \mapsto y_{w_{i+1}}^{i+1}$$

   is order-preserving.

   a. Prove that P has a fixed point free order-preserving self-map iff $FPFG(P)$ has a solution.
     Hint. Use Exercise 13-6.
   b. Prove that $(1, 1)$-consistency can be enforced on $FPFG(P)$ by only considering pairs of
     variables x_i and x_{i+1}.
   c. Prove that, after enforcing $(1, 1)$-consistency, either the resulting network is empty or
     backtracking with the natural variable ordering $x_1, \ldots, x_h$ will produce a solution without
     ever having to backtrack.
   d. Count the required steps to obtain the run-time bound. If your algorithm takes longer
     than the stated bound, think about refining the algorithm.
   e. Assess the storage requirements to obtain the space bound.

13-9. Let P be a finite ordered set and let P_k be the set of elements of rank k. The ordered set P_{ps}
is obtained as follows.

- Replace every adjacency of an element $p_k \in P_k$ with an element $p_{k+j} \in P_{k+j}$ with
  $j \geq 2$ with a chain (a "string of pearls") $p_k < p_{k+1} < \cdots < p_{k+j-1} < p_{k+j}$, where the
  $p_{k+1}, \ldots, p_{k+j-1}$ are new elements to be added to P.
- Add only the indicated comparabilities plus those forced by transitivity.

Call P_{ps} the **pearl-strung version** of P.
We define the **edge-width** e of an ordered set P to be the width of the pearl-strung
version P_{ps}.
Prove that for each $e > 0$ there is a polynomial p_e and an algorithm that, for ordered sets
of edge-width $\leq e$, decides in $\leq p_e(|P|)$ steps if P has a fixed point free order-preserving
self-map.

13.2 The Number of Graded Ordered Sets

Throughout, we will use the set $\{1, \ldots, n\}$ as the ground set that carries all the orders
involved. To abbreviate notation, we will denote the set $\{1, \ldots, n\}$ as $[n]$.

Definition 13.5. *Let $G_{n,k}$ be the set of graded ordered sets with ground set $[n]$ and
up to k grade levels. Also let $g_{n,k} := |G_{n,k}|$.*

By the above definition, $g_{n,n}$ is the number of all graded ordered sets with ground
set $[n]$, $g_{n,1} = 1$, $g_{n,k} \leq g_{n,k+1}$ for all k, and $g_{n,k} = g_{n,n}$ for $k \geq n$. The natural
way to build a graded ordered set is to assign each point a grade level and then
to insert adjacencies only between points in adjacent grade levels. Although this
process can generate the same ordered set in different ways, the number of ordered
sets generated this way will be important in the future.

Definition 13.6. *Define $C_{n,k}$ as the set of $(k + 1)$-tuples $(X_1, \ldots, X_k, \leq)$ obtained
as follows. $X_1, \ldots, X_k$ are k disjoint, not necessarily nonempty subsets of $[n]$ such
that $\bigcup_{i=1}^{k} X_k = [n]$. $\leq$ is an order on $[n]$ such that, for $j \in \{1, \ldots, k - 1\}$, the only
adjacencies in $\leq$ are between elements of X_j and X_{j+1}. We let $c_{n,k} := |C_{n,k}|$.*

Interestingly enough, it is now possible to compute the $c_{n,k}$ directly and to obtain
from them the numbers $g_{n,k}$. This is because every set in $C_{n,k}$ can be envisioned as
built in two ways: On one hand, there is the straightforward way described above,
and, on the other hand, it can be built as a certain union of graded ordered sets as
described in the proof below.

Lemma 13.7 (See [162]).   *Let* $\begin{pmatrix} n \\ n_1,\ldots,n_k \end{pmatrix} = \dfrac{n!}{n_1!n_2!\cdots n_k!}$ *be the multinomial coefficient. Then*

$$c_{n,k} = \sum_{\substack{n_1+n_2+\cdots+n_k=n \\ n_i \geq 0}} \begin{pmatrix} n \\ n_1,\ldots,n_k \end{pmatrix} 2^{n_1 n_2 + n_2 n_3 + \cdots + n_{k-1}n_k}$$

$$= \sum_{\substack{s_1+s_2+\cdots+s_k=n \\ s_i \geq 0}} \begin{pmatrix} n \\ s_1,\ldots,s_k \end{pmatrix} g_{s_1,1} g_{s_2,2} \cdots g_{s_k,k}.$$

Proof. To prove that the first sum is equal to $c_{n,k}$, we proceed as follows. Note that the sizes $n_1,\ldots,n_k$ of the sets $X_1,\ldots,X_k$ in the partition of $[n]$ are nonnegative numbers that must add up to n. Hence $C_{n,k}$ is the pairwise disjoint union

$$C_{n,k} = \bigcup_{\substack{n_1+n_2+\cdots+n_k=n \\ n_i \geq 0}} \{(X_1,\ldots,X_k,\leq) : (\forall 1 \leq j \leq k)|X_j| = n_j\}.$$

This means that we are done if we can find the sizes of the sets in the above union. For any given sequence of sizes $n_1,\ldots,n_k$, the multinomial coefficient $\begin{pmatrix} n \\ n_1,\ldots,n_k \end{pmatrix}$ gives the number of ways to assign the elements of $[n]$ to disjoint sets $X_1,\ldots,X_k$ with $|X_j| = n_j$. For every such partition of $[n]$, there are $2^{n_{j-1}n_j}$ ways to insert comparabilities between X_{j-1} and X_j. Because all combinations are possible, we have

$$\left| \{(X_1,\ldots,X_k,\leq) : (\forall 1 \leq j \leq k)|X_j| = n_j\} \right|$$

$$= \begin{pmatrix} n \\ n_1,\ldots,n_k \end{pmatrix} 2^{n_1 n_2 + n_2 n_3 + \cdots + n_{k-1}n_k},$$

which proves the first equality.

To prove that the second sum is equal to $c_{n,k}$, we proceed as follows. For each $(X_1,\ldots,X_k,\leq)$ in $C_{n,k}$, partition $[n]$ as follows. Let S_k be the set of all numbers m that, in $([n],\leq)$, are connected to an element of X_1. Let $S'_j := [n] \setminus \bigcup_{i=j+1}^{k} S_i$. Let S_j be the set of all numbers m that are connected to an element of X_{k-j+1} in $\left(S'_j, \leq |_{S'_j \times S'_j} \right)$. Then each $(S_j, \leq |_{S_j \times S_j})$ is a graded ordered set with at most j grade levels and ground set S_j. This means that $C_{n,k}$ is the following pairwise disjoint union.

$$C_{n,k} = \bigcup_{\substack{s_1 + s_2 + \cdots + s_k = n \\ s_i \geq 0}} \left\{ (X_1, \ldots, X_k, \leq) : [n] = \bigcup_{j=1}^{k} S_j; \right.$$

$$|S_i| = s_i; \ S_i \cap S_j = \emptyset \text{ for } i \neq j; S_j \subseteq \bigcup_{i=k-j+1}^{k} X_i;$$

if $S_j \neq \emptyset$, then $S_j \cap X_{k-j+1} \neq \emptyset$;

$\leq |_{S_j \times S_j}$ is a graded order with at most j grade levels;

elements of distinct S_i are not $\leq$ −adjacent;

$$X_i = \bigcup_{j=1}^{i} \left\{ m \in S_j : g_{S_j}(m) = i - j + 1 \right\} \right\}.$$

Again, we need to compute the sizes of the sets within the disjoint union. For any given sequence of sizes $s_1, \ldots, s_k$ there are $\binom{n}{s_1, \ldots, s_k}$ ways to assign the elements of $[n]$ to disjoint sets $S_1, \ldots, S_k$ with $|S_j| = s_j$. For each S_j, there are $g_{s_j,j}$ ways to order S_j in the desired fashion. This proves the second equality. ∎

Lemma 13.7 provides an infinite set of nonlinear equations with which we can compute the $g_{n,k}$ from the $c_{n,k}$. How do you solve such a system of equations? Although there are no techniques that work in general, in this case, the algebra of formal power series will allow us to solve the problem. Generating functions are central for this type of argument.

Definition 13.8. *Let $\{a_n\}_{n=0}^{\infty}$ be a sequence of numbers. Then the **generating function** of $\{a_n\}_{n=0}^{\infty}$ is the formal power series $\sum_{n=0}^{\infty} \frac{a_n}{n!} x^n$.*

Essentially the generating function is a formal power series (the radius of convergence can be and often is zero) such that the nth formal derivative at $x = 0$ is a_n. If the equations we try to solve have the right pattern, the generating function can become a very powerful tool as we will see shortly. First, another lemma.

Lemma 13.9. *If $T(x) = \sum_{n=0}^{\infty} t_n x^n$ and $t_0 = 1$, then*

$$\frac{1}{T(x)} = 1 + \sum_{n=1}^{\infty} \left[\sum_{\substack{n_1 + n_2 + \cdots + n_l = n \\ l > 0, \ n_i > 0}} (-1)^l t_{n_1} t_{n_2} \cdots t_{n_l} \right] x^n.$$

Proof. Clearly the zeroth coefficient of the formal product of the formal power series for $T(x)$ and $\frac{1}{T(X)}$ is 1. For $n \geq 1$, we obtain that the nth coefficient is

$$
t_n + \sum_{k=0}^{n-1} t_k \left[\sum_{\substack{n_1 + n_2 + \cdots + n_l = n - k \\ l > 0, \ n_i > 0}} (-1)^l t_{n_1} t_{n_2} \cdots t_{n_l} \right]
$$

$$
= t_n + \sum_{\substack{n_1 + n_2 + \cdots + n_l = n \\ l > 0, \ n_i > 0}} (-1)^l t_{n_1} t_{n_2} \cdots t_{n_l}
$$

$$
+ \sum_{\substack{n_1 + n_2 + \cdots + n_l + n_{l+1} = n \\ l > 0, \ n_i > 0}} (-1)^l t_{n_1} t_{n_2} \cdots t_{n_l} t_{n_{l+1}} = 0. \qquad \blacksquare
$$

We now define the following generating functions.

Definition 13.10 (See [162]).   *Let* $C_k(x) := \sum_{n=0}^{\infty} \frac{c_{n,k}}{n!} x^n$, $G_k(x) := \sum_{n=0}^{\infty} \frac{g_{n,k}}{n!} x^n$, *and*

$$
D_k(x) := \frac{1}{C_k(x)} =: \sum_{n=0}^{\infty} \frac{d_{n,k}}{n!} x^n.
$$

With these generating functions, we can finally prove enumeration formulas for graded sets. The number of graded ordered sets with n elements is obtained as $g_{n,n}$.

Theorem 13.11. *Klarner's enumeration of graded orders* (see [162]). *For any natural numbers* $n, k \geq 2$, *we have the following*

$$
d_{n,k} = \sum_{\substack{n_1 + n_2 + \cdots + n_l = n \\ l > 0, \ n_i > 0}} (-1)^l \binom{n}{n_1, \ldots, n_l} c_{n_1,k} c_{n_2,k} \cdots c_{n_l,k}
$$

$$
g_{n,k} = \sum_{j=0}^{n} \binom{n}{j} c_{j,k} d_{n-j,k-1}
$$

Proof. By Lemma 13.7, we have that $C_k(x) = G_1(x) G_2(x) \cdots G_k(x)$. (Simply notice that, in a product of formal power series, the nth coefficient is the sum of the products of all coefficients whose powers add up to n.) Therefore $G_1(x) = C_1(x)$ and, for $k > 1$, we have $G_k(x) = \frac{C_k(x)}{C_{k-1}(x)} = C_k(x) \frac{1}{C_{k-1}(x)}$.

The formula for the $d_{n,k}$ is a direct consequence of Lemma 13.9. The formula for the $g_{n,k}$ follows from the formula for the nth coefficient of a product of formal power series.   ∎

Exercises

13-10. The multinomial theorem.

    a. Prove the multinomial theorem, that is, prove that, if $a_1, \ldots, a_k \in \mathbb{R}$ and $n \in \mathbb{N}$, then

$$(a_1 + \cdots + a_k)^n = \sum_{n_1 + \cdots + n_k = n, n_i \geq 0} \binom{n}{n_1, \ldots, n_k} a_1^{n_1} \cdots a_k^{n_k},$$

      where the $\binom{n}{n_1, \ldots, n_k} := \dfrac{n!}{n_1! \cdots n_k!}$ are multinomial coefficients.

    b. Prove that any multinomial coefficient $\binom{n}{n_1, n_2, \ldots, n_k}$ is bounded by k^n.

13.3   The Asymptotic Number of Graded Ordered Sets

When there is no exact enumeration formula for a quantity, such as the number of orders that can be placed on $[n]$, it is natural to ask how large the quantity is for large values of a given parameter, such as the size n here. This question is also natural when a precise, but complex, formula, such as Theorem 13.11, is available. Such approximations are the subject of asymptotic enumeration, where we try to find value to which a quantity is asymptotically equal.

Definition 13.12. *Two sequences $\{a_n\}_{n \in \mathbb{N}}$ and $\{b_n\}_{n \in \mathbb{N}}$ of real numbers are called* **asymptotically equal** *(in symbols $a_n \sim_{\text{as}} b_n$) iff* $\lim\limits_{n \to \infty} \dfrac{a_n}{b_n} = 1$.

The asymptotic value of the number of ordered sets with ground set $[n]$ will be provided in Theorem 13.23. A standard approach to finding asymptotic values for the size of a class of objects is to first establish an asymptotic value for a large subclass and then show that all other subclasses are small in comparison. This approach is taken to prove Theorem 13.23 in [165, 166]. However this proof is very technical.[1] Thus we will merely exemplify the described approach by proving an asymptotic formula for the number of graded ordered sets in Theorem 13.13. Asymptotically, the graded ordered sets are the largest class of ordered sets. The gap we leave is that we will not show that the class of nongraded ordered sets is

---

[1] Indeed, one of the authors of [165, 166] considers it "horrible."

small in comparison. For the proof of Theorem 13.23, consider [57, 165, 166]. Open Questions 5 and 6 at the end of this chapter are asking for simpler ways to prove the result.

Theorem 13.13. *The number of graded ordered sets with ground set $[n]$ is asymptotically equal to $2^{\frac{n^2}{4}+\frac{3}{2}n+O(\sqrt{n})}$.*

The estimate that we will prove is not as exact as it could be, because I did not think the technical details needed for the exact bound would illuminate the subject any more. We will spend the rest of this section proving Theorem 13.13 in a series of lemmas. All of these lemmas, except for the first one, are analytical in nature. This is not surprising, because we are dealing with an analytical notion.

Convention 13.14. *For the remainder of this section, inequalities that do not hold for all n are inequalities that hold for large n, that is, for all n beyond a certain N_0. Moreover, divisions will be performed regardless of whether the result is an integer or not. The small perturbations effected by rounding to integer values have no influence on the asymptotic results. Thus, to keep the presentation uncluttered, these parts of the computation are suppressed.*

In the above fashion, notation will be a little more economical. The following lemma is the main combinatorial tool to obtain the estimate. The upper bound derived here will generically be called "the upper bound" in what follows.

Lemma 13.15. *The number of graded ordered sets with exactly k nonempty grade levels is bounded below by*

$$\sum_{\substack{n_1 + n_2 + \cdots + n_k = n \\ n_i > 0}} \binom{n}{n_1, \ldots, n_k} \left(2^{n_1} - n\right)^{n_2} \left(2^{n_2} - n\right)^{n_3} \cdots \left(2^{n_{k-1}} - n\right)^{n_k}$$

and it is bounded above by

$$\sum_{\substack{n_1 + n_2 + \cdots + n_k = n \\ n_i > 0}} \binom{n}{n_1, \ldots, n_k} 2^{n_1 n_2 + n_2 n_3 + \cdots n_{k-1} n_k}.$$

Proof. To construct all graded ordered sets with exactly k grade levels and ground set $[n]$ we proceed as follows. Partition the ground set into k nonempty levels and then insert covering relations between adjacent levels. This procedure will produce all graded ordered sets with k levels. We will find upper and lower bounds on the number of structures that can be produced in this fashion.

First note that $\binom{n}{n_1, \ldots, n_k} = \dfrac{n!}{n_1! \cdots n_k!}$ is the number of possible distributions of n elements into k "levels," each having n_i elements. With the sums running over all possible choices $n_1, \ldots, n_k$, we cover all possible distributions of n elements into k bins so that no bin is empty. To verify the upper bound, notice that there can be at most $2^{n_i n_{i+1}}$ connections between the ith and the $(i+1)$st "level." The upper bound counts a few orders more than once. For example, some disconnected graded ordered sets can be generated in several ways here.

The lower bound is less than the number of graded orders such that, in the ith level $(1 < i \leq k)$, each element has at least one lower cover in the $(i-1)^{\text{st}}$ level and such that no two elements have the same lower covers. Indeed, when building such a set, for the jth element of level i for which we choose a set of upper bounds, we have $2^{n_{i-1}} - 1 - (j-1) \geq 2^{n_{i-1}} - n$ choices for a set of lower covers. Thus there are at least $(2^{n_{i-1}} - n)^{n_i}$ possible ways to connect the ith and the $(i-1)^{\text{st}}$ level. Because no two elements on the same level have the same lower covers and all elements are above an element on level 1, no order is doubly counted. This establishes the lower bound.

Note that some summands will even turn out negative for the lower bound. Because we are looking for a lower bound, this is of no consequence. ∎

Lemma 13.16. *The number of graded ordered sets with three grade levels is asymptotically bounded above by the quantity* $2^{\frac{n^2}{4} + \frac{3}{2}n + O(\sqrt{n})}$ *and it is bounded below by the quantity* $2^{\frac{n^2}{4} + \frac{3}{2}n - O(\log_2(n))}$.

Proof. For the lower bound, notice that, in the sum from Lemma 13.15, the term for $n_1 = \dfrac{n}{4}, n_2 = \dfrac{n}{2},$ and $n_3 = \dfrac{n}{4}$ is very large:

$$\binom{n}{\frac{n}{4}, \frac{n}{2}, \frac{n}{4}} \left(2^{\frac{n}{4}} - n\right)^{\frac{n}{2}} \left(2^{\frac{n}{2}} - n\right)^{\frac{n}{4}} \sim_{\text{as}} \binom{n}{\frac{n}{2}}\binom{\frac{n}{2}}{\frac{n}{4}} 2^{\frac{n}{4}\frac{n}{2} + \frac{n}{2}\frac{n}{4}}$$

$$\geq 2^{n - O(\log_2(n))} 2^{\frac{n}{2} - O(\log_2(n))} 2^{\frac{n^2}{4}} = 2^{\frac{n^2}{4} + \frac{3}{2}n - O(\log_2(n))}.$$

The first asymptotic equality is because $\left(1 - \dfrac{n}{2^{\frac{n}{4}}}\right)^{\frac{n}{2}} \to 1$ as $n \to \infty$ and $\left(1 - \dfrac{n}{2^{\frac{n}{2}}}\right)^{\frac{n}{4}} \to 1$ as $n \to \infty$. The inequality holds because $n\binom{n}{\frac{n}{2}} \geq 2^n$.

To obtain the upper bound, we argue as follows. For $n_1 + n_2 + n_3 = n$, we have that $n_1 n_2 + n_2 n_3 = n_2(n - n_2)$. For the part of the upper bound in Lemma 13.15 in which $\left|\dfrac{n}{2} - n_2\right| \geq \sqrt{n}$, this is maximized for $n_2 = \dfrac{n}{2} \pm \sqrt{n}$. Thus, in this part of the sum, we can bound the power of 2 with a term that is independent of the summation. Now notice (see Exercise 13-10a) that the sum of multinomial coefficients is $\displaystyle\sum_{n_1 + \cdots + n_k = n, n_i \geq 0} \binom{n}{n_1, n_2, \ldots, n_k} = k^n$. Hence

$$\sum_{\substack{n_2 \le \frac{n}{2} - \sqrt{n} \text{ or} \\ n_2 \ge \frac{n}{2} + \sqrt{n} \\ n_1 + n_2 + n_3 = n}} \binom{n}{n_1, n_2, n_3} 2^{n_2(n-n_2)} \le 3^n 2^{\frac{n^2}{4} - n} = 2^{\frac{n^2}{4} + n(\log_2 3 - 1)} \le 2^{\frac{n^2}{4} + n}.$$

The main part of the sum is estimated as follows.

$$\sum_{\substack{\frac{n}{2} - \sqrt{n} \le n_2 \le \frac{n}{2} + \sqrt{n} \\ n_1 + n_2 + n_3 = n}} \binom{n}{n_1, n_2, n_3} 2^{n_1 n_2 + n_2 n_3}$$

$$= \sum_{\substack{\frac{n}{2} - \sqrt{n} \le n_2 \le \frac{n}{2} + \sqrt{n} \\ n_1 + n_2 + n_3 = n}} \binom{n}{n_2} \binom{n - n_2}{n_1} 2^{n_2(n_1 + n_3)}$$

$$\le n 2^n 2^{\frac{n}{2} + \sqrt{n}} 2^{\frac{n^2}{4}} = 2^{\frac{n^2}{4} + \frac{3}{2}n + \sqrt{n} + \log_2(n)}.$$

∎

We now (almost) know the asymptotic size of what turns out to be the largest class of graded ordered sets. The remainder of the proof of Theorem 13.13 consists of a series of arguments that show that all other classes of graded ordered sets are of negligible size compared to this class. Thus, in the following "negligible," will mean "negligible in comparison to the class of graded sets with three grade levels."

Lemma 13.17. *The numbers of all graded ordered sets with one or two grade levels are negligible for large n.*

Proof. There is only one graded ordered set with one grade level. For the graded ordered sets with two levels notice that $\sum_{n_1=1}^{n-1} \binom{n}{n_1} 2^{n_1(n-n_1)} \le 2^{\frac{n^2}{4} + n}$. ∎

Lemma 13.18. *The number of graded ordered sets with four grade levels is negligible for large n.*

Proof. First we estimate the parts of the upper bound with $n_1 \ge n^{\frac{9}{10}}$ or (by symmetry) $n_4 \ge n^{\frac{9}{10}}$.

$$\sum_{\substack{n_1 + \cdots + n_4 = n \\ n_i > 0; \\ n_1 \ge n^{\frac{9}{10}}}} \binom{n}{n_1, n_2, n_3, n_4} 2^{n_1 n_2 + n_2 n_3 + n_3 n_4}$$

$$= \sum_{\substack{n_1 + \cdots + n_4 = n \\ n_i > 0; \\ n_1 \geq n^{\frac{9}{10}}}} \binom{n}{n_1, n_2 + n_4, n_3} \binom{n_2 + n_4}{n_4} 2^{n_1(n_2 + n_4) + (n_2 + n_4)n_3} 2^{-n_1 n_4}$$

$$\leq \sum_{\substack{n_1 + n_2 + n_3 = n \\ n_i > 0; \\ n_1 \geq n^{\frac{9}{10}}}}' \binom{n}{n_1, n_2, n_3} 2^{n_1 n_2 + n_2 n_3} \sum_{n_4 = 1}^{n} \binom{n}{n_4} (2^{-n_1})^{n_4}$$

$$\leq 2^{\frac{n^2}{4} + \frac{3}{2}n + O(\sqrt{n})} \left(\left(1 + 2^{-n^{\frac{9}{10}}} \right)^n - 1 \right)$$

$$= 2^{\frac{n^2}{4} + \frac{3}{2}n + O(\sqrt{n})} \left(\left(\left(1 + \frac{1}{2^{n^{\frac{9}{10}}}} \right)^{2^{n^{\frac{9}{10}}}} \right)^{\frac{n}{2^{n^{\frac{9}{10}}}}} - 1 \right)$$

$$\leq 2^{\frac{n^2}{4} + \frac{3}{2}n + O(\sqrt{n})} \left(e^{\left(\frac{n}{2^{n^{\frac{9}{10}}}} \right)} - 1 \right)$$

$$\leq 2^{\frac{n^2}{4} + \frac{3}{2}n + O(\sqrt{n})} \frac{2n}{2^{n^{\frac{9}{10}}}},$$

where the last estimate shows that the sum is negligible compared to the number of graded orders with three levels. For $n_1, n_4 \leq n^{\frac{9}{10}}$, we have

$$\binom{n}{n_1, n_2, n_3, n_4} \leq \binom{n}{n_1}\binom{n}{n_4}\binom{n}{n_2} \leq n^{n^{\frac{9}{10}}} n^{n^{\frac{9}{10}}} 2^n = 2^{n + 2n^{\frac{9}{10}} \log_2(n)}.$$

Therefore we are done, because the remaining sum in the upper bound is at most $2^{\frac{n^2}{4} + n + 2n^{\frac{9}{10}} \log_2(n)}$. ∎

The above arguments clearly show one of the great benefits of asymptotic enumeration. Because the estimates only need to hold for large enough n, we can sometimes use rather crude inequalities and still achieve the result.[2] As you can also see in the previous proofs, the term $2^{n_1 n_2 + n_2 n_3 + \cdots + n_{k-1} n_k}$ in the estimate of Lemma 13.15 plays a dominating role. We will therefore investigate this term in more detail. The maximization of the exponent subject to the constraints

---

[2]However, this benefit can come back to haunt us if we are interested in the rate of convergence. We will not address this issue in this text.

$\sum_{i=1}^{k} n_i = n$ and $n_i \geq 1$ can be an epic exercise in the use of Lagrange multipliers. There is a quicker combinatorial way to obtain estimates. The following lemma shows that, the more clustered the elements in the levels are, the larger an exponent we have in our upper bound estimate.

Lemma 13.19. *Let* $k \geq 4, F(x_1, \ldots, x_k) := \displaystyle\sum_{i=1}^{k-1} x_i x_{i+1}$ *and* $n_1, \ldots, n_k \in \mathbb{N}$. *Then there is a permutation* $\sigma : [k] \to [k]$ *and a* $j \in \{2, \ldots, k-1\}$ *such that:*

1. *For* $1 \leq i \leq j-1$, *we have* $n_{\sigma(i)} \leq n_{\sigma(i+1)}$.
2. *For* $j \leq i \leq k-1$, *we have* $n_{\sigma(i)} \geq n_{\sigma(i+1)}$.
3. $F(n_1, \ldots, n_k) \leq F(n_{\sigma(1)}, \ldots, n_{\sigma(k)})$.
4. *For* $k \geq 4$, *we have*

$$F(n_{\sigma(1)}, \ldots, n_{\sigma(k)})$$

$$\leq F\left(1, \ldots, 1, n_{\sigma(j-1)} + n_{\sigma(j+1)} - 1, n_{\sigma(j)} + \sum_{i \neq j-1, j, j+1} (n_{\sigma(i)} - 1), 1, \ldots, 1\right),$$

where the term with $n_{\sigma(j)}$ *occurs at the earliest in the third position and at the latest in the* $(k-1)^{\text{st}}$ *position.*

Proof. Let σ be a permutation such that part 3 holds and such that the value $F(n_{\sigma(1)}, \ldots, n_{\sigma(k)}) \geq F(n_1, \ldots, n_k)$ is maximal. To prove parts 1 and 2, we consider a sequence of assumptions that each leads to a contradiction.

First assume that there is a $j \in \{2, \ldots, k-1\}$ such that $n_{\sigma(j)} < n_{\sigma(1)}$. Choose $j \in \{2, \ldots, k-1\}$ such that $n_{\sigma(j)}$ is minimal. Then, with

$$\tau = (\tau(1), \ldots, \tau(k)) := (\sigma(j), \sigma(1), \ldots, \sigma(j-1), \sigma(j+1), \ldots, \sigma(k)),$$

we have

$$F(n_{\sigma(1)}, \ldots, n_{\sigma(k)})$$

$$= \sum_{i=1}^{k-1} n_{\sigma(i)} n_{\sigma(i+1)}$$

$$= n_{\sigma(j-1)} n_{\sigma(j)} + n_{\sigma(j)} n_{\sigma(j+1)} - n_{\sigma(1)} n_{\sigma(j)} - n_{\sigma(j+1)} n_{\sigma(j-1)}$$

$$\quad + \sum_{i \neq j-1, j} n_{\sigma(i)} n_{\sigma(i+1)} + n_{\sigma(1)} n_{\sigma(j)} + n_{\sigma(j+1)} n_{\sigma(j-1)}$$

$$= n_{\sigma(j-1)}(n_{\sigma(j)} - n_{\sigma(j+1)}) + n_{\sigma(j)}(n_{\sigma(j+1)} - n_{\sigma(1)}) + F(n_{\tau(1)}, \ldots, n_{\tau(k)})$$

$$< n_{\sigma(j)}(n_{\sigma(j)} - n_{\sigma(j+1)}) + n_{\sigma(j)}(n_{\sigma(j+1)} - n_{\sigma(j)}) + F(n_{\tau(1)}, \ldots, n_{\tau(k)})$$

$$= F(n_{\tau(1)}, \ldots, n_{\tau(k)}),$$

contradicting the choice of σ. Hence, we can assume that $n_{\sigma(1)} \leq n_{\sigma(i)}$, and, by symmetry, $n_{\sigma(k)} \leq n_{\sigma(i)}$, for all $i \in \{2, \ldots, k-1\}$.

Now assume that there is a $j \in \{2, \ldots, k-1\}$ such that $n_{\sigma(j-1)} > n_{\sigma(j)} < n_{\sigma(j+1)}$. Find $l \in \{2, \ldots, j-1\}$ such that $n_{\sigma(l-1)} \leq n_{\sigma(j)} \leq n_{\sigma(l)}$. Then, with

$$\tau = (\sigma(1), \ldots, \sigma(l-1), \sigma(j), \sigma(l), \ldots, \sigma(j-1), \sigma(j+1), \ldots, \sigma(k)),$$

we have

$$
\begin{aligned}
&F(n_{\sigma(1)}, \ldots, n_{\sigma(k)}) \\
&= n_{\sigma(j-1)} n_{\sigma(j)} + n_{\sigma(j)} n_{\sigma(j+1)} + n_{\sigma(l-1)} n_{\sigma(l)} \\
&\quad - n_{\sigma(l-1)} n_{\sigma(j)} - n_{\sigma(l)} n_{\sigma(j)} - n_{\sigma(j+1)} n_{\sigma(j-1)} + F(n_{\tau(1)}, \ldots, n_{\tau(k)}) \\
&= n_{\sigma(j+1)} (n_{\sigma(j)} - n_{\sigma(j-1)}) + n_{\sigma(l)} (n_{\sigma(l-1)} - n_{\sigma(j)}) \\
&\quad + n_{\sigma(j)} (n_{\sigma(j-1)} - n_{\sigma(l-1)}) + F(n_{\tau(1)}, \ldots, n_{\tau(k)}) \\
&< n_{\sigma(j)} (n_{\sigma(j)} - n_{\sigma(j-1)}) + n_{\sigma(j)} (n_{\sigma(l-1)} - n_{\sigma(j)}) \\
&\quad + n_{\sigma(j)} (n_{\sigma(j-1)} - n_{\sigma(l-1)}) + F(n_{\tau(1)}, \ldots, n_{\tau(k)}) \\
&= F(n_{\tau(1)}, \ldots, n_{\tau(k)}),
\end{aligned}
$$

contradicting the choice of σ.

Similarly (see Exercise 13-11), we can prove that there is no $j \in \{2, \ldots, k-1\}$ such that $n_{\sigma(j-1)} > n_{\sigma(j)} = n_{\sigma(j+1)} = \cdots = n_{\sigma(j+m)} < n_{\sigma(j+m+1)}$. Thus, by maximality of $F(n_{\sigma(1)}, \ldots, n_{\sigma(k)})$, there is a $j \in [k]$ so that $n_{\sigma(1)}, \ldots, n_{\sigma(j)}$ is nondecreasing and $n_{\sigma(j)}, \ldots, n_{\sigma(k)}$ is nonincreasing. To finish the proof of 1 and 2, we have to show that $j \notin \{1, k\}$.

Note that $j \in \{1, k\}$ would mean that $n_{\sigma(i)}$ is nonincreasing or nondecreasing. Because, for all $i \in \{2, \ldots, k-1\}$, we have $n_{\sigma(1)} \leq n_{\sigma(i)}$ and $n_{\sigma(k)} \leq n_{\sigma(i)}$, this would imply that $n_{\sigma(i)}$ is constant. Hence, in case $j \in \{1, k\}$, we can simply pick a different $j \in \{2, \ldots, k-1\}$ and 1 and 2 both hold.

For part 4, note that, for $i < j-1$ and $i > j+1$, we have

$$n_{\sigma(i)} n_{\sigma(i+1)} \leq (n_{\sigma(i)} - 1)(n_{\sigma(j+1)} + n_{\sigma(j-1)} - 1) + n_{\sigma(i+1)}.$$

Thus all of these terms, which are lost by going from the left side of the inequality to the right side, are replaced by corresponding larger terms. Moreover, the terms $n_{\sigma(j-1)} n_{\sigma(j)} + n_{\sigma(j)} n_{\sigma(j+1)}$ on the left occur on the right side as $(n_{\sigma(j-1)} + n_{\sigma(j+1)} - 1) n_{\sigma(j)} + n_{\sigma(j)}$. The term $n_{\sigma(j+1)} n_{\sigma(j+2)}$ on the left (if it exists) has the upper bound $(n_{\sigma(j-1)} + n_{\sigma(j+1)} - 1) n_{\sigma(j+2)} + n_{\sigma(j+2)}$ on the right. The $(k-2)$ "-1"s that were neglected above are picked up by the $(k-4)$ "$+1$"s that are generated by multiplying the "$+1$"s on the sides and by the term $(n_{\sigma(j-1)} + n_{\sigma(j+1)} - 1)$ (if this term is ≥ 2) that is multiplied by the 1 to its left. If $(n_{\sigma(j-1)} + n_{\sigma(j+1)} - 1) < 2$, then all n_i except for $n_{\sigma(j)}$ are 1 and the statement is trivial. ∎

Lemma 13.20. *Let* $k \geq 4$, *let* $F(x_1, \ldots, x_k) := \sum_{i=1}^{k-1} x_i x_{i+1}$ *and let the numbers*

$n_1, \ldots, n_k \in \mathbb{N}$ *be such that* $\sum_{i=1}^{k} n_i = n$. *Then the following hold.*

1. $F(n_1, \ldots, n_k) \leq \dfrac{n^2}{4} + \left(2 - \dfrac{k}{2}\right) n + \left(1 - \dfrac{k}{2}\right)^2 - 2.$

2. *For* $k = 6$ *and all* $n_i \geq 2$, *we have* $F(n_1, \ldots, n_6) \leq \dfrac{n^2}{4} - 2n + 8.$

3. *For* $k = 5$ *and all* $n_i \geq 2$, *we have* $F(n_1, \ldots, n_5) \leq \dfrac{n^2}{4} - n + 1.$

Proof. To prove 1, note that, by part 4 of Lemma 13.19, we have

$$F(n_1, \ldots, n_k) \leq F\left(1, \ldots, 1, \frac{n}{2} - \frac{k}{2} + 1, \frac{n}{2} - \frac{k}{2} + 1, 1, \ldots, 1\right)$$

$$= \left(\frac{n}{2} - \frac{k}{2} + 1\right)^2 + n - k + 2 + k - 4$$

$$= \frac{n^2}{4} + \left(2 - \frac{k}{2}\right) n + \left(1 - \frac{k}{2}\right)^2 - 2.$$

The proofs of 2 and 3 are similar, so we only prove 3. Part 2 and some details of part 3 are left as Exercise 13-12. Note that an argument similar to part 4 of Lemma 13.19 gives the first estimate below.

$$F(n_1, \ldots, n_5) \leq F\left(2, 2, \frac{n}{2} - 3, \frac{n}{2} - 3, 2\right)$$

$$= \left(\frac{n}{2} - 3\right)^2 + 2n - 12 + 4$$

$$= \frac{n^2}{4} - n + 1.$$

∎

Lemma 13.21. *The number of graded ordered sets with five or six grade levels is negligible for large n.*

Proof. We only prove the estimate for the number of five leveled graded ordered sets. By part 3 of Lemma 13.20, we have that

$$\sum_{i=1}^{4} n_i n_{i+1} \leq \frac{n^2}{4} - n + 1$$

if all n_i are at least 2. This implies, via the multinomial theorem, that

$$\sum_{\substack{n_1+\cdots+n_5=n \\ n_i>1}} \binom{n}{n_1,\ldots,n_5} 2^{n_1 n_2+\cdots+n_4 n_5} \le 5^n 2^{\frac{n^2}{4}-n+1} \le 2^{-\frac{n}{10}} 2^{\frac{n^2}{4}+\frac{3}{2}n},$$

which is negligible. This leaves us to estimate the part of the sum in which one of the n_i is equal to 1. By part 3 of Lemma 13.19, this part of the sum is bounded by the sum for which $n_1 = 1$, which is in turn estimated by

$$\sum_{\substack{n_1+\cdots+n_5=n \\ n_i>0;\,n_1=1}} \binom{n}{n_1,n_2,n_3,n_4,n_5} 2^{n_1 n_2+n_2 n_3+n_3 n_4+n_4 n_5}$$

$$\le \sum_{\substack{1+n_2+\cdots+n_5=n \\ n_i>0}} \binom{n}{1,n_2,n_3,n_4,n_5} 2^{n_2+n_2 n_3+n_3 n_4+n_4 n_5}$$

$$\le n \sum_{\substack{1+n_2+\cdots+n_5=n \\ n_i>0}} \binom{n}{n_2,n_3+1,n_4,n_5} 2^{n_2(n_3+1)+(n_3+1)n_4+n_4 n_5}$$

$$\le n \sum_{\substack{n_1+\cdots+n_4=n \\ n_i>0}} \binom{n}{n_1,n_2,n_3,n_4} 2^{n_1 n_2+n_2 n_3+n_3 n_4}$$

$$\le n2^{\frac{n^2}{4}+\frac{3}{2}n+O(\sqrt{n})} \frac{2n}{2^{n\frac{9}{10}}},$$

by the estimates in the proof of Lemma 13.18. Again, this is negligible. The estimate for the number of graded ordered sets with six grade levels is obtained in similar fashion. ∎

Lemma 13.22. *The number of graded ordered sets with more than six grade levels is negligible for large n.*

Proof. Use Lemma 13.20, part 1 to estimate the power of 2 with a term that is independent of the summation. The summation of the multinomial coefficients yields k^n. Thus we estimate the upper bound for the number of graded ordered sets with k levels by $2^{\log_2(k)n+\frac{n^2}{4}+(2-\frac{k}{2})n+(1-\frac{k}{2})^2-2}$. For all $k \ge 7$, this quantity is bounded by $2^{-\frac{n}{10}} 2^{\frac{n^2}{4}+\frac{3}{2}n}$ for large enough n. ∎

This concludes the proof of Theorem 13.13. We finally state the exact asymptotic values for the number of ordered sets with ground set $[n]$.

Theorem 13.23 (For a proof, see [57, 165, 166]).   *Let P_n be the number of ordered sets with ground set $[n]$. For $n \to \infty$ we have, for odd n:*

$$P_n^{(\text{odd})} \underset{\text{as}}{\sim} \left(\frac{2}{\pi}\right)^{\frac{1}{2}} \left(\sum_{x=-\infty}^{\infty} 2^{-x^2}\right) 2^{\frac{n^2}{4}+\frac{3}{2}n-\frac{1}{2}\log n}$$

and, for even n:

$$P_n^{(\text{even})} \underset{\text{as}}{\sim} \left(\frac{2}{\pi}\right)^{\frac{1}{2}} \left(\sum_{x=-\infty}^{\infty} 2^{-\left(x^2-\frac{1}{2}\right)^2}\right) 2^{\frac{n^2}{4}+\frac{3}{2}n-\frac{1}{2}\log n}.$$

Exercises

13-11. Show that, in the proof of Lemma 13.19, there is no index $j \in \{2, \dots, k-1\}$ such that we have $n_{\sigma(j-1)} > n_{\sigma(j)} = n_{\sigma(j+1)} = \cdots = n_{\sigma(j+m)} < n_{\sigma(j+m+1)}$.

13-12. Finishing Lemma 13.20.

    a. Fill in the remaining detail in the proof of part 3 of Lemma 13.20 by producing an argument similar to part 4 of Lemma 13.19.

    b. Prove part 2 of Lemma 13.20.

13-13. Finish the proof of Lemma 13.21 by proving an upper bound on the number of graded ordered sets with six levels.

13-14. Prove that the sizes of the following classes of ordered sets (with ground set $[n]$) are asymptotically negligible.

    a. Decomposable ordered sets.

    b. Ordered sets that have an irreducible point.
       Hint. Consider the sets treated in Lemma 13.16.

    c. Ordered sets that have a retractable point.
       Hint. Consider the sets treated in Lemma 13.16.

    d. Ordered sets that have the fixed point property.
       Hint. Consider the sets treated in Lemma 13.16 and show that most contain a four-crown-tower.

13-15. Improve the estimate of $2^{\frac{n^2}{4}+\frac{3}{2}n+O(\sqrt{n})}$ for the number of graded ordered sets in Theorem 13.13 to $2^{\frac{n^2}{4}+\frac{3}{2}n+O(\log_2(n))}$. The easy part is to estimate the number of graded ordered sets with three levels and $\left|\frac{n}{2}-n_2\right| \le \log_2(n)$. The hard part is the estimate of that number for $\sqrt{n} \ge \left|\frac{n}{2}-n_2\right| \ge \log_2(n)$.

13.4   The Number of Nonisomorphic Ordered Sets

In Section 13.3, we found the asymptotic number of all possible graded ordered
sets with ground set $[n]$ and we stated the asymptotic number of all (graded and
nongraded) ordered sets with ground set $[n]$. We did not take into account that,
although two ordered sets might not be equal, they could still be isomorphic.
For most purposes, isomorphic structures are not considered to be different from
each other. Thus it makes sense to ask for the number of "truly different," that is,
nonisomorphic, ordered sets with n elements.

Unfortunately, counting nonisomorphic structures is often harder than counting
all structures. One way to tackle this problem is to first count all structures and then
determine which fraction of them is large enough to contain one representative from
each isomorphism class. Because there are $n!$ permutations for an n-element set,
this fraction is at least $\frac{1}{n!}$ times the number of all structures. The following theorem
shows that, for ordered sets, this fraction is (asymptotically) indeed $\frac{1}{n!}$.

Theorem 13.24 (See [228], Corollary 2.3). *Let $\mathcal{O}_n$ be the number of ordered sets
with ground set $[n]$ and let $[\mathcal{O}]_n$ be the number of nonisomorphic ordered sets with
n elements. Then*

$$\frac{\mathcal{O}_n}{n![\mathcal{O}]_n} \to 1 \qquad \text{as } n \to \infty.$$

Proof. This proof is an adaptation of the proof of the Main Lemma in [228] for the
case of ordered sets.

First note that $\mathcal{O}_n \le n![\mathcal{O}]_n$. Finding an upper bound for $n![\mathcal{O}]_n$ that grows at
the same speed as $\mathcal{O}_n$ will complete the proof. We first derive another expression
for $n![\mathcal{O}]_n$.

Let $\mathcal{P}_n$ be the set of ordered sets with ground set $[n]$ and let $[\mathcal{P}]_n$ be the set
of isomorphism classes of ordered sets with ground set $[n]$. Then $|\mathcal{P}_n| = \mathcal{O}_n$ and
$|[\mathcal{P}]_n| = [\mathcal{O}]_n$. For any permutation π of $[n]$, let $F(\pi)$ be the number of ordered sets
in $\mathcal{P}_n$ for which π is an automorphism. For each ordered set $([n], \le) \in \mathcal{P}_n$, there are
exactly $|\mathrm{Aut}([n], \le)|$ permutations π such that π is an automorphism for $([n], \le)$.
With σ_n denoting the set of permutations of $[n]$, we compute the following.

$$\sum_{\pi \in \sigma_n} F(\pi) = \sum_{([n],\le)\in\mathcal{P}_n} |\mathrm{Aut}([n], \le)|$$

$$= \sum_{[([n],\le)]\in[\mathcal{P}]_n} \sum_{([n],\le')\in[([n],\le)]} |\mathrm{Aut}([n], \le')|$$

Now consider the sum $\sum_{([n],\le')\in[([n],\le)]} |\mathrm{Aut}([n], \le')|$. Fix a representative
$([n], \le)$ of $[([n], \le)]$. Any ordered set $([n], \le')$ in $[([n], \le)]$ is obtained from $([n], \le)$
by permuting the elements of $[n]$ and defining the order $\le'$ as follows. If σ is
a permutation of n elements, then $p \le' q$ iff $\sigma^{-1}(p) \le \sigma^{-1}(q)$. There are $n!$

ways to permute the elements of $([n], \leq)$. Sometimes distinct permutations do not lead to distinct orders. Let $([n], \leq')$ be an ordered set obtained from $([n], \leq)$ as indicated above and let $\sigma_1, \ldots, \sigma_m$ be all distinct permutations that can be used to construct $([n], \leq')$ from $([n], \leq)$ as indicated. Then $|\text{Aut}([n], \leq')| \geq m$, because the identity and $\sigma_2 \circ \sigma_1^{-1}, \ldots, \sigma_m \circ \sigma_1^{-1}$ are distinct automorphisms of $([n], \leq')$. Moreover, $|\text{Aut}([n], \leq')| \leq m$, because, for each automorphism τ of $([n], \leq')$, we must have that $\tau = \tau \circ \sigma_1 \circ \sigma_1^{-1} = \sigma_k \circ \sigma_1^{-1}$ for some k. Therefore $\sum_{([n], \leq') \in [([n], \leq)]} |\text{Aut}([n], \leq')| = n!$ and we conclude the following

$$
\sum_{\pi \in \sigma_n} F(\pi) = \sum_{[([n], \leq)] \in [\mathcal{P}]_n} \sum_{([n], \leq') \in [([n], \leq)]} |\text{Aut}([n], \leq')|
$$

$$
= \sum_{[([n], \leq)] \in [\mathcal{P}]_n} n! = n![\mathcal{O}]_n
$$

To finish the proof, we will show that $\sum_{\pi \in \sigma_n} F(\pi) \leq \mathcal{O}_n \left(1 + 2^{-\frac{n}{8}} \right)$ for large enough n. As in Section 13.3, inequalities that do not hold for all n are to be seen as holding for large enough n. Because $F(\text{id}_n) = \mathcal{O}_n$, we must show that the contribution from the nontrivial permutations is negligible. To do this, we partition the remaining permutations into the following two sets.

$$
X := \left\{ \pi \in \sigma_n \setminus \{\text{id}_n\} : \pi \text{ has more than } \frac{9n}{10} \text{ fixed points} \right\},
$$

$$
Y := \left\{ \pi \in \sigma_n \setminus \{\text{id}_n\} : \pi \text{ has at most } \frac{9n}{10} \text{ fixed points} \right\}.
$$

We start with the permutations in X. Let $\pi \in X$. We find an overestimate for the number of ordered sets on $[n]$ for which π is an automorphism as follows. Let A be a set that contains exactly one element from every nontrivial cycle of π. Let B contain the remaining elements of the nontrivial cycles of π. Let $a := |A|, b := |B|$ and note that $a \leq b$.

Let $\leq$ be an order relation on $[n]$ such that π is an automorphism of $([n], \leq)$. Then B and $[n] \setminus B$ are ordered subsets with the induced order relations. Conversely, given order relations on B and on $[n] \setminus B$, we claim that there are at most $2^{2|A| \cdot |B|}$ order relations on $[n]$ that induce these orders and that are such that π is an automorphism. Indeed, with restrictions to B and $[n] \setminus B$ as given, there are $2|A||B|$ possible comparabilities between elements of A and B. Thus there are $2^{2|A||B|}$ possible sets of comparabilities between A and B. The comparabilities between elements of $A \cup B$ and the fixed points of π are dictated by π (see Exercise 13-16). Thus $F(\pi)$ is bounded by the number of orders on $[n] \setminus B$ times the number of orders on B times $2^{2|A||B|}$. With the result of Exercise 13-17 and $f \geq \frac{9n}{10}$ denoting the number of fixed points of π, we obtain the following.

$$F(\pi) \leq 2^{\frac{(f+a)^2}{4}+\frac{3}{2}(f+a)+O(\log_2(f+a))} 2^{\frac{b^2}{4}+\frac{3}{2}b+O(\log_2(b))} 2^{2ab}$$

$$\leq 2^{\frac{1}{4}\left(f+\left(\frac{a+b}{2}\right)\right)^2+\frac{3}{2}n+\frac{1}{4}\left(\frac{a+b}{2}\right)^2+2\left(\frac{a+b}{2}\right)^2+O(\log_2(b))+O(\log_2(f+a))}$$

Because π has at least $\frac{9n}{10}$ fixed points, we have that $a + b \leq \frac{1}{10}n$. Moreover $a + b \geq 2$, because the smallest possible cycle has two elements. Finally, there are $\binom{n}{a+b}$ ways to choose the $a + b$ points that are not fixed and there are $(a + b)!$ ways to permute them. Thus, with $i := a + b$ we obtain the following.

$$\sum_{\pi \in X} F(\pi) \leq \sum_{i=2}^{\lfloor \frac{n}{10} \rfloor} \binom{n}{i} i! 2^{\frac{1}{4}\left(n-\frac{i}{2}\right)^2+\frac{3}{2}n+\frac{9}{4}\left(\frac{i}{2}\right)^2+O(\log_2(n))}$$

$$\leq \sum_{i=2}^{\lfloor \frac{n}{10} \rfloor} 2^{i\log_2(n)+\frac{1}{4}n^2+\frac{3}{2}n-\frac{1}{2}n\frac{i}{2}+\frac{5}{2}\left(\frac{i}{2}\right)^2+O(\log_2(n))}$$

$$\leq \sum_{i=2}^{\lfloor \frac{n}{10} \rfloor} 2^{\frac{1}{4}n^2+\frac{3}{2}n+\frac{i}{2}\left(2\log_2(n)+\frac{5}{4}i-\frac{1}{2}n\right)+O(\log_2(n))}$$

$$\leq 2^{\log_2\left(\frac{n}{10}\right)} 2^{\frac{1}{4}n^2+\frac{3}{2}n+\frac{n}{20}\left(2\log_2(n)+\frac{1}{8}n-\frac{1}{2}n\right)+O(\log_2(n))}$$

$$\leq \mathcal{O}_n 2^{-\frac{1}{80}n^2} \leq \mathcal{O}_n 2^{-\frac{1}{4}n}$$

This leaves us with the estimation for Y. Let $\pi \in Y$. Choose A and B to be disjoint sets of points that are not fixed by π such that $|B| = \lfloor \frac{n}{20} \rfloor \leq \frac{n}{20}$ and such that, for each $x \in B$, there is exactly one element $y \in A$ such that y is in the same cycle of π as x. Then $|A| \leq |B|$ and $[n] \setminus (A \cup B)$ also contains points that are not fixed by π.

Nonetheless, the number of orders such that π is an automorphism can be bounded similar to what was done above. Again let $a := |A|$ and $b := |B|$. There are $2^{\frac{1}{4}(n-b)^2+\frac{3}{2}(n-b)+O(\log_2(n-b))}$ ways to order $[n] \setminus B$, $2^{\frac{1}{4}b^2+\frac{3}{2}b+O(\log_2(b))}$ ways to order B, and at most 2^{2ab} possible ways to impose orders between points of A and points of B. By choice of A and B, the remaining order relations are dictated by the fact that π is supposed to be an automorphism (see Exercise 13-16). Thus

$$F(\pi) \leq 2^{\frac{(n-b)^2}{4}+\frac{3}{2}(n-b)+O(\log_2(n-b))} 2^{\frac{b^2}{4}+\frac{3}{2}b+O(\log_2(b))} 2^{2ab}$$

$$\leq 2^{\frac{n^2}{4}+\frac{3}{2}n-\frac{1}{2}nb+\frac{b^2}{4}-\frac{3}{2}b+\frac{b^2}{4}+\frac{3}{2}b+2b^2+O(\log_2(n-b))+O(\log_2(b))}$$

$$= 2^{\frac{n^2}{4}+\frac{3}{2}n-\frac{1}{2}nb+\frac{5}{2}b^2+O(\log_2(n))}$$

$$= 2^{\frac{n^2}{4}+\frac{3}{2}n+\frac{1}{2}b(5b-n)+O(\log_2(n))}$$

$$\leq 2^{\frac{n^2}{4}+\frac{3}{2}n+\frac{1}{2}\frac{n}{20}\left(\frac{n}{4}-n\right)+O(\log_2(n))}$$

$$= 2^{\frac{n^2}{4}+\frac{3}{2}n-\frac{3}{160}n^2+O(\log_2(n))}.$$

With i denoting the number of points not fixed by π, there are again $\binom{n}{i}$ ways to choose the points not fixed by π and $i!$ ways to permute them. Thus

$$\sum_{\pi \in Y} F(\pi) \leq \sum_{i=\left\lfloor \frac{n}{10} \right\rfloor}^{n} \binom{n}{i} i! 2^{\frac{n^2}{4} + \frac{3}{2}n - \frac{3}{160}n^2 + O(\log_2(n))}$$

$$\leq \sum_{i=\left\lfloor \frac{n}{10} \right\rfloor}^{n} 2^{i \log_2(n)} 2^{\frac{n^2}{4} + \frac{3}{2}n - \frac{3}{160}n^2 + O(\log_2(n))}$$

$$\leq 2^{\log_2(n) + n \log_2(n) + \frac{n^2}{4} + \frac{3}{2}n - \frac{3}{160}n^2 + O(\log_2(n))} \leq \mathcal{O}_n 2^{-\frac{1}{4}n}.$$

Adding our estimates now proves the result. ∎

The ratio of $\frac{1}{n!}$ between isomorphism classes and all structures can only be realized if most structures have no symmetry, which can be formalized as follows.

Corollary 13.25. *Let $\mathcal{R}_n$ be the number of rigid orders on $[n]$ and let $[\mathcal{R}]_n$ be the number of nonisomorphic rigid orders on an n-element set. Then, for $n \to \infty$, we have that $\frac{\mathcal{R}_n}{\mathcal{O}_n} \to 1$ and $\frac{[\mathcal{R}]_n}{[\mathcal{O}]_n} \to 1$.*

Proof. Suppose, for a contradiction, that $\liminf_{n \to \infty} \frac{[\mathcal{R}]_n}{[\mathcal{O}]_n} = c < 1$. Denote the number of nonrigid orders on $[n]$ by $\mathcal{F}_n$ and let $[\mathcal{F}]_n$ be the number of nonisomorphic nonrigid orders on $[n]$. Every nonrigid ordered set has at least one nontrivial automorphism. This means $\mathcal{F}_n \leq \frac{n!}{2}[\mathcal{F}]_n$. This in turn implies that, for a sequence of numbers n that goes to infinity, there are $d_n \leq \frac{c+1}{2} < 1$ such that the following hold. (Note that $\mathcal{R}_n = n![\mathcal{R}]_n$.)

$$\mathcal{O}_n = \mathcal{R}_n + \mathcal{F}_n \leq n![\mathcal{R}]_n + \frac{1}{2}n![\mathcal{F}]_n = d_n n![\mathcal{O}]_n + (1 - d_n)\frac{1}{2}n![\mathcal{O}]_n$$

$$= \frac{1 + d_n}{2} n![\mathcal{O}]_n \leq \frac{3 + c}{4} n![\mathcal{O}]_n.$$

This is a contradiction to Theorem 13.24. Therefore $\liminf_{n \to \infty} \frac{[\mathcal{R}]_n}{[\mathcal{O}]_n} = 1$ and hence $\lim_{n \to \infty} \frac{[\mathcal{R}]_n}{[\mathcal{O}]_n} = 1$. Finally, $\lim_{n \to \infty} \frac{\mathcal{R}_n}{\mathcal{O}_n} = \lim_{n \to \infty} \frac{\mathcal{R}_n}{\mathcal{O}_n} \frac{n![\mathcal{O}]_n}{n![\mathcal{R}]_n} = \lim_{n \to \infty} \frac{n![\mathcal{O}]_n}{\mathcal{O}_n} = 1.$ ∎

Exercises

13-16. Let $\leq$ be an order on $[n]$ such that the permutation $\pi : [n] \to [n]$ is an automorphism. Let $b \in [n]$ be an element of a nontrivial cycle of π. Let $a \in [n]$ be an element of the same cycle, say, $a = \pi^k(b)$. Let $m > k$ be the smallest natural number so that $b = \pi^m(b)$. Let $x \in [n]$ be arbitrary. Prove that $b \leq x$ iff $a \leq \pi^k(x)$.

13-17. Let f, a, b be positive numbers such that $a \leq b$ and $f \geq \frac{1}{2}(b - a)$.

Prove that $(f + a)^2 + b^2 \leq \left(f + \left(\frac{a+b}{2}\right)\right)^2 + \left(\frac{a+b}{2}\right)^2$.

Remarks and Open Problems

1. For more on enumeration, see [80]. For more on generating functions, see [324].
2. For precise formulas for the number of nonisomorphic graded ordered sets with n elements, see [163].
3. The numbers of different (nonisomorphic) orders for an n-element ordered set, up to isomorphism to a certain threshold were given (chronologically) in [37, 44, 55, 82, 134]. In each paper, different approaches were used to obtain more numbers, with the latest version of nauty and Traces (see [203]) probably being the state of the art in enumeration. The numbers that are currently known are the following (see [37]).

n	nonisomorphic orders	total number of orders on $[n]$
1	1	1
2	2	3
3	5	19
4	16	219
5	63	4,231
6	318	130,023
7	2,045	6,129,859
8	16,999	431,723,379
9	183,231	44,511,042,511
10	2,567,284	6,611,065,248,783
11	46,749,427	1,396,281,677,105,899
12	1,104,891,746	414,864,951,055,853,499
13	33,823,827,452	171,850,728,381,587,059,351
14	1,338,193,159,771	98,484,324,257,128,207,032,183
15	68,275,077,901,156	77,567,171,020,440,688,353,049,939
16	4,483,130,665,195,087	83,480,529,785,490,157,813,844,256,579
17		122,152,541,250,295,322,862,941,281,269,151
18		241,939,392,597,201,176,602,897,820,148,085,023

4. Aside from graded ordered sets, there are other classes of ordered sets for which exact formulas for the number of n-element orders *in this class* are available. Among these classes are series parallel ordered sets, interval ordered sets, and tiered ordered sets. An overview of these classes is given in [80].
5. Find a proof for Theorem 13.23 that is easier than the proof in the literature. The proof of Theorem 13.13 together with [57] allows us to establish the bounds of Theorem 13.23 for the number of graded ordered sets. All that "remains"

would be to show that the number of nongraded ordered sets is asymptotically negligible. A step towards this goal might be to first solve Problem 6 below and then use sharp bounds on the number of graded sets to establish bounds on the number of nongraded sets.

6. Prove or disprove that the number of graded ordered sets with four grade levels is asymptotically $2^{\frac{n^2}{4}+\frac{5}{4}n+O(\log_2(n))}$. Prove or disprove that the number of graded ordered sets with $k \geq 5$ grade levels is asymptotically equal to $2^{\frac{n^2}{4}+\left(1-\frac{k-5}{2}\right)n+O(\log_2(n))}$. These conjectures are motivated by the fact that most graded ordered sets with three grade levels have structure $\frac{n}{4} - \frac{n}{2} - \frac{n}{4}$. To me, this suggests that most graded ordered sets with $k > 3$ levels will have three large levels that look like the typical graded set of size $n - k + 3$ and all other levels contain exactly one element. I am not sure about the degree of difficulty of this conjecture. An approach as given in Section 13.3 appears promising, but it's probably quite tedious.

7. For asymptotic formulas for the number of ordered sets of bounded *width*, see [35].

8. What is the asymptotic number of $(n+2)$-element lattices? There are asymptotic upper and lower bounds, see [80], but there is no asymptotic expression.

9. Corollary 13.25 is an example of a "zero-one-law." Many combinatorial properties (to be precise, those that can be stated using a sentence in first-order logic) are so that, as the size of the underlying structure gets large, the probability that a randomly chosen structure has the property (rigidity, in the case of Corollary 13.25) approaches 1, or this probability approaches 0. This result can be found in [84] or also in Section II.2, Theorem 6 in [26]. In a sense, this is not surprising. For ordered sets, we have seen that, for large n, almost all ordered sets are graded with three levels. The largest class of ordered sets is thus very homogeneous.

10. Knowing the number of ordered sets with a given ground set can help answer certain questions without actually constructing the answer.

   For example, the number of linear extensions of an ordered set is a number between 1 and $n!$. Because the number of nonisomorphic ordered sets grows faster than $n!$, there must be nonisomorphic ordered sets with the same number of linear extensions. S. Felsner showed me this application of the "probabilistic method," which provided a negative answer to my simple-minded question if reconstruction of the number of linear extensions would solve the reconstruction problem.

   As another example, an ordered set and its dual have the same number of endomorphisms. Via an argument similar to the above, there must be two nonisomorphic and non-dually isomorphic ordered sets P and Q of the same size such that $|\text{End}(P)| = |\text{End}(Q)|$. However, note that, by Exercise 12-37, no two distinct non-dual ordered sets can have the same *set* of endomorphisms.

   For more work of a probabilistic nature in ordered sets, see, for example, Chapter 7 in [311].

11. Let $m, n \in \mathbb{N}$ be given and let the union $T := \bigcup_{a=1}^{n} \{1, \ldots, m\}^a$ be ordered by $(x_1, \ldots, x_k) \le (y_1, \ldots, y_l)$ iff $k \le l$ and, for all $i \in \{1, \ldots, k\}$, we have $x_i = y_i$. Let $c_1, \ldots, c_b$ be elements of T.

    Find estimates for the number of elements of $S := T \setminus \bigcup_{i=1}^{b} \uparrow c_i$.

    The above can be used to describe the tree that is searched by a search algorithm such as backtracking or forward checking (see Section 5.4). Insights into the above question may allow us to estimate run times of search algorithms.

    The above description of the search space is from [326].

12. Another entity related to enumeration in ordered sets is the Möbius function. For more on this function, see [178, 261].

13. To consider average time complexity, see, for example [113], we need to know how the underlying structures are distributed. Thus refinements of the results in this chapter that provide tight upper bounds on the speed of convergence could be interesting in showing that, for certain problems, the behavior on the class of ordered sets of height 2 dominates average time complexity considerations.

Appendix A
Some Algebraic Topology

This appendix is an introduction to the basic notions of algebraic topology that are needed for Sections 9.6 and 9.7. To keep the new vocabulary limited, we will not use the language of category theory here, though those versed in it will easily be able to identify the functors, etc., behind the results. Most of the standard concepts from algebraic topology have been taken from [295], Chapter 4, Sections 1–4. Proofs that were omitted are either short or a reference to the literature is given. For a more current introduction, consider [124].

Notation A.1. *We know how to assign to an ordered set P a graph $G_C(P)$, the comparability graph, see Definition 6.6. We also know how to assign to a graph G a simplicial complex K_G, see Example 9.13. Finally, we know who to assign to a simplicial complex K a topological space $|K|$, see Definition 9.23. In this appendix, we will learn how to assign to a simplicial complex K a chain complex $C(K)$, see Definition A.4. All these notions allow constructions that naturally arise in their respective settings. These constructions can then also be executed for structures induced by other structures: For example, we can consider the homology complex for a chain complex induced by a simplicial complex induced by a comparability graph of an ordered set. To reduce the amount of notation and without spelling out all the necessary definitions, which would be a large task indeed, we will sometimes use language formally "out of turn": For example, we will talk about the homology complex of an ordered set, rather than about the lengthy description above. We shall also abbreviate symbols accordingly: For example, we will write $H(P)$ rather than $H(C(K(G_C(P))))$. This practice should not cause confusion, as we will always assume that the construction is performed for the appropriate induced structure.*

© Springer International Publishing 2016
B. Schröder, *Ordered Sets*, DOI 10.1007/978-3-319-29788-0

A.1   Chain Complexes

The key idea to using homology in fixed point theory is to connect algebra with simplicial complexes. In this section, we will translate/embed simplicial complexes and their morphisms into the theory of chain complexes, which is the first step. Throughout this section, let us consider a four chain $P = \{0, 1, 2, 3\}$ with its natural order as an example. The chain P, its comparability graph and the clique complex of the comparability graph, which is also called the "P-chain complex," are pictured in Figure A.1.

Definition A.2. *A **chain complex** $C = (\{C_n\}_{n \in \mathbb{Z}}, \{\partial_q\}_{q \in \mathbb{Z}})$ is an ordered pair of a family $\{C_n\}_{n \in \mathbb{Z}}$ of abelian groups and a family of functions $\partial_q : C_q \to C_{q-1}$, called **boundary maps**, such that $\partial_q \partial_{q+1} = 0$ for all $q \in \mathbb{Z}$.*
   *C is called **finitely generated** iff $C_q = 0$ for all but finitely many q and all nonzero C_q have a finite set of generators.*

   To obtain a chain complex from a simplicial complex, we consider the topological realization of the simplicial complex. For the four chain in Figure A.1 this is the tetrahedron given in Figure A.2. All simplicial complexes in this appendix are assumed to be finite.

Definition A.3. *Let $S = \{v_0, \dots, v_q\}$ be a finite set of points. Two linear orders $v_{j_0} < v_{j_1} < \cdots < v_{j_q}$ and $v_{k_0} < v_{k_1} < \cdots < v_{k_q}$ are called **equivalently oriented** iff the permutation σ such that $\sigma \circ (k_0, \dots, k_q) = (j_0, \dots, j_q)$ is even. (Equivalent*

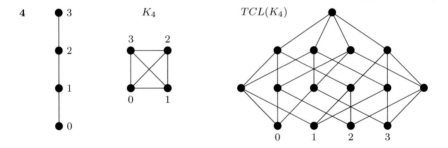

Fig. A.1 The four chain **4**, the corresponding comparability graph K_4, and the corresponding clique complex visualized as $TCL(K_4)$

Fig. A.2 The topological realization of a four chain, which is a tetrahedron. The orientations of the "hidden" boundary pieces (triangles) of the tetrahedron are indicated. The front triangle is positively oriented (counterclockwise)

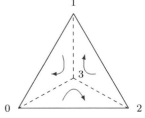

orientation is an equivalence relation and it has two equivalence classes for $q > 0$.)
The equivalence class of $v_{j_0} < v_{j_1} < \cdots < v_{j_q}$ will be denoted by $[v_{j_0}, v_{j_1}, \ldots, v_{j_q}]$
and will be called an **orientation** *of S. If $v_0, \ldots, v_q$ are vertices of a simplex and*
$v_i = v_j$ for some $i \neq j$, then we let $[v_0, \ldots, v_q] := 0$.

Although the orientations of the geometric objects we visualize are certainly
subject to some choices, the connections between higher-dimensional and lower-
dimensional objects via the boundary map are determined by the following.

Definition A.4. *Let $K = (V, S)$ be a simplicial complex. For $q \in \mathbb{Z}$ let $C_q(K)$*
be the free abelian group generated by the orientations of the q-simplices with
different orientations of the same simplex being additive inverses of each other.
Let the function $\partial_q^K : C_q(K) \to C_{q-1}(K)$ be the homomorphism defined on the
generators by

$$\partial_q^K([v_0, \ldots, v_q]) = \sum_{i=0}^{q} (-1)^i [v_0, \ldots, \widehat{v_i}, \ldots, v_q],$$

where, as is often customary, the hat indicates that the vertex under the hat is to
be dropped. (We will show that ∂_q^K is well-defined via this definition.) With this
definition, $C(K) := (\{C_q(K)\}_{q \in \mathbb{Z}}, \{\partial_q^K\}_{q \in \mathbb{Z}})$ is a chain complex, called the **oriented**
chain complex of K.

The orientations of $\partial_2(\{0, 1, 2, 3\})$ are indicated in Figure A.2. Note that the
boundaries of the solid tetrahedron (the triangular faces) are oriented in such a way
that their boundaries (sides) in turn are traversed once in each direction by anyone
who travels the boundaries of the triangles in the direction indicated in the triangle.
From my (most certainly limited) view of this subject, the motivation for this particular definition of the boundary operator seems to lie deeply in considerations of
differential geometry somewhere near Stokes' theorem. Algebraically, the following
proof shows that the right way to line up alternating signs is what makes the oriented
chain complex a chain complex.

Proof that the oriented chain complex really is a chain complex. To see that
the ∂_q^K are well-defined, note that a transposition of adjacent elements v_j, v_{j+1} and
a subsequent transposition of elements v_k, v_{k+1} (in the new indexing after the first
transposition) in $[v_0, \ldots, v_q]$ do not affect the right-hand side of the definition of
$\partial_q^K([v_0, \ldots, v_q])$. Every even permutation is a composition of an even number of
transpositions of adjacent elements. Hence this shows that the functions ∂_q^K are well-
defined.

To prove that $C(K) = (\{C_q(K)\}_{q \in \mathbb{Z}}, \{\partial_q^K\}_{q \in \mathbb{Z}})$ truly is a chain complex, we must
show that $\partial_{q-1}^K \partial_q^K = 0$ for all $q \geq 1$. (For $q < 1$ the maps $\partial_{q-1}^K \partial_q^K$ map into the group
$C_{q-1}(K) = \{0\}$.) To do this, we can limit ourselves to investigating the action of
$\partial_{q-1}^K \partial_q^K$ on the generators.

$$\partial^K_{q-1}\partial^K_q([v_0,\ldots,v_q]) = \partial^K_{q-1}\left(\sum_{i=0}^{q}(-1)^i[v_0,\ldots,\widehat{v_i},\ldots,v_q]\right)$$

$$= \sum_{i=0}^{q}(-1)^i\partial^K_{q-1}([v_0,\ldots,\widehat{v_i},\ldots,v_q])$$

$$= \sum_{i=0}^{q}(-1)^i\left(\sum_{j=0}^{i-1}(-1)^j[v_0,\ldots,\widehat{v_j},\ldots,\widehat{v_i},\ldots,v_q]\right.$$

$$\left. + \sum_{j=i+1}^{q}(-1)^{j-1}[v_0,\ldots,\widehat{v_i},\ldots,\widehat{v_j},\ldots,v_q]\right)$$

$$= \sum_{i=0}^{q}\sum_{j=0}^{i-1}(-1)^{i+j}[v_0,\ldots,\widehat{v_j},\ldots,\widehat{v_i},\ldots,v_q]$$

$$+ \sum_{i=0}^{q}\sum_{j=i+1}^{q}(-1)^{i+j-1}[v_0,\ldots,\widehat{v_i},\ldots,\widehat{v_j},\ldots,v_q]$$

$$= \sum_{i=1}^{q}\sum_{j=0}^{i-1}(-1)^{i+j}[v_0,\ldots,\widehat{v_j},\ldots,\widehat{v_i},\ldots,v_q]$$

$$+ \sum_{j=1}^{q}\sum_{i=0}^{j-1}(-1)^{i+j-1}[v_0,\ldots,\widehat{v_i},\ldots,\widehat{v_j},\ldots,v_q]$$

$$= 0.$$

∎

As with any of the objects we have defined so far, we are interested in the natural maps between these objects. For chain complexes, the natural maps are the chain maps.

Definition A.5. *A* ***chain map*** *(see Figure A.3) $f : C \to C'$ from the chain complex $C = (\{C_q\}_{q\in\mathbb{Z}}, \{\partial_q\}_{q\in\mathbb{Z}})$ to the chain complex $C' = (\{C'_n\}_{n\in\mathbb{Z}}, \{\partial'_q\}_{q\in\mathbb{Z}})$ is a family $f = \{f_q\}_{q\in\mathbb{Z}}$ of group homomorphisms $f_q : C_q \to C'_q$ such that, for all $q \in \mathbb{Z}$, we have $f_{q-1} \circ \partial_q = \partial'_q \circ f_q$.*

Chain maps can be induced by simplicial maps as shown below.

Proposition A.6. *Let $K = (V,\mathcal{S})$ and $K' = (V',\mathcal{S}')$ be simplicial complexes and let $f : K \to K'$ be a simplicial map. Then the function $f^{\mathrm{ch}} : C(K) \to C(K')$ defined by $f^{\mathrm{ch}}_q([s_0,\ldots,s_q]) := [f(s_0),\ldots,f(s_q)]$ on the generators and extended in the natural fashion is a chain map.*

Fig. A.3 A chain map

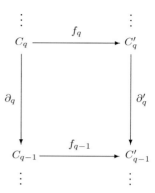

Proof. First note that f_q^{ch} is well-defined. Indeed, an even permutation of the $s_0, \ldots, s_q$ translates into an even permutation of the $f(s_0), \ldots, f(s_q)$.

Now we must show that $\partial_q^{K'} \circ f_q^{\mathrm{ch}} = f_{q-1}^{\mathrm{ch}} \circ \partial_q^{K}$. Yet this is quite trivial, because, for all generators, we have the following.

$$f_{q-1}^{\mathrm{ch}} \circ \partial_q^{K}([v_0, \ldots, v_q]) = f_{q-1}^{\mathrm{ch}}\left(\sum_{i=0}^{q}(-1)^i[v_0, \ldots, \widehat{v_i}, \ldots, v_q]\right)$$

$$= \sum_{i=0}^{q}(-1)^i[f(v_0), \ldots, \widehat{f(v_i)}, \ldots, f(v_q)]$$

$$= \partial_q^{K'}([f(v_0), \ldots, f(v_q)])$$

$$= \partial_q^{K'} f_q^{\mathrm{ch}}([v_0, \ldots, v_q]).$$

∎

We thus have embedded/translated the objects that we care about, simplicial complexes and their morphisms, into the theory of chain complexes and chain maps. The next step is to use algebraic tools to obtain fixed point results.

A.2 The Lefschetz Number

As a possible motivation for the Lefschetz number of a map, note that the entries on the diagonal of a matrix tell "how much" of a vector that is multiplied with the matrix will still point in the original direction. The trace of a matrix is the sum of the entries on the diagonal. Thus, if we have a large trace, then lots of vectors have images with large components in the original direction. This simplistic description will be made more precise in the following. We forego the proof that the trace of the matrix representation of each f_q that makes up a chain map is unique, though.

Definition A.7. *Let* $C = (\{C_q\}_{q\in\mathbb{Z}}, \{\partial_q\}_{q\in\mathbb{Z}})$ *be a finitely generated chain complex and let* $f : C \to C$ *be a chain map. The* **Lefschetz number** $\Lambda(f)$ *of* f *is defined to be*

$$\Lambda(h) := \sum_{q\in\mathbb{N}}(-1)^q Tr(f_q),$$

where Tr denotes the trace.

As was tried to motivate above, the Lefschetz number of a map can be used as a sufficient criterion for the existence of fixed points or cliques or simplices.[1] Let $f : V \to V$ be a simplicial map on the finite simplicial complex $K = (V, \mathcal{S})$. If $\Lambda(f) \neq 0$, then there is a $q \in \mathbb{N}$ such that $Tr(f_q) \neq 0$. Thus there is a q-dimensional simplex $S = \{v_0, \ldots, v_q\} \in \mathcal{S}$ such that $f_q([v_0, \ldots, v_q]) = k[v_0, \ldots, v_q] + h$ for some $k \in \mathbb{Z} \setminus \{0\}$ and h a combination of the generators other than S. However, because

$$f_q([v_0, \ldots, v_q]) = [f(v_0), \ldots, f(v_q)]$$

this means

$$f_q([v_0, \ldots, v_q]) \in \{[v_0, \ldots, v_q], -[v_0, \ldots, v_q]\}.$$

Hence we have that $f[S] = S$, that is, f has a fixed simplex. Now if K is a clique complex and f is induced by a graph endomorphism, this means that there is a clique C such that $f[C] = C$. If K is the clique complex of the comparability graph of the finite ordered set P and f is an order-preserving map on P, then f maps a chain to itself, which naturally means that f has a fixed point.

A.3 (Integer) Homology

Here, specifically in Lemma A.12, is "where the miracle of algebraic topology occurs" (at least for our purposes). Through Lemma A.12, many Lefschetz numbers become computable and thus a host of combinatorially surprising fixed point theorems enters the theory.

Lemma A.8. *Let* K *be a simplicial complex and let* L *be a subcomplex of* K. *Then* $C_q(L)$ *is a subgroup of* $C_q(K)$ *and the quotient group* $C_q(K, L) := C_q(K)/C_q(L)$ *is well-defined. Moreover* $C(K, L) := (\{C_q(K, L)\}_{q\in\mathbb{Z}}, \{\widetilde{\partial}_q\}_{q\in\mathbb{Z}})$ *is a chain complex with boundary map*

$$\widetilde{\partial}_q : C_q(K, L) \to C_{q-1}(K, L); \qquad (c + C_q(L)) \mapsto \partial_q(c) + C_{q-1}(L).$$

∎

---

[1]J.D. Farley showed me the following short argument.

Definition A.9. *Let* $C = (\{C_q\}_{q\in\mathbb{Z}}, \{\partial_q\}_{q\in\mathbb{Z}})$ *be a chain complex. We will call the subgroups* $Z_q(C) := \ker(\partial_q)$ *the subgroups of cycles of* C *and the subgroups* $B_q(C) := \partial_{q+1}[C_{q+1}]$ *the subgroups of boundaries of* C. *Two cycles whose difference is a boundary are called homologous. The homology groups are the quotient groups*

$$H_q(C) := Z_q(C)/B_q(C).$$

Lemma A.10 (See [295], Chapter 4, Section 1, Theorem 1).   *Let a chain complex* $C = (\{C_q\}_{q\in\mathbb{Z}}, \{\partial_q\}_{q\in\mathbb{Z}})$ *be given. Then*

$$Z(C) := (\{Z_q(C)\}_{q\in\mathbb{Z}}, \{\partial_q|_{Z_q(C)}\}_{q\in\mathbb{Z}}),$$

$$B(C) := (\{B_q(C)\}_{q\in\mathbb{Z}}, \{\partial_q|_{B_q(C)}\}_{q\in\mathbb{Z}})$$

are chain complexes. If we define

$$\partial_q^* : H_q(C) \to H_{q-1}(C); \qquad (c + B_q(C)) \mapsto \partial_q(c) + B_{q-1}(C),$$

then ∂_q^* *is a group homomorphism and* $H(C) := (\{H_q(C)\}_{q\in\mathbb{Z}}, \{\partial_q^*\}_{q\in\mathbb{Z}})$ *is a chain complex. We will call these complexes the* **cycle**, **boundary**, *and* **homology complex**.

∎

Lemma A.11 (See [295], Chapter 4, Section 1, Theorem 1). *Let chain complexes* $C = (\{C_q\}_{q\in\mathbb{Z}}, \{\partial_q\}_{q\in\mathbb{Z}})$ *and* $C' = (\{C'_q\}_{q\in\mathbb{Z}}, \{\partial'_q\}_{q\in\mathbb{Z}})$ *be given and let the function* $f : C \to C'$ *be a chain map. Let*

$$f_q^* : H_q(C) \to H_q(C'); \qquad (c + B_q(C)) \mapsto f_q(c) + B_q(C').$$

Then all f_q^* *are well-defined and* f^* *is a chain map.* ∎

Lemma A.12 (See [295], Chapter 4, Section 7, Theorem 6). *Let* C *be a finitely generated chain complex and let* $f : C \to C$ *be a chain map. Then*

$$\Lambda(f) = \Lambda(f^*).$$

Proof. First note that $f_q[Z_q(C)] \subseteq Z_q(C)$ and $f_q[B_q(C)] \subseteq B_q(C)$, because, for any $x \in Z_q(C)$, we have

$$\partial_q f_q(x) = f_{q-1}\partial_q(x) = f_{q-1}(0) = 0,$$

while, for any $y \in B_q(C)$, there is an $x \in C_{q+1}$ with $y = \partial_{q+1}(x)$ and thus

$$f_q(y) = f_q(\partial_{q+1}(x)) = \partial_{q+1}f_{q+1}(x) \in B_q(C).$$

Now, because $Z_q(C)$ is isomorphic to $H_q(C) \oplus B_q(C)$ and $C_q/Z_q(C)$ is isomorphic to $B_{q-1}(C)$ (with ∂_q being the isomorphism), we have

$$Tr(f_q|_{Z_q(C)}) = Tr(f_q^*) + Tr(f_q|_{B_q(C)}),$$
$$Tr(f_q) = Tr(f_q|_{Z_q(C)}) + Tr(f_{q-1}|_{B_{q-1}(C)}).$$

Adding the two equations, multiplying with $(-1)^q$, and summing over q give the equality of the Lefschetz numbers. ∎

We thus have translated Lefschetz numbers of chain maps in the chain complex associated with a simplicial complex into Lefschetz numbers of their associated maps in the homology complex. The advantage in this translation is that the homology groups tend to be smaller and more manageable than the original groups in the oriented chain complex. The zeroth homology group can be described geometrically.

Definition A.13. *Let $K = (V, S)$ be a simplicial complex. A **path** is a subset $\{p_0, \ldots, p_n\}$ such that, for $k \in \{1, \ldots, n\}$, the set $\{p_{k-1}, p_k\}$ is a simplex of K. A **component** of K is a maximal full subcomplex C such that any two vertices of C are in a path. K is called **connected** iff K has exactly one component.*

Lemma A.14. *Let $K = (V, S)$ be a finite simplicial complex and let k be the number of components of K. Then $H_0(K)$ is isomorphic to $\mathbb{Z}^k$.*

Proof. The generators for $B_0(K)$ are the differences $v - w$, where v, w are vertices of K with $\{v, w\} \in S$. Thus if u, z are vertices in the same component of K, then $u - z \in B_0(K)$. On the other hand if $u - z \in B_0(K)$, then $u - z$ must be the sum of finitely many generators of $B_0(K)$ and thus u and z must be in the same component of K.

Because $Z_0(K) = C_0(K)$ is generated by the vertex set of K, this implies that any two vertices x, y in the same component of K satisfy $x + B_0(K) = y + B_0(K)$. On the other hand, if x' and y' are not in the same component of K, then we have $x' + B_0(K) \neq y' + B_0(K)$. Hence $H_0(K)$ has exactly k generators with no further conditions to be satisfied. That is, it is isomorphic to $\mathbb{Z}^k$. ∎

In particular, this means that, if K is connected, then $H_0(K)$ is isomorphic to $\mathbb{Z}$. If this is the only interesting homology group of K, we will say that K is acyclic. Acyclicity is the key property in using homology to prove fixed point theorems so far, see Lemma A.16 below.

Definition A.15. *Let $C = (\{C_q\}_{q \in \mathbb{Z}}, \{\partial_q\}_{q \in \mathbb{Z}})$ be a chain complex. Then C is called **acyclic** iff*

$$H_q(C) = \begin{cases} \{0\}; & \text{if } q \neq 0, \\ \mathbb{Z}; & \text{if } q = 0. \end{cases}$$

Fig. A.4 The truncated face lattice of a triangulation of the projective plane. (By Example 2.2 in [12], this ordered set is acyclic)

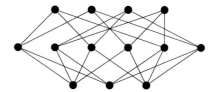

Lemma A.16. *Suppose $K = (V, S)$ is a finite acyclic simplicial complex. Then, for every simplicial map $f : V \to V$, we have that $f^* = \mathrm{id}_{\mathbb{Z}}$ and thus $\Lambda(f) = 1$. Hence, for every simplicial map $f : K \to K$, there is a simplex $\sigma \in S$ such that $f[\sigma] = \sigma$.*

∎

Exercises

A-1. Compute the homology groups of the clique complex of a path of length n.
A-2. Compute the homology groups of the clique complex of a cycle of length n.
A-3. Show that the ordered set in Figure A.4 has an order-preserving self-map f so that $\mathrm{Fix}(f)$ is not acyclic.

A.4   A Homological Reduction Theorem

After the above excursion into algebraic topology, we want to re-connect with combinatorial properties that are sufficient to assure acyclicity. On one hand, for combinatorial results such as the fixed point property, it seems more natural to use combinatorial sufficient conditions. On the other hand, the combinatorial conditions may be easier to check than acyclicity, despite formally being "stronger" than acyclicity. The following is a translation of the results in [204] from their original version for modulo 2 homology and graphs to integer homology and simplicial complexes.

Acyclicity means that homology groups are trivial. Thus we will need lemmas about cancelations.

Lemma A.17 (See [204], Lemma 5). *Let $K = (V, S)$ be a simplicial complex, let $v \in V$ be a vertex, and let $[v, v_1^1, \dots, v_q^1]$, $[v, v_1^2, \dots, v_q^2]$, $\dots$, $[v, v_1^t, \dots, v_q^t]$, $[v_0^{t+1}, v_1^{t+1}, \dots, v_q^{t+1}]$, $\dots$, $[v_0^p, v_1^p, \dots, v_q^p]$ be q-dimensional oriented simplices in the chain complex associated with K such that no v_i^j is equal to v. If v occurs in no oriented simplex of*

$$\partial_q \left(\sum_{j=1}^{t} [v, v_1^j, \dots, v_q^j] + \sum_{j=t+1}^{p} [v_0^j, v_1^j, \dots, v_q^j] \right),$$

then we have that

$$\partial_{q-1}\left(\sum_{j=1}^{t}[v_1^j,\ldots,v_q^j]\right)=0.$$

Proof. In case $q \leq 1$, there is nothing to prove, because the boundary of a sum of zero-dimensional simplices is 0. Thus, in the following, we assume that $q \geq 2$. Let $[w_1,\ldots,w_{q-1}]$ be a $(q-2)$-dimensional oriented simplex that occurs in one of the $(q-1)$-dimensional simplices $[v_1^j,\ldots,v_q^j]$, with $j \in \{1,\ldots,t\}$. Then $[v,w_1,\ldots,w_{q-1}]$ occurs once (positive or negative) in $\partial_q([v,v_1^j,\ldots,v_q^j])$. However, by assumption, we have that $[v,w_1,\ldots,w_{q-1}]$ does not occur in the image

$$\partial_q\left(\sum_{j=1}^{t}[v,v_1^j,\ldots,v_q^j]+\sum_{j=t+1}^{p}[v_0^j,v_1^j,\ldots,v_q^j]\right).$$

Therefore $[v,w_1,\ldots,w_{q-1}]$ occurs in an even number, say, $2k$, of the $[v,v_1^j,\ldots,v_q^j]$, where $j \in \{1,\ldots,t\}$. Moreover it occurs in such a way that, in k of the boundaries $\partial_q([v,v_1^j,\ldots,v_q^j])$, the summand $[v,w_1,\ldots,w_{q-1}]$ is positive and, in the other k boundaries, it is negative. If the simplex $[v,w_1,\ldots,w_{q-1}]$ is a positive summand of $\partial_q([v,v_1^j,\ldots,v_q^j])$, then the simplex $[w_1,\ldots,w_{q-1}]$ is a negative summand of $\partial_{q-1}([v_1^j,\ldots,v_q^j])$ and vice versa. Thus $[w_1,\ldots,w_{q-1}]$ occurs in

$$\partial_{q-1}\left(\sum_{j=1}^{t}[v_1^j,\ldots,v_q^j]\right)$$

k times as a positive summand and k times as a negative summand. Because $[w_1,\ldots,w_{q-1}]$ was arbitrary, the conclusion follows.   ∎

We are now in a position to prove the main theorem about homology in this appendix.

Theorem A.18 (See [204], Theorem 6). *Let $K = (V,S)$ be a simplicial complex, let $q \geq 1$, and let $v \in V$ be such that $H_q(N(v) \setminus \{v\}, Lk(v)) = \{0\}$. Then the following hold.*

1. *If $q = 1$ and $H_0(N(v) \setminus \{v\}, Lk(v)) = \mathbb{Z}$, then $H_1(K) = H_1(K[V \setminus \{v\}])$.*
2. *If $q \geq 2$ and $H_{q-1}(N(v) \setminus \{v\}, Lk(v)) = \{0\}$, then $H_q(K) = H_q(K[V \setminus \{v\}])$.*

Proof. Let $q \geq 1$ and assume that $H_q(N(v) \setminus \{v\}, Lk(v)) = \{0\}$. For $q = 1$, assume that we have $H_0(N(v) \setminus \{v\}, Lk(v)) = \mathbb{Z}$. For $q \geq 2$, assume that we have $H_{q-1}(N(v) \setminus \{v\}, Lk(v)) = \{0\}$.

First, we show that every homology class of q-dimensional cycles has a representative that does not contain v. To do this, we first consider the case $q \geq 2$.

Let

$$C := \sum_{j=1}^{t} [v, v_1^j, \dots, v_q^j] + \sum_{j=t+1}^{p} [v_0^j, v_1^j, \dots, v_q^j]$$

be a representative of the homology class $C + B_q(K)$, with $v_i^j \neq v$ and $\partial_q(C) = 0$. If $t = 0$, we are done, so we can assume that $t \geq 1$. By Lemma A.17,

$$\partial_{q-1}\left(\sum_{j=1}^{t} [v_1^j, \dots, v_q^j]\right) = 0$$

in K and hence this equation holds in $(N(v) \setminus \{v\}, Lk(v))$, too. Because $H_{q-1}(N(v) \setminus \{v\}, Lk(v)) = \{0\}$, there are q-dimensional simplices $[w_0^1, \dots, w_q^1], \dots, [w_0^s, \dots, w_q^s]$ in $(N(v) \setminus \{v\}, Lk(v))$ such that

$$\partial_q\left(\sum_{i=1}^{s} [w_0^i, \dots, w_q^i]\right) = \sum_{j=1}^{t} [v_1^j, \dots, v_q^j].$$

But then

$$\partial_{q+1}\left(\sum_{i=1}^{s} [v, w_0^i, \dots, w_q^i]\right) = \sum_{i=1}^{s} [w_0^i, \dots, w_q^i] - \sum_{j=1}^{t} [v, v_1^j, \dots, v_q^j].$$

Thus

$$\partial_q\left(\sum_{i=1}^{s} [w_0^i, \dots, w_q^i] + \sum_{j=t+1}^{p} [v_0^j, v_1^j, \dots, v_q^j]\right)$$

$$= \partial_q\left(\sum_{i=1}^{s} [w_0^i, \dots, w_q^i] - \sum_{j=1}^{t} [v, v_1^j, \dots, v_q^j]\right)$$

$$= \partial_q\partial_{q+1}\left(\sum_{i=1}^{s} [v, w_0^i, \dots, w_q^i]\right) = 0.$$

Thus

$$C' := \sum_{i=1}^{s} [w_0^i, \dots, w_q^i] + \sum_{j=t+1}^{p} [v_0^j, v_1^j, \dots, v_q^j]$$

forms a q-dimensional cycle with vertices in $V \setminus \{v\}$. Finally, C and C' are homologous, because

$$C' - C = \sum_{i=1}^{s} [w_0^i, \dots, w_q^i] + \sum_{j=t+1}^{p} [v_0^j, v_1^j, \dots, v_q^j]$$

$$- \left(\sum_{j=1}^{t} [v, v_1^j, \dots, v_q^j] + \sum_{j=t+1}^{p} [v_0^j, v_1^j, \dots, v_q^j] \right)$$

$$= \sum_{i=1}^{s} [w_0^i, \dots, w_q^i] - \sum_{j=1}^{t} [v, v_1^j, \dots, v_q^j] = \partial_{q+1} \left(\sum_{i=1}^{s} [v, w_0^i, \dots, w_q^i] \right).$$

We have thus shown that C' is a representative of the homology class of C that does not contain any occurrences of v.

Now we consider the case $q = 1$. Let

$$C := \sum_{j=1}^{t} (-1)^{s_j} [v, v_1^j] + \sum_{j=t+1}^{p} [v_0^j, v_1^j]$$

be a representative of the homology class $C + B_q(G)$, with $v_i^j \neq v$ and $\partial_q(C) = 0$. Then t must be even, say, $t = 2k$ and k of the s_j must be 0, the rest 1, because otherwise the boundary of C contains a multiple of $[v]$. Because, by assumption, $H_0(N(v) \setminus \{v\}, Lk(v)) = \mathbb{Z}$, we have, by Lemma A.14, that the link complex $(N(v) \setminus \{v\}, Lk(v))$, and hence the induced full subcomplex $K[N(v) \setminus \{v\}]$, is connected. However then we infer that

$$\sum_{j=1}^{t} (-1)^{s_j+1} [v_1^j] + B_0(K[N(v) \setminus \{v\}]) = 0 + B_0(K[N(v) \setminus \{v\}]),$$

because, in a connected simplicial complex, any sum of differences of zero-dimensional simplices is the boundary of a sum of 1-dimensional simplices (just use the paths that connect the two terms in each difference). The case $q = 1$ is now finished just like the case $q \geq 2$ (see Exercise A-4).

The remainder of the proof is the same for $q = 1$ and for $q \geq 2$.

We claim that, for any q-cycles C, C' that do not contain v, we have

$$C + B_q(K) = C' + B_q(K) \text{ iff } C + B_q(K[V \setminus \{v\}]) = C' + B_q(K[V \setminus \{v\}]).$$

Provided the claim holds, an isomorphism from $H_q(K)$ to $H_q(K[V \setminus \{v\}])$ would be the map that maps every homology class $\mathcal{C}$ in $H_q(K)$ to the corresponding homology class in $H_q(K[V \setminus \{v\}])$ that is defined by the representatives of $\mathcal{C}$ that do not contain v. Hence, establishing the claim finishes the proof.

To prove the claim, first note that the direction "$\Leftarrow$" is trivial. For the other direction, let $C + B_q(K) = 0 + B_q(K)$ with

$$C = \sum_{j=1}^{p} [v_0^j, v_1^j, \ldots, v_q^j]$$

and $v_i^j \neq v$ for all i, j. Then

$$C = \partial_{q+1} \left(\sum_{i=1}^{t} [v, w_1^i, \ldots, w_q^i, w_{q+1}^i] + \sum_{i=t+1}^{p} [w_0^i, w_1^i, \ldots, w_q^i, w_{q+1}^i] \right),$$

with $w_j^i \neq v$ for all i, j. By Lemma A.17,

$$\sum_{i=1}^{t} [w_1^i, \ldots, w_q^i, w_{q+1}^i]$$

is a q-cycle and, because $H_q(N(v) \setminus \{v\}, Lk(v)) = \{0\}$, there must be $(q+1)$-dimensional simplices in $(N(v) \setminus \{v\}, Lk(v))$ such that

$$\sum_{i=1}^{t} [w_1^i, \ldots, w_q^i, w_{q+1}^i] = \partial_{q+1} \left(\sum_{i=1}^{l} [u_0^i, \ldots, u_q^i, u_{q+1}^i] \right).$$

But then

$$\partial_{q+1} \left(\sum_{i=1}^{l} [u_0^i, \ldots, u_q^i, u_{q+1}^i] + \sum_{i=t+1}^{p} [w_0^i, w_1^i, \ldots, w_q^i, w_{q+1}^i] \right)$$

$$= \partial_{q+1} \left(\sum_{i=1}^{l} [u_0^i, \ldots, u_q^i, u_{q+1}^i] \right) + \partial_{q+1} \left(\sum_{i=t+1}^{p} [w_0^i, w_1^i, \ldots, w_q^i, w_{q+1}^i] \right)$$

$$= \sum_{i=1}^{t} [w_1^i, \ldots, w_q^i, w_{q+1}^i] + \partial_{q+1} \left(\sum_{i=t+1}^{p} [w_0^i, w_1^i, \ldots, w_q^i, w_{q+1}^i] \right)$$

$$+ \partial_{q+1} \left(\sum_{i=1}^{t} [v, w_1^i, \ldots, w_q^i, w_{q+1}^i] \right) - \partial_{q+1} \left(\sum_{i=1}^{t} [v, w_1^i, \ldots, w_q^i, w_{q+1}^i] \right)$$

$$= C + \sum_{i=1}^{t} [w_1^i, \ldots, w_q^i, w_{q+1}^i] - \partial_{q+1} \left(\sum_{i=1}^{t} [v, w_1^i, \ldots, w_q^i, w_{q+1}^i] \right)$$

$$= C + \sum_{i=1}^{t} [w_1^i, \ldots, w_q^i, w_{q+1}^i]$$

$$- \left[\sum_{i=1}^{t} [w_1^i, \ldots, w_q^i, w_{q+1}^i] + \sum_{j=1}^{q} (-1)^j \sum_{i=1}^{t} [v, w_1^i, \ldots, \widehat{w_j^i}, \ldots, w_q^i, w_{q+1}^i] \right]$$

$$= C + \sum_{j=1}^{q} (-1)^j \sum_{i=1}^{t} [v, w_1^i, \ldots, \widehat{w_j^i}, \ldots, w_q^i, w_{q+1}^i] = C,$$

with the last sum being zero because

$$\sum_{i=1}^{t} [w_1^i, \ldots, w_q^i, w_{q+1}^i]$$

is in $B_q(N(v) \setminus \{v\}, Lk(v))$ and hence in $B_q(K[N(v) \setminus \{v\}])$. We have shown that $C \in B_q(K[V \setminus \{v\}])$. For equal nonzero homology classes of K, the above shows that their difference is zero in $K[V \setminus \{v\}]$. This finishes the proof. ∎

Theorem A.18 allows us to compute the homology group $H_q(K)$ of the simplicial complex K from the homology group $H_q(K[V \setminus \{v\}])$, provided that the homology groups $H_q(N(v) \setminus \{v\}, Lk(v))$ and $H_{q-1}(N(v) \setminus \{v\}, Lk(v))$ of the link complex are trivial. Although Theorem A.18 is our main tool for work with acyclicity in ordered sets, note that Theorem A.18 is a theorem about homology in general, as it does not place hypotheses on $H_q(K)$ and $H_q(K[V \setminus \{v\}])$.

Corollary A.19. *Let $K = (V, S)$ be a finite simplicial complex and let $v \in V$ be such that the link complex $(N(v) \setminus \{v\}, Lk(v))$ is acyclic. Then, for all $q \in \mathbb{N} \cup \{0\}$, we have $H_q(K) = H_q(K[V \setminus \{v\}])$. In particular, $K[V \setminus \{v\}]$ is acyclic iff K is acyclic.*

Proof. By Theorem A.18, for all $q \geq 1$, we have $H_q(K) = H_q(K[V \setminus \{v\}])$. Because $(N(v) \setminus \{v\}, Lk(v))$ is acyclic, we have that $(N(v) \setminus \{v\}, Lk(v))$ is connected. This means that $K[V \setminus \{v\}]$ has the same number of components as K. Hence, by Lemma A.14, we have $H_0(K[V \setminus \{v\}]) = H_0(K)$. ∎

The above, together with the removal of escamotable points as discussed in Section 9.6, gives a proof of Baclawski and Björner's result on truncated non-complemented lattices (see Theorem 9.44) that is entirely algebraic. Contractibility of the neighborhood was needed in [12] to guarantee removability of the point without affecting the homology. The contractibility of escamotable graphs was a nice consequence, which, however, is stronger than needed. We have seen here that the notion of a weakly escamotable vertex (see Definition 9.32) is strong enough for our purposes.

Unfortunately, there is no analogue of Corollary A.19 with "acyclic" replaced by "fixed clique property": In the comparability graph of the ordered set in Figure 1.3, the comparability graph of $\downarrow 11 \setminus \{11\}$ has the fixed clique property (in fact, it is acyclic), but the ordered set obtained by removing the vertex 11 does not even have the fixed point property. For the converse, the situation may be slightly better in that I do not have a counterexample. However, the set in part b) of Figure 1.1 has the fixed point property (though its comparability graph does not have the fixed clique property), and if we attach an additional point whose lower covers are exactly the points i and k, then we obtain an ordered set that does not have the fixed point property although the strict lower bounds of the new point form an acyclic ordered set.

We conclude this appendix by connecting dismantlability to acyclicity.

Lemma A.20. *Call a graph* $G = (V, E)$ **dismantlable** *iff either* G *has only one vertex or there is a vertex* $v \in V$ *such that*

- *There is a* $w \in V \setminus \{v\}$ *such that* $N(v) \subseteq N(w)$, *and*
- $G[V \setminus \{v\}]$ *is dismantlable.*

Every finite dismantlable graph is acyclic.

Proof. This is a proof by induction on $n = |V|$. For $n = 1$, there is nothing to prove. For the induction step $\{1, \ldots, n\} \to (n + 1)$, let G have $n + 1$ vertices and let v, w be as in the definition of dismantlable graphs. Then $N(v) \setminus \{v\}$ contains a point w that is adjacent to all other points in $N(v) \setminus \{v\}$. Thus $G[N(v) \setminus \{v\}]$ is dismantlable and hence, by induction hypothesis, acyclic. Moreover, $G[V \setminus \{v\}]$ is dismantlable (see Exercise A-10) and hence, by induction hypothesis, acyclic. Thus, by Corollary A.19, G is acyclic. ∎

Exercises

A-4. Finish the case $q = 1$ in the first part of the proof of Theorem A.18.

A-5. Investigating the hypotheses of Theorem A.18.

    a. Prove that the simplicial complex $K_1(V_1, S_1)$ with vertices $V_1 = \{1, 2, 3\}$ and simplices $S_1 = \mathcal{P}(\{1, 2, 3\}) \setminus \{\emptyset\}$ is acyclic.

    b. Prove that the simplicial complex $K_1(V_2, S_2)$ with vertices $V_2 = \{1, 2, 3, 4\}$ and simplices $S_2 = \mathcal{P}(\{1, 2, 3, 4\}) \setminus \{\emptyset, \{1, 2, 3, 4\}\}$ is not acyclic.

    c. Show that, in the hypotheses of Theorem A.18, the link complex $(N(v) \setminus \{v\}, Lk(v))$ cannot be replaced with the full subcomplex $K(N(v) \setminus \{v\})$ of the center-deleted neighborhood.

A-6. Define connectedly collapsible graphs and prove that they are acyclic.

A-7. Define dismantlable simplicial complexes and prove that they are acyclic.

A-8. Define connectedly collapsible simplicial complexes and prove that they are acyclic.

A-9. Call an ordered set acyclic iff its chain complex is acyclic.

    a. Give an example of an acyclic ordered set P that contains a point $x \in P$ so that $P \setminus \{x\}$ is not acyclic.

    b. Give an example of an ordered set P that is not acyclic and so that, for every point $x \in P$, the card $P \setminus \{x\}$ is acyclic.

c. Show that, for the acyclic ordered set in Figure A.4, for every point $x \in P$, the card $P \setminus \{x\}$ is not acyclic.

A-10. Prove that every simplicial retract of a dismantlable graph is dismantlable.

A-11. It is a consequence of [124], Section 2.1, Exercise 11, that, if K is an acyclic simplicial complex and A is a simplicial retract of K, then A is acyclic, too. Use this fact to prove that an ordered set of height 1 is acyclic iff it contains no crowns.

Remarks and Open Problems

This appendix can only shed the smallest of lights on the vast subject of algebraic topology. The arguments were held entirely algebraic, which may not be as intuitive as possible. However, it allowed the leanest possible complete presentation of the results that we need. For further guidance into the use of algebraic topology, consider the papers [12] and [48] as well as standard texts on algebraic topology such as [124, 295]. One-point reduction methods are also discussed in [15].

The first arguments by Baclawski and Björner that link algebraic topology to fixed points in ordered sets, see [12], Theorems 1.1, 1.2; [48], Théorème 1.1, actually prove Lefschetz type theorems that state that, for an order-preserving map $f : P \to P$, the Euler characteristic of the set Fix(f) of fixed points of f is equal to the Lefschetz number of f. The **Euler(–Poincaré) characteristic** $\chi(C)$ of C is defined to be

$$\chi(C) := \sum_{q \in \mathbb{N}} (-1)^q \rho(C_q),$$

where ρ denotes the rank. It can also be seen as the Lefschetz number of the identity function.

With these results, we also gain an insight into the structure of Fix(f) (also see Section 8.2.2). For brevity's sake, we did not pursue these arguments. We will be satisfied with the existence of a fixed point. Note, however, that Example 2.2 in [12] (see Figure A.4) provides an acyclic ordered set and an order-preserving self-map whose fixed point set is not acyclic (see Exercise A-3).

The main result in this appendix is that, if a simplicial complex is acyclic, then it has the fixed simplex property. I am not aware of any other sufficient conditions for the fixed simplex property that involve homology. If there are such conditions, they cannot solely be conditions on the homology groups: Let B_n be a truncated Boolean lattice on $n \geq 3$ elements, that is, an ordered set isomorphic to $\mathcal{P}(\{1, \ldots, n\}) \setminus \{\emptyset, \{1, \ldots, n\}\}$. Then B_n is the truncated face lattice of a simplicial complex whose zeroth homology group is $\mathbb{Z}$, whose $n - 2^{\text{nd}}$ homology group is $\mathbb{Z}$, and whose other homology groups are trivial. For any finite sequence of distinct natural numbers $3 \leq n_1 < n_2 < \ldots < n_m$ and natural numbers $k_1, \ldots, k_m$, construct an ordered set by, for each $j \in \{1, \ldots, m\}$, taking k_j copies of B_{n_j} in which one minimal element is colored, say, red, and identifying all red elements into one element. The resulting ordered set is the truncated face lattice of a simplicial complex whose zeroth homology group is $\mathbb{Z}$, whose $n_j - 2^{\text{nd}}$ homology group is $\mathbb{Z}^{k_j}$ and whose other homology groups are trivial. Clearly any B_{n_j} is a simplicial retract, so that this complex does not have the fixed simplex property. The same construction using truncated face lattices with the fixed simplex property and whose zeroth homology group is $\mathbb{Z}$, whose $n - 2^{\text{nd}}$ homology group is $\mathbb{Z}$, and whose other homology groups are trivial yields a simplicial complex whose zeroth homology group is $\mathbb{Z}$ and whose $n_j - 2^{\text{nd}}$ homology group is $\mathbb{Z}^{k_j}$ and whose other homology groups are trivial. So, given that, for any nontrivial configuration of homology groups, there are examples with and without the fixed simplex property, are there conditions that can be added to conditions on the homology groups to imply the fixed simplex property?

A major open problem in fixed point theory was to find entirely combinatorial proofs of the results that so far depend on algebraic topology. The recent paper [11] provides a combinatorial proof for many of these results. We present it in Appendix B, using the language of discrete Morse theory. It stands to reason that a constructive understanding of the key counting argument in Appendix B should lead to further insights into the theory of fixed simplices for simplicial complexes.

Appendix B
Some Discrete Morse Theory

Theorem 9.44, first published in [12], baffled researchers in ordered sets for about 30 years. The proof that uses algebraic topology to establish the fixed point property can of course be verified, and it is presented here in Appendix A and in Section 9.7. However, with the result being a combinatorial result, there was a desire for a combinatorial proof. Baclawski had such a proof in the mid-1990s and he did share preprints, but the proof was only published recently in [11]. The argument in [11] essentially uses discrete Morse theory, albeit with different language, as it was developed from scratch. (Such parallel developments are not uncommon, though they become rarer and more impressive as the complexity of the arguments increases.) In this appendix, we present the argument from [11] in the language of discrete Morse theory. For a fundamental and very readable introduction to discrete Morse theory, consider [96]. For a recasting of discrete Morse theory in terms of ordered sets, consider chapter 11 of [172].

B.1   Discrete Morse Functions

We start with simplicial complexes and a standard notation that simplifies the ways in which the size/dimension of the simplex is specified.

Definition B.1. *We will denote the dimension of a simplex as an exponent in parentheses. That is, when $K = (V, S)$ is a simplicial complex and $\alpha \in S$ is a d-dimensional simplex, then we will also write $\alpha^{(d)}$ instead of α to indicate the dimension of the simplex.*

All simplicial complexes in this appendix are assumed to be finite. Simplices are ordered by inclusion. A discrete Morse function on a simplicial complex is now a function that is "generally increasing as the dimension increases," but the definition allows for the occasional occurrence of "defects," if you will.

© Springer International Publishing 2016
B. Schröder, *Ordered Sets*, DOI 10.1007/978-3-319-29788-0

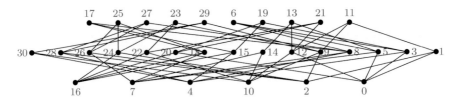

Fig. B.1 A simplicial complex (see [11], Example 16) on six vertices with a discrete Morse function's values indicated next to the simplices, which are depicted as points in an ordered set. (To obtain the sets that make up the simplicial complex, assign distinct singletons to the minimal elements and then use unions for suprema)

Definition B.2. *Let $K = (V, S)$ be a simplicial complex. A function $\phi : S \to \mathbb{R}$ is called a **discrete Morse function** iff, for every d-dimensional simplex $\alpha^{(d)} \in S$, the following two conditions hold.*

1. $\left| \{ \beta^{(d+1)} > \alpha^{(d)} : \phi \left(\beta^{(d+1)} \right) \leq \phi \left(\alpha^{(d)} \right) \} \right| \leq 1.$
2. $\left| \{ \beta^{(d-1)} < \alpha^{(d)} : \phi \left(\beta^{(d-1)} \right) \geq \phi \left(\alpha^{(d)} \right) \} \right| \leq 1.$

As a first example of a simplicial complex with a discrete Morse function, consider Figure B.1. By Exercise B-1, the function given in the picture is indeed a discrete Morse function. By Exercise B-2, a discrete Morse function can be made injective without losing essential properties, so that it is no accident that the values in Figure B.1 are a set of consecutive integers.

It turns out that the "defects," if they occur, can only occur at one end. That is, either there is an upper cover that does not have a strictly larger ϕ-value or there is a lower cover that does not have a strictly smaller ϕ-value, but not both.

Proposition B.3. *Let $K = (V, S)$ be a simplicial complex, let $\phi : S \to \mathbb{R}$ be a discrete Morse function, and let $\alpha^{(d)} \in S$ be a d-dimensional simplex. Then at most one of the two numbers $\left| \{ \beta^{(d+1)} > \alpha^{(d)} : \phi(\beta^{(d+1)}) \leq \phi(\alpha^{(d)}) \} \right|$ and $\left| \{ \beta^{(d-1)} < \alpha^{(d)} : \phi(\beta^{(d-1)}) \geq \phi(\alpha^{(d)}) \} \right|$ is equal to 1.*

Proof. Suppose, for a contradiction, that $\alpha^{(d)} \in S$ is a simplex so that there is a $\beta^{(d+1)} > \alpha^{(d)}$ so that $\phi(\beta^{(d+1)}) \leq \phi(\alpha^{(d)})$ and so that there is a $\gamma^{(d-1)} < \alpha^{(d)}$ so that $\phi(\gamma^{(d-1)}) \geq \phi(\alpha^{(d)})$. Then $\gamma^{(d-1)}$ has two fewer elements than $\beta^{(d+1)}$, which means that there is exactly one simplex $\delta^{(d)} \neq \alpha^{(d)}$ so that we have the containments $\gamma^{(d-1)} < \delta^{(d)} < \beta^{(d+1)}$. Because ϕ is a discrete Morse function, we have $\phi(\beta^{(d+1)}) > \phi(\delta^{(d)})$ and $\phi(\gamma^{(d-1)}) < \phi(\delta^{(d)})$. However, this would imply the contradiction $\phi(\alpha^{(d)}) \geq \phi(\beta^{(d+1)}) > \phi(\delta^{(d)}) > \phi(\gamma^{(d-1)}) \geq \phi(\alpha^{(d)})$.   ∎

Similar to calculus/analysis, we define critical simplices to be simplices that are candidates for being the place where a Morse function assumes its largest or smallest value. Note that, as Exercise B-3 shows, the analogy is not complete.

Definition B.4. *Let $K = (V, S)$ be a simplicial complex, let $\phi : S \to \mathbb{R}$ be a discrete Morse function, and let $\alpha^{(d)} \in S$ be a d-dimensional simplex. Then α*

*is called a **critical simplex** iff we have* $\left|\{\beta^{(d+1)} > \alpha^{(d)} : \phi(\beta) \leq \phi(\alpha)\}\right| = 0$ *and* $\left|\{\beta^{(d-1)} < \alpha^{(d)} : \phi(\beta) \geq \phi(\alpha)\}\right| = 0.$

Proposition B.3 shows that the noncritical simplices of a discrete Morse function are naturally matched in pairs: Match each simplex with the unique simplex above/below it for which the Morse function does not grow/shrink as we go up/down. These pairs can be considered as vectors that point from the lower-dimensional simplex to the higher-dimensional simplex. Of course, such pairings can be considered in general. The language for upper and lower simplices is from [11]. The notation $\nearrow$ for the corresponding upper and lower simplices is used only in this appendix.

Definition B.5. *Let* $K = (V, \mathcal{S})$ *be a simplicial complex. Then a **discrete vector field** on K is a collection $\mathcal{P}$ of pairs $\{\alpha^{(d)} < \beta^{(d+1)}\}$ of simplices of K so that each simplex is in at most one pair in $\mathcal{P}$. For each pair $\{\alpha^{(d)} < \beta^{(d+1)}\}$, we call $\alpha^{(d)}$ a **lower simplex** and we call $\beta^{(d+1)}$ an **upper simplex**. A simplex that is neither an upper nor a lower simplex will be called a **critical simplex**. Moreover, we denote $\alpha \nearrow := \beta$ and we denote $\nearrow \beta := \alpha$ and we let the sets of lower, upper, and critical simplices be denoted $\mathcal{L}(\mathcal{P}), \mathcal{U}(\mathcal{P}),$ and $\mathcal{C}(\mathcal{P})$, respectively.*

The vector field that naturally arises from the pairs of noncritical simplices of a discrete Morse function is called its gradient field. (Though, formally, it is more like the negative gradient field, as the function stagnates or decreases as we go from the start of the vector to its end.)

Definition B.6. *Let* $K = (V, \mathcal{S})$ *be a simplicial complex and let* $\phi : \mathcal{S} \to \mathbb{R}$ *be a discrete Morse function. Then every noncritical simplex $\alpha^{(d)} \in \mathcal{S}$ is either contained in a pair $\{\alpha^{(d)} < \beta^{(d+1)}\}$ with $\phi(\alpha^{(d)}) \geq \phi(\beta^{(d+1)})$ or it is contained in a pair $\{\beta^{(d-1)} < \alpha^{(d)}\}$ with $\phi(\beta^{(d-1)}) \geq \phi(\alpha^{(d)})$. The collection of these pairs is called the **gradient vector field** $\nabla \phi$ of ϕ.*

Discrete Morse functions are now similar to potential functions for vector fields in that there are no closed paths along which the function will decrease.

Definition B.7. *Let* $K = (V, \mathcal{S})$ *be a simplicial complex and let $\mathcal{P}$ be a discrete vector field on K. A $\mathcal{P}$-**path** is a sequence of simplices*

$$\alpha_0^{(d)}, \beta_0^{(d+1)}, \alpha_1^{(d)}, \beta_1^{(d+1)}, \ldots, \alpha_r^{(d)}, \beta_r^{(d+1)}, \alpha_{r+1}^{(d)}$$

so that, for each $i \in \{0, \ldots, r\}$, we have

$$\left\{\alpha_i^{(d)}, \beta_i^{(d+1)}\right\} \in \mathcal{P} \qquad \text{and} \qquad \beta_i^{(d+1)} > \alpha_{i+1}^{(d)} \neq \alpha_i^{(d)}.$$

*The simplex α_0 is called the **source** of the path and the simplex α_{r+1} is called the **target** of the path.*

*A $\mathcal{P}$-path is a **nontrivial closed $\mathcal{P}$-path** iff $r \geq 1$ and $\alpha_0^{(d)} = \alpha_{r+1}^{(d)}$.*

Theorem B.8. *Let $K = (V, \mathcal{S})$ be a simplicial complex and let $\mathcal{P}$ be a discrete vector field on K. Then $\mathcal{P}$ is the gradient field of a discrete Morse function iff there are no nontrivial closed $\mathcal{P}$-paths.*

Proof. For the "$\Rightarrow$"-direction, let $\mathcal{P}$ be the gradient field of the discrete Morse function ϕ and suppose, for a contradiction, that

$$\alpha_0^{(d)}, \beta_0^{(d+1)}, \alpha_1^{(d)}, \beta_1^{(d+1)}, \ldots, \alpha_r^{(d)}, \beta_r^{(d+1)}, \alpha_{r+1}^{(d)}$$

is a nontrivial closed $\mathcal{P}$-path. Then, by Definition B.6 and Proposition B.3, we have

$$\phi\left(\alpha_0^{(d)}\right) \geq \phi\left(\beta_0^{(d+1)}\right) > \phi\left(\alpha_1^{(d)}\right) \geq \phi\left(\beta_1^{(d+1)}\right) > \cdots$$
$$> \phi\left(\alpha_r^{(d)}\right) \geq \phi\left(\beta_r^{(d+1)}\right) > \phi\left(\alpha_{r+1}^{(d)}\right) = \phi\left(\alpha_0^{(d)}\right),$$

a contradiction.

We will prove the "$\Leftarrow$"-direction by induction on $|\mathcal{S}|$, with the base step being trivial. For the induction step, let $K = (V, \mathcal{S})$ be a simplicial complex, assume that the result holds for simplicial complexes with fewer than $|\mathcal{S}|$ simplices and let $\mathcal{P}$ be a discrete vector field that has no nontrivial closed $\mathcal{P}$-paths.

First consider the case that K has a maximal critical simplex μ. In this case, apply the induction hypothesis to $K' = (V, \mathcal{S} \setminus \{\mu\})$ to obtain a discrete Morse function $\phi' : \mathcal{S} \setminus \{\mu\} \to \mathbb{R}$ whose gradient field is $\mathcal{P}$. Extend the discrete Morse function ϕ' on K' to a discrete Morse function $\phi : \mathcal{S} \to \mathbb{R}$ on K by setting $\phi(\mu)$ to some value that is larger than the largest value of ϕ'. Then the gradient field of ϕ is $\mathcal{P}$.

This leaves the case in which all maximal simplices are noncritical. Let $d + 1$ be the largest dimension any simplex of K can have and let

$$\alpha_0^{(d)}, \beta_0^{(d+1)}, \alpha_1^{(d)}, \beta_1^{(d+1)}, \ldots, \alpha_r^{(d)}, \beta_r^{(d+1)}, \alpha_{r+1}^{(d)}$$

be a $\mathcal{P}$-path that cannot be extended to a longer $\mathcal{P}$-path by adding simplices on either end. Because there are no nontrivial closed $\mathcal{P}$-paths, all simplices in the path are distinct and $\alpha_0^{(d)}$ has no further upper covers. Apply the induction hypothesis to

$$K' = \left(V, \mathcal{S} \setminus \left\{\alpha_0^{(d)}, \beta_0^{(d+1)}\right\}\right)$$

to obtain a discrete Morse function

$$\phi' : \mathcal{S} \setminus \left\{\alpha_0^{(d)}, \beta_0^{(d+1)}\right\} \to \mathbb{R}$$

whose gradient field is

$$\mathcal{P} \setminus \left\{\{\alpha_0^{(d)} < \beta_0^{(d+1)}\}\right\}.$$

Let M be some value that is larger than the largest value of ϕ'. Extend the Morse function ϕ' on K' to a Morse function $\phi : S \to \mathbb{R}$ on K by setting $\phi(\alpha_0^{(d)})$ and $\phi(\beta_0^{(d+1)})$ equal to M. Then, because $\alpha_0^{(d)}$ has no further upper covers (and neither does $\beta_0^{(d+1)}$), the gradient field of ϕ is $\mathcal{P}$.   ∎

Exercises

B-1. Prove that the function indicated in Figure B.1 is a discrete Morse function. Indicate the lower, upper, and critical simplices.

B-2. Let $K = (V, S)$ be a simplicial complex and let $\phi : S \to \mathbb{R}$ be a discrete Morse function. Prove that there is an injective discrete Morse function $\psi : S \to \mathbb{R}$ so that $\nabla \psi = \nabla \phi$.

B-3. Let $K = (V, S)$ be a simplicial complex and let $\phi : S \to \mathbb{R}$ be a discrete Morse function.

    a. Prove that any simplex on which ϕ assumes its minimum is a critical simplex and a singleton.

    b. Prove that, if ϕ assumes its maximum on exactly one simplex μ, which happens to be maximal in the containment order, then μ is a critical simplex.

    c. Give an example of a simplicial complex and a discrete Morse function that has exactly one critical simplex.

    d. Explain how a discrete Morse function can assume its maximum at a noncritical simplex.

B.2   Collapsing Schemes

The key to a combinatorial proof of Theorem 9.44 is called a pseudo cone structure in [11]. For the purpose of this translation, I decided to call the corresponding Morse function a collapsing scheme.[1]

Definition B.9. *Let $K = (V, S)$ be a simplicial complex and let $\phi : S \to \mathbb{R}$ be a discrete Morse function. Then ϕ is called a **collapsing scheme** iff ϕ has exactly one critical simplex.*

Proposition B.10. *Let $K = (V, S)$ be a simplicial complex and let $\phi : S \to \mathbb{R}$ be a collapsing scheme. Then the unique critical simplex of ϕ is a singleton $\{v\}$ and $\phi(\{v\}) < \phi(\alpha)$ for all $\alpha \in S$*

Proof. See Exercise B-3a.   ∎

The reason we call the functions in Definition B.9 collapsing schemes lies in their connection to the idea of elementary collapses in Whitehead's sense (see [321]). The following definitions are taken from [292].

Definition B.11. *Let $K = (V, S)$ be a simplicial complex.*

---

[1] Terminology suggested by Russ Woodroofe.

1. An **end** is a pair $\{\alpha^{(d-1)}, \beta^{(d)}\} \subseteq S$ of simplices so that β is the unique simplex that strictly contains α. The dimension of the end is the dimension d of β. In this situation, α is also called a **free face**.
2. The subcomplex L of K is obtained from K by an **elementary collapse** iff there is an end $e = \{\alpha, \beta\}$ so that $L = (V', S \setminus \{\alpha, \beta\})$, where $V' = V$ if α has more than one vertex and $V' = V \setminus \alpha$ otherwise. We also write $K(e) := L$.
3. A **sequence of elementary collapses** in K is a sequence $\mathbf{e} = (e_1, \ldots, e_n)$ so that e_1 is an end in $K_1 := K$ and so that, for $i = 2, \ldots, n$, e_i is an end of $K_i := K_{i-1}(e_{i-1})$. In this case, we say that K collapses to $K(\mathbf{e}) = K_n(e_n)$.
4. K is called **collapsible** (in Whitehead's sense, see [321]) iff K collapses to one of its vertices.

Theorem B.12 (Also see [95], Theorem 3.3, or, [96], Lemma 2.6). *Let $K = (V, S)$ be a simplicial complex with an odd number of simplices $2n + 1 := |S|$. Then there is a collapsing scheme $\phi : S \to \mathbb{R}$ iff K is collapsible.*

Proof. Both directions are proved by induction on n.

For the direction "$\Rightarrow$," the base case $n = 0$ is trivial. Now, for the induction step, let $K = (V, S)$ be a simplicial complex with $|S| = 2n + 1 \geq 3$ simplices and let $\phi : S \to \mathbb{R}$ be a collapsing scheme. Let $d + 1$ be the largest possible dimension of a simplex in K. Then all $d+1$-dimensional simplices in K are upper simplices. Let $\alpha^{(d)}$ be a lower d-dimensional simplex so that, $\phi\left(\alpha^{(d)}\right) \geq \phi\left(\gamma^{(d)}\right)$ for all lower simplices $\gamma^{(d)} \in S$. Let $\beta^{(d+1)} := \alpha^{(d)} \nearrow$. Then $\phi\left(\alpha^{(d)}\right) \geq \phi\left(\beta^{(d+1)}\right)$, because, otherwise, $\beta^{(d+1)}$ would not be an upper simplex. Now suppose, for a contradiction, that there is another simplex $\eta^{(d+1)} > \alpha^{(d)}$ with $\eta^{(d+1)} \neq \beta^{(d+1)}$. Then, because $\eta^{(d+1)} \neq \beta^{(d+1)} = \alpha^{(d)} \nearrow$, we have $\phi\left(\alpha^{(d)}\right) < \phi\left(\eta^{(d+1)}\right)$. By choice of $\alpha^{(d)}$, this implies, for all lower d-dimensional simplices $\gamma^{(d)}$, that $\phi\left(\eta^{(d+1)}\right) > \phi\left(\alpha^{(d)}\right) \geq \phi\left(\gamma^{(d)}\right)$, contradicting the fact that $\eta^{(d+1)}$ must be an upper simplex. Choose $\alpha_{2n+1} := \alpha^{(d)}$ and $\alpha_{2n} := \beta^{(d+1)} = \alpha_{2n+1} \nearrow$. Let $S_{n-1} := S \setminus \{\alpha_{2n}, \alpha_{2n+1}\}$. If α_{2n+1} is a singleton, let $V_{n-1} := V \setminus \alpha_{2n+1}$, otherwise let $V_{n-1} := V$. Note that $K_{n-1} := (V_{n-1}, S_{n-1})$ is a simplicial complex and that $\phi|_{S_{n-1}}$ is a collapsing scheme on K_{n-1}. The result thus follows from the induction hypothesis.

For the direction "$\Leftarrow$," the base case $n = 0$ is trivial, too. For the induction step, let $K = (V, S)$ be a simplicial complex with $|S| = 2n + 1 \geq 3$ so that, for each $k \in \{1, \ldots, n\}$, the only strict upper bound of α_{2k+1} in $S_k := \{\alpha_1, \ldots, \alpha_{2k+1}\}$ is α_{2k}. If α_{2n+1} is a singleton, let $V_{n-1} := V \setminus \alpha_{2n+1}$, otherwise let $V_{n-1} := V$. Let $K_{n-1} := (V_{n-1}, S_{n-1})$. By induction hypothesis, there is a collapsing scheme $\phi_{n-1} : S_{n-1} \to \mathbb{R}$. Let $M := \max \{\phi_{n-1}(\alpha) : \alpha \in S_{n-1}\}$. Define

$$\phi(\alpha) := \begin{cases} \phi_{n-1}(\alpha); & \text{if } \alpha \in S_{n-1}, \\ M + 2; & \text{if } \alpha = \alpha_{2n+1}, \\ M + 1; & \text{if } \alpha = \alpha_n. \end{cases}$$

Then $\phi : S \to \mathbb{R}$ is a collapsing scheme.   ∎

Remark B.13. It is very tempting to hope that existence of a collapsing scheme can be checked in polynomial time: "Simply" remove ends until this is no longer possible. However, in this case, the situation is fundamentally different from results like Theorems 4.25 and 4.31. There are collapsible complexes that can be collapsed onto non-collapsible complexes. For an explicit example, see [17]. This example also shows that we cannot define unique cores for the removal of escamotable points. In fact, the decision problem whether a given simplicial complex of dimension ≥ 3 is collapsible is NP-complete (see [306] for a recent proof).   □

The fact that, in the truncated face lattice $\mathcal{S}$, a collapsing scheme allows the simplices to be removed in such a way that every simplex (viewed as a vertex in the chain complex of $\mathcal{S}$) is weakly escamotable implies the following connection to homology.

Corollary B.14. *A simplicial complex with a collapsing scheme is so that $\mathcal{S}$, viewed as a truncated lattice with the containment order, is acyclic.*

Proof. Use Corollary A.19 and the fact that, in every elementary collapse, we remove points whose neighborhoods are dismantlable and hence acyclic.   ∎

The converse of Corollary B.14 is not true, see Exercise B-11. Although Theorem B.12 completely characterizes simplicial complexes with a collapsing scheme, because of Remark B.13, it is still instructive to consider some examples of collapsing schemes.

Definition B.15. *A simplicial complex $K = (V, \mathcal{S})$ is called a **cone** iff there is a $v \in V$ so that, for every $\alpha \in \mathcal{S}$, we have that $\alpha \cup \{v\} \in \mathcal{S}$.*

Proposition B.16. *Every cone has a collapsing scheme.*

Proof. Define

$$\phi(\alpha) := \begin{cases} |\alpha|; & \text{if } v \notin \alpha, \\ |\alpha| - 1; & \text{if } v \in \alpha. \end{cases}$$

∎

Aside from cones, there is a recursive way to build further simplicial complexes that have a collapsing scheme. Note the similarity of Theorem B.17 below with Theorem 4.12 and Corollary A.19. In each of these different settings, we use conditions on the structure obtained through removal of a point and on the neighborhood of the point to obtain information about the whole structure.

Theorem B.17 (See [11], Theorem 11). *Let $K = (V, \mathcal{S})$ be a simplicial complex and let $a \in V$ be so that both the full subcomplex $K[V \setminus \{a\}]$ and the link complex $(N(a) \setminus \{a\}, Lk(\{a\}))$ have a collapsing scheme. Then K has a collapsing scheme.*

Proof. Let ϕ_1 be the collapsing scheme on $K[V \setminus \{a\}]$ and let ϕ_2 be the collapsing scheme on $(N(a) \setminus \{a\}, Lk(\{a\}))$. Let $\{w\}$ be the unique critical simplex in the link complex $(N(a) \setminus \{a\}, Lk(\{a\}))$, let $M := \max\{\phi_1(\alpha) : \alpha \subseteq V \setminus \{a\}, \alpha \in S\}$, and (see Proposition B.10) let $m := \phi_2(\{w\}) = \min\{\phi_2(\alpha) : \alpha \cup \{a\} \in S\}$. Define

$$\phi(\alpha) := \begin{cases} \phi_1(\alpha); & \text{if } a \notin \alpha, \\ \phi_2(\alpha \setminus \{a\}) - m + M + 1; & \text{if } a \in \alpha, \alpha \neq \{a\}, \\ M + 1; & \text{if } \alpha = \{a\}. \end{cases}$$

Then $\phi : S \to \mathbb{R}$ is a collapsing scheme.   ∎

Following [11], we will call simplicial complexes that are constructed recursively as suggested in Theorem B.17 link reducible.

Definition B.18 (See [11], Definition 13). *Let $K = (V, S)$ be a simplicial complex. Then K is called* **link reducible** *iff either K consists of a singleton set or there is an $a \in V$ so that $K[V \setminus \{a\}]$ and $(N(a) \setminus \{a\}, Lk(\{a\}))$ are link reducible.*

Clearly, by Theorem B.17, every link-reducible simplicial complex has a collapsing scheme. The simplicial complex in Figure B.1 shows that there are simplicial complexes with a collapsing scheme that are not link reducible (see Exercise B-7).

Regarding the connection to Theorem 4.12, Exercise B-4 shows that, if an ordered set P is connectedly collapsible, then its P-chain complex is link reducible. The converse direction is not true, that is, there are more ordered sets with a link-reducible P-chain complex than there are connectedly collapsible ordered set. We will see that the ordered sets in both classes have the fixed point property. For connectedly collapsible ordered sets, this follows from Theorem 4.12. For ordered sets with a chain complex that has a collapsing scheme, the combinatorial proof will take the remainder of this appendix.

Exercises

B-4. Let P be a connectedly collapsible ordered set.

   a. Prove that the P-chain complex of P is link reducible.
   b. Give an example of an ordered set that is not connectedly collapsible and whose P-chain complex is link reducible.
      Hint. Figure 4 in [182] is one place to find one, but there are other places, too.

B-5. Let $K = (V, S)$ be a simplicial complex and let $a \in V$.

   a. Prove that the following are equivalent.

   i. K has a collapsing scheme $\phi : S \to \mathbb{R}$ so that $\min \phi|_{\uparrow a} > \max \phi|_{K[V \setminus \{a\}]}$.
   ii. K has a collapsing scheme $\phi : S \to \mathbb{R}$ so that $\phi|_{K[V \setminus \{a\}]}$ is a collapsing scheme.
   iii. $K[V \setminus \{a\}]$ has a collapsing scheme and $(N(a) \setminus \{a\}, Lk(\{a\}))$ has a collapsing scheme.

   b. Give an example of a simplicial complex $K = (V, S)$ and a vertex $a \in V$ so that K has a collapsing scheme and $(N(a) \setminus \{a\}, Lk(\{a\}))$ does not.

B-6. Let $K = (V, \mathcal{S})$ be a simplicial complex and let $\phi : \mathcal{S} \to \mathbb{R}$ be a collapsing scheme. The collapsing scheme ϕ is called sorted iff ϕ is injective and, for all $\alpha, \beta \in \mathcal{S}$, with $|\alpha| = |\beta| = 1$ and $\phi(\alpha) < \phi(\beta)$, and for all $\gamma \in \uparrow \alpha \setminus \uparrow \beta$ we have that $\phi(\gamma) < \phi(\beta)$. Prove that K is link reducible iff K has a sorted collapsing scheme.

B-7. Existence of a collapsing scheme does not imply link reducibility.

    a. Prove that the simplicial complex in Figure B.1 is not link reducible.

    b. Prove that the discrete Morse function indicated in Figure B.1 is a collapsing scheme.

B-8. Link reducibility of the P-chain complex of the face lattice.

    a. Let $n \in \mathbb{N}$. Prove that the P-chain complex of $\mathcal{P}(\{1, \ldots, n\}) \setminus \{\emptyset\}$, ordered by set inclusion, is link reducible.
    Hint. Use Exercise B-4a.

    b. Let $n \in \mathbb{N}$. Prove that the P-chain complex of $\mathcal{P}(\{1, \ldots, n\}) \setminus \{\emptyset, \{1, \ldots, n\}, \{1, \ldots, n-1\}\}$, ordered by set inclusion, is link reducible.

    c. Let P be an ordered set and let $p \in P$ be so that the P-chain complex of $\updownarrow p \setminus \{p\}$ and the P-chain complex of $P \setminus \{p\}$ are both link reducible. Prove that the P-chain complex of P is link reducible.

    d. Let $(K, \mathcal{S})$ be a simplicial complex with a collapsing scheme. Prove that the P-chain complex of $\mathcal{S}$, ordered by inclusion, is link reducible.

    e. Show that the P-chain complex of the fixed point set in Exercise 8-16c is link reducible.

    f. Let $K = (V, \mathcal{S})$ be a simplicial complex so that there is a sequence of elementary collapses that reduces $K = (V, \mathcal{S})$ to $L = (W, \mathcal{T})$. Prove that, in $C = (\mathcal{S}, TCL(\mathcal{S}))$, there is an enumeration $\sigma_1, \ldots, \sigma_m$ of the elements of $\mathcal{S} \setminus \mathcal{T}$ so that, for $i = 1, \ldots, m$, we have that the link complex of σ_i in $C[\mathcal{S} \setminus \{\sigma_1, \ldots, \sigma_{i-1}\}]$ is link reducible.
    Note. Together with the example in [17], this shows that successive removal of vertices with link-reducible links does not lead to a unique core, even for chain complexes of ordered sets.

B-9. Show that, for the chain complex of the ordered set in Figure 1.3, there is a sequence of elementary collapses that ends in (a simplicial complex whose set of simplices is isomorphic to) a 12-crown.

B-10. Use Theorem B.12 and Exercises B-4 and B-6 to give an alternative proof for Exercise 4-28.

B-11. Prove that the chain complex of the ordered set in Figure A.4 does not have a collapsing scheme.
    Note. It can be shown that the chain complex of this ordered set is acyclic. Hence the above shows that the converse of Corollary B.14 is not true.

B-12. The connection to [11]. The following is Definition 3 in [11]. Let $K = (V, \mathcal{S})$ be a simplicial complex. A **pseudo cone** structure on K consists of

    • A partition of $\mathcal{S} \cup \{\emptyset\}$ into upper and lower simplices.
    • A partial order $\preceq$ on the simplices so that the following hold.

    – For each upper simplex $\sigma^{(d)}$, the set $\{\tau^{(d-1)} \in \mathcal{S} : \tau^{(d-1)} < \sigma^{(d)}\}$ has a $\preceq$-smallest element, which is a lower simplex denoted $\gamma(\sigma)$.
    – The function γ is a bijection from the set of upper simplices to the set of lower simplices.

    Prove that there is a pseudo cone structure on K iff K has a collapsing scheme.

B-13. Elementary collapses versus the fixed simplex property.

    a. Prove that there are simplicial complexes K and L so that $L = K(e)$ is obtained from K via an elementary collapse, L has the fixed simplex property, and K does not.
    Hint. Use the P-chain complex K of the ordered set in Figure 9.5. Prove that $K[P \setminus \{a\}]$ can be obtained from K via a sequence of elementary collapses. Also see Exercise 9-34.

b. Give explicit examples of simplicial complexes K and L so that $L = K(e)$ is obtained from K via an elementary collapse, L has the fixed simplex property, and K does not. *Hint.* Find where the property changes in the sequence of simplicial complexes in the previous hint.

B.3 Straightening Formulas

A typical way to establish the fixed point property for a finite ordered set is to prove that, for every order-preserving self-map, there is a point that is comparable to its image. However, there are many ordered sets for which it is not obvious whether such a point exists for every order-preserving self-map. It would be nice if there were methods that can be applied to certain points that are not comparable to their images, and which "somehow" will still lead to a fixed point. Obviously, any method used for such points will not be as straightforward as in Theorem 3.32. The notion of a hit from [11] (see Definition B.24) provides such a method, by counting the ways in which simplices can be mapped and not be fixed. We present the first tools needed for the counting argument in this section, translating the ideas from [11] to discrete Morse functions.

The gradient field of a discrete Morse function allows for a variation on the idea of a $\mathcal{P}$-path, which will be exploited in the future. We must be careful with the language, as we will work with paths that are definitely related to the vector field $\mathcal{P}$, but they are not necessarily $\mathcal{P}$-paths as in Definition B.7.

Definition B.19. *Let $K = (V, \mathcal{S})$ be a simplicial complex and let $\mathcal{P}$ be a discrete vector field on K. A $\mathcal{P}$-straightening is a sequence of simplices $\alpha_0, \alpha_1, \ldots, \alpha_r$ so that the following hold.*

1. *For each $i \in \{0, \ldots, \lceil \frac{r}{2} \rceil - 1\}$, we have that $\{\alpha_{2i}, \alpha_{2i+1}\} \in \mathcal{P}$.*
2. *For each $i \in \{1, \ldots, \lfloor \frac{r}{2} \rfloor\}$, we have that $\alpha_{2i-1} > \alpha_{2i} \neq \alpha_{2i-2}$ and the sizes satisfy $|\alpha_{2i-1}| - 1 = |\alpha_{2i}|$.*
3. *The simplices $\alpha_0, \ldots, \alpha_r$ are distinct.*
4. *α_r is not a lower simplex.*

*The simplex α_0 is called the **source** of the straightening and the simplex α_r is called the **target** of the straightening. The set of targets is defined to be*

$$\mathcal{T}(\mathcal{P}) := \mathcal{S} \setminus \mathcal{L}(\mathcal{P}) = \mathcal{U}(\mathcal{P}) \cup \mathcal{C}(\mathcal{P}).$$

The definition of the set of targets is justified, because, for any critical or upper simplex γ, the "sequence" γ is a $\mathcal{P}$-straightening. Moreover, note that, if $\mathcal{P}$ is the gradient field of a discrete Morse function, then condition 3 of Definition B.19 is satisfied when conditions 1 and 2 are satisfied.

Ultimately, we want to count the number of $\mathcal{P}$-straightenings from $f(\alpha)$ to an upper cover of α whenever $|f(\alpha)| = |\alpha|$. Hence, the idea of a straightening formula is natural.

Definition B.20 (See [11], Definition 8). *Let $K = (V, S)$ be a simplicial complex, let $\mathcal{P}$ be a discrete vector field on K, and let $\sigma \in S$ be a simplex. For every simplex τ that is not a lower simplex, let $k_{\sigma \to \tau}$ be the number of $\mathcal{P}$-straightenings with source σ and target τ. The **straightening formula** for σ, denoted $\mathrm{str}(\sigma)$, is defined to be the function $\mathrm{str}(\sigma) : \mathcal{T}(\mathcal{P}) \to \mathbb{N} \cup \{0\}$ given by*

$$\mathrm{str}(\sigma) := \sum_{\tau \in \mathcal{T}(\mathcal{P})} k_{\sigma \to \tau} \mathbf{1}_\tau,$$

where $\mathbf{1}_\tau[\alpha]$ is equal to 1 iff $\alpha = \tau$ and it is equal to zero otherwise. That is, $\mathrm{str}(\sigma)[\tau]$ is the number of $\mathcal{P}$-straightenings with source σ and target τ.

Proposition B.21 (See [11], Theorem 9). *Let $K = (V, S)$ be a simplicial complex, let $\mathcal{P} = \nabla\phi$ be the gradient field of the discrete Morse function $\phi : S \to \mathbb{R}$ on K, and let $\sigma \in S$ be a simplex.*

1. *If $\sigma \in \mathcal{U}(\nabla\phi) \cup \mathcal{C}(\nabla\phi)$, then $\mathrm{str}(\sigma) = \mathbf{1}_\sigma$.*
2. *If $\sigma \in \mathcal{L}(\nabla\phi)$, then*

$$\mathrm{str}(\sigma) = \mathbf{1}_{\sigma \nearrow} + \sum_{\lambda \prec \sigma \nearrow,\ \lambda \neq \sigma} \mathrm{str}(\lambda).$$

Proof. Let $\sigma \in S$ be a simplex. If $\sigma \in \mathcal{U}(\nabla\phi) \cup \mathcal{C}(\nabla\phi)$, then the only $\nabla\phi$-straightening with source σ is the path consisting of the one simplex σ. Hence, in this case, $\mathrm{str}(\sigma) = \mathbf{1}_\sigma$.

For the remainder of the argument, we can assume that σ is a lower simplex. By definition of $\nabla\phi$-straightenings, every $\nabla\phi$-straightening that starts at σ starts with the simplices $\sigma, \sigma \nearrow$. Hence the sequence $\sigma, \sigma \nearrow$ is the only $\nabla\phi$-straightening that starts at σ and has two simplices. All other $\nabla\phi$-straightenings that start at σ have at least three simplices, $\sigma, \sigma \nearrow, \lambda$, where λ is a lower cover of $\sigma \nearrow$ that is not equal to σ. Let $\sigma, \sigma \nearrow, \lambda, \dots, \tau$ be a $\nabla\phi$-straightening that starts at σ and has at least three simplices. Then $\lambda, \dots, \tau$ is a $\nabla\phi$-straightening that starts at λ. The converse is easily established, and the formula follows. ∎

Although, for $d > 0$, there can be many $\mathcal{P}$-straightenings starting at the same d-dimensional source, for sources with exactly one element, the situation is simpler.

Lemma B.22. *Let $K = (V, S)$ be a simplicial complex, let $\mathcal{P} = \nabla\phi$ be the gradient field of the discrete Morse function $\phi : S \to \mathbb{R}$ on K, and let $\sigma \in S$ be a lower simplex with $|\sigma| = 1$. Then there is a unique longest $\nabla\phi$-straightening that starts at σ and ends at a simplex with one element.*

Proof. For every lower simplex α with $|\alpha| = 1$, the simplex $\alpha \nearrow$ has exactly two lower covers, namely, α and $(\alpha \nearrow) \setminus \alpha$. Hence every $\nabla\phi$-straightening that starts at α and ends at a simplex with one element that is not α must start with the simplices $\alpha, (\alpha \nearrow), (\alpha \nearrow) \setminus \alpha$.

For every lower simplex σ with $|\sigma| = 1$, let r_σ be the largest number r so that there is a $\nabla\phi$-straightening $\sigma = \alpha_0, \dots, \alpha_r$. By the above, for all lower simplices σ with $|\sigma| = 1$, we have $r_\sigma \geq 2$. Moreover, if there is a σ so that $r_\sigma = 2k$, then, for $i = 1, \dots, k$, there are simplices τ_i so that $r_{\tau_i} = 2i$.

Let σ be a lower simplex with $|\sigma| = 1$ and $r_\sigma = 2k$. We will prove the claim by induction on r_σ. The base step $k = 1$ (that is, $r_\sigma = 2$) follows from the first paragraph. For the induction step $k \to k + 1$, note that every $\nabla\phi$-straightening that starts at σ and ends at a simplex with one element that is not σ must start with the simplices $\sigma, (\sigma \nearrow), (\sigma \nearrow) \setminus \sigma$. By induction hypothesis, there is a unique longest $\nabla\phi$-straightening that starts at $(\sigma \nearrow) \setminus \sigma$ and ends at a simplex with one element. This proves the result. ∎

Proposition B.23 (See [11], Theorem 10). *Let* $K = (V, S)$ *be a simplicial complex, let* $\mathcal{P} = \nabla\phi$ *be the gradient field of the discrete Morse function* $\phi : S \to \mathbb{R}$ *on* K, *and let* $\rho, \tau \in S$ *be simplices with* $|\rho| = |\tau|$. *Then the following holds for the sum*

$$A(\rho, \tau) := \sum_{\sigma \preceq \rho} \mathrm{str}(\sigma)[\tau].$$

1. *If* ρ *is an upper simplex, then* $A(\rho, \tau)$ *is even.*
2. *If* ρ *is a lower simplex with* $|\rho| \geq 2$ *and all simplices* σ *with* $|\sigma| = |\rho| \geq 2$ *and* $\phi(\tau) \leq \phi(\sigma) < \phi(\rho)$ *are noncritical, then* $A(\rho, \tau)$ *is even.*
3. *If* ρ *is a lower simplex,* $|\rho| = |\tau| = 1$ *and* τ *is the endpoint of the unique longest* $\nabla\phi$-*straightening starting at* ρ, *then* $A(\rho, \tau)$ *is one.*
4. *If* ρ *is a lower simplex,* $|\rho| = |\tau| = 1$ *and* τ *is not the endpoint of the unique longest* $\nabla\phi$-*straightening starting at* ρ, *then* $A(\rho, \tau)$ *is zero.*

Proof. For part 1, let ρ be an upper simplex. By Proposition B.21, we obtain the following.

$$\sum_{\sigma \preceq \rho} \mathrm{str}(\sigma) = \mathbf{1}_\rho + \sum_{\sigma \prec \rho} \mathrm{str}(\sigma)$$

$$= \mathbf{1}_\rho + \mathrm{str}(\nearrow \rho) + \sum_{\sigma \prec \rho,\, \sigma \neq \nearrow \rho} \mathrm{str}(\sigma)$$

$$= \mathbf{1}_\rho + \mathbf{1}_{(\nearrow \rho)\nearrow} + \sum_{\lambda \prec (\nearrow \rho)\nearrow,\, \lambda \neq \nearrow \rho} \mathrm{str}(\lambda) + \sum_{\sigma \prec \rho,\, \sigma \neq \nearrow \rho} \mathrm{str}(\sigma)$$

$$= \mathbf{1}_\rho + \mathbf{1}_\rho + \sum_{\lambda \prec \rho,\, \lambda \neq \nearrow \rho} \mathrm{str}(\lambda) + \sum_{\sigma \prec \rho,\, \sigma \neq \nearrow \rho} \mathrm{str}(\sigma)$$

$$= 2 \left(\mathbf{1}_\rho + \sum_{\sigma \prec \rho,\, \sigma \neq \nearrow \rho} \mathrm{str}(\sigma) \right)$$

Hence, for any simplex τ with $|\rho| = |\tau|$, the sum $A(\rho, \tau)$ is even.
For part 2, let ρ be a lower simplex with $|\rho| \geq 2$. Then the following holds.

$$2\mathrm{str}(\rho) + \sum_{\mu \prec \rho \nearrow} \sum_{\sigma \prec \mu} \mathrm{str}(\sigma)$$

$$= \mathrm{str}(\rho) + \mathrm{str}(\rho) + \sum_{\sigma \prec \rho} \mathrm{str}(\sigma) + \sum_{\mu \prec \rho \nearrow, \, \mu \neq \rho} \sum_{\sigma \prec \mu} \mathrm{str}(\sigma)$$

$$= \mathrm{str}(\rho) + \mathbf{1}_{\rho \nearrow} + \sum_{\lambda \prec \rho \nearrow, \, \lambda \neq \rho} \mathrm{str}(\lambda) + \sum_{\sigma \prec \rho} \mathrm{str}(\sigma) + \sum_{\mu \prec \rho \nearrow, \, \mu \neq \rho} \sum_{\sigma \prec \mu} \mathrm{str}(\sigma)$$

$$= \mathbf{1}_{\rho \nearrow} + \sum_{\sigma \preceq \rho} \mathrm{str}(\sigma) + \sum_{\mu \prec \rho \nearrow, \, \mu \neq \rho} \sum_{\sigma \preceq \mu} \mathrm{str}(\sigma)$$

Because, for every simplex $\sigma < \rho \nearrow$ with $|\sigma| = |\rho \nearrow| - 2$, there are exactly two simplices $\mu_1 \neq \mu_2$ so that $\sigma \prec \mu_1, \mu_2 \prec \rho \nearrow$, every term in the double sum $\sum_{\mu \prec \rho \nearrow} \sum_{\sigma \prec \mu} \mathrm{str}(\sigma)$ on the left side occurs exactly twice. Hence, every value of the function in the above computation is even. Because $|\rho| = |\tau|$, we have that $\tau \neq \rho \nearrow$ and hence $\mathbf{1}_{\rho \nearrow}[\tau] = 0$. Therefore $\sum_{\sigma \preceq \rho} \mathrm{str}(\sigma)[\tau]$ is even iff $\sum_{\mu \prec \rho \nearrow, \, \mu \neq \rho} \sum_{\sigma \preceq \mu} \mathrm{str}(\sigma)[\tau]$ is even.

Let $\mu \prec \rho \nearrow$ be so that $\mu \neq \rho$. Then $\phi(\mu) < \phi(\rho \nearrow) \leq \phi(\rho)$. If $\sum_{\sigma \preceq \mu} \mathrm{str}(\sigma)[\tau] > 0$, then $\mathrm{str}(\mu)[\tau] > 0$ or there is a $\sigma_0 \prec \mu$ so that $\mathrm{str}(\sigma_0)[\tau] > 0$. In case $\mathrm{str}(\mu)[\tau] > 0$, we have $\phi(\mu) > \phi(\tau)$ or $\mu = \tau$. In case there is a $\sigma_0 \prec \mu$ so that $\mathrm{str}(\sigma_0)[\tau] > 0$, we have $\phi(\sigma_0) \geq \phi(\tau)$. If $\phi(\mu) > \phi(\sigma_0)$, then $\phi(\mu) > \phi(\tau)$. If $\phi(\mu) \leq \phi(\sigma_0)$, then $\mu = \sigma_0 \nearrow$ and, because there is a $\nabla\phi$-straightening that starts at σ_0, goes through μ and ends at τ, we obtain $\phi(\mu) > \phi(\tau)$ or $\mu = \tau$. Hence, if $\sum_{\sigma \preceq \mu} \mathrm{str}(\sigma)[\tau] > 0$, then $\phi(\mu) > \phi(\tau)$ or $\mu = \tau$, which, in particular, means $\phi(\rho) > \phi(\mu) \geq \phi(\tau)$.

Let d be the dimension of ρ and of τ. Note that $\rho \neq \tau$, because ρ is a lower simplex. First consider the case that ρ is a d-dimensional lower simplex for which $\phi(\rho)$ is the smallest possible among d-dimensional lower simplices. If the sum $\sum_{\mu \prec \rho \nearrow, \, \mu \neq \rho} \sum_{\sigma \preceq \mu} \mathrm{str}(\sigma)[\tau] = 0$, then $\sum_{\sigma \preceq \rho} \mathrm{str}(\sigma)[\tau]$ is even. Otherwise, for each $\mu \prec \rho \nearrow, \mu \neq \rho$ so that $\sum_{\sigma \preceq \mu} \mathrm{str}(\sigma)[\tau] > 0$, we have $\phi(\rho) > \phi(\mu) \geq \phi(\tau)$. By hypothesis, μ cannot be critical, so that, because there is no d-dimensional lower simplex whose ϕ-value is smaller than $\phi(\rho)$, μ must be an upper simplex. By part 1, $\sum_{\sigma \preceq \mu} \mathrm{str}(\sigma)[\tau]$ is even. Hence $\sum_{\mu \prec \rho \nearrow, \, \mu \neq \rho} \sum_{\sigma \preceq \mu} \mathrm{str}(\sigma)[\tau]$ is even.

Thus $\sum_{\sigma \preceq \rho} \mathrm{str}(\sigma)[\tau]$ is even for the d-dimensional lower simplices ρ for which $\phi(\rho)$ is the smallest possible among d-dimensional lower simplices. The result now follows inductively for all d-dimensional simplices ρ for which there is no critical simplex γ of the same dimension so that $\phi(\rho) > \phi(\gamma) \geq \phi(\tau)$: If $\phi(\rho)$ is not the smallest possible among d-dimensional lower simplices, then, for all $\mu \prec \rho \nearrow$ so that $\mu \neq \rho$, we have that $\sum_{\sigma \preceq \mu} \mathrm{str}(\sigma)[\tau] = 0$ or μ is an upper simplex with even $\sum_{\sigma \preceq \mu} \mathrm{str}(\sigma)[\tau]$ or μ is a lower simplex with $\phi(\mu) < \phi(\rho)$ and, by induction hypothesis, $\sum_{\sigma \preceq \mu} \mathrm{str}(\sigma)[\tau]$ is even.

For parts 3 and 4, let ρ be a lower simplex and let $|\rho| = |\tau| = 1$. Then $\rho \neq \tau$ and the sum $A(\rho, \tau)$ is the number of $\nabla\phi$-straightenings starting at ρ and ending at τ. By Lemma B.22 and because τ must be a critical simplex, the number of $\nabla\phi$-straightenings starting at ρ and ending at τ is one iff τ is the endpoint of the unique longest $\nabla\phi$-straightening starting at ρ and ending at a simplex with one element, zero if not.   ∎

Exercises

B-14. Let $K = (V, S)$ be a simplicial complex and let $\mathcal{P}$ be a discrete vector field on K. Prove that, if a $\mathcal{P}$-straightening contains a critical simplex, then this simplex is the last simplex of the straightening.

B-15. Let $K = (V, S)$ be a simplicial complex, let $\mathcal{P}$ be a discrete vector field on K, and let γ be a critical simplex that is not strictly contained in another simplex. Prove that there is no $\mathcal{P}$-straightening that starts at a simplex $\alpha \neq \gamma$ and that ends at γ.

B-16. Let $K = (V, S)$ be a simplicial complex, let $\mathcal{P}$ be a discrete vector field on K, and let γ be a critical simplex. Prove that, if $\alpha_0, \dots, \alpha_r = \gamma$ is a $\mathcal{P}$-straightening that ends at γ, then $|\alpha_0| = |\gamma|$.

B.4   Hits and Fixed Simplices

We are now ready to define hits and use them to prove that simplicial complexes that (simplicially, see Definition B.11) collapse to a vertex have the fixed simplex property. Note that the definition of a hit uses the fact that the existence of $\nabla\phi$-straightenings from one simplex to another gives another ordering of the simplices that is not the containment order.

Definition B.24 (See [11], Definition 28). *Let $K = (V, S)$ be a simplicial complex, let $\phi : S \to \mathbb{R}$ be a discrete Morse function, and let $f : V \to V$ be a simplicial self-map. A **hit** is an ordered pair $(\alpha^{(d)}, \beta)$ so that*

1. $\alpha^{(d)} < \beta^{(d+1)}$ or $\alpha^{(d)} = \beta^{(d)}$,
2. $|f(\alpha)| = |\alpha|$,
3. $\mathrm{str}(f(\alpha))[\beta] > 0$. *(In particular, this means that β is an upper simplex or a critical simplex.)*

*The **multiplicity** of the hit $(\alpha^{(d)}, \beta)$ is equal to $\mathrm{str}(f(\alpha))[\beta]$, which is the number of $\nabla\phi$-straightenings from $f(\alpha)$ to β. If (α, β) is not a hit, we also say that its multiplicity is zero.*

Proposition B.25 (See [11], Proposition 29). *Let $K = (V, S)$ be a simplicial complex, let $\phi : S \to \mathbb{R}$ be a discrete Morse function, and let $f : V \to V$ be a simplicial self-map. If α is a fixed simplex of f, then exactly one of the following holds.*

1. α is an upper simplex or a critical simplex and (α, α) is a hit.
2. α is a lower simplex and $(\alpha, \alpha \nearrow)$ is a hit.

Moreover, if (α, β) is a hit and α is a fixed simplex, then β is an upper simplex or a critical simplex and $\beta \in \{\alpha, \alpha \nearrow\}$.

Proof. Clearly, α is either upper, lower or critical. Parts 1 and 2 follow directly from the definition.

For the "Moreover," part, let (α, β) be a hit so that $f(\alpha) = \alpha$. If α is an upper simplex or a critical simplex, then $\mathrm{str}(f(\alpha)) = \mathrm{str}(\alpha) = \mathbf{1}_\alpha$. Hence, in this case, the only possible value for β is α.

If α is a lower simplex, then, $\alpha = f(\alpha)$ implies $\mathrm{str}(\alpha)[\beta] = \mathrm{str}(f(\alpha))[\beta] > 0$. Because (α, β) is a hit and $\mathrm{str}(\alpha)[\alpha] = 0$, we must have that $\beta^{(d+1)} > \alpha^{(d)}$. However, the only upper cover β of α so that $\mathrm{str}(\alpha)[\beta] > 0$ is $\beta = \alpha \nearrow$.   ∎

Proposition B.26 (See [11], Theorem 31, part (1)). *Let $K = (V, S)$ be a simplicial complex, let $\phi : S \to \mathbb{R}$ be a discrete Morse function, and let $f : V \to V$ be a simplicial self-map. For every simplex β with $|\beta| \geq 2$ so that there are no critical simplices γ with $|\gamma| = |\beta|$ and $\phi(\gamma) \geq \phi(\beta)$, the number of hits (counted with multiplicity) that have β as their second component is even. For every critical simplex β with $|\beta| = 1$, the number of hits (counted with multiplicity) that have β as their second component is 1 iff β is the target of the unique longest $\nabla\phi$-straightening starting at $f(\beta)$ and ending at a simplex with one element, and it is zero otherwise.*

Proof. We start with $\ell := |\beta| \geq 2$. If β is not the second component of a hit, then there is nothing to prove. So let β be a simplex so that there is a simplex α so that (α, β) is a hit. Because f is a simplicial map, we have $|f(\beta)| \leq \ell$. If $|f(\beta)|$ were strictly smaller than $\ell - 1$, we would have $|f(\alpha)| \leq |f(\beta)| < \ell - 1$ and $\mathrm{str}(f(\alpha))[\beta]$ could not be nonzero. Hence $|f(\beta)| \in \{\ell - 1, \ell\}$.

First consider the case that $|f(\beta)| = \ell - 1 \geq 1$. Because f is a simplicial map, there are exactly two lower covers, call them λ_1 and λ_2, of β so that we have $|f(\lambda_i)| = \ell - 1 = |\lambda_i|$. Moreover, again because f is a simplicial map, we have $f(\lambda_1) = f(\lambda_2) = f(\beta)$. Hence the multiplicity of the hit (λ_1, β) equals that of the hit (λ_2, β) and the total number of hits (α, β) (counted with multiplicity) is even.

Second, consider the case that $|f(\beta)| = \ell \geq 2$. Because f is a simplicial map, f is bijective on β. Hence $\{\lambda : \lambda \preceq f(\beta)\} = \{f(\alpha) : \alpha \preceq \beta\}$ and then

$$\sum_{\alpha \preceq \beta} \mathrm{str}(f(\alpha))[\beta] = \sum_{\lambda \preceq f(\beta)} \mathrm{str}(\lambda)[\beta].$$

The former sum is the number of hits (counted with multiplicity) with second component β. The latter sum is even by Proposition B.23 with $\rho = f(\beta)$ and $\tau = \beta$.

Finally, let β be a critical simplex with $|\beta| = 1$. If β is not the second component of a hit, then there is nothing to prove. So let β be a simplex so that there is a simplex α so that (α, β) is a hit. Then we have $\alpha = \beta$, $|f(\alpha)| = 1$ and we have that

$\mathrm{str}(f(\beta))[\beta] = \mathrm{str}(f(\alpha))[\beta] > 0$. By Lemma B.22, this implies $\mathrm{str}(f(\beta))[\beta] = 1$, which is the multiplicity of (α, β). ∎

Proposition B.27 (See [11], Theorem 31, part (2)). *Let $K = (V, S)$ be a simplicial complex, let $\phi : S \to \mathbb{R}$ be a discrete Morse function, and let $f : V \to V$ be a simplicial self-map. For every simplex α, the number of hits (counted with multiplicity) that have α as a first component is odd iff α is a fixed simplex of f.*

Proof. For any $\alpha, \rho \in S$ with $|\alpha| = |\rho|$, we define

$$Q(\alpha, \rho) := \sum_{\beta \supseteq \alpha} \mathrm{str}(\rho)[\beta].$$

Note that, in case $|\alpha| = |f(\alpha)|$, $Q(\alpha, f(\alpha))$ is the number of hits (counted with multiplicity) that have α as a first component.

We will first prove that, for all $\alpha \in S$, we have that $Q(\alpha, \alpha) = 1$. If α is an upper simplex or a critical simplex, then $\mathrm{str}(\alpha) = \mathbf{1}_\alpha$ and hence $Q(\alpha, \alpha) = 1$. If α is a lower simplex, then $\mathrm{str}(\alpha)[\alpha \nearrow] = 1$. Now let $\beta \in S$ be so that $\mathrm{str}(\alpha)[\beta] > 0$ and $\beta \supseteq \alpha$. Then $|\beta| = |\alpha| + 1$ and $\phi(\nearrow \beta) > \phi(\lambda)$ for all $\lambda \prec \beta, \lambda \neq \nearrow \beta$. On the other hand, because there is a $\nabla\phi$-straightening from α to β, we have that $\alpha \neq \nearrow \beta$ would imply that $\phi(\alpha) > \phi(\nearrow \beta)$, which cannot be, because $\alpha \prec \beta$. Hence $\alpha = \nearrow \beta$ and hence $\beta = \alpha \nearrow$. Thus $Q(\alpha, \alpha) = 1$.

Now we will prove that, for all distinct $\alpha, \rho \in S$, we have that $Q(\alpha, \rho)$ is even. The proof is by induction on $\sum_{\tau \in S} \mathrm{str}(\rho)[\tau]$. In the base step, $\sum_{\tau \in S} \mathrm{str}(\rho)[\tau] = 1$, ρ cannot be a lower simplex, because, by Proposition B.21, the sum is greater than 1 for lower simplices. Hence ρ is an upper simplex or a critical simplex, which implies that $\mathrm{str}(\rho) = \mathbf{1}_\rho$. Because $|\rho| = |\alpha|$ and $\rho \neq \alpha$, for all $\beta \supseteq \alpha$, we have $\mathrm{str}(\rho)[\beta] = \mathbf{1}_\rho[\beta] = 0$.

For the induction step, let ρ be so that $\sum_{\tau \in S} \mathrm{str}(\rho)[\tau] = n > 1$, and assume that the result has been proved for all simplices ρ' so that $\sum_{\tau \in S} \mathrm{str}(\rho')[\tau] < n$. Because $n > 1$, ρ must be a lower simplex. By Proposition B.21, we have that $\mathrm{str}(\rho) = \mathbf{1}_{\rho\nearrow} + \sum_{\lambda \prec \rho\nearrow, \lambda \neq \rho} \mathrm{str}(\lambda)$. Let $\ell := |\rho \nearrow|$ and let $\lambda_1 = \rho, \lambda_2, \ldots, \lambda_\ell$ be the lower covers of $\rho \nearrow$. Then

$$Q(\alpha, \rho) = \begin{cases} 1 + \sum_{j=2}^{\ell} Q(\alpha, \lambda_j); & \text{if } \alpha \subseteq \rho \nearrow, \\ \sum_{j=2}^{\ell} Q(\alpha, \lambda_j); & \text{if } \alpha \not\subseteq \rho \nearrow. \end{cases}$$

In case $\alpha \subseteq \rho \nearrow$, α is equal to one of the λ_j, say, $\alpha = \lambda_2$. Then $\alpha \neq \lambda_j$ for $j \geq 3$ and, by induction hypothesis, $Q(\alpha, \rho) = 1 + 1 + \sum_{j=3}^{\ell} Q(\alpha, \lambda_j)$, which is even.

In case $\alpha \not\subseteq \rho \nearrow$, α is not equal to any of the λ_j. Hence, by induction hypothesis, all $Q(\alpha, \lambda_j)$ are even and so is $Q(\alpha, \rho)$.

Thus $Q(\alpha, \rho)$ is odd iff $\alpha = \rho$. Hence $Q(\alpha, f(\alpha))$, the number of hits (counted with multiplicity) with first component α, is odd iff $\alpha = f(\alpha)$. ∎

Theorem B.28 (See [11], Corollary 33). *Let $K = (V, S)$ be a simplicial complex so that there is a collapsing scheme $\phi : S \to \mathbb{R}$. Then every simplicial self-map $f : V \to V$ has a fixed simplex.*

Proof. Let $f : V \to V$ be a simplicial self-map. By Proposition B.26, because there are no critical simplices with 2 or more elements, for every simplex β with $|\beta| > 1$, there is an even number of hits (counted with multiplicity) with second component β. Moreover, if $\{v\}$ is the unique critical simplex of ϕ, then every $\nabla\phi$-straightening that ends at a singleton simplex ends at $\{v\}$ and there is a unique $\nabla\phi$-straightening from $f(\{v\})$ to $\{v\}$. Hence the number of hits (counted with multiplicity) whose second component is a simplex β with $|\beta| = 1$ is 1. Overall, this means that the number of hits (counted with multiplicity) is odd.

Therefore, there must be a simplex α so that the number of hits (counted with multiplicity) with first component α is odd. Hence, by Proposition B.27, α is a fixed simplex of f. ∎

Exercises

B-17. Let $K = (V, S)$ be a simplicial complex so that there is a discrete Morse function $\phi : S \to \mathbb{R}$ with exactly three critical simplices, one of which, call it γ, has two elements and two of which, call them η and ν, have one element. Moreover, assume that $\phi(\gamma)$ is the largest value of ϕ on the two element simplices and that $\phi(\nu)$ is the largest value of ϕ on the one-element simplices. Prove that K has the fixed simplex property.

Remarks and Open Problems

1. Can the results here be extended to prove results on the fixed simplex property for simplicial complexes that only collapse onto another non-singleton simplicial complex (that may not be acyclic)?
2. Is there a nontrivial application of Exercise B-17? Functions as in Exercise B-17 can be constructed from collapsing schemes: Attach an element whose link is a singleton. So the question is if there are simplicial complexes that have such a function and no collapsing scheme. I think an argument as in Theorem B.12 should show that the answer is "no," but Exercise B-17 still is a good exercise to review the argument that provides Theorem B.28.
3. Can the results presented here be translated into results for ordered sets that are not truncated lattices?
4. Let P be an ordered set whose chain complex has a collapsing scheme and let $f : P \to P$ be order-preserving. Conjecture 34 in [11] states that $\text{Fix}(f)$ should have a collapsing scheme, too, but the following example by M. Kukieła shows that this is not true.

Let K be a 3-dimensional simplicial complex that has a collapsing scheme, but which can also be collapsed onto a non-collapsible subcomplex L of K. (For an explicit example of such a simplicial complex, see [17].) Now let X be the simplicial complex that is built of two copies of the complex K, say, K and K', glued by their common subcomplex L. Because X collapses to K (collapse the K' part to L) and K is collapsible, the complex X is collapsible. However, let f be the simplicial map of X that is constant on L and that maps the vertices of K that are outside of L to their corresponding vertices in K' and vice versa. The fixed simplices of f form the subcomplex L and L is not collapsible. To produce an example of an ordered set with a chain complex that has a collapsing scheme, take the truncated face lattice of X and the order-preserving map that is induced by f. Now the P-chain complex of P is link reducible (remove vertices in the order in which the corresponding simplices in X would be removed in a collapsing scheme) and hence it has a collapsing scheme.

References

1. Abian, A. (1971). Fixed point theorems of the mappings of partially ordered sets. *Rendiconti del Circolo Mathematico di Palermo, 20*, 139–142.
2. Abian, S., & Brown, A. B. (1961). A theorem on partially ordered sets with applications to fixed point theorems. *Canadian Journal of Mathematics, 13*, 78–82.
3. Adámek, J., Herrlich, H., & Strecker, G. (1990). *Abstract and concrete categories.* New York: Wiley.
4. Aeschlimann, R., & Schmidt, J. (1992). Drawing orders using less ink. *Order, 9*, 5–13.
5. Aho, A. V., Hopcroft, J. E., & Ullman, J. D. (1983). *Data structures and algorithms.* Reading: Addison-Wesley.
6. Alvarez, L. (1965). Undirected graphs realizable as graphs of modular lattices. *Canadian Journal of Mathematics, 17*, 923–932.
7. Arditti, J. C. (1976). Graphes de comparabilité et dimension des ordres. *Notes de Recherches*, CRM 607. Centre de Recherches Mathématiques Université de Montréal.
8. Arditti, J. C., & Jung, H. A. (1980). The dimension of finite and infinite comparability graphs. *Journal of the London Mathematical Society, 21*(2), 31–38.
9. Bacchus, F., & Grove, A. (1995). On the forward checking algorithm. In *Principles and practices in constraint programming (CP-95). Lecture notes in computer science* (Vol. 976, pp. 292–309). Berlin: Springer. Available at http://www.cs.toronto.edu/~fbacchus/on-line.html
10. Baclawski, K. (1977). Galois connections and the Leray spectral sequence. *Advances in Mathematics, 25*, 191–215.
11. Baclawski, K. (2012). A combinatorial proof of a fixed point property. *Journal of Combinatorial Theory, Series A, 119*, 994–1013.
12. Baclawski, K., & Björner, A. (1979). Fixed points in partially ordered sets. *Advances in Mathematics, 31*, 263–287.
13. Balof, B., & Bogart, K. (2003). Simple inductive proofs of the Fishburn and Mirkin theorem and the Scott–Suppes theorem. *Order, 20*, 49–51.
14. Bandelt, H. J., & van de Vel, M. (1987). A fixed cube theorem for median graphs. *Discrete Mathematics, 67*, 129–137.
15. Barmak, J. (2011). *Algebraic topology of finite topological spaces and applications. Lecture notes in mathematics* (Vol. 2032). Heidelberg: Springer.
16. Bélanger, M. F., Constantin, J., & Fournier, G. (1994). Graphes et ordonnés démontables, propriété de la clique fixe. *Discrete Mathematics, 130*, 9–17.

© Springer International Publishing 2016
B. Schröder, *Ordered Sets*, DOI 10.1007/978-3-319-29788-0

17. Benedetti, B., & Lutz, F. (2013). The dunce hat in a minimal non-extendably collapsible 3-ball. Electronic Geometry Model No. 2013.10.001. Available at http://www.eg-models. de/models/Polytopal_Complexes/2013.10.001/_direct_link.html and at http://arxiv.org/abs/ 0912.3723

18. Bergman, C., McKenzie, R., & Nagy, Sz. (1982). How to cancel a linearly ordered exponent. *Colloquia Mathematica Societatis János Bolyai, 21*, 87–94.

19. Berman, J., & Köhler, P. (1976). Cardinalities of finite distributive lattices. *Mitteilungen aus dem Mathematics Seminar Giessen, 121*, 103–124.

20. Bhavale, A. N. (2013). Enumeration of Certain Algebraic Systems and Related Structures. Ph.D. thesis, University of Pune.

21. Birkhoff, G. (1967). *Lattice theory* (3rd ed.). Providence: AMS Colloquium Publications XXV.

22. Björner, A. (1981). Homotopy type of posets and lattice complementation. *Journal of Combinatorial Theory, Series A, 30*, 90–100.

23. Blyth, T. (2005). *Lattices and ordered algebraic structures*. London: Springer.

24. Bogart, K. P. (1994). Intervals and orders: What comes after interval orders? In V. Bouchitté & M. Morvan (Eds.), *Orders, algorithms and applications (Proceedings of the ORDAL '94 in Lyon). Lecture notes in computer science* (Vol. 831, pp. 13–32). Berlin: Springer.

25. Bogart, K. P., Freese, R., Kung, J. P. S. (Eds.). (1990). *The Dilworth theorems: Selected papers of Robert P. Dilworth*. Boston: Birkhäuser.

26. Bollobás, B. (1978). *Extremal graph theory*. London/New York/San Francisco: Academic.

27. Bollobás, B. (1979). *Graph theory. Graduate texts in mathematics* (Vol. 63). New York: Springer.

28. Bondy, J. A., & Hemminger, R. L. (1976). Graph reconstruction – A survey. *Journal of Graph Theory, 1*, 227–268.

29. Bordat, J. P. (1992). Sur l'algorithmique combinatoire d'ordres finis. Thèse de Docteur d'état, Université Montpellier II.

30. Bouchitté, V., & Habib, M. (1989). The calculation of invariants for ordered sets. In I. Rival (Ed.), *Algorithms and order* (pp. 231–279). Dordrecht: Kluwer Academic.

31. Bouchitté, V., & Morvan, M. (1994). *Orders, algorithms and applications (Proceedings of the International Workshop ORDAL '94 in Lyon). Lecture notes in computer science* (Vol. 831). Berlin: Springer.

32. Brightwell, G. (1988). Linear extensions of infinite posets. *Discrete Mathematics, 70*, 113–136.

33. Brightwell, G. (1989). Semiorders and the $\frac{1}{3}$-$\frac{2}{3}$ conjecture. *Order, 5*, 369–380.

34. Brightwell, G. (1993). On the complexity of diagram testing. *Order, 10*, 297–303.

35. Brightwell, G., & Goodall, S. (1990). The number of partial orders of fixed width. *Order, 15*, 315–337.

36. Brightwell, G. R., Felsner, S., & Trotter, W. T. (1995). Balancing pairs and the cross product conjecture. *Order, 12*, 327–349.

37. Brinkmann, G., & McKay, B. (2002). Posets on up to 16 points. *Order, 19*, 147–179.

38. Brown, R. (1982). The fixed point property and Cartesian products. *The American Mathematical Monthly*, November issue, 654–678.

39. Brualdi, R. A., Jung, H. C., & Trotter, W. T. (1994). On the poset of all posets on *n* elements. *Discrete Applied Mathematics, 50*, 111–123.

40. Carl, S., & Heikkilä, S. (1992). An existence result for elliptic differential inclusions with discontinuous nonlinearity. *Nonlinear Analysis, 18*, 471–479.

41. Carl, S., & Heikkilä, S. (2011). *Fixed point theory in ordered sets and applications; from differential and integral equations to game theory*. New York: Springer.

42. Caspard, N., Leclerc, B., & Monjardet, B. (2012). *Finite ordered sets, concepts, results and uses*. New York: Cambridge University Press.

43. Chang, C., Jonsson, B., & Tarski, A. (1964). Refinement properties for relational structures. *Fundamenta Mathematicae, 55*, 249–281.

44. Chaunier, C., & Lygerōs, N. (1992). The number of orders with thirteen elements. *Order, 9,* 203–204.
45. Chen, P. C. (1992). Heuristic sampling: A method for predicting the performance of tree searching programs. *SIAM Journal on Computing, 21,* 295–315.
46. Clay Mathematics Institute web site. (2015). http://www.claymath.org/millennium-problems
47. Cohn, D. L. (1980). *Measure theory.* Boston: Birkhäuser.
48. Constantin, J., & Fournier, G. (1985). Ordonnés escamotables et points fixes. *Discrete Mathematics, 53,* 21–33.
49. Cook, S. A. (1971). The complexity of theorem-proving procedures. In *Proceedings of the Third Annual ACM Symposium on Theory of Computing* (pp. 151–158). New York: Association for Computing Machinery.
50. Corominas, E. (1990). Sur les ensembles ordonnés projectifs et la propriété du point fixe. *Comptes Rendus de l'Académie des Sciences Paris Série I, 311,* 199–204.
51. Cousot, P., & Cousot, R. (1979). Constructive versions of Tarski's fixed point theorems. *Pacific Journal of Mathematics, 82,* 43–57.
52. Cousot, P., & Cousot, R. (1979). A constructive characterization of the lattices of all retractions, preclosure, quasi-closure and closure operators on a complete lattice. *Portugaliae Mathematica, 38,* 185–198.
53. Crapo, H. H. (1982). Ordered sets: Retracts and connections. *Journal of Pure and Applied Algebra, 23,* 13–28.
54. Crawley, P., & Dilworth, R. P. (1973). *Algebraic theory of lattices.* Englewood Cliffs: Prentice Hall.
55. Culberson, J. C., & Rawlins, G. J. E. (1991). New results from an algorithm for counting posets. *Order, 7,* 361–374.
56. Davey, B., & Priestley, H. (1990). *Introduction to lattices and order.* Cambridge: Cambridge University Press.
57. Davidson, J. L. (1986). Asymptotic enumeration of partial orders. *Congressus Numerantium, 53,* 277–286.
58. Davis, A. C. (1955). A characterization of complete lattices. *Pacific Journal of Mathematics, 5,* 311–319.
59. Debruyne, R., & Bessière, C. (1997). Some practicable filtering techniques for the constraint satisfaction problem. In *Proceedings of the 15th International Joint Conference on Artificial Intelligence (IJCAI)* (pp. 412–417).
60. Dechter, R. (1992). Constraint networks. In S. Shapiro (Ed.), *Encyclopedia of artificial intelligence* (pp. 276–284). New York: Wiley.
61. Dechter, R., & Pearl, J. (1989). Tree clustering for constraint networks. *Artificial Intelligence, 38,* 353–366.
62. Dieudonné, J. (1960). *Foundations of modern analysis.* New York/London: Academic.
63. Dilworth, R. P. (1950). A decomposition theorem for partially ordered sets. *Annals of Mathematics, 51,* 161–166.
64. Dilworth, R. P. (1990). Chain partitions in ordered sets. In K. P. Bogart, R. Freese, & J. P. S. Kung (Eds.), *The Dilworth theorems: Selected papers of Robert P. Dilworth* (pp. 1–6). Boston: Birkhäuser.
65. Donalies, M., & Schröder, B. (2000). Performance guarantees and applications for Xia's algorithm. *Discrete Mathematics, 213,* 67–86 (Proceedings of the Banach Center Minisemester on Discrete Mathematics, Week on Ordered Sets).
66. Dreesen, B., Poguntke, W., & Winkler, P. (1985). Comparability invariance of the fixed point property. *Order, 2,* 269–274.
67. Duffus, D. (1984). Automorphisms and products of ordered sets. *Algebra Universalis, 19,* 366–369.
68. Duffus, D., & Goddard, T. (1996). The complexity of the fixed point property. *Order, 13,* 209–218.
69. Duffus, D., Kierstead, H. A., & Trotter, W. T. (1991). Fibres and ordered set coloring. *Journal of Combinatorial Theory, Series A, 58,* 158–164.

70. Duffus, D., Łuczak, T., Rödl, V., & Ruciński, A. (1998). Endomorphisms of partially ordered sets. *Combinatorics, Probability and Computing, 7*, 33–46.
71. Duffus, D., Poguntke, W., & Rival, I. (1980). Retracts and the fixed point problem for finite partially ordered sets. *Canadian Mathematical Bulletin, 23*, 231–236.
72. Duffus, D., & Rival, I. (1978). A logarithmic property for exponents of partially ordered sets. *Canadian Journal of Mathematics, 30*, 797–807.
73. Duffus, D., & Rival, I. (1981) A structure theory for ordered sets. *Discrete Mathematics, 35*, 53–118.
74. Duffus, D., Rival, I., & Simonovits, M. (1980). Spanning retracts of a partially ordered set. *Discrete Mathematics, 32*, 1–7.
75. Duffus, D., Rödl, V., Sands, B., & Woodrow, R. (1992). Enumeration of order-preserving maps. *Order, 9*, 15–29.
76. Duffus, D., Sands, B., Sauer, N., & Woodrow, R. E. (1991). Two-coloring all two-element maximal antichains. *Journal of Combinatorial Theory, Series A, 57*, 109–116.
77. Duffus, D., & Sauer, N. (1987). Fixed points of products and the strong fixed point property. *Order, 4*, 221–231.
78. Dushnik, B., & Miller, E. W. (1941). Partially ordered sets. *American Journal of Mathematics, 63*, 600–610.
79. Edelman, P. (1979). On a fixed point theorem for partially ordered sets. *Discrete Mathematics, 15*, 117–119.
80. El-Zahar, M. (1989). Enumeration of ordered sets. In I. Rival (Ed.), *Algorithms and order. NATO advanced science institute series c: Mathematical and physical sciences* (pp. 327–352). Dordrecht: Kluwer Academic.
81. Erné, M. (1981). Open question on p. 843 of [246].
82. Erné, M., & Stege, K. (1991). Counting finite posets and topologies. *Order, 8*, 247–265.
83. Ewacha, K., Li, W., & Rival, I. (1991). Order, genus and diagram invariance. *Order, 8*, 107–113.
84. Fagin, R. (1976). Probabilities on finite models. *Journal of Symbolic Logic, 41*, 50–58.
85. Farley, J. D. (1993). The uniqueness of the core. *Order, 10*, 129–131.
86. Farley, J. D. (1995). The number of order-preserving maps between fences and crowns. *Order, 12*, 5–44.
87. Farley, J. D. (1997). Perfect sequences of chain-complete posets. *Discrete Mathematics, 167/168*, 271–296.
88. Farley, J. D. (1997). The fixed point property for posets of small width. *Order, 14*, 125–143.
89. Farley, J. D., & Schröder, B. (2001). Strictly order-preserving maps into $\mathbb{Z}$, II: A 1979 problem of Erné. *Order, 18*, 381–385.
90. Felsner, S., & Trotter, W. T. (2000). Dimension, graph and hypergraph coloring. *Order, 17*, 167–177.
91. Fishburn, P. C. (1985). *Interval orders and interval graphs: A study of partially ordered sets* New York: Wiley.
92. Fishburn, P. C., & Trotter, W. T. (1999). Geometric containment orders: A survey, *Order, 15*, 168–181.
93. Fofanova, T., Rival, I., & Rutkowski, A. (1994). Dimension 2, fixed points and dismantlable ordered sets. *Order, 13*, 245–253.
94. Fofanova, T., & Rutkowski, A. (1987). The fixed point property in ordered sets of width two. *Order, 4*, 101–106.
95. Forman, R. (1998). Morse theory for cell complexes. *Advances in Mathematics, 134*, 90–145.
96. Forman, R. (2002). A user's guide to discrete Morse theory. *Séminaire Lotharingien de Combinatoire, 48*. Article B48c
97. Fouché, W. (1996). Chain partitions of ordered sets. *Order, 13*, 255–266.
98. Freese, R., Ježek, J., & Nation, J. (1995). *Free lattices. Mathematical surveys and monographs* (Vol. 42). Providence: American Mathematical Society
99. Freuder, E. C. (1982). A sufficient condition for backtrack-free search. *Journal of the ACM, 29*, 24–32.

100. Frías-Armenta, M. E., Neumann-Lara, V., & Pizaña, M. A. (2004). Dismantlings and iterated clique graphs. *Discrete Mathematics, 282*, 263–265.
101. Gallai, T. (1967). Transitiv orientierbare graphen. *Acta Mathematica Hungarica, 18*, 25–66.
102. Ganter, B. (2013). *Diskrete mathematik: Geordnete mengen.* Berlin/Heidelberg: Springer.
103. Garey, M. R., & Johnson, D. S. (1979). *Computers and intractability: A guide to the theory of NP-completeness.* San Francisco: Freeman.
104. Gaschnig, J. (1978). Experimental case studies of backtrack vs. (Waltz)-type vs. new algorithms for satisfying assignment problems. In *Proceedings of the Second Canadian Conference on Artificial Intelligence*, Toronto (pp. 268–277).
105. Ghouilà-Houri, A. (1962). Caractérisation des graphes nonorientés dont on peut orienter les arêtes de manière à obtenir le graphe d'une relation d'ordre. *Comptes Rendus de l'Académie des Sciences Paris, 254*, 1370–1371.
106. Gibson, P., & Zaguia, I. (1998). Endomorphism classes of ordered sets, graphs and lattices. *Order, 15*, 21–34.
107. Gikas, M. (1986). Fixed points and structural problems in ordered sets. Ph.D. dissertation, Emory University.
108. Gilmore, P., & Hoffman, J. (1963). A characterization of comparability graphs and of interval graphs. *Canadian Journal of Mathematics, 16*, 539–548.
109. Golumbic, M. (1980). *Algorithmic graph theory and perfect graphs.* New York: Academic.
110. Gottlob, G. (2012). On minimal constraint networks. *Artificial Intelligence, 191/192*, 42–60.
111. Grant, K., Nowakowski, R., & Rival, I. (1995). The endomorphism spectrum of an ordered set. *Order, 12*, 45–55.
112. Grätzer, G. (1978). *General lattice theory.* Basel: Birkhäuser.
113. Gurevich, Y., & Shelah, S. (1987). Expected computation time for Hamiltonian path problem. *SIAM Journal on Computing, 16*(3), 486–502.
114. Gysin, R. (1977). Dimension transitiv orientierbarer Graphen. *Acta Mathematica Academiae Scientiarum Hungaricae, 29*, 313–316.
115. Habib, M., Morvan, M., Pouzet, M., & Rampon, J.-X. (1993). Interval dimension and MacNeille completion. *Order, 10*, 147–151.
116. Habib, M., Morvan, M., & Rampon, J.-X. (1993). On the calculation of transitive reduction-closure of orders. *Discrete Mathematics, 111*, 289–303.
117. Halmos, P. R. (1974). *Naive set theory. Undergraduate texts in mathematics.* New York: Springer.
118. Han, C.-C., & Lee, C.-H. (1988). Comments on Mohr and Henderson's path consistency algorithm. *Artificial Intelligence, 36*, 125–130.
119. Haralick, R. M., & Elliott, G. L. (1980). Increasing tree search efficiency for constraint satisfaction problems. *Artificial Intelligence, 14*, 263–313.
120. Harzheim, E. (2005). *Ordered sets.* New York: Springer.
121. Hashimoto, J. (1948). On the product decomposition of partially ordered sets. *Mathematica Japonica, 1*, 120–123.
122. Hashimoto, J. (1951). On direct product decomposition of partially ordered sets. *Annals of Mathematics, 54*(2), 315–318.
123. Hashimoto, J., & Nakayama, T. (1950). On a problem of G. Birkhoff. *Proceedings of the American Mathematical Society, 1*, 141–142.
124. Hatcher, A. (2002). *Algebraic topology.* Cambridge: Cambridge University Press. Available at http://www.math.cornell.edu/~hatcher/AT/ATpage.html
125. Hazan, S. (1992). The Projection Property for Orders and Triangle-Free Graphs. Ph.D. dissertation, Vanderbilt University.
126. Hazan, S. (1996). On triangle-free projective graphs. *Algebra Universalis, 35*(2), 185–196.
127. Hazan, S., & Neumann-Lara, V. (1995). Fixed points of posets and clique graphs. *Order, 13*, 219–225.
128. Hazan, S., & Neumann-Lara, V. (1998). Two order invariants related to the fixed point property. *Order, 15*, 97–111.

129. Heikkilä, S. (1990). On fixed points through a generalized iteration method with applications to differential and integral equations involving discontinuities. *Nonlinear Analysis, 14*, 413–426.

130. Heikkilä, S. (1990). On differential equations in ordered Banach spaces with applications to differential systems and random equations. *Differential Integral Equations, 3*, 589–600.

131. Heikkilä, S., & Lakhshmikantham, V. (1994). *Monotone iterative techniques for discontinuous nonlinear differential equations.* New York: Marcel Dekker.

132. Heikkilä, S., & Lakhshmikantham, V. (1995). On mild solutions of first order discontinuous semilinear differential equations in Banach spaces. *Applicable Analysis, 56*, 131–146.

133. Heikkilä, S., Lakhshmikantham, V., & Sun, Y. (1992). Fixed point results in ordered normed spaces with applications to abstract and differential equations. *Journal of Mathematical Analysis and Applications, 163*, 422–437.

134. Heitzig, J., & Reinhold, J. (2000). The number of unlabeled orders on fourteen elements. *Order, 17*, 333–341.

135. Hell, P., & Nešetril, J. (2004). *Graphs and homomorphisms. Oxford lecture series in mathematics and its applications* (Vol. 28). Oxford: Oxford University Press.

136. Heuser, H. (1983). *Lehrbuch der analysis, Teil 2.* Stuttgart: B. G. Teubner.

137. Hewitt, E., & Stromberg, K. R. (1965). *Real and abstract analysis. Springer graduate texts in mathematics* (Vol. 25). Berlin/Heidelberg: Springer.

138. Hiraguchi, T. (1955). On the dimension of orders. *Science Reports of Kanazawa University, 4*, 1–20.

139. Höft, H., & Höft, M. (1976). Some fixed point theorems for partially ordered sets. *Canadian Journal of Mathematics, 28*, 992–997.

140. Höft, H., & Höft, M. (1988). Fixed point invariant reductions and a characterization theorem for lexicographic sums. *Houston Journal of Mathematics, 14*(3), 411–422.

141. Höft, H., & Höft, M. (1991). Fixed point free components in lexicographic sums with the fixed point property. *Demonstratio Mathematica, XXIV*, 294–304.

142. Hogg, T., Huberman, B., & Williams, C. (1996). Phase transitions and the search problem. *Artificial Intelligence, 81*, 1–15.

143. Hopcroft, J., & Karp, R. (1973). A $n^{\frac{5}{2}}$ algorithm for maximum matching in bipartite graphs. *SIAM Journal on Computing, 2*, 225–231.

144. Hughes, J. (2004). The Computation and Comparison of Decks of Small Ordered Sets. MS thesis, Louisiana Tech University.

145. Ille, P. (1993). Recognition problem in reconstruction for decomposable relations. In N. W. Sauer et al. (Eds.), *Finite and infinite combinatorics in sets and logic* (pp. 189–198). Dordrecht: Kluwer Academic.

146. Ille, P., & Rampon, J.-X. (1997). Reconstruction of posets with the same comparability graph. *Journal of Combinatorial Theory (B), 74*, 368–377.

147. Jachymski, J. (2007). The contraction principle for mappings on a metric space with a graph. *Proceedings of the American Mathematical Society, 136*, 1359–1373.

148. Jawhari, E., Misane, D., & Pouzet, M. (1986). Retracts: Graphs and ordered sets from the metric point of view. In I. Rival (Ed.), *Combinatorics and ordered sets. Contemporary mathematics* (Vol. 57, pp. 175–226). Providence: American Mathematical Society.

149. Jeavons, P., Cohen, D., & Pearson, J. (1998). Constraints and universal algebra. *Annals of Mathematics and Artificial Intelligence, 24*, 51–67.

150. Jónsson, B. (1982). Arithmetic of ordered sets. In I. Rival (Ed.), *Ordered sets* (pp. 3–41). Dordrecht: D. Reidel.

151. Jónsson, B., & McKenzie, R. (1982). Powers of partially ordered sets: Cancellation and refinement properties. *Mathematica Scandinavica, 51*, 87–120.

152. Kahn, J., & Saks, M. (1984). Balancing poset extensions. *Order, 1*, 113–126.

153. Kelly, D. (1984). Unsolved problems: Removable pairs in dimension theory. *Order, 1*, 217–218.

154. Kelly, D. (1985). Comparability graphs. In I. Rival (Ed.), *Graphs and order* (pp. 3–40). Dordrecht: D. Reidel.
155. Kelly, D., & Trotter, W. T. (1982). Dimension theory for ordered sets. In I. Rival (Ed.), *Ordered sets* (pp. 171–211). Dordrecht: D. Reidel.
156. Kelly, P. J. (1957). A congruence theorem for trees. *Pacific Journal of Mathematics, 7,* 961–968.
157. Kierstead, H., & Trotter, W. T. (1991). A note on removable pairs. In Y. Alavi et al. (Eds.), *Graph theory, combinatorics and applications* (Vol. 2, pp. 739–742). New York: Wiley.
158. Kimble, R. (1973). Extremal Problems in Dimension Theory for Partially Ordered Sets. Ph. D. dissertation, MIT.
159. Kinoshita, S. (1953). On some contractible continua without fixed point property. *Fundamenta Mathematicae, 40,* 96–98.
160. Kisielewicz, A. (1988). A solution of Dedekind's problem on the number of isotone Boolean functions. *Journal für die Reine und Angewandte Mathematik, 386,* 139–144.
161. Kislitsin, S. S. (1968). Finite partially ordered sets and their associated sets of permutations. *Matematicheskiye Zametki, 4,* 511–518.
162. Klarner, D. (1969). The number of graded partially ordered sets. *Journal of Combinatorial Theory, 6,* 12–19.
163. Klarner, D. (1970). The number of classes of isomorphic graded partially ordered sets. *Journal of Combinatorial Theory, 9,* 412–419.
164. Kleitman, D. J., & Markowsky, G. (1975). On Dedekind's problem: The number of isotone Boolean functions II. *Transactions of the American Mathematical Society, 213,* 373–390.
165. Kleitman, D. J., & Rothschild, B. L. (1970). The number of finite topologies. *Proceedings of the American Mathematical Society, 25,* 276–282.
166. Kleitman, D. J., & Rothschild, B. L. (1975). Asymptotic enumeration of partial orders on a finite set. *Transactions of the American Mathematical Society, 205,* 205–220.
167. Knaster, B. (1928). Un théorème sur les fonctions d'ensembles. *Annales de la Société Polonaise de Mathématique, 6,* 133–134.
168. Knuth, D. E. (1975). Estimating the efficiency of backtrack programs. *Mathematics of Computation, 29,* 121–136.
169. Köbler, J., Schöning, U., & Torán, J. (1993). *The graph isomorphism problem: Its structural complexity. Progress in theoretical computer science.* Boston: Birkhäuser.
170. Kondrak, G., & van Beek, P. (1997). A theoretical evaluation of selected backtracking algorithms. *Artificial Intelligence, 89,* 365–387.
171. Korshunov, A. (1977). On the number of monotone Boolean functions (in Russian). *Problemy Kibernetiki, 38,* 5–108.
172. Kozlov, D. (2008). *Combinatorial algebraic topology.* New York: Springer.
173. Kratsch, D., & Rampon, J.-X. (1994). A counterexample about poset reconstruction. *Order, 11,* 95–96.
174. Kratsch, D., & Rampon, J.-X. (1994). Towards the reconstruction of posets. *Order, 11,* 317–341.
175. Kratsch, D., & Rampon, J.-X. (1996). Width two posets are reconstructible. *Discrete Mathematics, 162,* 305–310.
176. Kukieła, M., & Schröder, B. (2012). A 2-antichain that is not contained in any finite retract. *Algebra Universalis, 67,* 59–62.
177. Kumar, V. (1992). Algorithms for constraint satisfaction problems – A survey. *AI magazine, 13,* 32–44.
178. Kung, J. P. S. (1999). Möbius inversion. *Encyclopedia of Mathematics.* Available at https://www.encyclopediaofmath.org/index.php/M%C3%B6bius_inversion
179. Langley, L. J. (1995). A recognition algorithm for orders of interval dimension 2, ARIDAM VI and VII (New Brunswick, NJ, 1991/1992). *Discrete Applied Mathematics, 60,* 257–266.
180. Larose, B. (1995). Minimal automorphic posets and the projection property. *International Journal of Algebra and Computation, 5,* 65–80.

181. Larrión, F., Neumann-Lara, V., & Pizaña, M. (2004). Clique divergent clockwork graphs and partial orders. *Discrete Applied Mathematics, 141*, 195–207.
182. Larrión, F., Pizaña, M., & Villaroel-Flores, R. (2008). Contractibility and the clique graph operator. *Discrete Mathematics, 308*, 3461–3469.
183. Li, B. (1990). All retraction operators on a lattice need not form a lattice. *Journal of Pure and Applied Algebra, 67*, 201–208.
184. Li, B. (1993). The core of a chain complete poset with no one-way infinite fence and no tower. *Order, 10*, 349–361.
185. Li, B. (1996). The ANTI-order for caccc posets – Part I. *Discrete Mathematics, 158*, 173–184.
186. Li, B. (1996). The ANTI-order for caccc posets – Part II. *Discrete Mathematics, 158*, 185–199.
187. Li, B., & Milner, E. C. (1992). The PT order and the fixed point property. *Order, 9*, 321–331.
188. Li, B., & Milner, E. C. (1993). A chain complete poset with no infinite antichain has a finite core. *Order, 10*, 55–63.
189. Li, B., & Milner, E. C. (1997). The ANTI-order and the fixed point property for caccc posets. *Discrete Mathematics, 175*, 197–209.
190. Li, B., & Milner, E. C. (1997). Isomorphic ANTI-cores of caccc posets. *Discrete Mathematics, 176*, 185–195.
191. Lindenstrauss, J., & Tzafriri, L. (1973). *Classical Banach spaces. Springer lecture notes in mathematics* (Vol. 338). New York: Springer
192. Liu, W.-P., & Wan, H. (1993). Automorphisms and isotone self-maps of ordered sets with top and bottom. *Order, 10*, 105–110.
193. Lonc, Z., & Rival, I. (1987). Chains, antichains and fibres. *Journal of Combinatorial Theory, Series A, 44*, 207–228.
194. Lovász, L. (1967). Operations with structures. *Acta Mathematica Hungarica, 18*, 321–328.
195. Lovász, L. (1971). On the cancellation law among finite relational structures. *Periodica Mathematica Hungarica, 1*, 145–156.
196. Lubiw, A. (1981). Some NP-complete problems similar to graph isomorphisms. *SIAM Journal of Computing, 10*, 11–21.
197. Luks, E. M. (1982). Isomorphism of graphs of bounded valence can be tested in polynomial time. *Journal of Computer and System Science, 25*, 42–65.
198. Mackworth, A. K. (1992). Constraint satisfaction. In S. Shapiro (Ed.), *Encyclopedia of artificial intelligence* (pp. 284–293). New York: Wiley.
199. Maltby, R. (1992). A smallest-fibre-size to poset-size ratio approaching $\frac{8}{15}$. *Journal of Combinatorial Theory (A), 61*, 328–330.
200. Maróti, M., & Zádori, L. (2012). Reflexive digraphs with near unanimity polymorphisms. *Discrete Mathematics, 312*, 2316–2328.
201. McGregor, J. J. (1979). Relational consistency algorithms and their application in finding subgraph and graph isomorphisms. *Information Sciences, 19*, 229–250.
202. McKay, B. (1997). Small graphs are reconstructible. *The Australasian Journal of Combinatorics, 15*, 123–126.
203. McKay, B. D., & Piperno, A. (2013). Practical graph isomorphism, II. *Journal of Symbolic Computation, 60*, 94–112.
204. McKee, T., & Prisner, E. (1996). An approach to graph-theoretic homology. In Y. Alavi et al. (Eds.), *Combinatorics, graph theory, and algorithms. Proceedings of the Eighth Quadrennial International Conference in Graph Theory, Combinatorics, Algorithms and Applications* (Vol. II, pp. 631–640).
205. McKenzie, R. (1999). Arithmetic of finite ordered sets: Cancellation of exponents I. *Order, 16*, 313–333.
206. McKenzie, R. (2000). Arithmetic of finite ordered sets: Cancellation of exponents II. *Order, 17*, 309–332.
207. Milner, E. C. (1990). Dilworth's decomposition theorem in the infinite case. In K. P. Bogart, R. Freese, & J. P. S. Kung (Eds.), *The Dilworth theorems: Selected papers of Robert P. Dilworth* (pp. 30–35). Boston: Birkhäuser.

208. Mitas, J. (1992). The Structure of Interval Orders. Doctoral dissertation, TH Darmstadt.
209. Mitas, J. (1995). Interval orders based on arbitrary ordered sets. *Discrete Mathematics, 144*, 75–95.
210. Mohr, R., & Henderson, T. C. (1986). Arc and path consistency revisited. *Artificial Intelligence, 28*, 225–233.
211. Müller, H., & Rampon, J.-X. (1997). Partial orders and their convex subsets. *Discrete Mathematics, 165/166*, 507–517.
212. Nadel, B. (1989). Constraint satisfaction algorithms. *Computational Intelligence, 5*, 188–224.
213. Nešetřil, J., & Rödl, V. (1977). Partitions of finite relational and set systems. *Journal of Combinatorial Theory (A), 17*, 289–312.
214. Nešetřil, J., & Rödl, V. (1984). Combinatorial partitions of finite posets and lattices. *Algebra Universalis, 19*, 106–119.
215. Nešetřil, J., & Rödl, V. (1987). Complexity of diagrams. *Order, 3*, 321–330.
216. Niederle, J. (2008). Forbidden retracts for finite ordered sets of width at most four. *Discrete Mathematics, 308*, 1774–1784.
217. Nowakowski, R. (1981). Open question on p. 842 of [246].
218. Nowakowski, R., & Rival, I. (1979). A fixed edge theorem for graphs with loops. *Journal of Graph Theory, 3*, 339–350.
219. Pelczar, A. (1961). On the invariant points of a transformation. *Annales Polonici Mathematici, XI*, 199–202.
220. Peles, M. A. (1963). On Dilworth's theorem in the infinite case. *Israel Journal of Mathematics, 1*, 108–109.
221. Pickering, D., & Roddy, M. (1992). On the strong fixed point property. *Order, 9*, 305–310.
222. Polat, N. (1995). Retract-collapsible graphs and invariant subgraph properties. *Journal of Graph Theory, 19*, 25–44.
223. Poston, T. (1971). Fuzzy Geometry. Ph.D. thesis, University of Warwick.
224. Pouzet, M. (1979). Relations non reconstructible par leurs restrictions. *Journal of Combinatorial Theory (B), 26*, 22–34.
225. Pretorius, L., & Swanepoel, C. (2000). Partitions of countable posets, papers in honour of Ernest J. Cockayne. *Journal of Combinatorial Mathematics and Combinatorial Computing, 33*, 289–297.
226. Priestley, H. A., & Ward, M. P. (1994). A multipurpose backtracking algorithm. *Journal of Symbolic Computation, 18*, 1–40.
227. Prisner, E. (1992). Convergence of iterated clique graphs. *Discrete Mathematics, 103*, 199–207.
228. Prömel, H. J. (1987). Counting unlabeled structures. *Journal of Combinatorial Theory, Series A, 44*, 83–93.
229. Prosser, P. (1993). Hybrid algorithms for the constraint satisfaction problem. *Computational Intelligence, 9*, 268–299.
230. Provan, J. S., & Ball, M. O. (1983). The complexity of counting cuts and of computing the probability that a graph is connected. *SIAM Journal on Computing, 12*, 777–788.
231. Purdom, P. W. (1978). Tree size by partial backtracking. *SIAM Journal on Computing, 7*, 481–491.
232. Quackenbush, R. (1986). Unsolved problems: Dedekind's problem. *Order, 2*, 415–417.
233. Quilliot, A. (1983). Homomorphismes, points fixes, rétractions et jeux de poursuite dans les graphes, les ensembles ordonnés et les espaces métriques. Thèse de doctorat d'état, Univ. Paris VI.
234. Quilliot, A. (1983). An application of the Helly property to the partially ordered sets. *Journal of Combinatorial Theory (A), 35*, 185–198.
235. Quilliot, A. (1985). On the Helly property working as a compactness criterion for graphs. *Journal of Combinatorial Theory (A), 40*, 186–193.
236. Rabinovitch, I. (1978). The dimension of semiorders, *Journal of Combinatorial Theory (Series A), 25*, 50–61.

237. Rabinovitch, I. (1978). An upper bound on the dimension of interval orders. *Journal of Combinatorial Theory (Series A), 25*, 68–71.
238. Rabinovitch, I., & Rival, I. (1979). The rank of a distributive lattice. *Discrete Mathematics, 25*, 275–279.
239. Ramachandran, S. (1981). On a new digraph reconstruction conjecture. *Journal of Combinatorial Theory Series B, 31*, 143–149.
240. Rampon, J.-X. (2005). What is reconstruction for ordered sets? *Discrete Mathematics, 291*, 191–233.
241. Ramsey, F. P. (1930). On a problem of formal logic. *Proceedings of the London Mathematical Society, 30*(2), 264–286.
242. Reiner, V., & Welker, V. (1999). A homological lower bound for order dimension of lattices. *Order, 16*, 165–170.
243. Reuter, K. (1989). Removing critical Pairs. *Order, 6*, 107–118.
244. Rival, I. (1976). A fixed point theorem for finite partially ordered sets. *Journal of Combinatorial Theory (A), 21*, 309–318.
245. Rival, I. (1982). The retract construction. In I. Rival (Ed.), *Ordered sets* (pp. 97–122). Dordrecht: Dordrecht-Reidel.
246. Rival, I. (Ed.). (1982). *Ordered sets*. Boston: Dordrecht-Reidel
247. Rival, I. (Ed.). (1984). *Graphs and order*. Boston: Dordrecht-Reidel.
248. Rival, I. (1984). The diagram. In I. Rival (Ed.), *Graphs and order* (pp. 103–136). Dordrecht: Dordrecht-Reidel.
249. Rival, I. (1985). Unsolved problems: The diagram. *Order, 2*, 101–104.
250. Rival, I. (1985). Unsolved problems: The fixed point property. *Order, 2*, 219–221.
251. Rival, I. (Ed.). (1986). Combinatorics and ordered sets. In *Proceedings of the AMS-IMS-SIAM Summer Research Conference at Humboldt State University, Contemporary Mathematics* (Vol. 57). Providence: American Mathematical Society.
252. Rival, I. (Ed.). (1989). *Algorithms and order*. Dordrecht/Boston: Kluwer.
253. Rival, I., & Rutkowski, A. (1991). Does almost every isotone self-map have a fixed point? In *Extremal problems for finite sets. Bolyai mathematical society studies* (Vol. 3, pp. 413–422). Hungary: Viségrad.
254. Rival, I., & Zaguia, N. (Eds.). (1999). *ORDAL '96, Papers from the Conference on Orders Algorithms and Applications, Ottawa, 1996. Theoretical computer science* (Vol. 217). Amsterdam: Elsevier Science Publishers.
255. Roddy, M. (1994). Fixed points and products. *Order, 11*, 11–14.
256. Roddy, M. (2002). Fixed points and products: Width 3. *Order, 19*, 319–326.
257. Roddy, M. (2002). On an example of Rutkowski and Schröder. *Order, 19*, 365–366.
258. Roddy, M., & Schröder, B. (2005). Isotone relations revisited. *Discrete Mathematics, 290*, 229–248.
259. Rödl, V., & Thoma, L. (1995). The complexity of cover graph recognition for some varieties of finite lattices. *Order, 12*, 351–374.
260. Rossi, F., van Beek, P., & Walsh, T. (2006). *Handbook of constraint programming*. Amsterdam: Elsevier.
261. Rota, G.-C. (1964). On the foundations of combinatorial theory I: Theory of Möbius functions. *Zeitschrift für Wahrscheinlichkeitstheorie und verwandte Gebiete, 2*, 340–368.
262. Rutkowski, A. (1986). Cores, cutsets and the fixed point property. *Order, 3*, 257–267.
263. Rutkowski, A. (1986). The fixed point property for sums of posets. *Demonstratio Mathematica, 4*, 1077–1088.
264. Rutkowski, A. (1989). The fixed point property for small sets. *Order, 6*, 1–14.
265. Rutkowski, A., private communication.
266. Rutkowski, A., & Schröder, B. (1994). Retractability and the fixed point property for products. *Order, 11*, 353–359.
267. Sabin, D., & Freuder, E. C. (1994). Contradicting conventional wisdom in constraint satisfaction. In *Proceedings of the 11th European Conference on Artificial Intelligence*, Amsterdam (pp. 125–129).

268. Saks, M. (1985). Unsolved problems: Balancing linear extensions of ordered sets. *Order, 2,* 327–330.
269. Sands, B. (1985). Unsolved problems. *Order, 1,* 311–313.
270. Schröder, B. (1993). Fixed point property for 11-element sets. *Order, 10,* 329–347.
271. Schröder, B. (1995). On retractable sets and the fixed point property. *Algebra Universalis, 33,* 149–158.
272. Schröder, B. (1996). Fixed cliques and generalizations of dismantlability. In Y. Alavi et al. (Eds.), *Combinatorics, graph theory, and algorithms. Proceedings of the Eighth Quadrennial International Conference in Graph Theory, Combinatorics, Algorithms and Applications* (Vol. II, pp. 747–756).
273. Schröder, B. (1998). On cc-comparability invariance of the fixed point property. *Discrete Mathematics, 179,* 167–183.
274. Schröder, B. (2000). The uniqueness of cores for chain-complete ordered sets. *Order, 17,* 207–214.
275. Schröder, B. (2000). Reconstruction of the neighborhood deck of ordered sets. *Order, 17,* 255–269.
276. Schröder, B. (2001). Reconstruction of N-free ordered sets. *Order, 18,* 61–68.
277. Schröder, B. (2002). Examples on ordered set reconstruction. *Order, 19,* 283–294.
278. Schröder, B. (2003). On ordered sets with isomorphic marked maximal cards. *Order, 20,* 299–327.
279. Schröder, B. (2004). More examples on ordered set reconstruction. *Discrete Mathematics, 280,* 149–163.
280. Schröder, B. (2005). The automorphism conjecture for small sets and series parallel sets. *Order, 22,* 371–387.
281. Schröder, B. (2006). Examples of powers of ordered sets with the fixed point property. *Order, 23,* 211–219.
282. Schröder, B. (2007). *Mathematical analysis – A concise introduction.* Hoboken: Wiley.
283. Schröder, B. (2010). *Fundamentals of mathematics – An introduction to proofs, logic, sets and numbers.* Hoboken: Wiley.
284. Schröder, B. (2010). Pseudo-similar points in ordered sets. *Discrete Mathematics, 310,* 2815–2823.
285. Schröder, B. (2012). The fixed point property for ordered sets. *Arabian Journal of Mathematics, 1,* 529–547.
286. Schröder, B. (2015). Homomorphic constraint satisfaction problem solver. http://www.math.usm.edu/schroeder/software.htm
287. Schröder, B. (2015). The fixed vertex property for graphs. *Order, 32,* 363–377.
288. Schröder, B. (2015). The fixed point property for ordered sets of interval dimension 2 (submitted to order).
289. Schröder, B. (2016). The use of retractions in the fixed point theory for ordered sets. In M. Alfuraidan, & Q. H. Ansari (Eds.), *Fixed point theory and graph theory – Foundations and integrative approaches* (pp. 365–417). Amsterdam: Elsevier. Chapter in the Proceedings of the Workshop on Fixed Point Theory and Applications, King Fahd University of Petroleum and Minerals, December 2014, Dhahran, Saudi Arabia.
290. Schröder, B. (2016). Applications of dismantlability in analysis (in preparation).
291. Scott, D., & Suppes, P. (1958). Foundational aspects of theories of measurement. *Journal of Symbolic Logic, 23,* 113–128.
292. Segev, Y. (1994). Some remarks on finite 1-acyclic and collapsible complexes. *Journal of Combinatorial Theory, Series A, 65,* 137–150.
293. Smith, B., & Dyer, M. (1996). Locating the phase transition in binary constraint satisfaction problems. *Artificial Intelligence, 81,* 155–181.
294. Smithson, R. E. (1971). Fixed points of order-preserving multifunctions. *Proceedings of the American Mathematical Society, 28,* 304–310.
295. Spanier, E. H. (1966). *Algebraic topology.* New York: Springer.
296. Spinrad, J. (1988). Subdivision and lattices. *Order, 5,* 143–147.

297. Stanley, R. P. (1979). Balanced Cohen-Macauley complexes. *Transactions of the American Mathematical Society, 249*, 139–157.
298. Stockmeyer, P. K. (1977). The falsity of the reconstruction conjecture for tournaments. *Journal of Graph Theory, 1*, 19–25. Erratum: *Journal of Graph Theory, 62*, 199–200 (2009).
299. Stockmeyer, P. K. (1981). A census of non-reconstructible digraphs. I: Six related families. *Journal of Combinatorial Theory (B), 31*, 232–239.
300. Stockmeyer, P. K. (1988). Tilting at windmills, or My quest for nonreconstructible graphs; 250th anniversary conference on graph theory (Fort Wayne, IN, 1986). *Congressus Numerantium, 63*, 188–200.
301. Stong, R. E. (1966). Finite topological spaces. *Transactions of the American Mathematical Society, 123*, 325–340.
302. Sysło, M. (1984). A graph theoretic approach to the jump-number problem. In I. Rival (Ed.), *Graphs and order* (pp. 185–215). Boston: Dordrecht-Reidel.
303. Szpilrajn, E. (1930). Sur l'extension de l'ordre partiel. *Fundamenta Mathematicae, 16*, 386–389.
304. Szymik, M. (2015). Homotopies and the universal fixed point. *Order, 32*, 301–311.
305. Tancer, M. (2010). *d*-Collapsibility is NP-complete for $d \geq 4$. *Chicago Journal of Theoretical Computer Science, 2010*, 28 pp. Article 3
306. Tancer, M. (2012). Recognition of collapsible complexes is NP-complete. Available at http://arxiv.org/abs/1211.6254
307. Tarski, A. (1955). A lattice-theoretical fixpoint theorem and its applications. *Pacific Journal of Mathematics, 5*, 285–309.
308. Thakare, N., Pawar, M., & Waphare, B. (2002). A structure theorem for dismantlable lattices and enumeration. *Periodica Mathematica Hungarica, 45*, 147–160.
309. Trotter, W. T. (1975). Inequalities in dimension theory for posets. *Proceedings of the American Mathematical Society, 47*, 311–316.
310. Trotter, W. T. (1976). A forbidden subposet characterization of an order dimension inequality. *Mathematical Systems Theory, 10*, 91–96.
311. Trotter, W. T. (1992). *Combinatorics and partially ordered sets: Dimension theory*. Baltimore: Johns Hopkins University Press.
312. Trotter, W. T., Moore, J. I., & Sumner, D. P. (1976). The dimension of a comparability graph. *Proceedings of the American Mathematical Society, 60*, 35–38.
313. Tsang, E. (1993). *Foundations of constraint satisfaction*. New York: Academic. Out of print. Now available through Amazon books on demand, see http://www.bracil.net/edward/FCS.html
314. Tverberg, H. (1967). On Dilworth's theorem for partially ordered sets. *Journal of Combinatorial Theory, 3*, 305–306.
315. van Beek, P., & Dechter, R. (1995). On the minimality of row-convex constraint networks. *Journal of the ACM, 42*, 543–561.
316. von Rimscha, M. (1983). Reconstructibility and perfect graphs. *Discrete Mathematics, 47*, 79–90.
317. Wagon, S. (1993). *The Banach-Tarski paradox*. Cambridge: Cambridge University Press.
318. Walker, J. W. (1984). Isotone relations and the fixed point property for posets. *Discrete Mathematics, 48*, 275–288.
319. West, D. (1985). Parameters of partial orders and graphs: Packing, covering and representation. In I. Rival (Ed.), *Graphs and orders* (pp. 267–350). Dordrecht: Dordrecht-Reidel.
320. West, D. (1996). *Introduction to graph theory*. Upper Saddle River: Prentice Hall.
321. Whitehead, J. H. C. (1939). Simplicial spaces, nuclei and *m*-groups. *Proceedings of the London Mathematical Society, 45*(2), 243–327.
322. Wiedemann, D. (1991). A computation of the eighth Dedekind number. *Order, 8*, 5–6.
323. Wild, M. (1992). Cover-preserving order embeddings into Boolean lattices. *Order, 9*, 209–232.
324. Wilf, H. S. (1994). *Generating functionology*. New York: Academic.
325. Willard, S. (1970). *General topology*. Reading: Addison-Wesley.

326. Williams, C., & Hogg, T. (1994). Exploiting the deep structure of constraint problems. *Artificial Intelligence, 70*, 73–117.
327. Williamson, S. (1992). Fixed Point Properties in Ordered Sets. Ph. D. dissertation, Emory University.
328. Xia, W. (1992). Fixed point property and formal concept analysis. *Order, 9*, 255–264.
329. Yannakakis, M. (1982). On the complexity of the partial order dimension problem. *SIAM Journal of Algebra and Discrete Mathematics, 3*, 351–358.
330. Zádori, L. (1992). Order varieties generated by finite posets. *Order, 8*, 341–348.
331. Zádori, L. (1998). Characterizing finite irreducible relational sets. *Acta Scientiarum Mathematicarum (Szeged), 64*, 455–462.

Index

© Springer International Publishing 2016

B. Schröder, *Ordered Sets*, DOI 10.1007/978-3-319-29788-0